Introduction to Nitride Semiconductor Blue Lasers and Light Emitting Diodes

Introduction to Nitride Semiconductor Blue Lasers and Light Emitting Diodes

Edited by
Shuji Nakamura and Shigefusa F. Chichibu

TAYLOR & FRANCIS
Founded 1798

London and New York

First published 2000 by Taylor & Francis
11 New Fetter Lane, London EC4P 4EE

Simultaneously published in the USA and Canada
by Taylor & Francis

29 West 35th Street, New York, NY 10001

Taylor & Francis is an imprint of the Taylor & Francis Group

Printed and bound in Great Britain by Biddles Ltd, Guildford and King's Lynn

Publisher's Note

This book has been prepared from camera-ready copy provided by the authors.

British Library Cataloguing in Publication Data
A catalogue record for this book is available from the British Library

Library of Congress Cataloging in Publication Data
A catalogue record for this book has been requested

ISBN 0-7484-0836-3

Contributors

Daniel L. Barton	Sandia National Laboratories, P.O. Box 5800, Albuquerque, NM 87185-1081, USA
Shigefusa F. Chichibu	Institute of Applied Physics University of Tsukuba, 1-1-1 Tennodai, Tsukuba, Ibaraki 305-8573, Japan
Steven P. DenBaars	Departments of Materials, Electrical and Computer Engineering, University of California, Santa Barbara, CA 93106-5050, USA
Yoichi Kawakami	Department of Electronic Science and Engineering, Kyoto University, Yoshida-Honcho, Sakyo-ku, Kyoto 606-8051, Japan
Shuji Nakamura	Department of Research and Development, Nichia Chemical Industries Ltd. 491 Oka, Kaminaka, Anan, Tokushima 774-8601, Japan
Marek Osinski	Center for High Technology Materials, University of New Mexico, 1313 Goddard SE, Albuquerque, NM 87106, USA
Fernando A. Ponce	Department of Physics and Astronomy Arizona State University, P. O. Box 871504, Tempe, AZ 85287-1504, USA
Takayuki Sota	Department of Electrical, Electronics, and Computer Engineering, Waseda University, 3-4-1 Ohkubo, Shinjuku, Tokyo 169-8555, Japan
Masakatsu Suzuki	Central Research Laboratory, Matsushita Electric Industrial Co. Ltd. 3-4 Hikaridai, Seika-cho, Souraku-gun, Kyoto, 619-02, Japan
Takeshi Uenoyama	Central Research Laboratory, Matsushita Electric Industrial Co. Ltd. 3-4 Hikaridai, Seika-cho, Souraku-gun, Kyoto, 619-02, Japan
Chris G. Van de Walle	Xerox Palo Alto Research Center, 3333 Coyote Hill Road, Palo Alto, CA 94304, USA

Preface

Solid-state light-emitting diodes (LEDs) are the ultimate light source. The brightness and durability of solid-state LEDs make them ideal for use in displays and light sources, while semiconductor laser diodes (LDs) have been used in everything from optical communications systems to compact disk players. These applications have been limited, however, by the lack of materials that can efficiently emit blue or green light. For example, full-color displays require at least three primary colors, namely red, green, and blue, to reproduce the visible color spectrum. Such a combination is also needed to make solid-state white-light-emitting devices that would be more durable and consume less power consumption than conventional incandescent bulbs or fluorescent lamps. At present, incandescent bulb lamps and fluorescent lamps are used as light sources for many applications. However, these conventional light sources are old, traditional glass-vacuum-type light sources with poor reliability and durability and a low luminous efficiency. In the past, electronic circuits were made of bulky glass vacuum tubes in spite of poor reliability and durability. However, nowadays all electronic circuits are made with compact highly reliable solid-state semiconductor circuits. Thus, only light sources are not solid-state semiconductors but still made by old, traditional technology.

Recently, however, the development of nitride semiconductors opened the way to obtain all-solid-state semiconductor light sources. For lasers, the shorter the wavelength, the sharper the focused light diameter, which would increase the storage capacity of optical disks. Digital versatile disks (DVDs), which appeared on the market in 1996, rely on red AlInGaP semiconductor lasers and have a data capacity of about 4.7 gigabytes (Gbytes), compared to 0.65 Gbytes for compact disks (CDs) which use infrared lasers. By moving to bluish-purple wavelengths through the use of Nitride semiconductors, the storage capacity can be increased up to 15 Gbytes. Thus, nitride semiconductors are also indispensable for manufacturing semiconductor lasers of next generation DVDs.

The rapid progress of nitride semiconductor devices is extraordinary in comparison with that for other conventional III-V compound semiconductor devices which are made of GaAlAs, InGaAsP, or AlInGaP. For example, the first InGaN-based high-brightness blue LEDs have been put into the market in 1993. Then, InGaN-based quantum-well-structure high-brightness blue and green LEDs appeared in 1995. Now in 1999, these InGaN-based green and blue LEDs are used in many applications, such as LED full-color displays, traffic lights, and white lighting sources. They actually have higher reliability, longer lifetime, and lower power consumption, in comparison to incandescent sources, which will decrease the energy consumption and protect natural resources using the environmentally safe material of nitride semiconductors. Also, amber nitride-based LEDs were developed recently, and the characteristics of these devices are superior to those of conventional amber AlInGaP LEDs, especially for the intensity stability against the ambient temperature. Thus, the only remaining visible color which has not been developed using nitride semiconductors is red. The luminous efficiency of present InGaN-based green and blue LEDs are 30 and 5 lm/W. That of conventional incandescent bulb lamps is 10 lm/W. Thus, when we make white light using the presently available LEDs, a white LED lamp with a luminous efficiency of 30 lm/W can be fabricated, which is much higher than that (10 lm/W) of the conventional incandescent bulb lamps. Also, there is a great difference in the durability and reliability. The LEDs have an almost indefinite lifetime (longer than 100,000 hours) and superior reliability compared to that of conventional incandescent bulb lamps. Also, the luminous efficiency of blue and green LEDs is still increasing day by day as a result of continuous research efforts such that they will soon catch up with that of fluorescent lamps. Thus, conventional light sources should be replaced with LEDs from the viewpoint of saving energy and natural resources, because of their much longer lifetime, better durability, and higher efficiency. Bluish-purple InGaN-based lasers, which can have the shortest wavelength ever obtained, would have many applications and would cause a major impact in the optoelectronic industry, as did blue and green LEDs. The short wavelength of 400 nm (bluish-purple) means that the light can be focused more sharply, which would increase the storage capacity of optical disks. The first InGaN-based LD under room-temperature (RT) pulsed operation was reported in 1995. In 1996, the first RT continuous-wave (CW) operation was achieved. Then, a lifetime of 10,000 hours was attained in 1998 and LD distribution is started in 1999. The speed of the progress in the research and development of these InGaN-based light-emitting devices is extremely rapid.

The authors of the various chapters were carefully selected as the leaders in the development of the nitride semiconductors and devices. Chapter 1 gives an introductory supplying basic backgrounds in the area of nitride semiconductor blue LEDs and materials. Chapter 2 is for learning a basic lasing theory in semiconductors. Chapters 3 and 4 treat basic problems in nitride semiconductors and quantum well structures such as conductivity control, impurity problems, band offsets, microstructure of imperfection in crystals like threading dislocation, column structure, and polarity problem. In Chapter 5 discussed are emission mechanisms in nitrides which dominate bright emissions from InGaN qunatum wells such as quantum well excitons localized at the potential minima produced by the alloy compositional fluctuations in InGaN. In Chapters 6 and 7, degradation mechanisms of InGaN devices are discussed. Finally, current performance of blue LEDs and LDs are described.

This book is to be used as a textbook for postgraduate courses, professional engineers or scientists wishing to learn the latest progress in nitride semiconductors and related optical devices.

The editors wish to thank most warmly the authors and publishers who have contributed to this book. Voluntary help of T. Mizutani and M. Sugiyama in the editorial procedures is appreciated. We also wish to express our appreciation to our wives for continuous encouragements.

May 1, 1999

Nichia Chemical Industries — Shuji Nakamura
University of Tsukuba — Shigefusa F. Chichibu

Contents

Chapter 1
Basics Physics & Materials Technology of GaN LEDs and LDs

Steven P. DenBaars

1.1 INTRODUCTION

Recently, the Group-III nitride based semiconductors have emerged as the leading material for the production of blue LEDs and blue laser diodes. Nature has blessed the $(In_xGa_{1-x})_yAl_{1-y}N$ alloy system with a continuous and direct bandgap semiconductor alloy spanning ultraviolet to blue/green/yellow wavelengths. As we will learn in this chapter the direct bandgap semiconductor is perhaps the most efficient means to generate light from electricity. The purpose of this chapter is to review the basic physics of LEDs, materials selection, epitaxial growth, and describe the historical evolution of technologies employed in the manufacturing of LEDs and laser diodes.

1.1.1 Historical Evolution of LED Technology

The achievement of bright blue InGaN LEDs has basically led to a revolution in LED technology and opened up enormous new markets that were not accessible before. The remarkable advancements in GaN based LED materials depended on several key breakthroughs in materials synthesis and fabrication. The lack of p/n junctions in the Group-III nitrides and poor crystal quality slowed research for many decades. The use of InGaN/GaN double heterostructure (DH) in LEDs in 1994 by Nakamura (Nakamura *et al.*, 1994a) and achievement of p-doping in GaN by Akasaki (Amano *et al.*, 1990a) is widely credited with reigniting the III-V nitride system. It has taken over twenty years since the first optical pumped stimulated emission was observed in GaN crystal (Dingle *et al.*, 1994) and first LEDs (Pankove *et al.*, 1971) were fabricated to come to the recent realization of blue lasers. In the late 1980s remarkable progress for high brightness LEDs for displays occurred. This remarkable revolution in LED performance for the commercially important compound semiconductors materials system is illustrated in Figure 1.1. Notable progress is the development of high efficiency AlGaAs, AlGaInP, and AlGaInN alloys. In particular, the efficiency of p/n GaN LEDs increased rapidly after the first reports of p/n junctions by Akasaki in 1989. This was caused primarily by the higher efficiency afforded by heterojunction based LEDs.

The advances in GaN and GaAs based LEDs has enabled many new markets to be opened for LEDs. For many years solid state device researchers have dreamed of obtaining a bright blue light source. Only now can one use three LEDs to tune to any color in the visible range, or even use a single blue LED in combination with phosphors to make "white LEDs". This concept is particularily attractive because the solid state nature of semiconductor devices produce very high reliability. The average lifespan of an LED is of the order of tens of years (Craford, 1977). Full-color and white LEDs are now appearing in numerous applications ranging from outdoor TVs, traffic signals, scanners, flashlights, and automotive backlighting. Another potentially large application of GaN is in fabricating blue laser diodes (LDs) for extremely high-density optical storage systems. Because the storage density of optical compact discs (CDs) and digital video discs (DVDs) is inversely proportional to the square of the laser wavelength a 4-8 fold increase in capacity could be realized with short wavelength laser diodes. Future research is needed to commercialize GaN for optical storage, energy efficient lighting, communications, printing, projection TV, and even surgery. As GaN manufacturing volumes increase and costs decrease one can expect to see GaN LEDs and LDs in ever increasing applications wherever economical and reliable illumination is needed.

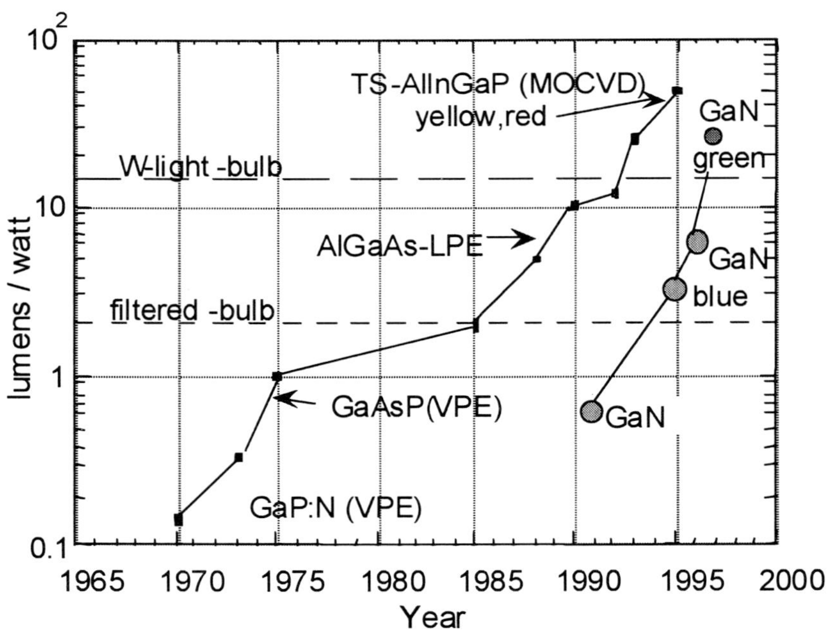

Figure 1.1 Evolution of LED performance (lumens/watt) for the commercially important materials.

1.2 BASIC PHYSICS OF LEDs: INJECTION LUMINESCENCE

In order to better understand GaN LEDs it is important to understand the physics governing light emission in semiconductors. Light emitting diodes (LEDs) are a class of diodes that emit spontaneous radiation under suitable forward bias conditions. LEDs are fabricated from semiconductor materials in which a p/n junction is fabricated. Injection electroluminescence (EL) is the most important mechanism for exciting the semiconductor materials. Under forward bias conditions, electrons are injected into the p-type semiconductor and holes are injected into the n-type material. Recombination of these minority carriers with the majority carriers at the p/n junction results in near bandgap radiation. The wavelength of the emitted light is governed by the energy bandgap of the semiconductor material, and is given as:

$$\lambda = \frac{1.24}{E_{gap}}[\mu m] \qquad (1.1)$$

A schematic representation of a radiative recombination in a forward bias p/n junction is shown in Figure 1.2. In GaN the p-type material is fabricated from Mg-doped GaN and the n-type material is silicon doped GaN. Prior to 1989 only materials emitting radiation covering the green through infrared spectrum existed. The first visible LEDs were red GaAsP emitters invented at General Electric by Holnyak and Bavacqua in 1962 (Holynak *et al.,* 1962). Blue was the missing primary color (red, green and blue) and as such full-color emission could not be achieved. Therefore with the advent of the GaN system the full-visible color spectrum is now accessible for the first time.

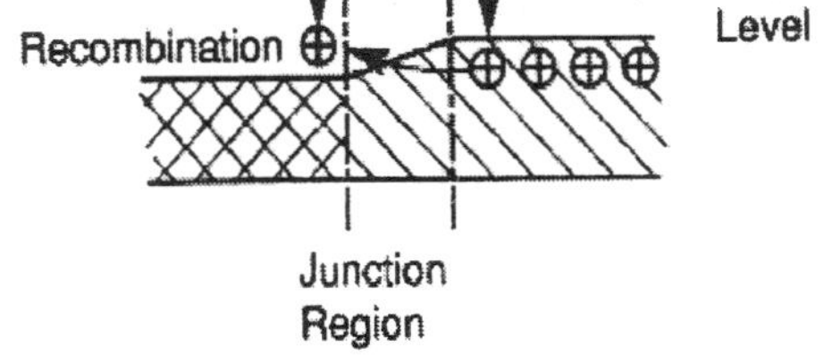

Figure 1.2 Schematic representation of a p/n junction in a light emitting diode under forward bias.

1.2.1 Direct and Indirect Band-Gap Material

Semiconductor LED materials can be divided into two different classes of band structure, exhibiting either direct or indirect bandgap behavior. In a direct bandgap semiconductor, electrons in conduction band minima and holes in the valence band maxima have the same momentum as shown in Figure 1.3. The probability of a band-to-band radiative transition for the direct bandgap semiconductor is high since momentum will be conserved upon recombination. In contrast, in an indirect bandgap semiconductor the electrons in the conduction band minima have different momentum from the holes in the valence band maxima, and therefore momentum is not conserved. In an indirect semiconductor an electron would have to change its momentum before recombining with a hole. A third component, a phonon, is required to participate in the recombination event for the electron to change its momentum and a transition to occur in the indirect gap semiconductor. This results in a significantly lower recombination probability for a band-to-band transition to occur in the indirect gap semiconductor. This fact is especially evident when comparing direct bandgap GaN to indirect SiC LEDs. GaN LEDs exhibit quantum efficiencies as high as 12%, which is significantly higher than the 0.02% observed in SiC LEDs.

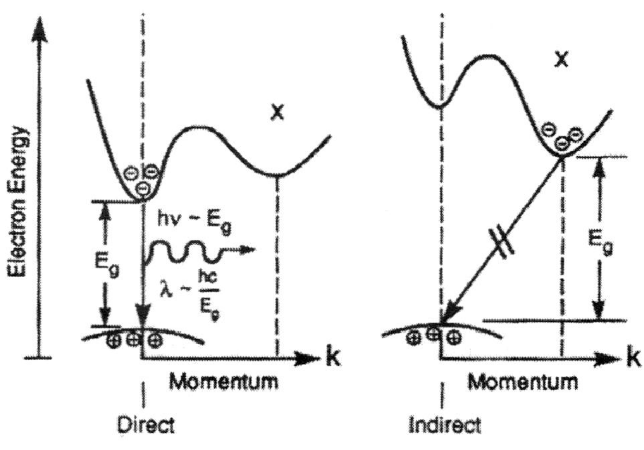

Figure 1.3 Energy band diagram of direct (GaN) and indirect bandgap (SiC) semiconductors. Electron energy is plotted versus momentum, k.

1.2.2 Radiative Recombination

Electron-hole recombination can be either radiative or non-radiative. These two recombination pathways can be considered as parallel processes occuring across the energygap of the semiconductor, as illustrated in Figure 1.4. Maximizing the probability for radiative recombination and decreasing the probability for non-radiative recombination paths can increase light output from an LED. Several mechanisms cause non-radiative recombination in semiconductors, among these are; auger recombination, recombination at defect sites, and multiphonon emission at deep impurity sizes (Henry *et al.*, 1976). Radiative transitions in semiconductors occur primarily by interband transitions, excitonic recombination, or recombination through impurity centers.

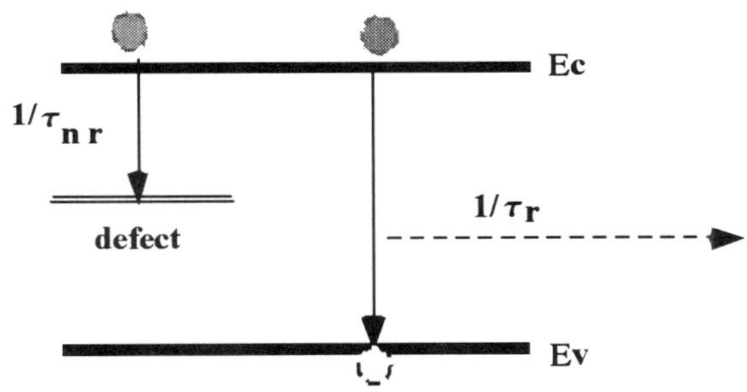

Figure 1.4 Schematic illustration of radiative and non-radiative trnasistion in a semiconductor.

Under continuous injection of carriers an excess carrier density will exist in the semiconductor. For charge neutrality to exist the excess electrons "ΔN" and excess holes "ΔP" must be equal after injection ceases; the excess carrier density has been found to decrease exponentially with time with a characteristic lifetime:

$$\Delta N = \Delta No \exp(-t/t), \tag{1.2}$$

where ΔN_o = initial excess carrier concentration, and

t = recombination lifetime.

The recombination lifetime consists of both a radiative lifetime and a non-radiative lifetime

$$\frac{1}{t} = \frac{1}{t_r} + \frac{1}{t_{nr}}. \tag{1.3}$$

The radiative recombination rate is defined with the excess carrier density and recombination lifetime as (Kressel *et al.*, 1977a):

$$R_r = \frac{\Delta n}{t_r}. \tag{1.4}$$

Likewise the non-radiative recombination rate:

$$R_{nr} = \frac{\Delta n}{t_{nr}}. \tag{1.5}$$

The internal radiative quantum efficiency is defined as the ratio of the rate of radiative recombination to the total recombination rate:

$$\eta_i = R_r/(R_r + R_{nr}) . \tag{1.6}$$

Therefore, the internal quantum efficiency can be expressed in terms of the radiative and non-radiative lifetimes.

$$n_i = \frac{1/t_r}{1/t_r + 1/t_{nr}} = \frac{1}{1 + 1/t_r/t_{nr}}. \tag{1.7}$$

By keeping t_r/t_{nr} small a maximum internal quantum efficiency can be obtained. This is done by making t_{nr} long compared to t_r. Growing materials with low defect densities and eliminating impurities that cause non-radiative deep levels are certain material properties that increase the quantum efficiency. Typical GaP and GaAs based LEDs must be manufacturer with lower than 10^4 defects cm^{-2} in order to achieve reasonable efficiencies. One apparent exception to this rule is that GaN LEDs still emit light efficiently despite exhibiting defect densities as high as $10^{10}cm^{-2}$ (Lester *et al.*, 1995). Using the principle of detailed balance, (Shockley, 1950) in which at equilibrium the rate of optical generation of electron-hole pairs is equal to their rate of radiative recombination, the spontaneous recombination rate for band-to-band emission can be expressed as:

$$R_{sp} = B_r np. \tag{1.8}$$

Where B_r (cm^3/s) is the radiative recombination coefficient. Under non-equilibrium conditions additional carriers are introduced into the material and the spontaneous recombination rate occurs. At high injection levels, the excess carrier density will substantially exceed the background concentration:

$$\Delta n > p_o + n_o, \tag{1.9}$$

so that the average carrier lifetime can be expressed as:

$$t_r \cong \frac{1}{B_r \Delta n}, \tag{1.10}$$

which occurs in the "bimolecular" recombination region. In the case of bimolecular recombination, the lifetime value is continually changing so t_r represents the average carrier lifetime. The other region for the lifetime occurs at low injection levels in which the excess carriers decay exponentially with time. In this region the lifetime can be determined by the background carrier concentration:

$$t_r \cong \frac{1}{B_r(p_o + n_o)}. \tag{1.11}$$

For n-type material $n_0 >> p_0$:

$$t_r \cong \frac{1}{B_r n_o}. \tag{1.12}$$

Likewise, for p-type material

$$t_r \cong \frac{1}{B_r p_o}. \tag{1.13}$$

For intrinsic material at low injection levels the longest possible lifetime is found

$$t_r \cong \frac{1}{2B_r n_i}. \tag{1.14}$$

Interface recombination is another factor that effects the effective lifetime and is taken into account by dividing the average non-radiative interface recombination velocity S by the active layer thickness d. The effective lifetime is then a combination of all the competing process lifetimes:

$$\frac{1}{t_{eft}} = \frac{1}{t_{nr}} + \frac{2S}{d} + B_r(n_o + p_o). \tag{1.15}$$

Since interfaces occur on each side of the active region in DH and SQW LED, two multiplies the interface recombination velocity.

1.2.3 External Quantum Efficiency

The best measure for LED efficiency is how much light is generated for the injected current. Because of the direct bandgap in GaN most carriers recombine through the radiative pathway and quantum efficiencies as high as 10% are achieved. Fundamentally, the external quantum efficiency can be expressed as the ratio of photons out per injected electron:

$$\eta_{ext} = (phonons\text{ - }out)/(phonons\text{ - }in) \tag{1.16}$$

The external quantum efficiency is product of material and optical coupling efficiency given as:

$$\eta_{ext} = \eta_{inj}\eta_i\eta_{opt}, \tag{1.17}$$

where η_{inj} is the injection efficiency and η_i is the light generation efficiency, and η_{opt} is the optical or light extraction efficiency. Since light generation is predominantly generated on the p-side of the diode for GaP (Kressel *et al.,* 1977c) and AlGaAs (Varon *et al.,* 1981), a high electron injection efficiency is typically desired. This dictates that high injection efficiency will be achieved if the electron current dominates over the hole current.

For a material to be an efficient visible light emitter it should meet three important criteria, namely (1) possess an efficient radiative pathway, (2) have an energy gap of appropriate width, (3) and be controllably doped both p and n-type with low resistivities. The most important materials system that has had great technological success in meeting these demands is the III–V semiconductors and their alloys. Table 1.1 summarizes the commercially important visible LED materials and the external quantum efficiencies achieved in each system currently achievable.

Table 1.1 Summary of Commercially Available Visible LED Materials

Material	Color Range	Peak Emission	Luminous Efficiency (lm/w)	External Quantum Eff. (%)	Growth Technique	Device Type
AlGaInP	Red	636nm	35	24%	MOCVD	DH
AlGaAs	Red	650 nm	8.0	16%	LPE	DH
AlGaInP	Amber	590nm	40	10%	MOCVD	DH
InGaN	Blue	470nm	10	11%	MOCVD	QW
InGaN	Green	520nm	34	10%	MOCVD	QW
InGaN	UV	372nm	0	7.5%	MOCVD	DH
InGaN	Amber	590nm	14	3.5%	MOCVD	QW
ZnSe	Blue	512 nm	17	5.3%	MBE	DH
GaP:Zn-O	Red	650 nm	4.0	5.0%	LPE	H
GaP:N	Yellow-Green	565 nm	1.8	0.3%	LPE	H
GaAsP:N	Yellow	590 nm	1.0	0.3%	VPE	H
SiC	Blue	470 nm	0.02	0.02%	CVD	H

DH = Double Heterostructure, H = Homojunction Device, QW = Quantum Well

1.2.4 Luminous Efficiency

The luminous efficiency is expressed in lumens per watt. The term luminous efficiency refers to the energy conversion efficiency of the light source in converting electrical power into light we can see, luminous flux. To calculate the luminous efficiency we multiple the radiant power efficiency by the luminous efficacy factor which is wavelength dependent and peaks at 683 lumens/watt at 555 nm. The CIE luminosity curve for the visible wavelength range is shown in Figure 1.5 below, the solid line representing the luminous efficiency for a 100% efficient emitter. As is evident from this curve GaN and AlGaInP LEDs is exceeding the efficiency of incandescent light sources in the red to green regions of the spectra. By further improving the external quantum efficiencies of the III-V semiconductors we can expect that LEDs will one day be the most efficient light sources available.

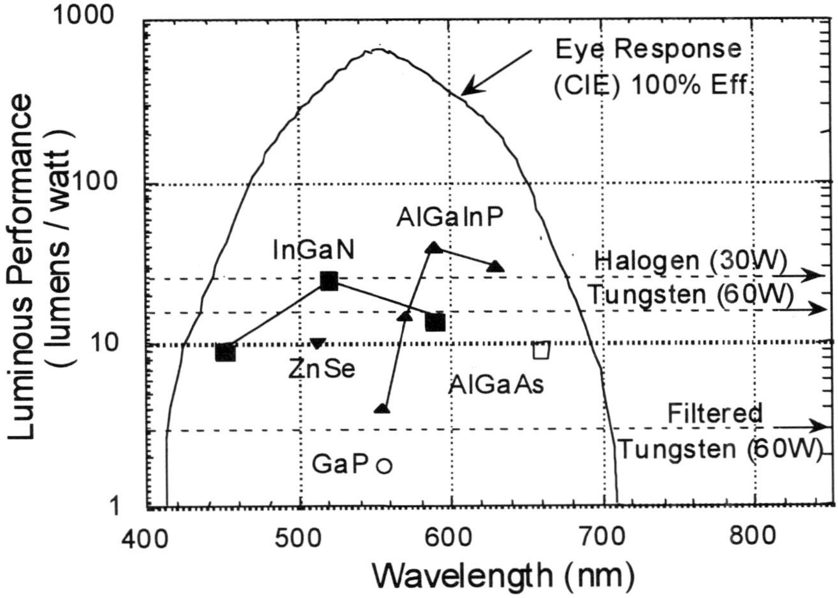

Figure 1.5 Luminous Efficiencies of LEDs over the visible range.

1.2.5 Injection Efficiency

The external efficiency is also influenced by the carrier injection efficiency into the active region. For injection into a p-semiconductor the injection efficiency is defined as the ratio of the electron current J_e to the total forward current J_f and can be expressed as:

$$n_{inj} = \frac{J_e}{J_e + J_h} = \frac{J_e}{J_f}. \tag{1.18}$$

From p-n junction theory (Shockley, 1950) we can define the injection efficiency in terms of relevant materials properties:

$$n_{inj} = \frac{D_n n L_p}{D_n n L_p + D_p p L_n}, \tag{1.19}$$

where D_n and D_p are the diffusion coefficients for electrons and holes, respectively. L_p and L_n are the minority carrier diffusion lengths, and n and p are the net electron and hole concentrations on the two sides of the junction. Typically in GaAs minority carrier diffusion lengths are one micron or more. In the nitrides the diffusion lengths are potentially much smaller. At thermal equilibrium the net electron and hole concentration can be expressed with the doping level of donors N_D and acceptors N_A:

$$n = n_i^2 / N_A, \tag{1.20}$$

and

$$p = n_i^2 / N_D. \tag{1.21}$$

Therefore to obtain a high injection efficiency we want to adjust the doping such that $N_D > N_A$. However, to optimize the optical efficiency the material must be kept transparent by light generated at the junction. Therefore free carrier absorption must be minimized by lowering the doping level in the heavily doped n-type material to below 10^{18} cm^{-3} (Varon *et al.*, 1981).

1.2.6 Heterojunction vs. Homojunction LED Materials

Both homojunction and heterojunction semiconductor structures can be employed for visible LEDs. By combining wide bandgap semiconductors with narrow bandgap semiconductor a "heterojunction" is formed and extremely desirable properties unattainable in homostructures become possible (Kroemer, 1963). In homojunction LEDs, the p and n materials are composed of the same energy gap semiconductor, while in heterojunction LEDs two materials of different bandgap are utilized. A typical GaAsP LED consists of a homojunction with a diffused p/n junction. Since the GaAsP is grown on an absorbing GaAs substrate only light emitted upwards from the junction escapes. This so called "bandgap engineering" (Capasso *et al.*, 1989) allows the LED designer to achieve higher luminous efficiencies than conventional homostructures. The two major benefits of heterostructure LEDs are the increased minority carrier injection efficiency and the ability to use wide-gap material. The wide-bandgap layers are transparent to the light generated in the narrow gap active layer and therefore do not contribute to photon reabsorption. The radiative efficiency in heterostructure LEDs is higher than homostructure LEDs because of the increased current injection efficiency at the heterojunction. In particular, since the radiative efficiency of the p-side is usually much higher than the n-side, higher efficiency devices are made by eliminating the minority carrier injection into the n-type material. By creating a heterostructure at the p/n interface a valence band discontinuity is formed which provides a hole potential barrier. By adding an additional heterostructure confinement at the other side of the active layer increases because the second heterointerface prevents the injected electrons from diffusing out of the narrow-gap layer.

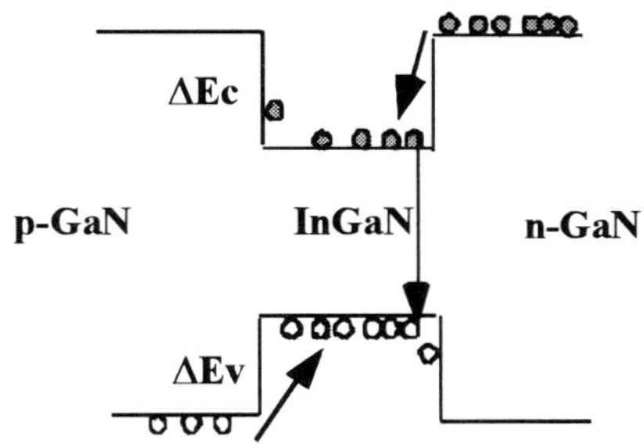

Figure 1.6 Energy band diagram for a heterojunction. LEDs in which the conduction band discontinuity Δ Ec and Valence band discontinuity ΔEv provide a barrier to confine carriers into active region.

Figure 1.6 illustrates the energy band diagram for heterojunction flat-band conditions where the applied bias just cancels the built in field. For heterojunctions the bandgap discontinuity affords a significant improvement in the injection efficiency. The conduction band discontinuity ΔEc and the valence band discontinuity ΔEv, produce a barrier to electron and hole flow, respectively. For a p-n heterojunction the ratio of electron to hole current can be expressed as (Casey *et al.*, 1978):

$$\frac{J_e}{J_h} = \frac{D_n n_{po} L_p}{D_p p_{no} L_n}, \tag{1.22}$$

where L_p and L_n are the minority carrier diffusion lengths in the p and n material, respectively. D_n and D_p are the electron and hole diffusion, respectively. From Fermi-Dirac statistics the electron-hole density product is a function of the materials energy gap:

$$np = n_i^2 = N_e N_v \exp(-E_g / kT). \tag{1.23}$$

Where N_c and N_v are the conduction and valence band effective density of states, respectively. Therefore the electron to hole injection ratio now becomes a function of the energy gap difference:

$$\frac{J_e}{J_h} = \frac{D_n L_p}{D_p L_n} \frac{n_o}{p_o} \frac{N_c^p n_v^p}{N_c^n n_v^n} \exp(\Delta E_g / kT). \tag{1.24}$$

For most compound semiconductor heterostructures, the exponential term in the above equation dominates and a high electron to hole injection efficiency can be achieved as long as the barrier height is sufficient. This nearly ideal single-sided injection of electrons into the active layer makes the double heterostructure GaN/InGaN, and AlGaInP/InGaP LEDs highly efficient. The energy gap differences also minimize the effect of the doping level on the injection efficiency, which dominated the homojunction LED efficiency. Selection of the wide gap material must take lattice matching into consideration to avoid forming dislocations near the interface that would act as non-radiative recombination centers and therefore significantly reduce the radiative efficiency.

1.2.7 Quantum Well LEDs

The quantum well (QW) LED is a special class of heterojunction LED in which the thickness of the active region is less than the deBroglie wavelength of the electron in the semiconducting materials. Basically the electron and hole are confined so tightly in space that discrete energy levels arise due to the wave-particle nature of the electron and hole. This effect causes electrons and holes in the subbands to recombine and emission is blue shifted to larger energies than the bulk DH LEDs. The details of emission in quantum structures are especially important in laser diodes and will be discussed in more detail in later chapters.

1.3 LED MATERIALS SELECTION

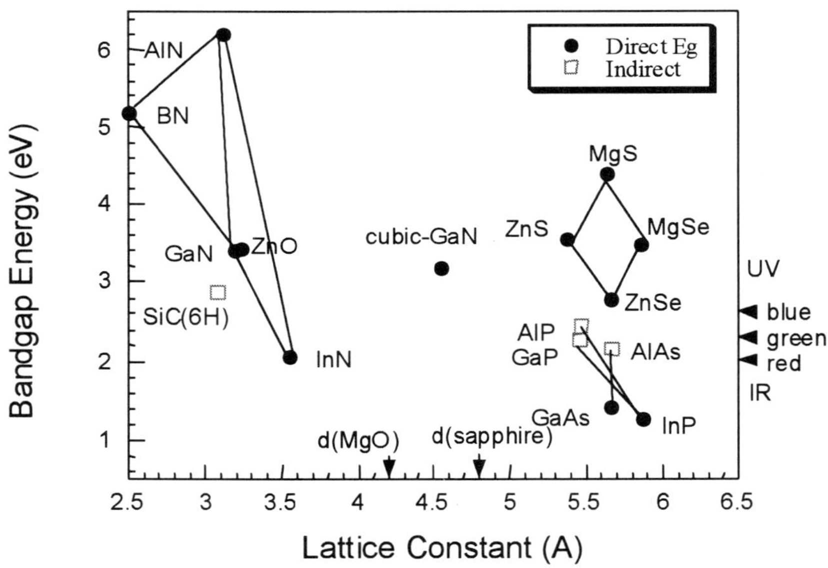

Figure 1.7 Bandgap energy versus bond length for various semiconductors. Note the general increase in bandgap as the bond length decreases.

1.3.1 Energy Band Structure/Lattice Constants

The band gap in the (Al,Ga,In)N based materials system ranges from 1.9eV (InN), 3.4 (GaN) to 6.2 eV (AlN). The band structure is currently thought to be a direct bandgap across the entire alloy range. Therefore, as illustrated in Figure 1.7, almost the entire visible range and deep UV wavelengths are spanned in the group-III nitride alloy system. This direct bandgap is especially fortuitous as it allows for high quantum efficiency light emitters to be fabricated in the group-III nitride system. Band bowing has not been accurately determined for all the alloys and is therefore not including in the bandgap diagram shown.

1.3.2 GaN Physical Properties

The Group-III nitrides possess several remarkable physical properties which make them particularly attractive for reliable solid state device applications. The wide bandgap materials possess low dielectric constants with high thermal conductivity pathways. As shown in Table 1.2 Group-III nitrides exhibit fairly high bond strengths and very high melting temperatures. The large bond strengths could possibly inhibit dislocation motion and improve reliability in comparison to other II-VI and III-V materials. In addition, the nitrides are resistant to chemical etching and should allow GaN based devices to be operated under harsh environments. These properties may lead to a device with superior reliability.

Table 1.2 Thermal Conductivities energy gaps, bond strengths and mobility of III-V nitrides (Slack, 1973, Sze, 1981 and Harrison, 1980).

	AlN	GaN	InN
K (W/m °K)	320	120	-
Tm (°K)	3500	2800	2150
PN_2 (KBar)	60	45	37
Bond(eV/bond)	2.88	2.2	1.93
Mobility(300K)	-	900cm^2/Vs	-

1.3.3 GaN based LED Structures

Two types of Blue LEDs are commercially available: Laterally contacted devices with 2.5 nm thick single quantum well (SQW) InGaN active regions on sapphire substrates from Nichia Chemical Industries, Ltd., and vertically contacted GaN/AlGaN double heterostructure (DH) devices on conducting SiC substrates by Cree Research Inc. The laterally contacted device is constructed from a low temperature GaN buffer layer first grown at 550°C, followed by a subsequent n-GaN layer, a 2.5 nm thick InGaN undoped active region, a p-$Al_{0.1}Ga_{0.9}N$ cladding, and Mg-doped p-GaN layer. Evaporating a p-type contact consisting of NiAu, and a thin NiAu current spreading layer completed the LED process. After chlorine reactive ion etching (RIE) of the mesa, TiAl was evaporated as n-contact metal. The SQW LEDs are extremely efficient currently, the best external efficiency achieved for the blue and green LEDs is 10% and 12% external quantum efficiencies, respectively (Nakamura *et al.*, 1997).The white LEDs are approximately 5% and have a luminous efficacy of 10 lumens/watt, which is becoming competitive with existing incandescent sources. In addition, newly developed InGaN amber LEDs are superior to conventional AlGaInP amber LEDs in terms of temperature performance. The wavelength shift as a function of temperature is much smaller in GaN than in GaAs based LEDs. Mukai showed a comparison of InGaN to AlGaInP based LEDs at elevated temperature 80°C in which the InGaN LED light output is decreased by only 20%, whereas the AlGaInP LED light output is down by 70% (Mukai *et al.*, 1998). This excellent temperature performance for GaN based LEDs in comparison to conventional

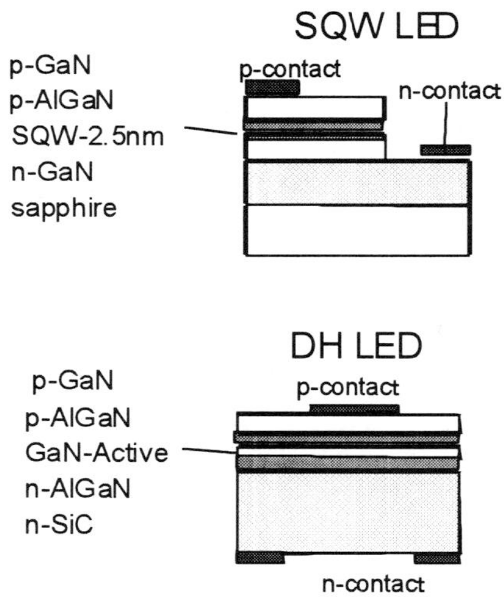

Figure 1.8 Cross-section of two types of blue GaN LEDs that are commercially available. Because of the increase optical extraction efficiency, a transparent substrate (TS) either sapphire or SiC is employed. The SQW device from Nichia uses lateral current conduction, while CREE uses a vertically conducting buffer layer.

GaAs based LEDs is also seen when comparing to GaP green and AlGaAs red LEDs. The InGaN yellow LEDs are not as bright as transparent substrate (TS) AlGaInP LEDs, but their performance rivals absorbing substrate AlGaInP LEDs. At 20 mA the new yellow LED is 4Candela and has a external efficiency of 3.3 % at 594 nm (Dingle *et al.*, 1994). With these achievements nitride LED's rank among the highest efficiency LED's on the market as shown in Table 1.1 The fundamental improvement in the material quality was attributed in part to the invention of a two-flow MOCVD reactor which features a vertical subflow and horizontal source flow (Nakamura, 1991a). This subflow reduces the boundary layer thickness, thus providing a higher concentration of active nitrogen species at the substrate surface. The evolution of LEDs to efficiency is comparable and in some cases exceeds tungsten light bulbs as shown in Figure 1.5. The bright blue/green GaN LEDs that have been achieved by Nichia are undergoing testing for traffic signal replacement (Nakamura, 1995b). This is a truly remarkable achievement given that the dislocation densities in these LEDs are approximately $2\text{-}10\times10^{10}cm^{-2}$. The high defect densities would normally severely reduce any radiative emission from GaAs based LEDs where dislocation densities of about $10^4/cm^{-2}$ limit the efficiency (Lester *et al.*, 1995).

The second type of LED structure is grown on SiC substrates and is made by Cree Research Inc. These devices feature a MOCVD grown GaN-AlGaN p-n junction grown on 6H-SiC substrates. SiC has the advantage of having a smaller lattice misfit to GaN and the possibility of providing n-doped substrates. Another advantage is that the vertical conduction pathway allows for a drop in replacement part similar to the other GaP and GaAs based structures which have the same vertical geometry. The high thermal conductivity of SiC substrates might also benefit the high temperature high output power applications of GaN LEDs. The GaN/SiC output power is lower than that available on sapphire and is commercialy available with 0.5 mW and external quantum efficiency of 1.0% (Cree Research Inc., 1995).

A key development in the achievement of high brightness blue LED device structures is the use of the ternary compound InGaN. High quality films of this alloy allowed the extension of GaN LEDs into the visible spectrum and the ability to use double heterostructure technology. The use of double heterostructure is very effective for obtaining good carrier confinement and single-sided injection into the active layer. Good quality InGaN was initially demonstrated by Yoshimoto (Yoshimoto *et al.*, 1991) using a high indium source flow and a high growth temperature of 800°C. At these temperatures the high indium source flow is necessary to maintain an overpressure of indium as indium is desorbing from the InGaN alloy. Nakamura *et al.* further improved the structural and optical quality of InGaN films using the two-flow MOCVD approach. In the first LEDs, it was reported that Co-doping with Zn and Si was found necessary to extend the LEDs into the blue and blue-green portion of the spectrum. Subsequently, growths of high indium composition in thin single quantum well (SQW) LEDS allowed direct transition in the blue, green and yellow spectral regions (Nakamura *et al.*, 1995a).

1.4 CRYSTAL GROWTH

Many techniques have been investigated for the crystal growth of III-V nitride thin films. Recently, metalorganic chemical vapor deposition (MOCVD) and molecular beam epitaxy (MBE) have grown the highest quality GaN based materials.

1.4.1 MOCVD Growth

MOCVD is non-equilibrium growth technique, which relies on vapor transport of the precursors and subsequent reactions of group-III alkyls and group-V hydrides in a heated zone. The MOCVD technique originated from the early research of Manasevit (1968) who demonstrated that triethylgallium (TEGa) and arsine deposited single crystal GaAs pyrolytically in an open tube cold-wall reactor. Manasevit and co-workers (Manasevit *et al.*, 1969 and 1979) subsequently expanded the use of this technique for the growth of $GaAs_{1-y}P_y$, $GaAs_{1-y}Sb_y$, and Al containing compounds. Composition and growth rate are controlled by precisely controlling the mass flow rate and dilution of the various components of the gas stream. The organometallic group III sources are either liquids, such as trimethylgallium (TMGa) and trimethylaluminum (TMAl), or solids such as trimethylindium (TMIn). The organometallic sources are stored in bubblers through which a carrier gas (typically hydrogen) flows. The bubbler temperature is to precisely control the vapor pressure over source material. The carrier gas will saturate with vapor from the source and transport the vapor to the heated substrate. The group V sources are most commonly gaseous hydrides, e.g. ammonia NH_3 for nitride growth. Dopant materials can be metal organic precursors (diethylzinc, DEZn, cyclopentadienylmagnesium, Cp_2Mg) or Hydrides (silane, or disilane Si_2H_6). The substrate usually rests on a block of graphite called a susceptor that can be heated by a RF coil, resistance heated, or radiantly by a strip heater. An important feature of the MOCVD process is that the walls are kept substantially colder than the heated interior substrate which minimizes wall deposits and reduces reactant depletion effects which hot-walls would cause. The basic MOCVD reaction describing the GaN deposition process is:

$$Ga(CH_3)_3(v) + NH_3(v) \rightarrow GaN(s) + 3CH_4(v)(l). \tag{1.25}$$

However, the details of the reaction are not well known and the intermediate reactions are thought to be complex. Further work is needed to understand the fundamentals of the crystal growth process.

Both atmospheric-pressure and low-pressure MOCVD reactors are employed by various researchers in the growth of GaN. Atmospheric pressure reactors are favored because of the high partial pressures of ammonia, or nitrogen containing precursor is achievable. The majority of the research groups in Japan have utilized atmospheric pressure for this reason. The breakthrough in bright blue LEDs was achieved by Nakamura *et al.* (1994b) using a modified MOCVD system. Nichia Chemistries Inc. has employed a novel two-flow approach that yielded excellent film quality. As shown in figure 1.9, in this reactor sources are supplied from a horizontal inlet, and a vertical subflow than drives the reactants to the growing film surface.

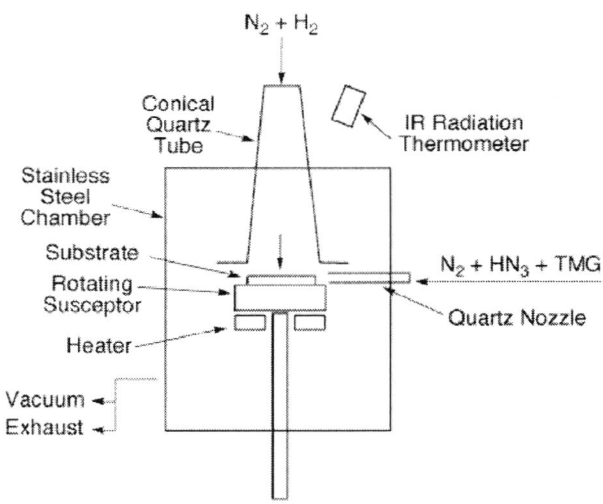

Figure 1.9 Two-flow MOCVD approach (After Nakamura *et al.*)

MOCVD Reactor designs for GaN growth must overcome problems presented by high growth temperatures, prereactions, and flow and film non-uniformity. Typically in GaN growth very high temperatures are required because of the high bond-strength of the N-H bond in ammonia precursors. This fact is compounded by the thermodynamic tendency of ammonia to prereact with the group-III metalorganic compounds to form non-volatile adducts. These factors contribute to the difficulties currently facing researchers in the design and scale-up of III-V nitride deposition systems. Much research activity is needed in the scale-up and understanding of the mechanism of gallium nitride growth by MOCVD.

1.4.2 MOCVD Systems for Production

Currently, several types of MOCVD reactor geometries are being developed for the mass production of GaN based materials and devices. Both atmospheric pressure and low-pressure systems are being produced by the major MOCVD equipment manufacturer (Aixtron GmbH, Emcore Corp., Nippon Sanso, and Thomas Swan, Ltd.). Three types of geometries are illustrated in Figure 1.10. This figure shows a high speed rotating disc reactor for LP-MOCVD, a closed space RDR for atmospheric pressure growth, and a two-flow horizontal flow pancake reactor. All three reactor designs are producing high quality GaN materials and it is not the intent of the author to judge one superior to the other. The benefits to each approach will be specific to the ultimate device and materials being grown. For the high speed vertical reactors from Emcore Corporation the main advantage is the large reactor size available (>6 wafers) and the thin uniform boundary layer caused

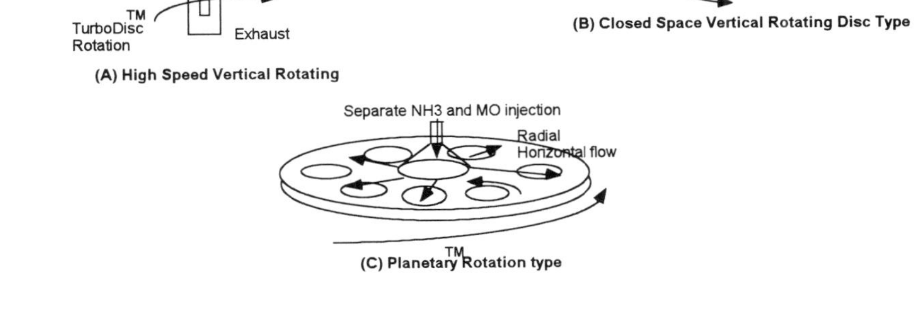

Figure 1.10 Three types of MOCVD production reactors for GaN. (a) High speed vertical rotating style (b) Closed space rotating disc type (c) Planetary rotation with radial horizontal flow.

by the high rotation speed (up to 1500 rpm). The closed space RDR has the benefit of atmospheric pressure operation because the low free height eliminates free convection. The two-flow horizontal planetary rotation™ reactor from Aixtron can also be operated at near atmospheric pressure and can accommodate large wafer volumes (>7 wafers). The selection of any reactor has to be carefully considered against factors such as: material quality, high throughput, reproducibility, maintenance, and source usage.

1.4.3 Molecular Beam Epitaxy (MBE)

Recently, many researchers have begun developing MBE for growth of the III-V nitrides. Several approaches have been investigated for supplying an atomic source of nitrogen. RF plasma and electron cyclotron resonance (ECR) microwave plasma sources are the two most successful techniques discovered to date (Lei *et al.*, 1991, Strite *et al.*, 1991 and Paisley *et al.*, 1989). In these systems the plasma source is used to crack molecular nitrogen. The plasma sources use a cylindrical cavity geometry to efficiently couple microwave energy into the nitrogen discharge area. The plasma stream is a complex mixture of atomic, molecular, and ionic N radicals. When using ECR sources a tradeoff between growth rate and ion damage has been observed (Lei *et al.*, 1991); under normal ECR use the flux of low energy reactive N species is so low that only low growth rates of 500Å/hr can be achieved. At higher microwave powers high growth rates can be achieved, but ion damage leading to deep levels and semi-insulating electrical properties is observed. Special care must be taken to limit the damage that high energy ionic species might cause if they from reach the surface. Mostakas *et al.*, (1993) have developed a apertured solenoid source which significantly reduced ion damage and enabled them to achieve the first GaN homojunction LED to be grown entirely by MBE (Molnar, 1995). Figure 1.11 shows one type of MBE nitride system that incorporates a nitrogen plasma source to achieve GaN materials. A major advantage of MBE for

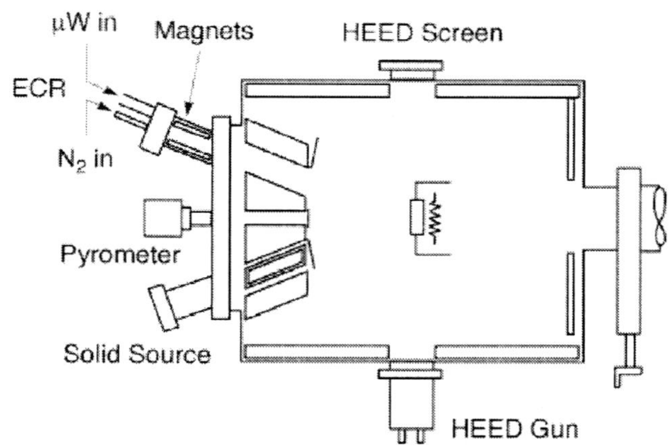

Figure 1.11 MBE nitride system using nitrogen plasma cracking.

nitride growth is the low growth temperature that can be achieved due to the atomic nitrogen source. This is in contrast to MOCVD which must employ high growth temperatures (>1000°C) to crack the ammonia molecules. The lower growth temperature should result in lower thermal stress upon cooling, less diffusion, and reduced alloy segregation. This is especially important in the AlGaN alloys which posses a large mismatch in the thermal expansion coefficient. Another advantage of MBE is that as-grown p-GaN films by MBE are not passivated due to the absence of atomic hydrogen species during the growth (Mostakas *et al.*, 1993).

1.4.4 Chloride Vapor Phase Epitaxy

Vapor Phase Epitaxy (VPE) is another technique being investigated for GaN growth In this technique Cl is used to transport the group-III species instead of organometallic sources. This has a distinct advantage in that large growth rates can be achieved with this technique over the previously described method. In fact Maruska and Tietjen (1969) grew the first high quality GaN thin films. In contrast to MOCVD, which is a non-equilibrium cold-wall reactor based technique, Chloride based VPE is an equilibrium based hot-wall reactor technique.

Very high growth rates of up to 40μm/hr can be achieved with the VPE technique. This has potential to generate thick GaN layers to actually form substrates. Akasaki *et al.* has used this technique to grow thick films on ZnO/Sapphire substrates and subsequently lift-off the layers and thus create a thick vapor deposited GaN "substrate". Considerable research is needed to control In segregation by VPE so that this technique can be applied to the InGaN alloys.

VPE relies on the gaseous transport of the constituent column III and column V sources to the deposition zone. Mass flow controllers achieve precise control of the molar flowrates of the two column V source. This allows independent control of the partial pressures that govern alloy composition grading. A high degree of flexibility in the composition and doping levels is achieved by accurate mass flow control. The versatile nature of VPE allows synthesis of the important III–V compounds such as GaN, ZnAs, GaP, InGaP, InAsP, and GaAsSb. However, the important ternary $Al_xGa_{1-x}As$ compounds and more importantly AlGaN compounds cannot be grown by VPE since the AlCl species reacts with the hot quartz walls resulting in unwanted impurities in the epitaxial film (Kressel, 1977a).

In contrast to MOCVD, Hydride-VPE is a reversible equilibrium based process in which the crystal growth is achieved by lowering the growth temperature of the substrate relative to the temperature of the reaction zone. For $GaAs_{1-x}P_x$ growth the Ga reservoir portion of the furnace is held at approximately 800°C. GaCl is formed by introducing a mixture of HCl and H_2 to the Ga reservoir. The reaction zone is kept at 1100°C to pyrolyze the hydrides NH_3. During deposition the substrate is kept at a lower temperature of approximately 750°C so that the vapor becomes supersaturated and deposition occurs.

The hydride-VPE process can be described by the following set of chemical reactions: gallium transport by chloride formation in the gallium zone:

$$2HCl(g) + 2Ga(l) \rightarrow 2GaCl(g) + H_2(g) \qquad T\sim 800°C, \qquad (1.26)$$

transport and reaction with the hydrides in the reaction zone:

$$GaCl(l) + NH_3(g) \rightarrow GaN\,(s) + HCl(g) + H_2(g) \qquad T{\sim}700°C. \qquad (1.27)$$

Ammonia is then thought to decompose at the surface and react with either GaCl or Ga to form GaN. However, further work on VPE reaction mechanisms is needed to clarify the reaction pathways. One of the main advantages of VPE is the ease of scale-up to a high capacity commercial process.

1.5 GROUP-III NITRIDE MATERIALS GROWTH ISSUES

Several major breakthroughs in the MOCVD growth of high quality GaN and InGaN enabled the fabrication of device quality group-III nitride materials. In this section we will discuss several of these key growth issues such as: (i) Substrates and the implementation of low temperature nucleation layers to provide a homogeneous nucleation on the sapphire substrates (Hiramatsu *et al.,* 1991), (ii) the achievement of p-type doping in GaN by overcoming the hydrogen passivation of acceptors (Amano *et al.*, 1989 and Nakamura *et al.*, 1992a), (iii) Growth of GaN, AlGaN, and InGaN with high optical and electrical quality and (iv) growth of GaN/InGaN heterostructures and quantum wells.

1.5.1 Substrates

Several problems in the epitaxial growth of nitrides originate from the non availability of single crystalline GaN substrates or other high quality single crystalline substrates with the same lattice parameters as GaN. For this reason, so far, most of the epitaxial growth of nitrides has been performed on sapphire or SiC substrates. In both cases, problems due to the lattice mismatch between the nitride epi-layer and the substrate (16% for sapphire and 3.5 % for SiC) have to be overcome. One of the major breakthroughs in the growth of device quality group-III nitride material was the implementation of nucleation layers. Using sapphire substrates, thin AlN or GaN nucleation layers deposited at temperatures between 500 and 750°C showed to remarkably improve the quality of the GaN film grown at temperatures above 1000°C (Akasaki, *et al.* 1989 and Nakamura, 1991b). In the case of SiC substrates, the growth is usually initiated with the deposition of a thin AlN nucleation layer at high temperatures (Weeks *et al.,* 1995).

By this means, GaN material of comparable quality on both types of substrates could be achieved. Since so far, most GaN growth has been performed on c-plane sapphire substrates, in the following section just the growth on c-plane sapphire will be discussed.

Additional substrate materials are currently being examined to determine if the properties of the GaN thin films can be enhanced by improved structural matching. From Figure 1.7 we can see that in addition to sapphire several other substrates offer potentially much better latticed and thermal matching. To this end, 6H-SiC, ZnO, and 3C-SiC, MgO are alternative substrate materials. ZnO has a wurtzite structure with lattice constants of (a=3.32Å, c=5.213Å) and thus offers a better structural match to the equilibrium wurtzite nitride. 3C-SiC and MgO are both cubic zinc-blende structures having better structural and thermal match to the nitrides than sapphire. 3C-SiC and MgO have cubic lattice constants of a=4.36Å and a=4.22Å, respectively. Although the nitrides are most commonly observed as the wurtzite (2H) polytype, they can also crystallizes in a metastable zinc-blende structure (a=4.52Å) when using non-equilibrium based growth techniques. The identification of a suitable substrate material that is lattice matched and thermally compatible with GaN wurtzite structure (a=3.19Å, c=5.185Å) will alleviate many of the difficulties associated with the deposition of device quality material.

1.5.2 Nucleation Layer technology

The deposition of a low temperature "buffer" layer of AlN on sapphire substrates was a key discovery in improving surface morphology and crystalline quality of GaN. Yoshida *et al.* (1983) was the first to recognize the importance of depositing a low temperature buffer layer. The surface morphology of the subsequent thick GaN layers can be markedly improved if thickness of the buffer layer is optimized. Subsequent work by Akasaki *et al.* (1988) resulted in significantly improved nitride films in which stimulated band-edge emission was observed (Amano *et al.*, 1990b). In this work it was reported that the AlN buffer layer relaxes the strain between GaN and sapphire, and promotes lateral growth of GaN. TEM results show that the buffer layer is highly dislocated and subsequent layers have a much lower dislocation density.

Structural characterization by transmission electron microscopy (TEM), atomic force microscopy (AFM), and high resolution X-ray diffractometry (XRD) revealed that the nitridation of the sapphire substrate plays an important role in controlling the defect structure and therefore dislocation densities in overgrown GaN films (Heying *et al.*, 1995). Based on these material studies, the understanding of X-ray rocking curves and their correlation with nucleation processes, material properties and defect structures in GaN films on sapphire was elucidated. It has been reported that changing the nitridation time of the sapphire surface before growth initiation results in a variation of the overall dislocation density between 10^{10} and $10^{8}cm^{-2}$. Based on these comprehensive structural characterizations, GaN thin films with record low dislocation densities of 1.7 x 10^{8} cm^{-2}.

1.5.3 Growth and doping of GaN

The best reported mobility for bulk GaN is by Nakamura *et al.* (1992b), in which a mobility of 900 at 300K and a low free carrier concentration of $2\times10^{16}cm^{-3}$ were observed. Exact defect density in these films is not known. In our studies the best films were nominally undoped 4 µm thick GaN films in which we observed 300 K mobilities of $780cm^{2}/Vs$ (n:300K = $6\times10^{16}cm^{-3}$). On these samples dislocation densities of $4\times10^{8}cm^{-2}$ for GaN on sapphire substrates was observed in cross-sectional TEM measurements (Kapolnek *et al.*, 1995). Even at excitation levels as low as 2.2 mW/cm^{2}, the 300K PL is dominated by the near band edge emission .

One of the key breakthroughs in the development of GaN technology was the achievement of p-doping using Mg and LEEBI treatment. Amano (Akasaki *et al.*, 1989) observed that under low electron beam irradiation (LEEBI) Mg-doped GaN exhibited much lower resistivity and the PL properties drastically improved. This achievement subsequently lead to the development of p/n GaN diodes with good turn on characteristics. Nakamura *et al.* (1993) built upon this fundamental breakthrough to achieve even higher p-doping and uniform activation of Mg by using high temperature thermal annealing under a nitrogen ambient. The passivation requires post growth treatment for MOCVD material to activate the dopants. During the growth, interstitial hydrogen is incorporated and a H-Mg acceptor complex forms which passivates the acceptor. This H-Mg bond can be broken by a high temperature annealing step under a inert environment. This work demonstrated that hydrogen compensation of Mg in the MOCVD growth of GaN was the principal problem that plagued previous researchers. High room temperature p-doping is further complicated by the high activation energy of Mg as the most commonly used dopant (270 meV) and the passivation of acceptors with hydrogen during CVD growth. The binding energies of dopants are dependent of the dielectric constant and effective mass of the material. The nitride system has low dielectric constant (GaN, $\epsilon(0)=9.5$) and large effective masses (GaN, $m_e=0.2m_0$, $m_h=0.75m_0$) resulting in large binding energies. This is especially pronounced in p-type doping, when comparing GaN to GaAs the acceptor levels are very deep because of the large hole mass. This has led to difficulties in high p-type doping. This is the result of two effects: high n-type background concentration compensating the p-dopant and the incomplete activation of the dopants at room temperature. The low p-type doping (typical values are $10^{17}cm^{-3}$) leads to high contact resistances and problems with current spreading. Further work on increasing the p-doping level and developing new p-dopants will result in substantial payoff in producing LEDs and lasers with lower operating voltages and higher power efficiencies.

Doping n-type is rather straightforward in GaN with silicon being the typical n-dopant. The as-grown material is a typical unintentionally n-type, which is widely believed to be due to intrinsic nitrogen vacancies. The Si donor lies just below the conduction band (Ea=15-25 meV). Therefore well-controlled n-type doping can be easily accomplished using silicon as the donor. The typically MOCVD precursor for n-type doping are silane (SiH_4) and disilane (Si_2H_6) which are typically diluted with hydrogen in the 200ppm range. Doping levels between 1×10^{17} to $2\times10^{19}cm^{-3}$ are easily achieved in the doping of GaN with silane.

1.5.4 Growth of AlGaN and AlGaN/GaN Heterostructures

High quality AlGaN films have been demonstrated by atmospheric pressure MOCVD as well as epitaxy performed under low pressure conditions. Parasitic reactions between TMAl and NH_3 are much more likely to occur than with TMGa. Two-flow-channel reactors with separate injection of the Group-III and Group-V precursors or reactor operation under low pressure conditions have been shown to successfully prevent excessive prereactions in the gas phase.

At growth temperatures below 1100°C, the mole fraction of aluminum in the AlGaN epitaxial layer was found to be almost directly proportional to the mole fraction of TMAl in the gas phase. At temperatures above 1100°C, the incorporation efficiency of gallium atoms decreased. This behavior was explained by a decreased sticking probability of Ga molecules at this high temperatures (Hirosawa *et al.*, 1993). K. Sayyah *et al.* (1986) reported on a decrease of the gallium incorporation efficiency with increasing TMAl flow and explained this behavior by an surface kinetic model and competitive adsorption. The same authors found an decrease of the gallium incorporation efficiency with increasing SiH_4 dopant flow also (Sayyah, *et al.*, 1986).

High quality AlGaN/GaN heterostructures are characterized by a very high mobility of the two dimensional electron gas at the interface (Kahn *et al.*, 1991) Values as high as 1500cm^2/Vs at room temperature have been achieved in the authors' laboratory (Wu *et al.*, 1996). The optical properties of AlGaN/GaN quantum wells (Kahn, *et al.*, 1990) were found to be determined by both quantum and strain related effects (Krishnankutty *et al.*, 1992a and 1992b). MOCVD AlN films showed a full width of half maximum of the (002) X-ray rocking curve peak as low as 97 arcseconds (Saxler *et al.*, 1994). AlN/GaN Superlattices of high structural and optical quality have also been fabricated by switched atomic layer MOCVD (Kahn *et al.*, 1993).

1.5.5 Growth of InGaN and InGaN/GaN Heterostructures

Growth of high quality InGaN is necessary to obtain good electrical and optical characteristics from LEDs. However, the growth of high quality InGaN has proven to be more difficult than GaN. InGaN growth needs to be performed at much lower temperatures than that of GaN, due to the low dissociation temperature of InN (Matsuoka *et al.*, 1988 and Koukitsu *et al.*, 1996). Furthermore, the decomposition of ammonia becomes less efficient with decreasing temperature due to the high kinetic barrier for breaking the nitrogen-hydrogen bonds. The growth of InGaN has to be performed at temperatures below 850°C because of the high volatility of indium at common GaN growth temperatures of above 1000°C. But even on InGaN layers grown at temperatures below 800°C, indium droplet formation was observed (Shimizu *et al.*, 1994).

The indium incorporation into the InGaN films strongly increases with decreasing growth temperature. Also, the relative indium incorporation coefficient k_{In} , defined as:

$$k_{\text{In}} = xs_{\text{In}} \times f_{\text{TMGa}} \,/\, (1\text{-}xs_{\text{In}}) \times f_{\text{TMIn}}. \tag{1.28}$$

(xs: In-In composition in the solid) increases with increasing growth rate at a given temperature, as shown in Figure 1.12. This indicates, that the incorporation of indium is limited by the evaporation of indium species from the surface. The tendency for evaporation decreases at lower temperature and/or increasing growth rate, when the indium species become trapped by the growing layer. But at growth temperatures below 760°C, high quality InGaN films could be obtained only by reducing the growth rate to values equal or lower than 3 Å/s in the authors' reactor.

These films showed intense band edge related luminescence at room temperature (Varon et al., 1981). The full width at half maximum (FWHM) of the X-ray diffraction peaks was 6.1 and 6.6 arcmin for layers containing 9% and 20% indium, respectively.

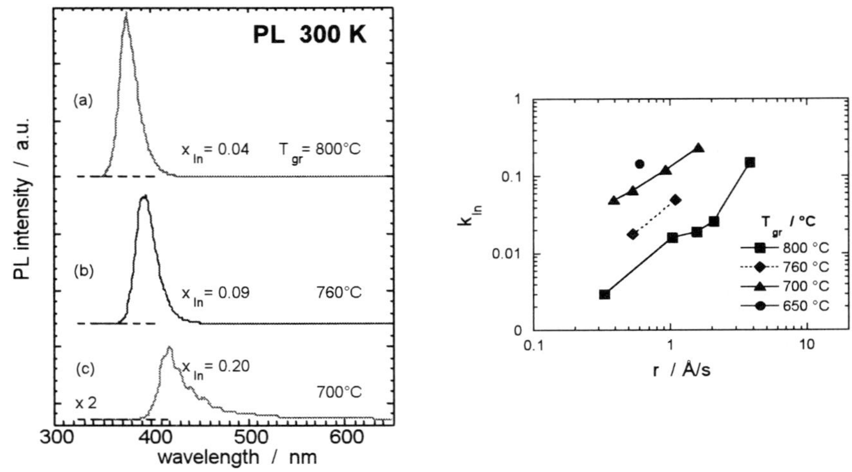

Figure 1.12 Effect of temperature on (a) PL wavelength and (b) Indium incorporation coefficient. Lower temperatures and higher growth rates result in higher indium incorporation.

1.6 CONCLUSIONS

The Group-III nitride semiconductors have emerged as the leading material for fabricating high efficiency and high reliability short wavelength emitters. MOCVD has emerged as the leading growth technology for depositing high quality GaN/InGaN heterostructure based devices. Well controlled doping, low background carrier concentrations, and high mobilities exceeding $800cm^2/Vs$ has been demonstrated by several research groups. The use of low temperature (500-600°C) GaN buffer layers and p-type GaN has resulted in high crystalline quality material on sapphire. Optimization of the MOCVD growth of InGaN/GaN based quantum structures has enabled high efficiency blue LEDs and laser diodes to be achieved. GaN based blue LEDs with external quantum efficiencies of 10% and 5mW output power at 20mA have recently been demonstrated, and are bright enough for full-color outdoor displays.

1.7 REFERENCES

Akasaki, I., Amano, H., 1988, *Appl. Phys. Lett.* **56**, p. 185.

Akasaki, I., Amano, H., Koide, Y., Hiramatsu, K. and Sawaki, N., 1989, *J. Cryst. Growth,* **89**, p. 209.

Amano, H., Sawaki, N., Akasaki, I. and Toyoda, Y., 1986, *Appl. Phys. Lett.* **48**, p. 353.

Amano, H., Kito, M., Hiramatsu, K. and Akasaki, I., 1989, *Jpn. J. Appl. Phys. Part II,* **28** (12), p. L2112.

Amano, H., Kitoh, M., Hiramatsu, K. and Akasaki, I., 1990a, *Gallium Arsenide and Related Compounds* 1989, (eds.) Ikoma, T., Watanabe, H., Bristol, *UKIOP*, p. 725.

Amano, H., Asaki, T. and Akasaki, I., 1990b, *Jpn. J. Appl. Phys. Lett.*, **29**, p. 205.

Capasso, F., Ripamonti, G., Tsang, W. T., Hutchinson, A. L. and Chu, S. N. G., 1989, *J. Appl. Phys.*, **65**(1), p. 388.

Casey, H.C. and Panish., M., 1978, *Heterostructure Lasers*, Academic Press, New York.

Craford, M.G., 1977, *IEEE Trans. Electron. Devices*, ED-**24**, p. 935.

Cree Research Inc., 1995, *Product Specification Sheet,* Super Blue LED C430-DH85.

Dingle, D., Shaklee, K.L., Leheny, R.F. and Zetterstrom, R.B., 1994, *Appl. Phys. Lett.*, **64**, p. 1377.

Harrison, W.A., 1980, *Elec. Structure and Properties of Solids Ed.* Freeman (San Francsico).

Henry, C.H. et al., 1976, *J. Luminescence*, **12/13**, p. 47.

Heying, B., Wu, X.H., Keller, S., Li, Y., Kapolnek, D., Keller, B., DenBaars, S.P. and Speck, J.S., 1995, *Appl. Phys. Lett.,* **68**(5) p. 643.

Hiramatsu, K., Itoh, S., Amano, H., Akasaki, I., Kuwano, N., Shiraishi, T. and Oki, K., 1991, *J. Cryst. Growth,* **115**, p. 628.

Hirosawa, K., Hiramatsu, K., Sawaki, N. and Akasaki, I., 1993, *Jpn. J. Appl. Phys.*, **32**, p. L1030.

Holynak, Jr. N. and Bavacqua, S.F., 1962, *Appl. Phys. Lett.*, **1**, p. 82.

Kahn, M.A., Skogman, R.A., Van Hove, J.M., Krishnankutty, S. and Kolbas, R.M., 1990, *Appl. Phys. Lett.,* **56**, p. 1257.

Kahn, M.A., Van Hove, J.M., Kuznia, J.N. and Olson, D.T., 1991, *Appl. Phys. Lett.,* **58**, p. 2408.

Kahn, M.A., Kuznia, J.N., Olson, D.T., George, T. and Pike, W.T., 1993, *Appl. Phys. Lett.*, **63**, p.3470.

Kapolnek, D., Wu, X.H., Heying, B., Keller, S., Keller, B.P., Mishra, U.K., DenBaars, S.P. and Speck, J.S., 1995, *Appl. Physics. Lett.*, **67** (11) p. 1541.

Koukitsu, A., Takahashi, N., Taki, T. and Seki, H., 1996, *Jpn. J. Appl. Phys.*, **35**, p. L673.

Kressel, M., 1977a, *Semiconductor Lasers and Heterojunction LEDs*, Academic Press, New York.

Kressel, M., *et al.*, 1977b, Academic Press, NY, p. 294.

Kressel, *M., et al.*, 1977c, Academic Press, NY, p. 295.

Krishnankutty, S. and Kolbas, R.M., Kahn, M.A., Kuznia, J.N., Van Hove, J.M. and Olson, D.T., J, 1992a, *Electron. Mat.*, **21**, p. 437.

Krishnankutty, S. and Kolbas, R.M., Kahn, M.A., Kuznia, J.N., Van Hove, J.M. and Olson, D.T., J, 1992b, *Electron. Mat.*, **21**, p. 609.

Kroemer, H., 1963, *Proc. IEEE*, **51**, p. 1782.

Lei, T., Fanciulli, M., Molnar, R.J., Moustakas, T.D., Graham, R.J. and Scanlon, J.D. , 1991, *Appl. Physics Lett.*, **59**, p. 94.

Lester, S.D., Ponce, F.A., Craford, M.G. and Steigerwald, 1995, D.A., *Appl. Phys. Lett,.* **66**, p. 1249.

Manasevit, H.M., 1968, *Appl. Phys. Lett.*, **12**, p. 156.

Manasevit, H.M. and Simpson, W.I., 1969, *Electrochem. Soc.*, **116**, p. 1725.

Manasevit, H.M. and Hess, K.L., 1979, *J. Electrochem. Soc.*, **126**, p. 2031.

Maruska, H.P. and Tietjen, J.J., 1969, *Appl. Physics Lett.* **15**, p. 327.

Matsuoka, T., Tanaka, H., Sasaki, T. and Katsui, A., 1988, *Inst. Phys. Conf. Ser., No 106, GaAs and related compounds*, chapter 3, p.141.

Molnar, R.J., Singh, R. and Mostakas,T., 1995, *Appl. Phys. Lett.,* **66**(3) p. 268.

Mostakas, T.D. and Molnar, R.J., 1993, *Mater. Res. Soc. Proc.*, **281**, p. 753.

Mukai, S., Narimatsu, H. and Nakamura, S., 1998, *Jpn. J. Appl. Phys. Part 2,* **37**, p. L479.

Nakamura., S., 1991a, *Jpn. J. Appl. Phys.,* **30**, p. 1620.

Nakamura., S., 1991b, *Jpn. J. Appl. Phys.,* **30**, p. L1705.

Nakamura., S., Iwasa N., Senoh M. and Mukai T., 1992a, *Jpn. J. Appl. Phys.,* **31**, p. 1258.

Nakamura., S., Mukai T. and Senoh M., J. 1992b, *Appl. Phys.,* **71**, p. 5543.

Nakamura., S., Senoh M. and Mukai T., 1993, *Jpn. J. Appl. Phys.*, **32**, p. L8.

Nakamura., S., Mukai T. and Senoh M., 1994a, *Appl. Phys. Lett.,* **64**, p. 1687.

Nakamura., S., Mukai T. and Senoh M., March 1994b, *Appl. Phys. Lett.,* **64**, p. 1687 .

Nakamura., S., Iwasa N., Senoh M., Nagahama, S., Yamada, T. and Mukai T., 1995a, *Jpn. J. Appl. Phys. Lett.,* **34**, p. L1332.

Nakamura., S., 1995b, private communication, Feb.

Nakamura., S. and Fasol, G., 1997, *The Blue Laser Diode*, Springer-Verlag, Heildelberg, Germany.

Paisley, M.J., Sitar, Z., Posthill, J.B. and Davis, R.F., 1989, *J. Vac. Sci. Technol.,* **A7**, p. 701.

Pankove, J.I., Miller, E.A. and Berkeyheiser, J.E., 1971, *RCA Rev.,* **32**, p. 383.

Saxler, A., Kung, P., Sun, C.J., Bigan, E. and Razeghi, M., 1994, *Appl. Phys. Lett.,* **64** , p. 339.

Sayyah, K., Chung, B.-C. and Gershenzon, M., 1986, *J. Cryst. Growth,* **77**, p. 424.

Shimizu, M., Hiramatsu K. and Sawaki, N., 1994, *J. Cryst. Growth*, **145**, p. 209.

Shockley, W., 1950, *Electrons and Holes in Semiconductors*, Van Nostrand Co., New York.

Slack, G.A., 1973, *Phys. Chem. Solids,* **34**, p. 321.

Strite, S., Ruan, J., Li, Z., Manning, N., Salavador, A., Chen, H., Smith, D.J., Choyke, W.J. and Morkoc, H., 1991, *J. Vac. Sci. Technol.,* **B9**, p. 1924.

Sze, S.M., 1981, *Physics of Semiconductor Devices*, p. 849.

Varon, J. *et al.* 1981, *IEEE Trans. Electron. Dev.*, **4**, p. 416.

Weeks, Jr. T. W., Bremser, M.D., Ailey, K.S., Carlson, E., Perry, W.G. and Davis, R.F., 1995, *Appl. Phys. Lett.* **67**, p. 401.

Wu, Y.-F., Keller, B.P., Keller, S., Kapolnek, D., Kozodoy, P., DenBaars, S.P. and Mishra, U.K., 1996, *Appl. Phys. Lett.* **69**, p. 1438.

Yoshida, S., Misawa, S. and Gonda, S., 1983, *J. Vac. Sci. Technol*, **B1**, p. 250.

Yoshimoto, N., Matsuoka, T., Sasaki, T. and Katsui, 1991, *A. Appl. Phys. Lett.* **59**, p. 2251.

Chapter 2
Theoretical Analysis of Optical Gain Spectra

Takeshi Uenoyama and Masakatsu Suzuki

2.1 INTRODUCTION

III-V nitrides are very attractive for Laser Diodes (LDs). Recently stimulated emission (Akasaki *et al.* 1995) and the pulse-lasing action (Nakamura *et al.* 1996) of the multiquantum well (MQW) laser diodes (LDs) by current injection at room temperature was reported. Furthermore, continuous-wave operation of MQW LDs with a lifetime of more than 10,000 hours has been achieved (Nakamura *et al.* 1997). Concerning the design of LDs, low threshold current density is one of the important requirements and it is tightly connected with the optical gain characteristics.

In this chapter, we study the optical gain spectra theoretically. Since LDs, where the population inversion is about to take place, have many carriers in the cavities, the optical gain must be investigated taking into account the many-body effects. Especially, the Coulomb interaction among the carriers plays important roles in the wide-gap semiconductors. Further, the GaN based LDs consist of the ternary layer of InGaN as the active layer in which is the compositional fluctuation of In content. It would cause the localized states and this is one of the issues to be clarified on how the localized states work in the laser oscillation. We have derived the formula of the optical gain using the linear response theory, based on the many-body states in Section 2.2. It is such a general formula that the mechanisms of the exciton state and the optical gain are treated on equal footing. The screening effect, bandgap renormalization and the Coulomb enhancement are also discussed there. In Section 2.3, we have studied the electronic properties of bulk GaN, AlN and GaN/AlGaN QW structures and discussed from a fundamental point of view, on the basis of the first-principles electronic band calculation and the ***k·p*** theory (Suzuki *et al.* 1995; Uenoyama and Suzuki 1995; Suzuki and Uenoyama 1996a, b, c, d, e). The practical approach of linking them provided the unknown physical parameters and made it possible to analyze the III-V nitrides and their QW structures. In Section 2.4, using the parameters, we concretely assess the optical gain spectra as the optical properties involving the valence band mixing, strain effects and many-body effects. As a result, a fundamental understanding of them was obtained, and it led to a guideline on material and device design with the III-V nitrides. Section 2.5 summarizes the chapter.

2.2 OPTICAL GAINS SPECTRA BY MANY-BODY APPROACH

2.2.1 Linear Response Theory

In this section, we mainly focus on the optical gain spectra using the linear response theory. Optical gain is the negative absorption coefficient which is proportional to the conductivity in terms of the recombination current. Let us derive the conductivity or susceptibility of the semiconductors.

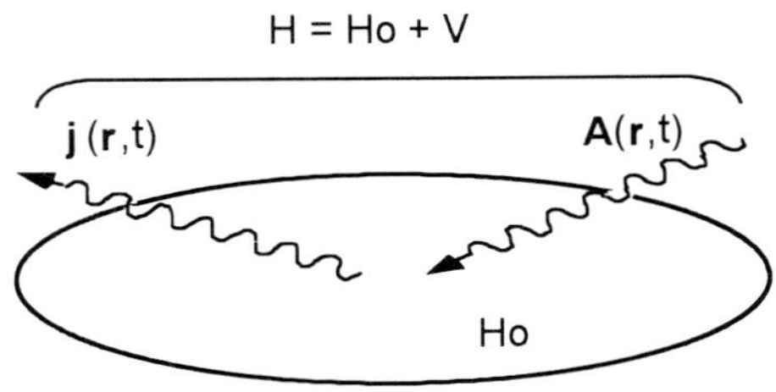

Figure 2.1 Image of the linear response.

Generally, in order to obtain some information from systems, we must apply the external fields, for example, electric field, magnetic field and so on, to the system. Then, we measure the response from the system against the external field, as shown in Figure 2.1. Before applying the external field, the system is maintained as an equilibrium state or a steady state. It might be a many-body state and can be specified by a certain Hamiltonian H_0. Suppose that the electric field, expressed by the vector potential $A(\boldsymbol{r},t)$, is applied to the equilibrium state as an external field, the system is no more in equilibrium and the current $j(\boldsymbol{r},t)$ would be induced. Then, we measure the current $j(\boldsymbol{r},t)$ as a response from the system. But the system is in non-equilibrium and should be described by the Hamiltonian H, given by

$$H = H_0 + V, \tag{2.1}$$

$$V = \frac{1}{c}\int d\boldsymbol{r} j(\boldsymbol{r})A(\boldsymbol{r},t),$$

where V is the energy, being composed by the external field $A(\boldsymbol{r},t)$ and the induced current $j(\boldsymbol{r},t)$. To evaluate the induced current, we use the density matrix operator at the non-equilibrium state. The time evolution of the density matrix operator is written by

$$\frac{\partial \rho}{\partial t} = i(\rho H - H\rho), \tag{2.2}$$

where H is the non-equilibrium Hamiltonian, including the external field. Then, we solve the equation of ρ within the linear approximation for the external filed $A(\boldsymbol{r},t)$. The density operator ρ is expressed as $\rho_0 + \rho_1$. ρ_0, is the density matrix operator at the equilibrium state and is dependent on the H_0. ρ_1 is caused by the external field and is the first order of the external field $A(\boldsymbol{r},t)$. Since ρ is time independent and commutes with H_0, the Equation (2.2) yields

$$\frac{\partial\tilde{\rho}_1}{\partial t} = i(\rho_0\tilde{V}(t) - \tilde{V}(t)\rho_0), \tag{2.3}$$

where

$$\tilde{\rho}_1 = e^{iH_0 t}\rho_1 e^{-iH_0 t},$$
$$\tilde{V}(t) = e^{iH_0 t}V(t)e^{-iH_0 t}.$$

Hereafter, the tilde of the operator means the unitary transformation like the interaction picture. $\tilde{\rho}$ is easily obtained by the integration of the Equation (2.3) and the induced current is calculated

$$\begin{aligned}\langle j(\boldsymbol{r},t)\rangle &= \mathrm{Tr}(\rho_1(t)j(\boldsymbol{r},t)) = \mathrm{Tr}(\tilde{\rho}_1\tilde{j}(\boldsymbol{r},t))\\ &= i\int_{-\infty}^{t} d\tau \mathrm{Tr}(\rho_0(\tilde{V}(\tau)\tilde{j}(\boldsymbol{r},t) - \tilde{j}(\boldsymbol{r},t)(\tilde{V}(\tau)))\\ &= \frac{i}{c}\int_{-\infty}^{t} d\tau\int d\boldsymbol{r}'\mathrm{Tr}(\rho_0(\tilde{j}(\boldsymbol{r}',\tau)\tilde{j}(\boldsymbol{r},t) - \tilde{j}(\boldsymbol{r},t)\tilde{j}(\boldsymbol{r}',\tau)))A(\boldsymbol{r}',\tau)\\ &= \frac{i}{c}\int_{-\infty}^{\infty} d\tau\int d\boldsymbol{r}'\chi(\boldsymbol{r},\boldsymbol{r}',t,t')A(\boldsymbol{r}',\tau)\end{aligned} \tag{2.4}$$

$$\chi(\boldsymbol{r},\boldsymbol{r}',t,t') = i\mathrm{Tr}(\rho_0(\tilde{j}(\boldsymbol{r}',t')\tilde{j}(\boldsymbol{r},t) - \tilde{j}(\boldsymbol{r},t)\tilde{j}(\boldsymbol{r}',t')))\theta(t-t'),$$

where χ is the susceptibility. Note that the *non-local* effect (Ishihara and Cho 1993) is considered in this approach since the field at $\boldsymbol{r}'$ affects the current density at $\boldsymbol{r}$.

χ is the retarded two-particle correlation function and is dependent on the time difference t-t'. Then its Fourier transformation is

$$\langle j(\boldsymbol{r},\omega)\rangle = \frac{1}{\mathrm{c}}\int d\boldsymbol{r}'\chi(\boldsymbol{r},\boldsymbol{r}',\omega)A(\boldsymbol{r}',\omega), \tag{2.5}$$
$$\chi(\boldsymbol{r},\boldsymbol{r}',\omega) = \int_{-\infty}^{\infty} d(t-t')e^{i\omega(t-t')}\chi(\boldsymbol{r},\boldsymbol{r}',t,t').$$

χ is the retarded function and is closely related to the temperature Green's function in terms of the analytical continuation. Then, we use the temperature Green's function instead of the retarded one since we can apply the perturbation theory (Mahan 1981).

The optical gain is the negative absorption coefficient $\alpha(\omega)$. The rate of the energy absorption in the semiconductor is given by the Joule heat,

$$Re\left[\int d\boldsymbol{r}\langle j(\boldsymbol{r},\omega)\rangle i\frac{\omega}{c}A^{*}(\boldsymbol{r},\omega)\right].$$

The absorption coefficient is given by the ratio between the absorbed energy and the input energy per unit length, as

$$\alpha(\omega)=Re\left(\frac{\int d\boldsymbol{r}\langle j(\boldsymbol{r},\omega)\rangle i\frac{\omega}{c}A^{*}(\boldsymbol{r},\omega)}{v_g\int d\boldsymbol{r}\frac{\varepsilon_0\omega^2}{4\pi_g^2}|A(\boldsymbol{r},\omega)|^2}\right)$$

$$=Im\left(\frac{4\pi\int d\boldsymbol{r}d\boldsymbol{r}'A(\boldsymbol{r}',\omega)\chi(\boldsymbol{r},\boldsymbol{r}',\omega)A^{*}(\boldsymbol{r},\omega)}{c\omega\bar{n}\int d\boldsymbol{r}|A(\boldsymbol{r},\omega)|^2}\right). \tag{2.6}$$

The denominator is the photon intensity, corresponding to the number of the photons. When the wavelength of the photon field is much longer than the mesoscopic scale, it can be treated as the dipole field. In this case, the absorption coefficient is simplified as

$$\alpha(\omega)\sim Im\left(\Gamma\frac{4\pi}{c\omega\bar{n}}\int d\boldsymbol{r}d\boldsymbol{r}'\chi(\boldsymbol{r},\boldsymbol{r}',\omega)\right), \tag{2.7}$$

where Γ and v_g are the confinement factor and the group velocity of the light in the media. $\bar{n}$ is the refractive index.

Consequently, the absorption coefficient is expressed by the current-current correlation function and its ensemble average is taken by the equilibrium states, although the induced current is evaluated by the non-equilibrium state. This is one of the salient features of the linear response theory. In the following sections, we concretely investigate the equilibrium state, described by the Hamiltonian H_0.

2.2.2 Screening Effects

The population inversion is necessary to realize laser oscillation. Concerning the semiconductor LDs, the population inversion takes place when many conduction electrons and the valence holes are created by photons or current injections. Therefore, we must consider the many-body effects to take a statistical average of the observables. In this section, we consider the simple two band model, composed of a single conduction band and a single valence band of the three dimensional electron gas. Since the wavenumber $\boldsymbol{k}$ is the good quantum number due to the translational invariance, the Hamiltonian H_0,the system is given by

$$\begin{aligned}H_0 &= \sum_{\boldsymbol{k}}\varepsilon_{\boldsymbol{k}}^{e}a_{\boldsymbol{k}}^{\dagger}a_{\boldsymbol{k}}+\frac{1}{2}\sum_{\boldsymbol{q}}v(\boldsymbol{q})\sum_{\boldsymbol{k}_1,\boldsymbol{k}_2}a_{\boldsymbol{k}_1}^{\dagger}a_{\boldsymbol{k}_2}^{\dagger}a_{\boldsymbol{k}_2-\boldsymbol{q}}a_{\boldsymbol{k}_1+\boldsymbol{q}}\\ &+\sum_{\boldsymbol{k}}\varepsilon_{\boldsymbol{k}}^{e}b_{\boldsymbol{k}}^{\dagger}b_{\boldsymbol{k}}+\frac{1}{2}\sum_{\boldsymbol{q}}v(\boldsymbol{q})\sum_{\boldsymbol{k}_1,\boldsymbol{k}_2}b_{\boldsymbol{k}_1}^{\dagger}b_{\boldsymbol{k}_2}^{\dagger}b_{\boldsymbol{k}_2-\boldsymbol{q}}b_{\boldsymbol{k}_1+\boldsymbol{q}}\\ &-\sum_{\boldsymbol{q}}v(\boldsymbol{q})\sum_{\boldsymbol{k}_1,\boldsymbol{k}_2}a_{\boldsymbol{k}_1}^{\dagger}b_{\boldsymbol{k}_2}^{\dagger}b_{\boldsymbol{k}_2-\boldsymbol{q}}a_{\boldsymbol{k}_1+\boldsymbol{q}},\end{aligned} \tag{2.8}$$

where ε^e_k and ε^h_k are the electron and the hole energies, respectively. $a^\dagger_k(a_k)$ and $b^\dagger_k(b_k)$ are the electron and hole creation (annihilation) operators at the state $\boldsymbol{k}$. $v(\boldsymbol{q})$ is the Coulomb interaction among the carriers. The last term expresses the direct Coulomb interaction between the electrons and holes. We ignore the exchange interaction between them because of the small coupling constant.

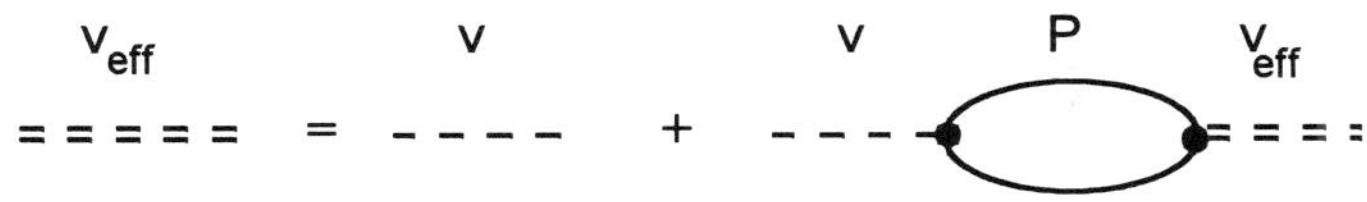

Figure 2.2 Feynman diagrams of screened potential.

When many conduction electrons and the valence holes exist, the bare Coulomb interaction $v(\boldsymbol{q})$ is screened by them through the many-body effect. Here we assume the quasi-equilibrium state where the chemical potentials of the electron and the holes are defined independently. The bare Coulomb interaction $v(\boldsymbol{q})$ is affected by the excited states in the conduction band and the valence band, respectively. Using the perturbation theory, the effective Coulomb interaction is given by

$$v_{eff}(\boldsymbol{q},i\omega) = v(\boldsymbol{q}) + v(\boldsymbol{q})\, P(\boldsymbol{q},i\omega)\, v_{eff}(\boldsymbol{q},i\omega), \tag{2.9}$$

$$P(\boldsymbol{q},i\omega) = \sum_{\boldsymbol{k}} \left(\frac{f(\varepsilon^e_{\boldsymbol{k}}) - f(\varepsilon^e_{\boldsymbol{k}+\boldsymbol{q}})}{i\omega + \varepsilon^e_{\boldsymbol{k}} - \varepsilon^e_{\boldsymbol{k}+\boldsymbol{q}}} + \frac{f(\varepsilon^h_{\boldsymbol{k}}) - f(\varepsilon^h_{\boldsymbol{k}+\boldsymbol{q}})}{i\omega + \varepsilon^h_{\boldsymbol{k}} - \varepsilon^h_{\boldsymbol{k}+\boldsymbol{q}}} \right), \tag{2.10}$$

where $P(\boldsymbol{q},i\omega)$ is the polarization, indicating the electron-hole excitation in the above and below the chemical potentials of the conduction band and the valence band, respectively. f is the Fermi-Dirac distribution function. Figure 2.2 shows the Feynman diagram of the Equation (2.9) and this corresponds to the Random Phase Approximation. Then, v_{eff} is

$$v_{eff}(\boldsymbol{q},i\omega) = \frac{v(\boldsymbol{q})}{\varepsilon(\boldsymbol{q},i\omega)}, \tag{2.11}$$

$$\varepsilon(\boldsymbol{q},i\omega) = 1 - v(\boldsymbol{q})\, P(\boldsymbol{q},i\omega). \tag{2.12}$$

The dielectric constant is modified by the free carriers in the continuum states and they cause the dynamic screening through the dielectric function $\varepsilon(\boldsymbol{q},i\omega)$.

Let us assess the dielectric function $\varepsilon(\boldsymbol{q},i\omega)$ in the two extreme cases to understand the screening effects.

Case 1 ($\omega = 0$, $\boldsymbol{q} \to 0$)
This is the static limit. The Equation (2.12) yields

$$\begin{aligned} \varepsilon(\boldsymbol{q},0) &= 1 - v(\boldsymbol{q})\, P(\boldsymbol{q},0) \\ &= 1 + v(\boldsymbol{q}) \left(\frac{\partial n_e}{\partial \mu_e} \Big|_{\varepsilon=\mu_e} + \frac{\partial n_h}{\partial \mu_h} \Big|_{\varepsilon=\mu_h} \right), \end{aligned} \tag{2.13}$$

where $n_e(n_h)$ and $\mu_e(\mu_h)$ are the electron (hole) density and chemical potential, respectively. In three dimensional case, $v(\boldsymbol{q}) = 4\pi e^2/(\varepsilon_0 q^2)$, then

$$\varepsilon(\boldsymbol{q},0) = 1 + \frac{\kappa^2}{q^2}, \tag{2.14}$$

where κ is the inverse screening length and the screened Coulomb potential $v_{eff}(\boldsymbol{r})$ in real space become the Yukawa type short range potential as

$$v_{eff}(\boldsymbol{r}) = \frac{1}{L^3}\sum_{\boldsymbol{q}} \frac{v(\boldsymbol{q})}{\varepsilon(\boldsymbol{q},0)} e^{i\boldsymbol{q}\cdot\boldsymbol{r}} = \frac{e^2}{\varepsilon_0 r} e^{-kr}. \tag{2.15}$$

Case 2 ($\omega \neq 0$, $\boldsymbol{q} \to 0$)

Since we can expand the distribution functions with respect to the energy difference for the small $\boldsymbol{q}$, the dielectric function $\varepsilon(\boldsymbol{q},i\omega)$ is given by

$$\varepsilon(\boldsymbol{q},i\omega) = 1 - \frac{\omega_{pl}^2(q)}{(i\omega)^2}, \tag{2.16}$$

$$\omega_{pl}^2(q) = v(\boldsymbol{q})q^2\left(\frac{n_e}{m_e^*} + \frac{n_h}{m_h^*}\right).$$

ω_{pl} is the sum of the electron and hole plasma frequencies and it becomes independent of the wavenumber. Note that in the two dimensional case, the plasma frequency is proportional to the square of the wavenumber.

2.2.3 Self-Energies of Electron Gas

The conduction electron and the valence hole interact with the other electrons and holes in the continuum states. Those interactions cause the energy modification to the single particle energy as

$$\epsilon_{\boldsymbol{k}}^{e,h} = \varepsilon_{\boldsymbol{k}}^{e,h} + \sum{}^{e,h}(\boldsymbol{k}, \varepsilon_{\boldsymbol{k}}^{e,h} + i\delta), \tag{2.17}$$

where $\Sigma^{e,h}$ is the self-energy of the electron or hole. The self-energy Σ can be evaluated by the perturbation theory, given by,

$$\sum{}^{e,h}(\boldsymbol{k},i\omega) = -\frac{1}{\beta}\sum_{\boldsymbol{q}}\sum_{i\omega'} v_{eff}(\boldsymbol{q},i\omega')G^{e,h}(\boldsymbol{k}+\boldsymbol{q},i\omega'), \tag{2.18}$$

where $G^e(\boldsymbol{k},i\omega)$ and $G^h(\boldsymbol{k},i\omega)$ are the electron and the hole temperature Green's functions, respectively. β is the inverse temperature. The self-energy is separated into the two terms as

$$\Sigma^{e,h} = \Sigma_{ex}^{e,h} + \Sigma_{ch}^{e,h}. \tag{2.19}$$

$\Sigma^{e,h}_{ex}$ is attributed from $G^{e,h}$, in other words, carrier distribution and $\Sigma^{e,h}_{ch}$ is from $\varepsilon(\boldsymbol{q},i\omega)$, i.e., plasmon. From the temperature Green's function formalism, the retarded function can be easily obtained by replacing the frequency as $i\omega \rightarrow \omega+i\delta$ in the temperature Green's function. Then, the self-energy is generally complex and the real part causes the energy shift of the particle. This energy shift decreases the electron and the hole energies. This red shift is called 'bandgap renormalization'. On the other hand, the imaginary part means the decay time of the quasi-particle. It corresponds to the relaxation time of the intraband transitions.

Since the dielectric function in Equation (2.12) is too complicated to calculate the self-energy practically, we use the simple plasmon-pole approximation. In this approximation, it is assumed that the dielectric function has a single pole in the frequency domain and its energy corresponds to the plasmon energy. Further, it satisfies the f-sum rules. Then the dielectric function can be written as

$$\frac{1}{\varepsilon(\boldsymbol{q},i\omega)} = 1+\frac{\omega_{pl}^2(q)}{(i\omega)^2-\omega_q^2(q)}, \tag{2.20}$$

where

$$\omega_q^2(q) = \omega_{pl}^2(1+\frac{q^2}{\kappa^2}). \tag{2.21}$$

This formula satisfies the two extreme cases, discussed in the previous section. When we sum over the frequency ω' in the Equation (2.18), the exchange self-energy and the Coulomb-hole self-energy are explicitly given by

$$\sum\nolimits_{ex}^{e,h}(\boldsymbol{k},i\omega) = -\sum_q v_{eff}(\boldsymbol{q},i\omega-\varepsilon_{\boldsymbol{k}+\boldsymbol{q}}^{e,h})f(\varepsilon_{\boldsymbol{k}+\boldsymbol{q}}^{e,h}), \tag{2.22}$$

$$\sum\nolimits_{ch}^{e,h}(\boldsymbol{k},i\omega) = \sum_q v(\boldsymbol{q})\frac{\omega_{pl}^2}{2\omega_q}\left[\frac{n_B(\omega_q)}{i\omega+\omega_q-\varepsilon_{\boldsymbol{k}+\boldsymbol{q}}}+\frac{n_B(\omega_q)+1}{i\omega-\omega_q-\varepsilon_{\boldsymbol{k}+\boldsymbol{q}}}\right], \tag{2.23}$$

where n_B is the Bose-Einstein distribution function. Note here that $\Sigma^{e,h}_{ex}$ depends on the charge distribution explicitly and that $\Sigma^{e,h}_{ch}$ depends on the charge density through the plasma frequency.

In this case, we only considered the carrier-carrier scattering which gives the plasmon in the dielectric function. When the carrier-phonon scattering is taken into account in the dielectric function, the broadening in terms of the scattering can be also obtained from the imaginary part of the self-energy. In general, the Coulomb-hole self-energy is much larger than the exchange one (Bauer and Ando 1986). Then, if we use the static approximation for the dielectric function before the summation of ω', the Coulomb-hole term is not taken into account and it would cause a severe error. Therefore we should adopt the quasi-static approximation (Haug *et al.* 1984) such that after summing over the frequency in the Equation (2.18), we make the carriers stay at the same states.

2.2.4 Coulomb Enhancement

In this section, we evaluate the electron-hole interaction in the optical gain/absorption coefficient, as general as possible. We again assume the two-band system as the electron and hole states. However, these states are labeled by the general quantum number $\boldsymbol{k}$ and it is not necessary as the wavenumber in the previous section. It might be the two dimensional wavenumbers, k_x, k_y and integer, which specifies the quantum well confinement. Since we use the effective mass approximation, the single particle wavefunctions are written by the

$$\psi^{e,h}(\boldsymbol{r}) = \phi^{e,h}(\boldsymbol{r}) u_{\boldsymbol{k}_0}^{e,h}(\boldsymbol{r}), \tag{2.24}$$

where $\phi^{e,h}(\boldsymbol{r})$ and $u^{e,h}{}_{\boldsymbol{k}0}(\boldsymbol{r})$ are the envelope function and periodic part of the Bloch function at $\boldsymbol{k}_0$ in k-space, respectively. Then, the current density operator $j(\boldsymbol{r})$ is expressed by

$$j(\boldsymbol{r}) = \sum_{\boldsymbol{k}_1,\boldsymbol{k}_2} a_{\boldsymbol{k}_1}^\dagger b_{\boldsymbol{k}_2}^\dagger \phi_{e,\boldsymbol{k}_1}^*(\boldsymbol{r}) \phi_{h,\boldsymbol{k}_2}^*(\boldsymbol{r}) w_0^* + h.c., \tag{2.25}$$

where $w_0 = \langle u^e | \boldsymbol{p} | u^h \rangle$. Since the integrals of the equations in Section 2.2.1 are on a mesoscopic scale corresponding to the scale of the envelope functions, the integration for the periodic parts has been done first. The current-current correlation function $\chi(\boldsymbol{r}, \boldsymbol{r}', i\omega)$ is given by

$$\chi(\boldsymbol{r},\boldsymbol{r}',i\omega) = -\int_0^\beta e^{i\omega\tau} \langle T j(\boldsymbol{r},\tau) j(\boldsymbol{r}',0) \rangle$$

$$= |w_0|^2 \sum_{\boldsymbol{k}_1,\boldsymbol{k}_2,\boldsymbol{k}'_1,\boldsymbol{k}'_2} \phi_{e,\boldsymbol{k}_1}^*(\boldsymbol{r}) \phi_{h,\boldsymbol{k}_2}^*(\boldsymbol{r}) \phi_{h,\boldsymbol{k}'_2}(\boldsymbol{r}') \phi_{e,\boldsymbol{k}'_1}(\boldsymbol{r}') \pi(\boldsymbol{k}_1,\boldsymbol{k}_2,\boldsymbol{k}'_1,\boldsymbol{k}'_2,i\omega), \tag{2.26}$$

$$\pi(\boldsymbol{k}_1,\boldsymbol{k}_2,\boldsymbol{k}'_1,\boldsymbol{k}'_2,i\omega) = \int_0^\beta e^{i\omega\tau} \langle T a_{\boldsymbol{k}_1}^\dagger(\tau) b_{\boldsymbol{k}_2}^\dagger(\tau) b_{\boldsymbol{k}'_2}(0) a_{\boldsymbol{k}'_1}(0) \rangle. \tag{2.27}$$

$\pi(\boldsymbol{k}_1, \boldsymbol{k}_2, \boldsymbol{k}_1', \boldsymbol{k}_2', i\omega)$ is the two-particle Green's function, which is expressed by the vertex function $\Gamma(\boldsymbol{k}_1, \boldsymbol{k}_2, \boldsymbol{k}_1', \boldsymbol{k}_2', i\omega)$ as

$$\pi(\boldsymbol{k}_1,\boldsymbol{k}_2,\boldsymbol{k}'_1,\boldsymbol{k}'_2,i\omega) = \pi^0(\boldsymbol{k}_1,\boldsymbol{k}_2,i\omega) \Gamma(\boldsymbol{k}_1,\boldsymbol{k}_2,\boldsymbol{k}'_1,\boldsymbol{k}'_2,i\omega), \tag{2.28}$$

where $\pi^0(\boldsymbol{k}_1, \boldsymbol{k}_2, i\omega)$ is the non-interacting Green's function, given by

$$\pi^0(\boldsymbol{k}_1,\boldsymbol{k}_2,i\omega) = \frac{1 - f(\epsilon_{\boldsymbol{k}_1}^e) - f(\epsilon_{\boldsymbol{k}_2}^h)}{i\omega - \epsilon_{\boldsymbol{k}_1}^e - \epsilon_{\boldsymbol{k}_2}^h}. \tag{2.29}$$

From the perturbation theory, the vertex function $\Gamma(\boldsymbol{k}_1, \boldsymbol{k}_2, \boldsymbol{k}_1', \boldsymbol{k}_2', i\omega)$ is obtained within the ladder approximation, as shown in Figure 2.3. The diagram yields

$$\Gamma(\boldsymbol{k}_1,\boldsymbol{k}_2,\boldsymbol{k}'_1,\boldsymbol{k}'_2,i\omega)$$

$$= \delta_{\boldsymbol{k}_1,\boldsymbol{k}'_1} \delta_{\boldsymbol{k}_2,\boldsymbol{k}'_2} + \sum_{\boldsymbol{k}''_1,\boldsymbol{k}''_2} V_{\boldsymbol{k}_1,\boldsymbol{k}_2,\boldsymbol{k}''_1,\boldsymbol{k}''_2} \frac{F(\boldsymbol{k}''_1,\boldsymbol{k}''_2)}{i\omega - \epsilon_{\boldsymbol{k}''_1}^e - \epsilon_{\boldsymbol{k}''_2}^h} \Gamma(\boldsymbol{k}''_1,\boldsymbol{k}''_2,\boldsymbol{k}'_1,\boldsymbol{k}'_2,i\omega), \tag{2.30}$$

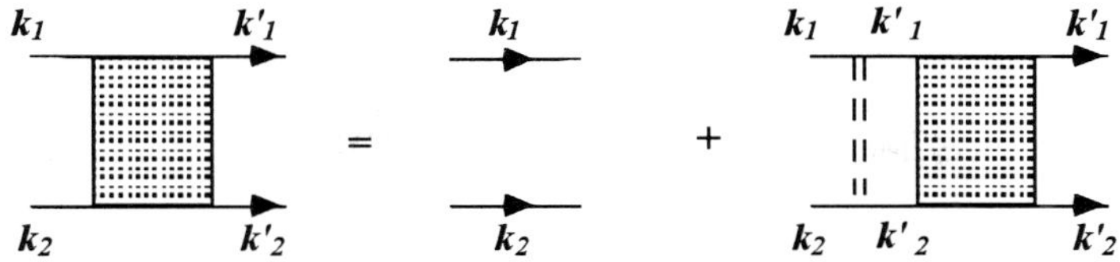

Figure 2.3 Feynman diagrams of the vertex parts.

where

$$v_{k_1,k_2,k'_1,k'_2} = \int dr_1 dr_2 \phi^*_{e,k_1}(r_1)\phi^*_{h,k_2}(r_2) v_{eff}(r_1 - r_2)\phi_{h,k'_2}(r_2)\phi_{e,k'_1}(r_1), \tag{2.31}$$

$$F(k_1,k_2) = 1 - f(\epsilon^e_{k_1}) - f(\epsilon^h_{k_2}). \tag{2.32}$$

v_{eff} is the static screened Coulomb potential and $F(k_1,k_2)$ is the phase space filling factor. Then, the two-particle Green's function satisfies the following integral equation:

$$\pi(k_1,k_2,k'_1,k'_2,i\omega) = \pi^0(k_1,k_2,i\omega)\delta_{k_1,k'_1}\delta_{k_2,k'_2}$$
$$+\pi^0(k_1,k_2,i\omega)\sum_{k''_1 k''_2} v_{k_1,k_2,k''_1,k''_2}\pi(k''_1,k''_2,k'_1,k'_2,i\omega). \tag{2.33}$$

We can solve the equation numerically, using the inverse matrix and obtain the current-current correlation function χ directly. On the other hand, in order to understand the exciton effect in the current-current correlation function from the physical point of view, we introduce the auxiliary function $\tilde{\pi}$ (r_1, r_2, r_1', r_2', $i\omega$), defined as

$$\tilde{\pi}(r_1,r_2,r'_1,r'_2,i\omega)$$
$$= \sum_{k_1,k_2,k'_1,k'_2} \phi^*_{e,k_1}(r_1)\phi^*_{h,k_2}(r_2)\phi_{h,k'_2}(r'_2)\phi_{e,k'_1}(r'_1)\pi(k_1,k_2,k'_1,k'_2,i\omega) \tag{2.34}$$
$$= \sum_{k_1,k_2,k'_1,k'_2} \frac{\phi_{h,k_2}(r_2)\phi_{e,k_1}(r_1)F(k_1,k_2)}{i\omega - \epsilon^e_{k''_1} - \epsilon^h_{k''_2}} \Gamma(k_1,k_2,k'_1,k'_2,i\omega)\phi^*_{e,k'_1}(r'_1)\phi^*_{h,k'_2}(r'_2).$$

Multiplying the operator $i\omega - h_e(r_1) - h_h(r_2)$, where $h_{e,h}(r)\phi_{e,h,k}(r) = \epsilon^{e,h}\phi_{e,h,k}(r)$, to the auxiliary function, the following differential equation is obtained instead of the integral Equation (2.33):

$$(i\omega - h_e(r_1) - h_h(r_2))\tilde{\pi}(r_1,r_2,r'_1,r'_2,i\omega)$$
$$+\int dr''_1\, dr''_2 \tilde{F}(r_1,r_2;r''_1,r''_2)v(r''_1 - r''_2)$$
$$\times \tilde{\pi}(r''_1,r''_2,r'_1,r'_2,i\omega) = \tilde{F}(r_1,r_2;r'_1,r'_2), \tag{2.35}$$

where

$$\widetilde{F}(\boldsymbol{r}_1,\boldsymbol{r}_2;\boldsymbol{r}'_1,\boldsymbol{r}'_2)=\sum_{\boldsymbol{k}_1,\boldsymbol{k}_2} F(\boldsymbol{k}_1,\boldsymbol{k}_2)\phi_{e,\boldsymbol{k}_1}(\boldsymbol{r}_1)\phi_{h,\boldsymbol{k}_2}(\boldsymbol{r}_2)\phi^*_{e,\boldsymbol{k}_1}(\boldsymbol{r}'_1)\phi^*_{h,\boldsymbol{k}_2}(\boldsymbol{r}'_2). \quad (2.36)$$

This differential equation can be solved from the eigenvalue problem as

$$(h_e(\boldsymbol{r}_1)+h_h(\boldsymbol{r}_2))\Psi_\lambda(\boldsymbol{r}_1,\boldsymbol{r}_2)-\int d\boldsymbol{r}'_1\, d\boldsymbol{r}'_2\widetilde{F}(\boldsymbol{r}_1,\boldsymbol{r}_2;\boldsymbol{r}_1',\boldsymbol{r}_2')v(\boldsymbol{r}_1'-\boldsymbol{r}_2')\Psi_\lambda(\boldsymbol{r}'_1,\boldsymbol{r}'_2,i\omega)$$
$$=E_\lambda\Psi_\lambda(\boldsymbol{r}_1,\boldsymbol{r}_2). \quad (2.37)$$

$\Psi_\lambda(\boldsymbol{r}_1, \boldsymbol{r}_2)$ is the general exciton wavefunction under the many-body ground state. Note that the Coulomb interaction between the electron and the hole becomes non-local and is factorized by the phase space filling factor. Therefore, it is hard to obtain the bound states of the electron-hole pairs when the population inversion occurs, i.e., $F(\boldsymbol{k}_1, \boldsymbol{k}_2) < 0$. Using the general exciton wavefunctions, the auxiliary function is expressed as

$$\widetilde{\pi}(\boldsymbol{r}_1,\boldsymbol{r}_2,\boldsymbol{r}'_1,\boldsymbol{r}'_2,i\omega)=\int d\boldsymbol{r}''_1\, d\boldsymbol{r}''_2\sum_\lambda\frac{\Psi_\lambda(\boldsymbol{r}_1,\boldsymbol{r}_2)\Psi^*{}_\lambda(\boldsymbol{r}''_1,\boldsymbol{r}''_2)\widetilde{F}(\boldsymbol{r}''_1,\boldsymbol{r}''_2;\boldsymbol{r}',\boldsymbol{r}')}{i\omega-E_\lambda}. \quad (2.38)$$

Since the current-current correlation function $\chi(\boldsymbol{r},\boldsymbol{r}',i\omega)$ is $\widetilde{\pi}(\boldsymbol{r}, \boldsymbol{r}, \boldsymbol{r}', \boldsymbol{r}', i\omega)$, the absorption spectra is given by

$$\alpha(\omega)=\frac{4\pi\,|\,w_0\,|^2}{c\omega n}\mathrm{Im}(\Pi(\omega+i\delta)), \quad (2.39)$$

$$\Pi(i\omega)=\sum_\lambda\frac{\int d\boldsymbol{r}d\boldsymbol{r}'d\boldsymbol{r}_1'd\boldsymbol{r}_2'A(\boldsymbol{r})\Psi_\lambda(\boldsymbol{r},\boldsymbol{r})\Psi^*{}_\lambda(\boldsymbol{r}'_1,\boldsymbol{r}'_2)\widetilde{F}(\boldsymbol{r}'_1,\boldsymbol{r}'_2;\boldsymbol{r}',\boldsymbol{r}')A^*(\boldsymbol{r}')}{(i\omega-E_\lambda)\int d\boldsymbol{r}\,|\,A(\boldsymbol{r})\,|^2}. \quad (2.40)$$

The Coulomb enhancement is included in the exciton wavefunction $\Psi_\lambda(\boldsymbol{r}_1, \boldsymbol{r}_2)$. When the electron-hole Coulomb interaction is ignored, $\Psi_\lambda(\boldsymbol{r}_1, \boldsymbol{r}_2)$ becomes $\phi^*_{e,k_1}(\boldsymbol{r}_1)\phi^*_{h,k_2}(\boldsymbol{r}_2)$ and the Equation (2.39) indicates the simple band-to-band transition. Since the quantum number is not limited within the wavenumber, the absorption by way of the localized states is also described. Further, as we treated the total wavenumber of the electron-hole pair as a finite number, this formula includes the effect of the super radiation.

Here we take the several special cases for the Equation (2.40).

1. Low density limit ($F(\boldsymbol{k}_1, \boldsymbol{k}_2)= 1$)

In this case, $\widetilde{F}(\boldsymbol{r}_1, \boldsymbol{r}_2; \boldsymbol{r}_1', \boldsymbol{r}_2', i\omega) = \delta(\boldsymbol{r}_1-\boldsymbol{r}_1')\,\delta(\boldsymbol{r}_2-\boldsymbol{r}_2')$. Then, we get:

$$\Pi(i\omega)=\sum_\lambda\frac{|\int d\boldsymbol{r}A(\boldsymbol{r})\Psi_\lambda(\boldsymbol{r},\boldsymbol{r})\,|^2}{(i\omega-E_\lambda)\int d\boldsymbol{r}\,|\,A(\boldsymbol{r})\,|^2}. \quad (2.41)$$

When we take the imaginary part of the $\Pi(\omega + i\delta)$, the general Elliott formula is obtained.

2. Dipole approximation

When wavelength of the photon field is much larger than the scale of the envelope wavefunctions, the vector field $A(\boldsymbol{r})$ can be treated as a constant in the integration of the numerator and the confinement factor should be multiplied instead as the Equation (2.7). If the system has the translational invariance, this integration of the matrix elements yields the selection rule. For example, in the uniform three dimensional system, the quantum number $\boldsymbol{k}$ is the wavenumber and the matrix element gives $\boldsymbol{k}_e = -\boldsymbol{k}_h$. Therefore, we just focus on the $\pi(\boldsymbol{k}, \boldsymbol{k}', \boldsymbol{k}, \boldsymbol{k}', i\omega)$ to evaluate the current-current correlation function.

Let us refer the relation with the Maxwell Bloch equation (Lindeberg and Koch 1988). In the derivation of the Maxwell Bloch equation, the time evolution of the $\langle a_k b_{-k}\rangle$ is evaluated. This bracket of the $\langle a_k b_{-k}\rangle$ means the average at the non-equilibrium state. It is expressed by

$$\left\langle a_k b_{-k}\right\rangle = \mathrm{Tr}(\rho_1 a_k b_{-k}) \propto \mathrm{Tr}(\rho_1 j_1). \tag{2.42}$$

This corresponds to the evaluation of the induced current in Section 2.2.1. Therefore, within the linear approximation with respect to the photon field, both procedures are same.

2.3 ELECTRONIC BAND STRUCTURES

2.3.1 Electronic Band Structures of Bulk GaN and AlN

GaN and AlN take stable hexagonal WZ structure (space group C^4_{6v}: $P6_3mc$) rather than meta-stable cubic ZB structure (space group T^2_d: $F\bar{4}3m$). Figure 2.4 shows the first Brillouin zones (BZs) of the WZ structures. The first-principles electronic band calculations for WZ structures have been performed by using the full-potential linearized augmented plane wave (FLAPW) method (Wimmer *et al.* 1981) within the local density functional approximation (LDA) (Gunnarson and Lundquvist 1976). In general, the nitrogen atom has such a strong electron affinity

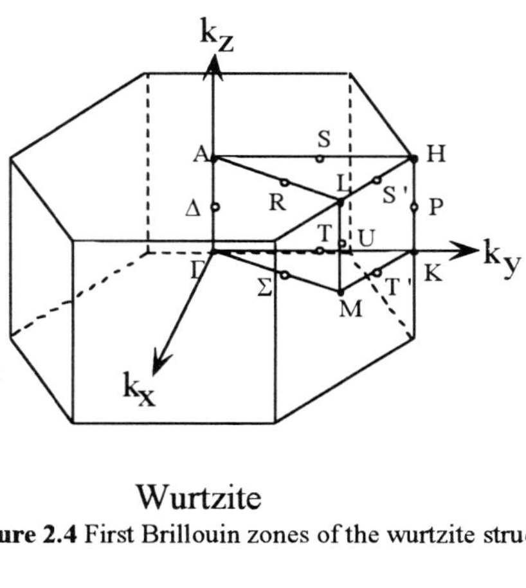

Figure 2.4 First Brillouin zones of the wurtzite structure.

that the valence charge densities of the III-V nitrides tend to be well localized at the nitrogen site. In such a case, a conventional pseudo-potential approach based on simple plane waves seems to be not adequate. Therefore, the FLAPW method where the plane wave is augmented by the atomic orbital, is adopted as the method of the first-principles band calculation. The FLAPW calculations are discussed in detail in reference (Suzuki *et al.* 1995).

Figure 2.5 shows the electronic band structure of WZ GaN. The spin-orbit interaction is included, and the energy dispersions are along high symmetry lines of the first BZ. As an example, the characteristics of WZ GaN are described. The energy bands consist of four parts which are three occupied bands and an unoccupied band. The lowest energy bands, which are localized between about 0.70 Ryd and 0.65 Ryd, mostly originate from the N 2*s* states. The upper side of these states are hybridized with the Ga 3*d* states. The second lowest energy bands are located between about 0.55 Ryd and 0.38 Ryd, where the lower energy side almost originates from the Ga 3*d* well-localized states, and the upper one consists of the hybridized states between the Ga 3*d* and the N 2*s* states. Due to this hybridization, the Ga 3*d* states must be included in the valence band states. In case of AlN, there is no hybridization between the cation *d* and the N 2*s* states, since the

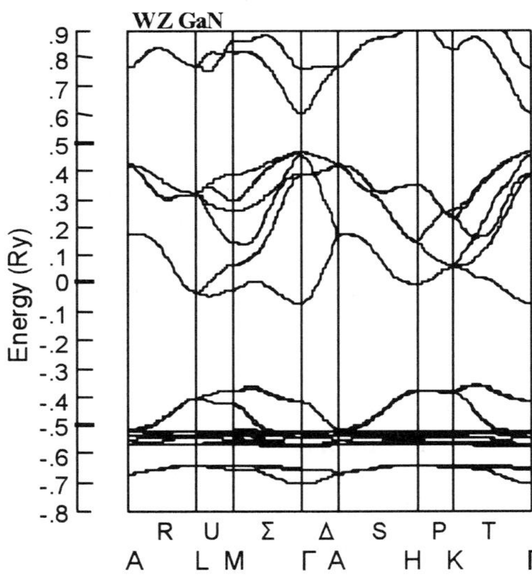

Figure 2.5 Electronic band structure of wurtzite GaN.

Al atom has no *d* electrons in the ground state. Consequently, the lowest energy band is combined with the upper side of the second lowest one, and these two bands become one narrow band. In the intermediate energy region, there are broad energy bands whose width is about 0.53 Ryd. The VBM is located at the top of these bands and at the Γ point. These energy bands arise from the N 2*s*, 2*p* and the Ga 4*s*, 4*p* states; however, the vicinity of the VBM almost consists of the N 2*p* states. The highest energy region separated by about 0.14 Ryd above the valence band is the conduction bands, and the CBM is also located at the Γ point. The vicinity of the CBM originates from the *s* states of N and Ga. Comparing the result of AlN with that of GaN, the main features in these two energy regions are similar, except the order of energy levels around the VBM.

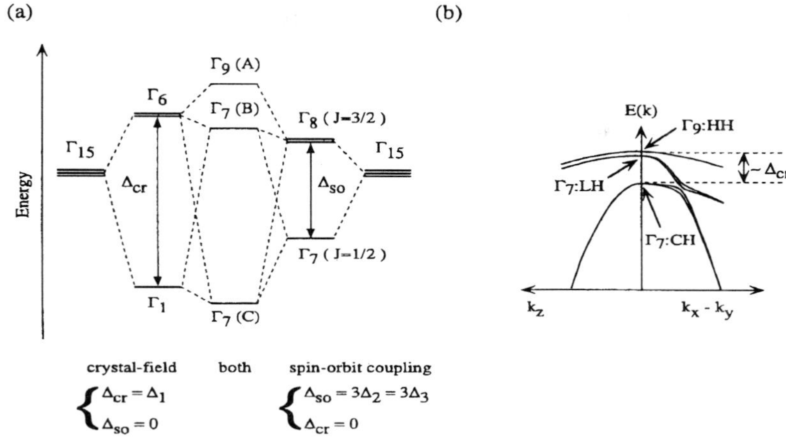

Figure 2.6 (a) Splitting at the top of valence bands of GaN under the influence of the crystal-field and the spin-orbit coupling. (b) shows the schematic band structure around the top of the valence band of the wurtzite GaN. The figures are not drawn to scale.

Next, we focus on the features of electronic structures near the VBM. We note that the electronic structures around the VBM of GaN in the WZ structure are different from conventional ZB III-V semiconductors. Figure 2.6 (a) shows the splitting at the top of the valence bands of GaN under the influence of the crystal-field and the spin-orbit interaction. Figure 2.6 (b) shows the schematic band structures around the VBM of WZ GaN. In the WZ structure, the VBM is split into two-fold and non-degenerate states even in the lack of the spin-orbit interaction. Using the irreducible representations of the group theory, the former state is labeled Γ_6, whose wave functions transform like x and y, and the latter one is labeled Γ_1, whose wave function transforms like z. The energy splitting between these two levels is induced by the hexagonal symmetry of the WZ structure, therefore, this splitting is called the crystal field splitting Δ_{cr}. The order of the two levels depends on the kind of materials, the ratio c/a of lattice constants and the internal parameter u. In case of GaN, the Γ_6 level is higher than the Γ_1 level, while in case of AlN, the Γ_1 level is higher. This difference between AlN and GaN is in agreement with the result by Rubio *et al.* (1993). Considering the spin-orbit interaction, the two-fold degenerate Γ_6 state is split into the Γ_9 and the Γ_7 levels, and the non-degenerate Γ_1 state is labeled Γ_7 as well. Then, the twoΓ_7 levels can be mixed by the spin-orbit interaction. The energy splitting among these three levels by the spin-orbit interaction is called a spin-orbit splitting. According to the FLAPW calculations, the spin-orbit splitting is much smaller than the crystal-field splitting in the III-V nitrides due to the weak spin-orbit interaction of the nitrogen atom. The hole effective masses of the three levels Γ_9, Γ_7 and Γ_7 have the large ***k***-dependence, as shown in Figure 2.6 (b), where we labeled three hole bands as HH (heavy), LH (light) and CH (crystal-field split-off), based on the feature in the k_x-k_y plane. The mass of the Γ_9 band is heavy along any $\boldsymbol{k}$ direction. On the other hand, the mass of the upper Γ_7 band is light in the k_x-k_y plane but is heavy along the k_z direction, and that of the lower Γ_7 band is the totall reverse. The lower Γ_7 band is split off from the Γ_9 and the upper Γ_7 bands even without the spin-orbit coupling. These features show that the mass anisotropy in the $\boldsymbol{k}$-space is not negligible when we carry out the material design or the characteristic analysis of the GaN-based QW devices like LDs.

Therefore, we must pay attention to it and treat the upper six valence bands in the analysis of the GaN-based QW structures.

2.3.2 Strain Effect on Electronic Band Structures

2.3.2.1 Biaxial and uniaxial strains in (0001) plane of wurtzite

Figure 2.7 shows the schematic band structure in the k_x-k_y plane around the VBM of bulk WZ GaN, (a) without a strain, (b) with a biaxial strain and (c) with a uniaxial strain in the (0001) plane (*c*-plane). It is found that each energy dispersion of HH, LH and CH bands is almost unchanged even under stress because the weak spin-orbit coupling makes the orbital character such as $|X\rangle$, $|Y\rangle$ and $|Z\rangle$ dominant. Under the condition of the compressive (tensile) biaxial strain, only the crystal-field splitting energy becomes effectively large (small) since the strained crystal still has the C_{6v} symmetry, which is the same as before. Then, the HH mass is still heavy, and the density of states (DOS) at the VBM is not so changed. This is very different from the ZB crystals, where the HH mass becomes light and the DOS decreases considerably since the T_d symmetry is changed to the D_{2d} one and the degeneracy is removed.

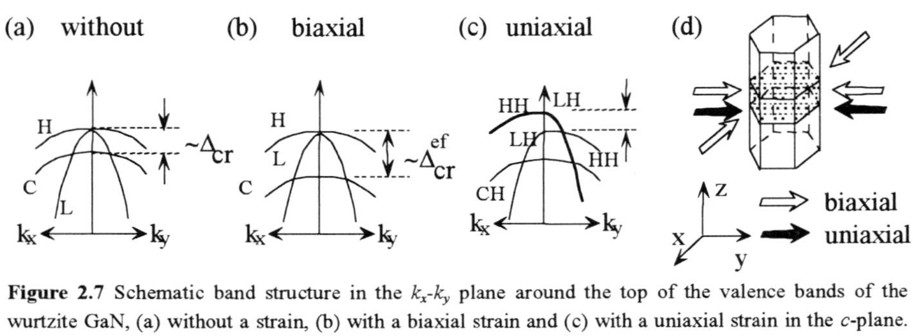

Figure 2.7 Schematic band structure in the k_x-k_y plane around the top of the valence bands of the wurtzite GaN, (a) without a strain, (b) with a biaxial strain and (c) with a uniaxial strain in the *c*-plane. (d) shows the direction of each strain.

On the other hand, the uniaxial strain in the *c*-plane gives the anisotropic energy splitting in the k_x-k_y plane since it causes the symmetry lowering from C_{6v} to C_{2v}. When the compressive (tensile) uniaxial strain along the *y*-direction, or the tensile (compressive) one along the *x*-direction, is induced, the HH band along the k_x direction as well as the LH band along the k_y direction moves to higher (lower) energy side. Because the only the $|Y\rangle$ becomes dominant in k_x-k_y plane. Thus, the

DOS at the VBM is largely reduced, compared with the conditions without a strain or with a biaxial strain.

2.3.3 *k·p* Theory for Wurtzite

Generally speaking, optical and transport properties are governed by the electronic band structures in close vicinity of the VBM and the CBM. Thus, the ***k·p*** theory is not only an appropriate method to describe the electronic structures around the VBM and the CBM, but also a convenient method to analyze the QW devices on the basis of the effective mass approximation. In the analyses of conventional ZB semiconductors, we frequently assume a parabolic band for the conduction bands, and the 4×4 and/or 6×6 Luttinger-Kohn Hamiltonians are used to describe the upper valence bands (Luttinger and Kohn 1955). In treating the valence bands together with the conduction bands, we often make use of the 8×8 Kane Hamiltonian (Kane 1957). For WZ materials, however, we must consider the hexagonal symmetry because the Luttinger-Kohn and Kane Hamiltonians are constructed under the condition of the cubic symmetry. Let us show the ***k·p*** and strain Hamiltonians for WZ structure.

The 8×8 Hamiltonian can be split into the 2×2 Hamiltonian $\mathbf{H}_{cc}$ for the CBM and 6×6 Hamiltonian $\mathbf{H}_{vv}$ for the VBM. The matrix elements of $\mathbf{H}_{cc}$ and $\mathbf{H}_{vv}$ are given by

$$\mathbf{H}_{\mathrm{cc}} = \begin{pmatrix} E_c & 0 \\ 0 & E_c \end{pmatrix}, \tag{2.43}$$

$$\mathbf{H}_{\mathrm{vv}} = \begin{pmatrix} F & 0 & -H^* & 0 & K^* & 0 \\ 0 & G & \sqrt{2}\Delta_3 & -H^* & 0 & K^* \\ -H & \sqrt{2}\Delta_3 & \lambda & 0 & I^* & 0 \\ 0 & -H & 0 & \lambda & \sqrt{2}\Delta_3 & I^* \\ K & 0 & I & \sqrt{2}\Delta_3 & G & 0 \\ 0 & K & 0 & I & 0 & F \end{pmatrix}, \tag{2.44}$$

where

$$\begin{aligned}
\mathrm{E_c} &= E_s^0 + \frac{\hbar^2 k_z^2}{2m_e^{//}} + \frac{\hbar^2 (k_x^2 + k_y^2)}{2m_e^{\perp}} + a_c(\varepsilon_{xx} + \varepsilon_{yy} + \varepsilon_{zz}),\\
F &= \Delta_1 + \Delta_2 + \lambda + \theta,\\
G &= \Delta_1 - \Delta_2 + \lambda + \theta,\\
H &= i(A_6 k_z + iA_7)(k_x + ik_y) + iD_6(\varepsilon_{zx} + i\varepsilon_{yz}),\\
I &= i(A_6 k_z - iA_7)(k_x + ik_y) + iD_6(\varepsilon_{zx} + i\varepsilon_{yz}),\\
K &= A_5(k_x^2 - k_y^2 + 2ik_x k_y) + D_5(\varepsilon_{xx} - \varepsilon_{yy} + 2i\varepsilon_{xy}),\\
\lambda &= E_p^0 + A_1 k_z^2 + A_2(k_x^2 + k_y^2) + D_1\varepsilon_{zz} + D_2(\varepsilon_{xx} + \varepsilon_{yy}),\\
\theta &= A_3 k_z^2 + A_4(k_x^2 + k_y^2) + D_3\varepsilon_{zz} + D_4(\varepsilon_{xx} + \varepsilon_{yy}).
\end{aligned}$$

E^0_s and E^0_p represent the energy levels of *s*-like and *p*-like functions at $\boldsymbol{k} = 0$, respectively. Δ_1 and $\Delta_{2,3}$ correspond to the crystal-field and spin-orbit splitting energies, respectively. A_i are the valence band parameters, corresponding to the Luttinger parameters γ_i in ZB crystals. Here we consider the strain effects in terms of the additional strain Hamiltonian which is the same form (the second rank tensor) as the $\boldsymbol{k{\cdot}p}$ Hamiltonian. a_c and D_i are the deformation potentials for the conduction and the valence bands, respectively. $\varepsilon_{ij}(i,j=x, y, z)$ are the *ij*-components of a strain tensor, where the diagonal terms ε_{ii} are defined as positive for tension. According to the FLAPW calculations and the group theory, the lowest conduction band state is an almost *s*-like state for the WZ structure as well as for the ZB structure, and it is more isotropic than the valence band states. Therefore, if the small anisotropy of the energy dispersions is neglected, namely, if the parabolic band is assumed, the electronic structure around the CBM is given by

$$E_c(\boldsymbol{k}) \simeq E^0_s + \frac{\hbar^2 \boldsymbol{k}^2}{2m^*_e} + a_c \varepsilon. \tag{2.45}$$

Then, we calculate the physical parameters by fitting the energy dispersions, obtained by the $\boldsymbol{k{\cdot}p}$ method and the first-principle calculations.

2.3.4 Physical Parameters

2.3.4.1 Crystal-Field and Spin-Orbit Splitting Energies

The calculated results of the crystal-field and spin-orbit splitting energies are shown in Table 2.1. The calculated values of Δ_{so} for WZ and ZB GaN are in good agreement with the experimentally observed values (Dingle *et al.* 1971[a] ; Chichibu *et al.* 1996[b] ; Flores *et al.* 1994[c]). On the other hand, the calculated value of Δ_{cr} for WZ GaN is in disagreement with the observed values, though the sign (Δ_{cr}>0) and the relative size ($\Delta_{cr} \mid > \mid \Delta_{so}|$) are consistent. The quantitative disagreement might be caused by the use of the ideal value $u = 3/8$ as an internal lattice parameter since the electronic structures around the VBM are so sensitive to strain effects. Further accurate measurements of the lattice constants using homogeneous strain-free crystals are desirous for more detailed comparison. Moreover, we must perform the first-principles calculations with lattice parameters free and investigate the dependence on the ratio c/a and the internal parameters u.

Table 2.1 Crystal-field and spin-orbit splitting energies (meV) of GaN and AlN.

			$\Delta_{cr(1)}$	Δ_{so}	Δ_2	Δ_3
wurtzite	GaN	calculated	72.9	15.6	5.2	5.2
		observed[a]	22(±2)	11(+5,-2)	—	—
		observed[b]	21	16	—	—
	AlN	calculated[c]	20.4	20.4	6.8	6.8

2.3.4.2 Luttinger-like Parameters

The calculated results of the Luttinger-like parameters are shown in Table 2.2. All the parameters are obtained within the quasi-cubic approximation. The values of A_i transformed from γ_i, with the help of the quasi-cubic approximation, are also listed. We note that the difference between the direct and the transformed results are not so large, although the latter is obtained by simply transforming without taking the relaxation of c/a and u into account. Therefore, the quasi-cubic approximation for the WZ structure is fairly good and useful. The similarity between the WZ and the ZB structures has been also investigated by the other researchers (Yeh *et al.* 1992), where it is reported that the trend in the difference of band gaps can be explained by the band folding and the symmetry changing. From our results, it is clarified that the quasi-cubic approximation for the WZ structure is valid in the III-V nitride system.

Table 2.2 The Luttinger-like valence band parameters of GaN and AlN. All values, except A_7 (Ryd·cm), are in units of $\hbar^2/2m_0$.

wurtzite	A_1	A_2	A_3	A_4	A_5	A_6	A_7
GaN calculated	–6.56	–0.91	5.65	–2.83	–3.13	–4.85	0.00
$\gamma \Rightarrow A_i$	–6.98	–0.56	6.42	–3.21	–2.90	–3.66	0.00
AlN calculated	–3.95	–0.27	3.68	–1.84	–1.95	–2.92	0.00
$\gamma \Rightarrow A_i$	–3.98	–0.26	3.72	–1.86	–1.63	–1.98	0.00
zincblende	γ_1	γ_2	γ_3				
GaN calculated	2.70	0.76	1.07				
AlN calculated	1.50	0.39	0.62				

2.3.4.3 Electron Effective Mass

The calculated electron effective masses are summarized in Table 2.3. The experimentally observed values for WZ GaN are also listed, compared with the theoretically derived values. The calculated values for WZ GaN are in good agreement with the observed values (Barker and Ilegems 1973[a]; Kosicki *et al.* 1970[b]; Dingle and Ilegems 1971[c]; Cunningham *et al.* 1972[d]; Rheinländer and Neumann 1974[e]; Pankove *et al.* 1975[f]; Meyer *et al.* 1995[g]; Drechsler *et al.* 1995[h]). As we expected, the anisotropy of the electron effective mass is quite small in the k space. According to the first principles calculations as well as the observed results, the band gap of WZ GaN is nearly equal to or slightly larger than that of ZB GaN.

2.3.4.4 Hole Effective Masses

The Luttinger-like parameters are ones to describe the hole effective masses, including the anisotropy in the k space. Thus, the feature of three hole bands around the VBM are discussed. The calculated hole masses are summarized in Table 2.4. According to the calculated results, the hole masses have the considerable anisotropy in k dependence, and the lowest conduction band is strongly coupled with only one hole band. Since the spin-orbit interaction in the

Table 2.3 Electron effective masses (m_0) of GaN and AlN. m_e^* denotes the average electron effective mass, and the other superscripts stand for the **k**-dependency.

wurtzite		m_e^*	$m_e^{\perp(\Sigma)}$	$m_e^{\perp(T)}$	$m_e^{\parallel(\Delta)}$
GaN	calculated	0.18	0.18	0.18	0.20
	observed[a]	0.20±0.02	0.20±0.02		0.20±0.06
	observed[b]	≈0.19	—	—	—
	observed[c]	0.25	—	—	—
	observed[d]	0.20~0.30	—	—	—
	observed[e]	0.27±0.06	—	—	—
	observed[f]	0.22±0.03	—	—	—
	observed[g]	0.236±0.005	—	—	—
	observed[h]	0.20±0.005	—	—	—
AlN	calculated	0.27	0.25	0.25	0.33

Table 2.4 Hole effective masses m_0 of GaN and AlN. m_{hh}, m_{lh} and m_{ch} (m_{sh}) denote heavy, light and crystal-field (spin-orbit) split-off hole masses, respectively, for the wurtzite (zincblende) structure.

wurtzite		m_{hh}	m_{lh}	m_{ch}	zincblende	m_{hh}	m_{lh}	m_{sh}
GaN	$k_x(\Sigma)$	1.61	0.14	1.04	$[k00](\Delta)$	0.86	0.86	0.17
	$k_y(\mathrm{T})$	1.44	0.15	1.03	$[kk0](\Gamma)$	4.89	0.84	0.15
	$k_z(\Delta)$	1.76	0.76	0.16	$[kkk](\Lambda)$	1.74	1.74	0.15
	observed[a]	$m^*_h \geq 0.6$						
	observed[b]	$m^*_h = 1 \sim 2$						
	observed[c]	$m^*_h = 0.8 \pm 0.2$						
AlN	$k_x(\Sigma)$	10.42	0.24	3.81	$[k00](\Delta)$	1.40	1.40	0.32
	$k_y(\mathrm{T})$	5.02	0.25	3.61	$[kk0](\Gamma)$	≫10	1.41	0.27
	$k_z(\Delta)$	3.53	3.53	0.25	$[kkk](\Lambda)$	3.72	3.72	0.25

III-V nitrides is quite small, $|X\rangle$, $|Y\rangle$ and $|Z\rangle$ are almost close to the eigenstates along the x, y and z direction, respectively. Then, the eigenstate along the any direction at the VBM is expressed by the linear combination of those states and it is coupled with the conduction band along the same direction. That is why only one hole band is coupled and the coupled hole mass becomes much lighter than the other two hole masses. In Table 2.4, the experimentally observed values (Kosicki *et al.* 1970[a]; Dingle and Ilegems 1971[b]; Pankove *et al.* 1975[c]) are also listed. Experimentally, the ***k***-dependence of hole masses has not been reported yet in detail. The calculated hole mass of the heavy hole band, which is the lowest in hole energy among the three hole bands, seems to be slightly heavier than the observed one. However, both the crystal-field and the spin-orbit splitting are so small that the observed hole mass may be the averaged value among them, namely the density of the states mass. Thus, we estimated the typical hole mass by averaging the ***k***-dependent masses with the weight of the star of ***k***. In averaging the hole masses, two (heavy and light) or three (heavy, light and crystal-field split-off) hole bands are considered. In both cases, the average hole mass m_h^* is about 0.95 ~ 1.10 m_0. It is consistent with the experimentally observed values.

2.3.5 Subband Structures of GaN/AlGaN Quantum Wells

In this section, the subband structures in the QWs are discussed. Their states are described by the direct product between the rapidly varying Bloch wave function and the slowly varying envelope function. In general, the latter is calculated within the effective mass approximation (Luttinger and Kohn 1955). In analysis of QW structures, a confining potential V(z) along the growth direction z is added and $\boldsymbol{k}_z$ becomes an operator in the $\boldsymbol{k}\cdot\boldsymbol{p}$ Hamiltonian. For the conduction band states, it is enough that only the lowest Γ_7 state is considered. For the valence band states, the upper six states must be explicitly treated in both WZ and ZB structure, because they are very close to one another due to the small spin-orbit and crystal-field splitting energies. In order to consider the mixing between these valence band states, the WZ 6×6 Hamiltonians should be used. The detail procedure was discussed in reference (Uenoyama and Suzuki 1995).

2.3.6 Subband in Wurtzite Quantum Wells

The valence subband structures in WZ GaN/$Al_{0.2}Ga_{0.8}N$ QWs, with the well lengths being (a) 30 Å and (b) 50 Å, are shown in Figure 2.8, where the strain effect by the lattice mismatch between GaN and $Al_{0.2}Ga_{0.8}N$ is neglected. As long as the $\boldsymbol{k}\cdot\boldsymbol{p}$ Hamiltonian is expanded within the second order of ***k***, the energy dispersions are isotropic in the k_x-k_y plane. This is very different from conventional ZB QWs and one of the remarkable features in WZ QWs. In the subband structure of WZ QWs, the energy bands can be labeled as HH_i, LH_i and CH_i as well as in WZ bulk GaN (see Figure 2.6 (b)). For any well length, the coupling between HH_i bands and the other kinds of bands is weak, and the HH_i bands are nearly parabolic. On the other hand, the LH_i bands are strongly coupled with the CH_i bands with a different parity, through the matrix elements H or I including $A_6 k_z k_\pm$. However, the feature of hole bands typically shown by the above three labels is almost unchanged even in the QWs because the crystal symmetry is still C_{6v} as well as in the bulk. This character originates from the hexagonal crystal symmetry and the weak spin-orbit coupling of the nitrogen atom. Thus, it also seems to be intrinsic in WZ nitride QWs and to be very different from conventional ZB GaAs/AlGaAs QWs. In case of conventional ZB QWs, the heavy hole mass becomes light and the light hole mass becomes heavy, compared with the bulk, due to the symmetry lowering.

Next, the well length dependence of the subband structure is discussed. Roughly speaking, the subband energies at the Γ point are inversely proportional to the bulk's hole masses along the confinement direction and to the square of the well length. This is generally called the quantum confinement effect. As for the bulk's masses along the k_z direction, the LH band is as heavy as the HH band and

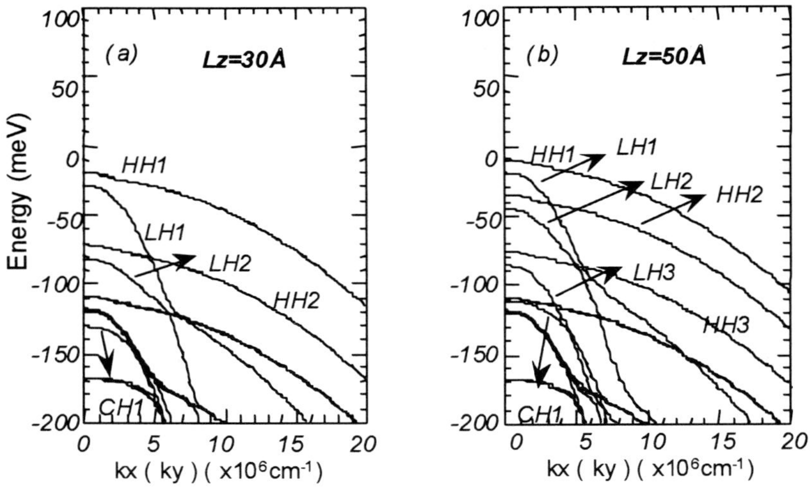

Figure 2.8 Valence subband structures of the unstrained wurtzite $GaN/Al_{0.2}Ga_{0.8}N$ single quantum well, with the well lengths L_z being (a) 30 Å and (b) 50 Å.

much heavier than the CH band. This is the reason why only CH_i bands are more split off from the HH_i and the LH_i bands in the QWs than in the bulk. In other words, the crystal-field splitting energy Δ_{cr} becomes effectively large due to the quantum confinement effect. For L_z = 30 Å, the CH_1 band is the only bound state among the CH_i bands, and the LH_2-CH_1 coupling causes the non-parabolicity of the LH_2 band. For L_z = 50 Å, the interval between the same labeled subbands becomes shorter than for L_z = 30 Å due to the quantum confinement effect, then the DOS around the VBM becomes a little larger. In WZ nitride QWs, thus, there is little well length dependence of the subband structure at the top of valence band due to very weak spin-orbit coupling and the unchanged symmetry. Therefore, the difference between the QWs with different well lengths is dependent on the number of the bound states alone.

Here, once again, we pay attention to the fact that the subband energy dispersions have similar characteristics to the bulk ones. The diagonal operators F and G in the $\boldsymbol{k{\cdot}p}$ Hamiltonian have the same parameter A_1+A_3 and Δ_1, and Δ_2 ~ 5meV is much smaller than the valence band offset ΔE_v ~ 100meV, so that $\phi^{hh}{}_m$ and $\phi^{lh}{}_m$ are almost the same eigenfunctions, i.e. $\langle\phi^{hh}{}_m|\phi^{lh}{}_{m'}\rangle \sim \delta_{m,\,m'}$. Here, the mixing between HH_i and LH_i with the same indices through $A_5k^2_\pm$ is very strong but meaningless because its effect is excluded by transforming the basis function to $|X\rangle$, $|Y\rangle$ and $|Z\rangle$, and any mixing other than them is almost zero. Therefore, the non-parabolicity of the energy dispersions is so weak. Then, the hole effective masses in the k_x-k_y plane are approximately given by $m_{hh} \sim -(A_2+A_4-A_5)^{-1}$ = 1.64 m_0, and $m_{lh} \sim -(A_2+A_4+A_5)^{-1}$ = 0.15m_0. These are the same as the bulk's hole masses. Generally, this property would be observed for the other III-V nitrides as well since the spin-orbit coupling of the nitrogen atom is negligible. As for the CH bands, the CH_1 is so weakly coupled with LH_2 through $A_6 k_z k_\pm$ that m_{ch} is about equal to A_2^{-1} and still heavy. This weak coupling modifies only the LH_2 band at $k \sim 6\times10^6$ cm^{-1} compared with the LH_1 and the LH_3 bands, due to the same parity of the envelope functions of the LH_1 and the LH_3 bands with that of the CH_1 band with respect to the $k_\pm$.

On the other hand, in the extreme case where Δ_2 is infinity, the hole effective masses in the plane of the quantum well are given by, $m_{hh} \sim -(A_2+A_4)^{-1} = 0.27m_0$, and $m_{lh} \sim -(A_2+A_4/3)^{-1} = 0.54m_0$. Note that m_{hh} becomes lighter than m_{lh}. This property has been observed in conventional ZB GaAs/AlGaAs QWs. Therefore, it is concluded that the decrease of the hole effective mass by the two dimensional carrier confinement is prevented by the small spin-orbit splitting energy.

2. 4 OPTICAL GAIN SPECTRA OF III-V NITRIDES LD STRUCTURES

2.4.1 Free Carrier Model

At first, we discuss the optical gain within free carrier model where Coulomb interaction is ignored. In this case, we can adopt the dipole approximation for the vector field and the two-particle Green's functions $\pi(k,k',i\omega)$ becomes non-interaction Green's function, expressed as

$$\pi(k,k',i\omega) = \pi^0(k,k,i\omega)\delta_{k,k'} = \frac{1-f_e(\varepsilon_k^e)-f_h(\varepsilon_k^h)}{i\omega-\varepsilon_k^e-\varepsilon_k^h},$$

Then, the optical gain is given by

$$g(\omega) = -\alpha(\omega) = \frac{2}{\bar{n}\omega c}\sum_k \frac{\hbar/\tau_{in}}{(\omega-\varepsilon_k^e-\varepsilon_k^h)^2+(\hbar/\tau_{in})^2} \times|w_0|^2\times\{f_e(\varepsilon_k^e-\mu_e)+f_h(\varepsilon_k^h-\mu_h)-1\}, \tag{2.46}$$

here ks are quantum numbers in a bulk, and c is the light velocity. $\bar{n}$ is the refractive index, as the value of which the observed value for WZ GaN ($\bar{n}$ = 2.67) (Ejder 1971) was used in both crystal structures. f_e and f_h are electron and hole distribution functions, respectively. μ_e and μ_h are the quasi-chemical potentials in the conduction and the valence bands, respectively. The calculations in this section were performed at room temperature (T = 298K). We assumed that the intraband relaxation time τ_{in} is the same as that of conventional ZB semiconductors, i.e., τ_{in} = 0.1 psec, though it is not sure for GaN.

In this subsection, we discuss the optical gain with respect to the quantum confinement and strain effects from the crystal symmetry point of view. The optical gain consists of three parts; the joint density of states (JDOS), the optical transition matrix elements and the occupation factor. Since the electron masses are much smaller than the hole masses, the JDOS is governed by the electron masses. In both the WZ and the ZB structures, they are not so different because of the *s*-like states at the CBM. Thus, the JDOS seem to be not so sensitive to the crystal symmetry. The optical transition matrix elements are obtained by sandwiching an optical polarization between the conduction and the valence band states. Since the splitting of the *p*-like states at the VBM is affected by the crystal-field, the optical transition matrix elements are strongly related to the crystal symmetry. As for the occupation factor, the positive sign indicates the population inversion. The quasi-chemical potential in f_h is dependent on the density of states (DOS) at the VBM, where the degeneracy is governed by the symmetry. Thus, the occupation factor is also affected by the crystal symmetry.

In the ZB case, the bulk's crystal symmetry is cubic, and there are three equivalent symmetry axes. The *p*-like six-fold degenerated VBM is separated into a four-fold degenerated Γ_8 state and a two-fold degenerated Γ_7 state by the spin-orbit coupling. When the QW structure is grown along the [001] direction, the crystal symmetry is changed to the tetragonal. Then, there are two equivalent symmetry axes, and the degeneracy of the Γ_8 state is removed. In general, the removal of the degeneracy causes the reduced DOS. Thus, it is expected that the DOS is reduced by using the QW structure. In the WZ case, the crystal symmetry of the QW structure grown along the [0001] direction is the same hexagonal as that of the bulk. Then, there is one symmetry axis, and the VBM is separated into three two-fold degenerated states, one Γ_9 state and two Γ_7 states. Since it makes no difference in the symmetry whether the QW structure is adopted, the band mixing is almost unchanged. Thus, adopting the QW structure is not so effective for reducing the DOS. This is one of the significant difference between the WZ and the ZB structures.

2.4.1.1 Optical Gain of Bulk GaN

In WZ GaN, the small spin-orbit coupling makes it possible to approximately express the three hole bands as follows,

$$|\Gamma_9^6(HH);\pm\frac{3}{2}\rangle = \frac{1}{\sqrt{2}}|X \pm iY,\pm\frac{1}{2}\rangle \tag{2.47}$$

$$|\Gamma_7^6(LH);\pm\frac{1}{2}\rangle \sim \frac{1}{\sqrt{2}}|X \pm iY,\mp\frac{1}{2}\rangle \tag{2.48}$$

$$|\Gamma_7^1(CH);\pm\frac{1}{2}\rangle \sim |Z,\pm\frac{1}{2}\rangle. \tag{2.49}$$

The basis $|\Gamma_i^j;\, m_J\rangle$ on the left-hand side indicates the eigenstate at the Γ point and the quantum number of the *z*-component of a total angular momentum operator J. The basis $|L,\, m_\sigma\rangle$ on the right-hand side indicates the orbital character and the quantum number of the *z*-component of a spin angular momentum operator σ_z. Along the k_z direction, the hole masses of the upper two bands are very heavy, and that of the only lowest CH band is light due to the strong coupling with the CBM through the $k_z p_z$ perturbation. Along the k_x direction, on the other hand, the CBM is strongly coupled with the mixed state

$$\frac{1}{\sqrt{2}}\left(|\Gamma_9^6;\pm\frac{3}{2}\rangle + |\Gamma_7^6;\mp\frac{1}{2}\rangle\right) \sim |X,\pm\frac{1}{2}\rangle,$$

and the hole mass of the only mixed band is light. Here, the lowest CH band is almost composed of the $|Z\rangle$ character, which yields the optical gain for the TM-mode. This is the reason why the TE-mode optical gain is dominant and why a large carrier density is needed to obtain the TM-mode optical gain in bulk WZ GaN.

In ZB GaN, the three hole bands can be expressed by

$$|\Gamma_8^{15}(HH);\pm\frac{3}{2}\rangle = \frac{1}{\sqrt{2}}|X \pm iY,\pm\frac{1}{2}\rangle \tag{2.50}$$

$$|\Gamma_8^{15}(LH);\pm\frac{1}{2}\rangle = \frac{1}{\sqrt{6}}(|X \pm iY,\mp\frac{1}{2}\rangle \mp 2|Z,\pm\frac{1}{2}\rangle) \tag{2.51}$$

$$|\Gamma_7^{15}(SH);\pm\frac{1}{2}\rangle = \frac{1}{\sqrt{3}}(|X \pm iY,\mp\frac{1}{2}\rangle \pm |Z,\pm\frac{1}{2}\rangle). \tag{2.52}$$

Since the hole masses of the upper two bands are very heavy along any $\boldsymbol{k}$ direction, the DOS at the VBM is larger than that in WZ GaN, as long as $\Delta_{cr} > \Delta_{so}$. Here, the LH band as well as the SH band includes the $|Z\rangle$ character, whose coefficients are independent of the spin-orbit splitting energy. Thus, the TM-mode optical gain shall start with the TE-mode one at the same carrier density. This is very different from WZ GaN. Figure 2.9 shows the maximum optical gain of bulk GaN. The transparent carrier density of the TE-mode in WZ GaN is lower than that in ZB GaN due to the small DOS at the VBM. However, the transparent carrier density of the TM-mode in WZ GaN is much higher than that in ZB GaN due to $\Delta_{cr} > \Delta_{so}$ and it is not observed in the range of Figure 2.9.

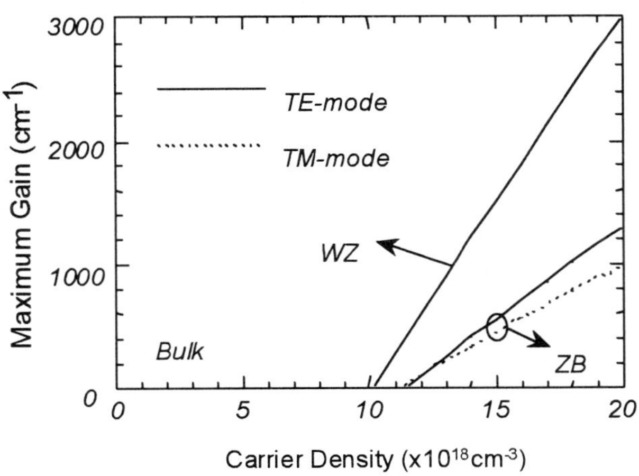

Figure 2.9 Optical gain of the wurtzite and the zincblende bulk GaN.

2.4.1.2 Optical Gain of Wurtzite Quantum Wells

When the carriers are confined in the QW, grown along the z direction, the wavenumber k_z is quantized as an integer index. Thus, the optical gain of the QW structure is given by

$$g(\omega) = \frac{2}{\bar{n}\omega c}\sum_{n,m}\sum_{k\perp}\frac{\hbar/\tau_{in}}{(\omega-\varepsilon^e_{n,k_\perp}-\varepsilon^h_{m,k_\perp})^2+(\hbar/\tau_{in})^2}$$
$$\times\left|w_0\langle\zeta^e_n\,|\,\zeta^h_m\rangle\right|^2\times\left\{f_e(\varepsilon^e_{n,k_\perp}-\mu_e)+f_h(\varepsilon^h_{m,k_\perp}-\mu_h)-1\right\} \tag{2.53}$$

where $m(n)$ represents the valence (conduction) subband label, and $k_\perp$ is the two-dimensional wave vector. Using the subband structures and the bulk's momentum matrix elements, we have calculated the optical gains with respect to the interband transition with the same wave vector k. The heterojunctions are perpendicular to the [0001] directions.

In the bulk WZ nitrides, the hybridization of the CH band with the HH and the

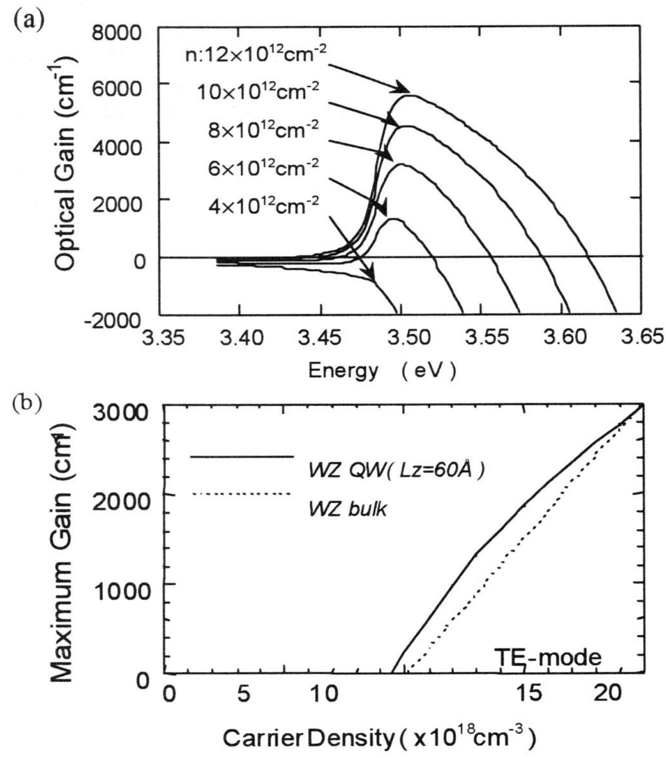

Figure 2.10 Optical gain spectra of the GaN/$Al_{0.2}Ga_{0.8}N$ quantum well for various carrier densities within free carrier model. The well length is 40 Å (b) Optical gain of the bulk GaN and $Ga_{0.2}Al_{0.8}N$ single quantum well with the well length L_z = 40 Å.

LH bands is negligible at the VBM, due to the small spin-orbit splitting energies. Thus, the eigenstates along the $k_x(k_y)$ direction can be approximately expressed as

$$HH\rangle \sim |Y(X)\rangle, |LH\rangle \sim |X(Y)\rangle, |CH\rangle \sim |Z(Z)\rangle.$$

In WZ QWs, the CH_i bands are more split off from the HH_i and LH_i bands. However, the band mixing is the same as that in the bulk GaN due to the same crystal symmetry as that of the bulk. Thus, the HH_1 mass is still heavy, and the DOS at the VBM is not so reduced. Then, the $|Z\rangle$-character, which yields the optical gain for the TM-mode, is very weak at the VBM. This is the reason why the optical gain for the TE-mode is dominant in WZ GaN-based QWs. The following discussion is limited to the TE-mode, and we take the optical polarization for the electric field as the *y*-direction in the calculations for TE-mode.

The optical gain spectra of the GaN/$Al_{0.2}Ga_{0.8}N$ quantum well with well length 40 Å are plotted in Figure 2.10 (a) for several carrier densities. In this calculation, the Luttinger like 6×6 Hamiltonian for the valence band were used to include the heavy hole and light hole band mixing. The effective parameters in the calculation were listed in Tables 2.1, 2.2, 2.3 and 2.5. It shows that about $5\times10^{12}cm^{-2}$ is necessary to obtain the gain. This value is two or three times larger compared with that of the conventional GaAs based LDs. This is due to the large density of states of the valence band. Note here that although several subband states were considered in the calculation, there are no features originated from the optical transitions other than the lowest conduction subband and highest valence subband transition for the wide energy range (~ 200 meV). It means that most of the characteristics of the optical gain are determined by the energy dispersions of those two bands.

The maximum optical gain of bulk WZ GaN and WZ GaN/$Al_{0.2}Ga_{0.8}N$ QWs with the well length L_z = 60 are shown in Figure 2.10 (b). As explained in the previous subsection, the threshold carrier density is not so reduced by adopting the QW structure, compared with the case of conventional ZB semiconductors. Figure2.11 shows the well length dependence of the maximum optical gain of WZ

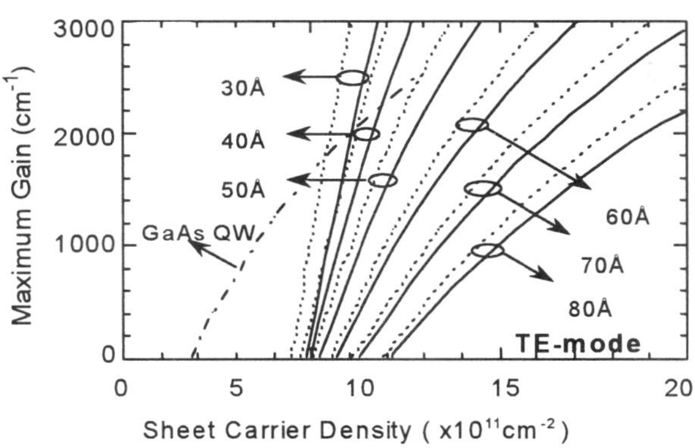

Figure 2.11 Optical gain of the wurtzite GaN/ $Al_{0.2}Ga_{0.8}N$ single quantum well, with the well lengths L_z being 30 ~ 80 Å. The solid and the dotted lines stand for the results without and with 0.5 % compressive biaxial strain, respectively. The calculated result for the zincblende GaAs/$Al_{0.3}Ga_{0.7}As$ single quantum wells with the well length L_z being 80 Å is shown by the dash-dotted line.

$GaN/Al_{0.2}Ga_{0.8}N$ QWs together with those of ZB $GaAs/Al_{0.3}Ga_{0.7}As$. In the unstrained QWs, as the well length becomes longer than 30 Å, more carrier density might be needed to obtain a sufficient optical gain. This is quite obvious from the feature of the subband structure. As the well length becomes longer, the interval between HH_1 and HH_2 bands becomes shorter, and the DOS around the VBM becomes larger. This is the reason why it is more difficult to realize the population inversion for longer well length. However, even for L_z= 30 Å, the threshold carrier density would be higher than ZB GaAs/AlGaAs QWs. According to the recent experimental report (Nakamura 1997), the threshold carrier density of InGaN MQW LDs is estimated to be $1 \sim 2 \times 10^{20}\ cm^{-3}$. On the other hand, the typical value of our calculated transparent carrier density of the unstrained GaN/AlGaN SQWs is $1 \sim 2 \times 10^{19}\ cm^{-3}$. Considering the loss of the injected carriers due to the irradiative recombination and the difference between MQWs and SQWs, our results seem to be reasonable. Further theoretical studies on structural dependence are necessary.

2.4.1.3 Optical Gain of Strained Wurtzite Quantum Wells

Figures 2.12 (a) shows the strain effect in the optical gain of WZ $GaN/Al_{0.2}Ga_{0.8}N$ QWs for various well lengths. In the compressive biaxial strained WZ QWs, the optical gain property is qualitatively improved for any well length, and the threshold carrier density would become lower than in the unstrained QWs. However, the reduction of the threshold carrier density is quantitatively not so effective. On the other hand, the tensile biaxial strain yields qualitatively negative effects on the threshold carrier density for any well length. Therefore, no biaxial strain comes up to our expectations of reducing the threshold carrier density for laser oscillation in WZ GaN/AlGaN QWs.

Here, on the analogy of the bulk, if it were possible to introduce a uniaxial strain into the *c*-plane of WZ QWs, it might cause much larger reduction of not only the DOS at the VBM but also the threshold carrier density. According to our calculated subband structures, any uniaxial strain in the *c*-plane reduces the DOS at the VBM due to the anisotropic energy splitting in the k_x-k_y plane. Considering the optical polarization, the useful uniaxial strain's directions are restricted to the following two cases. One is the compressive strain ($\varepsilon_{uni}{}^{com}$) parallel to the optical polarization for the electric field vector (**E**), that is the *y*-direction ($\varepsilon_{uni}{}^{com} \parallel \mathbf{E}$). The other is the tensile one ($\varepsilon_{uni}{}^{ten}$) perpendicular to it, that is the *x*-direction ($\varepsilon_{uni}{}^{ten} \perp \mathbf{E}$). Note that these two kinds of strains are in fact equivalent. Figure 2.12 (b) shows the schematic structure of the useful uniaxial strained WZ GaN/AlGaN QWs, corresponding to the former case. On the other hand, the useless uniaxial strain's directions are given by replacing the *y*-direction with the *x*-direction, namely, by changing the position of mirrors to the other lateral sides ($\varepsilon_{uni}{}^{com} \perp \mathbf{E}$, $\varepsilon_{uni}{}^{ten} \parallel \mathbf{E}$).

The calculated optical gains for uniaxial strained WZ $GaN/Al_{0.2}Ga_{0.8}N$ QWs are shown in Figure 2.12 (a), compared with those of biaxial strained QWs. If we induce the useful (useless) uniaxial strain, such as the compressive strain along the *y*-direction (*x*-direction), LH_i bands along the k_y (k_x) direction and HH_i bands along the k_x (k_y) direction move to the higher energy side. Then, the orbital component at the VBM becomes almost $|Y\rangle$- ($|X\rangle$-) character (see Figure 2.7.). In other words, such useful (useless) uniaxial strains selectively isolate the only $|Y\rangle$- ($|X\rangle$-) band at the VBM, which can (cannot) be coupled with the TE-polarized light. In that case,

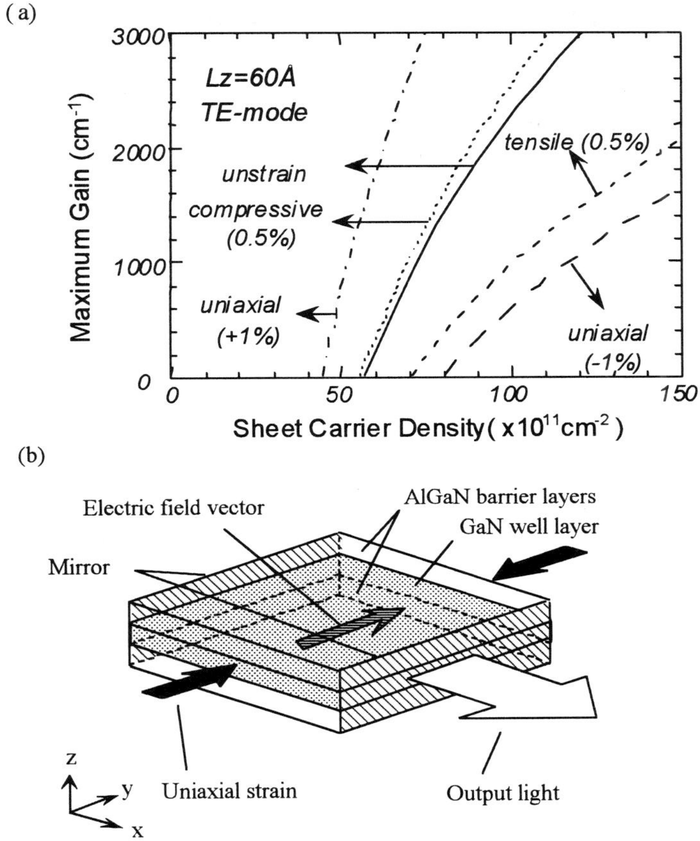

Figures 2.12 (a) Strain effect on optical gain of the wurtzite GaN/$Al_{0.2}Ga_{0.8}N$ quantum wells with the well lengths L_z being 60 Å. The solid, the dotted, the short-dashed, the dash-dotted and the long-dashed lines stand for the results without a strain, with ± 0.5 % compressive biaxial strains and with ± 1.0 % compressive uniaxial strains along the y-direction, respectively. (b) Schematic structure of the useful uniaxial strained wurtzite GaN/AlGaN single quantum wells.

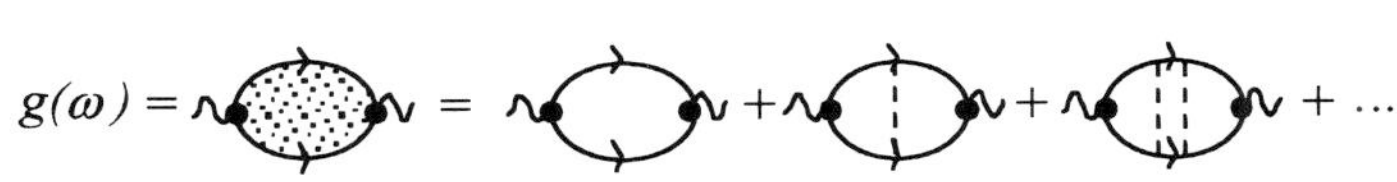

Figure 2.13 Ladder diagrams for the optical gain. The solid lines are Green's functions and dash lines are Coulomb interactions.

the recombination between the conduction band and valence band is also restricted and the recombination time would be longer. This is the reason why the above useful uniaxial strains cause not only the reduced DOS but also the reduced threshold carrier density. Furthermore, the reduced DOS is caused by the exclusion of the little contributive bands, which are HH_i bands along the k_y direction and LH_i bands along the k_x direction, to the optical gain for the TE-mode. Thus, the useful uniaxial strains give larger differential optical gain. On the other hand, the useless uniaxial strains cause the reduced DOS but the increased threshold carrier density and smaller differential optical gain. Therefore, if it were possible to introduce a uniaxial strain in the *c*-plane of WZ QWs, the threshold carrier density would be more efficiently reduced and the differential optical gain would be increased, as long as the relation between the uniaxial strain's direction and the optical polarization is suitable.

2.4.2 Coulomb Enhancement (Excitonic Effects) in the Optical Gain

In the previous section, the quantum confinement effect and strain effects in the optical gain were discussed within the free carrier model, i.e. no Coulomb interaction. One of salient features of wide-gap semiconductors is strong Coulomb interaction due to the small dielectric constant. Here we discuss how the Coulomb interaction plays in the optical gain spectra. In 1990, laser emission in (ZnCd)Se/ZnSe quantum wells at blue-green wavelength was observed under the resonant optical pumping into the lowest exciton resonance (Ding *et al.* 1992). At that time, the carrier density was small such that the population inversion at the band edge did not occur yet. Thus, excitonic effects in optical gain become an interest topic for wide-gap LDs, since the threshold current density could be reduced.

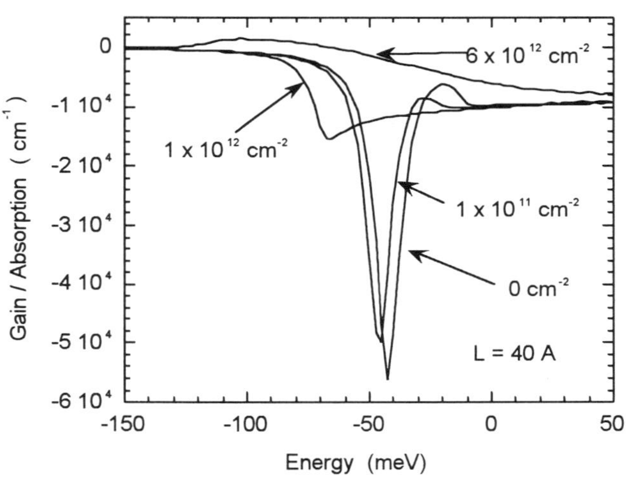

Figure 2.14 Optical gain spectra of the $GaN/Al_{0.2}Ga_{0.8}N$ quantum well for various carrier densities with Coulomb interaction with many-body effects. The well length is 40 Å

To incorporate the excitonic effect into the analysis of the optical gain, we must follow the procedures in Section 2.2.4 and evaluate the current-current correlation function, considering the Coulomb interaction among the carriers. Since the population inversion and the excitonic effects should be treated equally, we follow the ladder approximation as the many-body approach. Its diagram is shown in Figure 2.13. The bubble means the current-current correlation function and the upper solid line and lower one mean the electron and hole Green's function, respectively. The dash line shows the Coulomb interaction between the electron and hole. The first diagram of the right hand side in Figure 2.13 gives the optical gain within the free carrier model. The other diagrams include the Coulomb interactions, which means the photo-created electron hole pairs interact with each other through the Coulomb interaction before their recombinations. From the Equation (2.33), the summation of the infinity number of the diagrams can be expressed by the following recursion equation, so-called Bethe-Salpeter equation. (Haug *et al.* 1984)

$$\pi(k,k',i\omega) = \delta_{k,k'}\pi^0(k,k,i\omega) + \pi^0(k,k,i\omega)\sum_q V_s(q)\pi(k+q,k',i\omega). \tag{2.54}$$

This equation reproduces the Elliott's formula (Elliott 1957) for optical transition when there are no carriers in the conduction band and valence band as a ground state, in other word, corresponding to the two-body problem. When the many carriers exist near the population inversion, several many-body effects, for example screening effect and self-energy corrections, should be taken into account. To include them into the practical calculations, quasi-static approximation for the Coulomb interaction where the single plasmon-pole approximation was adopted. The detail was discussed in Section 2.2.4.

We show the optical gain spectra of the GaN/$Al_{0.2}Ga_{0.8}N$ quantum well in the two-band model in Figure 2.14 to clarify the effects caused by the Coulomb interaction. As we mentioned in the previous section, only the lowest conduction subband and highest valence subband are taken into account in the massess of the excitonic effects. The carrier density is changed from zero to $6\times10^{12}\text{cm}^{-2}$. The parameters used in the calculation were $m_e = 0.18\ m_0$, $m_h = 1.65\ m_0$ and $L = 40$ Å. The transparent carrier density is about $5\times10^{12}\text{cm}^{-2}$, which is the almost same as the one within free carrier model. The band edge absorption between the continuum states and the sharp exciton absorption can be observed clearly at the carrier density $n = 0$ from which the exciton binding energy can be estimated as about 40 meV. This value is so large that excitonic effects could be observed even at room temperature. Then, the binding energy and the absorption strength become reduced as increasing the carrier density. At $n = 1\times10^{12}\text{cm}^{-2}$, the band edge absorption is not observed clearly, but the excitonic absorption can still occur. However, when the optical gain appears, the excitonic absorption is completely gone. The reason is that the phase space filling factor $1-f_e(\varepsilon^e{}_k) - f_h(\varepsilon^h{}_k)$ kills the bound states at the population inversion. After the population inversion occurs, the electron-hole Coulomb interaction still enhances the oscillator strength, so-called Coulomb enhancement. But the enhancement factor is roughly 1.3 ~ 1.5 (Chow *et al.* 1994) against the free carrier model and the factor is less than that before the population inversion. Figure 2.15 shows the enhancement factor by comparing the gain spectra with/without the Coulomb interaction at the carrier density $n = 6\times10^{12}cm^{-2}$. The red-shift of the spectra is shown as increasing the carrier density, due to the band-gap renormalization. After all, the electron-hole Coulomb interaction or excitonic effect just enhance the oscillator strength of the transition and the exciton-like absorption is not obtained with the optical gain simultaneously. Furthermore, it becomes evident that the free carrier model is sufficient to discuss on the optical gain spectra other than the peak energy since the Coulomb enhancement factor is close to one. Therefore, the previous discussion of the quantum confinement and the strain effect in the optical gain are useful even when the Coulomb interaction is considered.

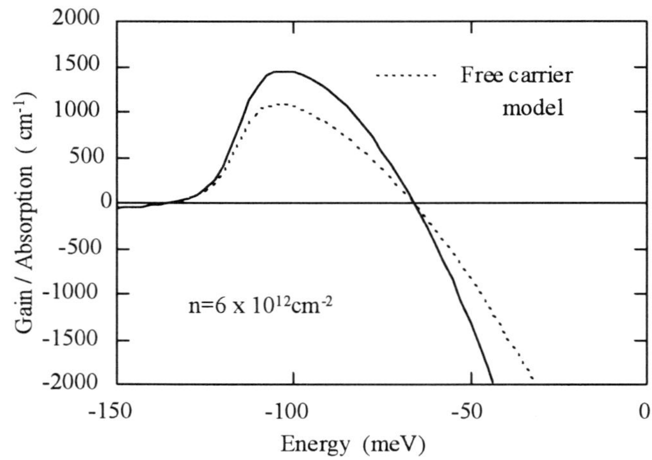

Figure 2.15 Optical gain spectra at the carrier density $n = 6\times10^{12}cm^{-2}$ with (solid line) / without (dash line) the Coulomb interaction.

2.4.3 Optical Gain with Localized States

Consequently, we could understand that the Coulomb interaction between the electron and the hole enhances the oscillator strength and that the enhancement factor decreases with increasing the carrier density. It means that the more frequently the electron and the hole scatter each other, the stronger the enhancement becomes. In the two-band model, there are few states near the band edge where the carriers can be scattered at the population inversion. Thus, the factor becomes small as 1.3. Then, we introduce another energy level, i.e. three-band model. In practice, most of the LD structures based on GaN have ternary wells in terms of InGaN. There are In compositional fluctuations, which would cause the localized states, and the radiative recombination attributed to the localized excitons by the compositional fluctuation were measured recently (Chichibu, *et al.* 1996, 1997; Narukawa, *et al.* 1997). Thus, it becomes important to study the relationship between the localized states and the optical gain. As we discussed in the previous section, the optical gain is enhanced by the Coulomb interaction, compared to the one within a free carrier model. This is so-called 'excitonic enhancement' or 'Coulomb enhancement'. Here we show another excitonic enhancement of the optical gain involved in the localized states. This enhancement is stronger than the conventional one in the previous section and it might play an important role to reduce the threshold carrier density.

The configuration of the quantum dot-like structures in the well is assumed to introduce the localized states, as shown in Figure 2.16 (a). The carriers are confined along the growth direction and the lateral one in the quantum disk with the radius R. The disks are assumed to be distributed randomly in the well and not to be correlated among them. Furthermore, only the holes are assumed to be trapped at the lowest subband in the disk, since the hole mass is much heavier than the electron mass. The wavefunction of the localized states might be given by

$$\psi_j(r,z)=\sqrt{\frac{\alpha}{\pi}}\exp\left(-\frac{\alpha}{2}(r-R_j)^2\right)\zeta_0^h(z), \tag{2.55}$$

where R_j is the position of the disk in the well and $\zeta_0{}^h(z)$ is the wavefunction of the highest valence subband state in the quantum well. α is the parameter which can be determined by the variational principle for the given radius R and potential depth of the disk. The energy diagram of the three-band model can be depicted schematically in Figure 2.16 (b). The Hamiltonian of the model including the Coulomb interaction is given by

$$H=\sum_k \varepsilon_k^e a_k^\dagger a_k+\sum_k \varepsilon_k^h b_k^\dagger b_k+\sum_{i=1}^{No}\varepsilon_i^d d_i^\dagger d_i+V^{ee}+V^{hh}+V^{eh}, \tag{2.56}$$

with

$$V^{ee}=\frac{1}{2}\sum_q\sum_{k_1,k_2}V(q)a_{k_1+q}^\dagger a_{k_2+q}^\dagger a_{k_2}a_{k_1},$$

$$\begin{aligned}V^{hh}=&\frac{1}{2}\sum_q\sum_{k_1,k_2}V(q)b_{k_1+q}^\dagger b_{k_2-q}^\dagger b_{k_2}b_{k_1}\\&+\frac{1}{2}\sum_q\sum_{j_1,j_2}V(q)e^{iq(R_{j1}-R_{j2})}e^{-\frac{q^2}{2\alpha}}d_{j_1}^\dagger d_{j_2}^\dagger d_{j_2}d_{j_1}\\&+\frac{1}{2}\sum_q\sum_{k_1,j_2}V(q)e^{ikR_{j2}}e^{-\frac{q^2}{4\alpha}}b_{k_1+q}^\dagger d_{j_2}^\dagger d_{j_2}b_{k_1}\\&+\frac{1}{2}\sum_q\sum_{j_1,k_2}V(q)e^{-ikR_{j1}}e^{-\frac{q^2}{4\alpha}}d_{j_1}^\dagger b_{k_2-q}^\dagger b_{k_2}d_{j_1},\end{aligned}$$

$$\begin{aligned}V^{eh}=&\sum_q\sum_{k_1,k_2}V(q)a_{k_1+q}^\dagger b_{k_2-q}^\dagger b_{k_2}a_{k_1}\\&+\sum_q\sum_{k_1,k_2,j}V(q)e^{i(k_2-q)R_j}e^{-\frac{(k_2-q)^2}{2\alpha}}\frac{2}{L}\sqrt{\frac{\pi}{\alpha}}a_{k_1+q}^\dagger d_j^\dagger b_{k_2}a_{k_1}\\&+\sum_q\sum_{k_1,k_2,j}V(q)e^{-ik_2R_j}e^{-\frac{k_2{}^2}{2\alpha}}\frac{2}{L}\sqrt{\frac{\pi}{\alpha}}a_{k_1+q}^\dagger b_{k_2-q}^\dagger d_j a_{k_1}\\&+\sum_q\sum_{k_1,j}V(q)e^{-iqR_j}e^{-\frac{q^2}{4\alpha}}a_{k_1+q}^\dagger d_j^\dagger d_j a_{k_1},\end{aligned}$$

where $\varepsilon^d_i,(d_i\ ,\ d_i^\dagger)$ are the energy level (annihilation and creation operators) of the localized states. N_0 is the number of the localized states. We assumed that the localized energy levels are equal to ε^d.

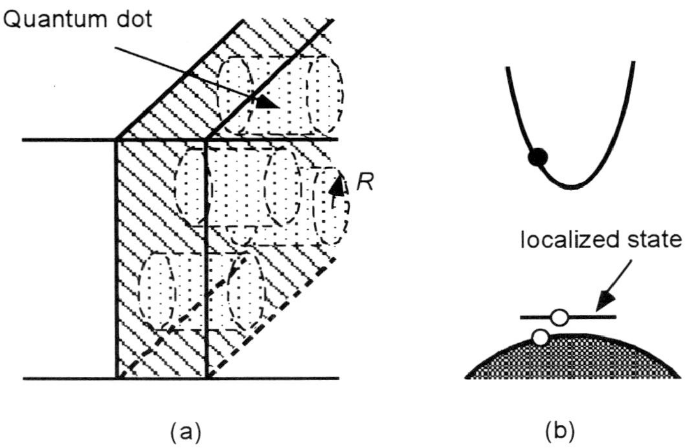

Figure 2.16 (a) Configuration of the quantum dot-like structures.
(b) Energy diagram of the three-band model.

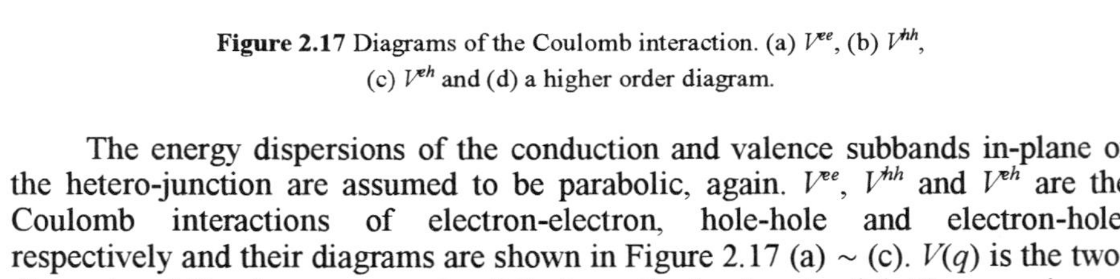

Figure 2.17 Diagrams of the Coulomb interaction. (a) V^{ee}, (b) V^{hh}, (c) V^{eh} and (d) a higher order diagram.

The energy dispersions of the conduction and valence subbands in-plane of the hetero-junction are assumed to be parabolic, again. V^{ee}, V^{hh} and V^{eh} are the Coulomb interactions of electron-electron, hole-hole and electron-hole, respectively and their diagrams are shown in Figure 2.17 (a) ~ (c). $V(q)$ is the two-dimensional Fourier components of the bare Coulomb potential. Here, we ignore the higher order Coulomb interactions with respect to the N_0/L^2, corresponding to the Figure 2.17 (d). There are two kinds of transitions; one is between the localized states and conduction subband states (channel-1 transition) and the other, between the conduction and valence subband states (channel-2 transition). The recombination current operator is expressed by

$$J = \sum_{k,j} w_{1,k,j} a_k d_j + \sum_k w_{2,k} a_k b_k + h.c., \tag{2.57}$$

where $w_{1,k,j}$ (= $\langle k|e/m{\cdot}p|v\psi_i\rangle\langle\zeta^e{}_0|\zeta^h{}_0\rangle$) and $w_{2,k,}$ (= $\langle ck|e/m{\cdot}p|vk\rangle\langle\zeta^e{}_0|\zeta^h{}_0\rangle$) are the optical transition matrix elements, respectively. The current-current correlation function is given by

$$\Pi(i\omega)=N_0\sum_{k,k',j}w_{1,k,j}\pi_{1,j}(k,k',i\omega)w^*_{1,k',j}+\sum_{k,k'}w_{2,k'}\pi_2(k,i\omega)w^*_{2,k}. \tag{2.58}$$

The two-particle Green's functions, $\pi_{1,j}(k,i\omega)$ and $\pi_2(k,i\omega)$ are obtained from the coupled Bethe-Salpeter equations (Uenoyama 1995)

$$\pi_{1,j}(k,k',i\omega)=\pi^0_{1,j}(k,k,i\omega)\left(\delta_{k,k'}+\sum_q V_s(q)(e^{-iqR_j}e^{-\frac{q^2}{4\alpha}}\pi_{1,j}(k+q,k',i\omega)\right.$$
$$\left.+e^{ikR_j}e^{-\frac{q^2}{2\alpha}}\frac{2}{L}\sqrt{\frac{\pi}{\alpha}}\pi_2(k+q,i\omega)\right),$$

$$\pi_2(k,k',i\omega)=\pi^0_2(k,k,i\omega)\left(\delta_{k,k'}+\sum_q V_s(q)(\sum_j(e^{-i(k+q)R_j}e^{-\frac{(k+q)^2}{2\alpha}}\right.$$
$$\left.\cdot\frac{2}{L}\sqrt{\frac{\pi}{\alpha}}\pi_{1,j}(k+q,\delta_{k,k'},i\omega)+\pi_2(k+q,i\omega))\right),$$

where $V_s(q)$ is the static screened Coulomb potential $V(q)/\varepsilon(q)$ and the single plasmon-pole approximation (Haug 1984) was adopted in the dielectric function $\varepsilon(q)$. $\pi^0{}_{1,j}(k,k,i\omega)$ and $\pi^0{}_2(k,k,i\omega)$ are the noninteracting two particle Green's functions, expressed as

$$\pi^0_{1,j}(k,k,i\omega)=\frac{1-f_e(\varepsilon^e_k)-f_h(\varepsilon^d_j)}{i\omega-\varepsilon^e_k-\varepsilon^d_j},\ \pi^0_2(k,k,i\omega)=\frac{1-f_e(\varepsilon^e_k)-f_h(\varepsilon^h_k)}{i\omega-\varepsilon^e_k-\varepsilon^h_k},$$

where f_e and f_h are the distribution functions for the electron and the hole, respectively.

We plot the optical gain spectra in Figure 2.18 (a) and (b), including the effects of the localized states and the Coulomb interaction. The carrier densities n and the localized state densities n_ℓ are (a) n = 2.5×10^{12} cm^{-2}, n_ℓ = 2×10^{12} cm^{-2} and (b) n = 6×10^{12} cm^{-2}, n_ℓ = 2×10^{12} cm^{-2}. The localized state energy was 88 meV from the top of the valence subband state at the radius R = 30 Å and the energy depth of the disk V_0 = 100 meV. The dash line shows the gain spectra without the coupling between the channel-1 and the channel-2 transitions. The band-gap renormalization is considered in each transition. The channel-1 and the channel-2 transition occur at ~ –160meV and ~ –100meV, respectively. At the carrier density n 2.5×10^{12} cm^{-2}, the population inversion occurs only at the channel-1 transition. At this time, the enhancement factor of the oscillator strength of the channel-1 transition is almost twice and this value is bigger than that of the conventional two-band model. When the carrier density is increased up to the n = 6×10^{12}cm^{-2}, the continuum states are also occupied and the population inversion occurs even at the channel-2 transition. At that time, the enhancement factor reduced to 1.1 at the channel-1 transition.

From that fact, it becomes convinced that to obtain the large enhancement factor at a certain transition, unoccupied states should exist close to the states causing the transition. On the other hand, instead of providing the unoccupied states to near the transition, decreasing temperature might be helpful. Because the Fermi surface becomes sharp against the energy, the unoccupied states increases near the Fermi energy. Then the oscillator strength of the transition near the Fermi energy would be enhanced like a Fermi-edge singularity.

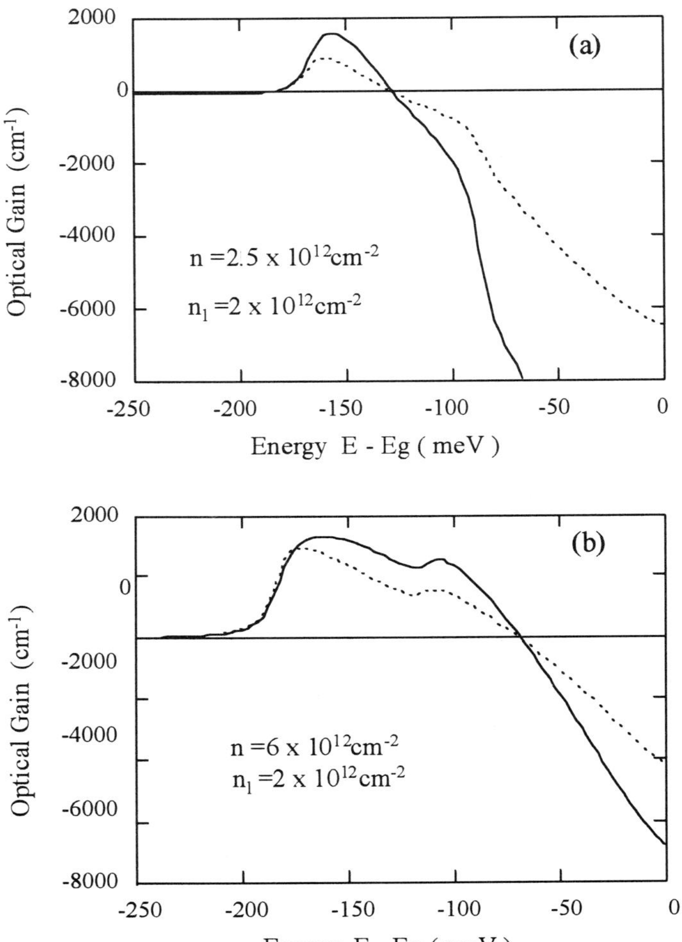

Figure 2.18 Optical gain spectra of the GaN/$Al_{0.2}Ga_{0.8}N$ quantum well for (a) $n = 2.5\times10^{12}$ cm^{-2}, $n_\ell = 2\times10^{12}$ cm^{-2} and (b) $n = 6\times10^{12}$ cm^{-2}, $n_\ell = 2\times10^{12}$ cm^{-2}. The localized level is -88 meV.

2.5 CONCLUSIONS

Electronic and optical properties of GaN/AlGaN quantum wells (QWs) as well as of bulk GaN and AlN were investigated on the basis of the first-principles calculation and the effective mass theory. The obtained information provides a fundamental understanding of the III-V nitrides and leads to a guideline on material and device design with them.

The electronic band structures of bulk GaN and AlN were calculated for the wurtzite (WZ) structure by the full-potential linearized augmented plane wave (FLAPW) method. The FLAPW band calculations for the deformed lattices were also performed, and the strain effects were investigated. In order to describe the electronic structure near the band edges, the effective mass Hamiltonian for the WZ structure was derived, taking the hexagonal C_{6v} symmetry into account. The physical parameters, such as effective masses, the Luttinger-like parameters, crystal-field and spin-orbit splitting energies, momentum matrix elements and deformation potentials, of WZ GaN and AlN were determined by reproducing the FLAPW band structures with the effective mass Hamiltonian. The parameters experimentally obtained so far for GaN are in agreement with our theoretical values.

The subband structures of GaN/AlGaN QWs were calculated on the basis of the effective mass approximation by using the physical parameters derived theoretically. In WZ GaN/AlGaN QWs, the subband dispersions are almost the same as the energy dispersions in bulk WZ GaN, and the heavy (light) hole mass is still heavy (light). This character originates from the hexagonal crystal symmetry and the weak spin-orbit coupling of the nitrogen atom, and it remains under biaxial stress. This is very different from conventional ZB GaAs/AlGaAs QWs, in which the heavy (light) hole mass becomes light (heavy) compared with bulk GaAs, due to the symmetry lowering from T_d to D_{2d}. Furthermore, the biaxial strain does not reduce the density of states (DOS) so much around the valence band maximum (VBM) in WZ GaN/AlGaN QWs. It was also found that the uniaxial strain in the c-plane of WZ QWs significantly reduced the DOS around the VBM, due to the symmetry lowering from C_{6v} to C_{2v}.

As the first step towards obtaining a guideline on the device design for the III-V nitride lasers, the optical gains with respect to the interband transition were evaluated for bulk GaN and GaN/AlGaN QWs. For the WZ structure, very heavy hole masses cause high threshold carrier density for laser oscillation, compared with conventional ZB III-V semiconductors. In WZ GaN/AlGaN QWs, the compressive biaxial strain can reduce the threshold carrier density, but it is not so effective quantitatively. It was suggested that the threshold carrier density would be considerably reduced if it were possible to introduce a uniaxial strain into the c-plane of WZ QWs. Furthermore, in the two-band model, the electron-hole Coulomb interaction enhances the oscillator strength at any carrier density but the optical gain and the excitonic absorption cannot be observed simultaneously. After the optical gain comes out, the Coulomb enhancement factor is around 1.3 and most of the characteristics of the gain spectra can be discussed within the free carrier model.

To increase the enhancement factor, we must provide the unoccupied states as final states of the Coulomb scattering. Introducing a localized level is one of the examples to do so. In this three-band model which includes the effects of the Coulomb interaction and the localized states, we have found a new excitonic enhancement of the optical gain involved the localized states. This enhancement is stronger than the conventional Coulomb enhancement in the two-band model. This result might lead to a possibility of low threshold carrier density.

The authors would like to thank Emeritus Professor A. Yanase of University of Osaka Prefecture for his helpful discussion and providing them with his FORTRAN source codes for the FLAPW band calculation.

2.6 REFFERENCES

Akasaki, I., Amano, H., Sota, S., Sakai, H., Tanaka, T. and Koike, M., 1995, *Jpn. J. Appl. Phys.*, **34**, p. L1517.

Barker, A.S. and Ilegems, M., 1973, *Phys. Rev. B,* **7**, p. 743.

Bauer, G.E.W. and Ando, T., 1986, *J. Phys.*, **C19**, p. 1537; *ibid.*,**C19**, p. 1553.

Chichibu, S., Shikanai, A., Azuhata, T., Sota, T., Kuramata, A., Horino, K. and Nakamura, S., 1996, *Appl. Phys. Lett.*, **68**, p. 3766.

Chichibu, S., Azuhata, T., Sota, T. and Nakamura, S., 1997, presented at *38th Electronic Material Conference, W-10, June 26-28, Santa Barbara, Appl. Phys. Lett.*, **69**, p. 4188 ; *ibid.*, **70**, p. 2822.

Cunningham, R.D., Brander, R.W., Kuee, N.D. and Wickender, D.K., 1972, *J. Luminescence,* **5**, p. 21.

Chow, W.W., Koch, S.W. and Sargent, III M., 1994, *Semiconductor-Laser Physics,* Springer-Verlag Berlin Heidelberg, p. 160.

Ding, J., Jeon, H., Ishihara, T., Hagerott, M. and Nurmikko, A.V., 1992, *Phys. Rev. Lett.*, **69**, p. 1707.

Dingle, R. and Ilegems, M., 1971, *Solid State Commun.*, **9**, p. 175.

Dingle, R., Sell, D.D., Stokowski, S.E. and Ilegems, M., 1971, *Phys. Rev. B*, **4**, p. 1211.

Drechsler, M., Hofmann, D.M., Meyer, B.K., Detchprohm, T., Amano, H. and Akasaki, I., 1995, *Jpn. J. Appl. Phys.*, **34**, p. L1178.

Ejder, E., 1971, *Phys. Status Solidi* A, **6**, p. 445.

Elliott, R.J., 1957, *Phys. Rev.*, **108**, p. 1384.

Flores, G.R., Contreras, H.N., Martinez, A.L., Powell, R.C. and Greene, J.E., 1994, *Phys. Rev. B,* **50**, p. 8433.

Gunnarson, O. and Lundquvist, B.I., 1976, *Phys. Rev. B,* **13**, p. 4274.

Haug, H., and Schmitt-Rink, S., 1984, *Prog. Quant. Electr.*, **9** , p. 3.

Ishihara, H. and Cho K., 1993, *Phys. Rev. B,* **48**, p. 7960.

Kane, E.O., 1957, *J. Phys. Chem. Solids,* **1**, p. 249.

Kosicki, B.B, Powell, R.J. and Burgiel, J.C., 1970, *Phys. Rev. Lett.*, **24**, p. 1421.

Lindberg, M. and Koch, S.W., 1988, *Phys. Rev. B,* **38**, p. 3342.

Luttinger, J.M. and Kohn, W., 1955, *Phys. Rev.*, **97**, p. 869.

Mahan, G.D., 1981, *Many-Particle Physics,* PLENUM PRESS, New York.

Meyer, B.K., Volm, D., Graber, A., Alt, H.C., Detchprohm, T., Amano, H. and Akasaki, I., 1995, *Solid State Commun.*, **95**, p. 597 .

Nakamura, S., 1997, *Tech. Rep. IEICE.* OPE96-**144**, LQE96-**142** 31 [in Japanese].

Nakamura, S., 1996, presented at *Materials Research Society Fall Meeting, N1.1, Dec.2-6,* Boston.

Nakamura, S., 1997, presented at *The 2nd Intl. Conf. on Nitride Semiconductors, S-1 Oct. 27-31,* Tokushima.

Narukawa, Y., Kawakami, Y., Fujita, Sz., Fujita, Sg. and Nakamura, S., 1997, *Phys. Rev.*, **55**, p. R1938; *Appl. Phys. Lett.*, **70**, p. 981.

Pankove, J.I., Bloom, S. and Harbeke, G., 1975, *RCA Rev.*, **36**, p. 163.

Rheinländer, B. and Neumann, H., 1974, *Phys. Stat. Sol. (b)*, **64**, p. K123.

Rubio, A., Corkill, J.L., Cohen, M.L., Shirley, E.L. and Louie, S.G., 1993, *Phys. Rev. B*, **48**, p. 11810.

Suzuki, M. and Uenoyama, T., 1996a, *Jpn. J. Appl. Phys.*, **35**, p. 543.

Suzuki, M. and Uenoyama, T., 1996b, *Jpn. J. Appl. Phys.*, **35**, p. 1420.

Suzuki, M. and Uenoyama, T., 1996c, *Jpn. J. Appl. Phys.*, **35**, p. L953.

Suzuki, M. and Uenoyama, T., 1996d, *Appl. Phys. Lett.*, **69**, p. 3378.

Suzuki, M. and Uenoyama, T., 1996e, *J. Appl. Phys.*, **80**, p. 6868.

Suzuki, M., Uenoyama, T. and Yanase, A., 1995, *Phys. Rev. B*, **52**, p. 8132.

Uenoyama, T., 1995, *Phys. Rev. B*, **51**, p. 10228.

Uenoyama, T. and Suzuki, M., 1995, *Appl. Phys. Lett.*, **67**, p. 2527.

Wimmer, E., Krakauer, H., Weinert, M. and Freeman, A.J., 1981, *Phys. Rev. B*, **24**, p. 864.

Yeh, C. , Lu, Z.W., Froyen, S. and Zunger, A., 1992, *Phys. Rev. B*, **46**, p. 10086.

Chapter 3
Electrical Conductivity Control

Chris G. Van de Walle

Proper control of electrical conductivity is essential in optoelectronic devices. In this chapter we discuss two aspects: Section 3.1 deals with doping of the *n*-type and *p*-type layers, and Section 3.2 describes heterostructure band offsets, which play an essential role in carrier confinement.

3.1 DOPING

Optoelectronic devices such as light-emitting diodes and laser diodes rely on carrier injection into the active region. Holes and electrons are injected on opposite sides of the device, and travel through various layers before recombining in the active region. Transport of holes through a layer requires that layer to be *p*-type doped (i.e., doped with acceptors); transport of electrons requires *n*-type doping (doping with donors).

Modifying the conductivity of a semiconductor through doping is one of the crucial elements of semiconductor device design. It has long been recognized that doping of wide-band-gap semiconductors often poses problems: typically, one type of doping is hard to achieve. In the case of GaN, *p*-type doping has always been difficult to achieve. In addition, as-grown GaN often exhibits a tendency for *n*-type conductivity. The general nature of this tendency, irrespective of growth technique, led to a belief that the *n*-type conductivity had to be *intrinsic* to the material, i.e., caused by *native defects*. Recent work has conclusively shown that native defects are *not* the cause of the *n*-type conductivity; instead, the conductivity can be attributed to unintentional incorporation of impurities.

Native defects are defects that are intrinsic to the crystal: vacancies (a missing atom), self-interstitials (an extra atom), or antisites (for instance, a gallium atom sitting on a nitrogen site in GaN). These defects usually introduce deep levels in the band gap of the semiconductor. Occasionally, the defect-induced level is near a band edge, and the defect can act as a dopant. Most often, however, native defects act as compensating centers; i.e., they counteract the electrical activity of the dopants.

In order to obtain optimal device behavior, doping levels need to be controlled and optimized. The mechanisms that can limit doping include: limited solubility of the dopant; large ionization energy of the dopant; compensation by native defects; compensation by foreign impurities; and compensation by alternative atomic configurations of the dopant. The latter refers to a situation where an impurity that acts as an acceptor when incorporated on a substitutional site may act as a donor when incorporated on an interstitial site.

In order to address these various doping-limiting mechanisms, it is essential to answer the following questions: What is the likelihood that various types of native defects and impurities will be incorporated? (For impurities, this obviously addresses the solubility.) What is the electrical activity of these centers; i.e., do they act as donors or acceptors, and what are their levels in the band gap? And, finally, what sources of compensation may occur? A theoretical formalism has been developed to address each of these issues, and Section 3.1.1 contains a description of the approach. In Section 3.1.2 we apply the formalism to the problem of *n*-type doping, and summarize the experimental status. Section 3.1.3. addresses the same issues for *p*-type doping.

3.1.1 Theory of Native Defects and Impurities

When considering a native defect or impurity, we usually want to know its concentration. Indeed, a defect that only occurs in very low concentrations does not pose much of a problem. Dopant impurities, on the other hand, need to be incorporated in high enough concentrations to have the desired effect on conductivity. Theoretically, the concentration of an impurity or native defect is determined by its formation energy, a quantity that expresses how much energy it costs to incorporate the defect or impurity in the crystal. Here we present an overview of the theoretical approach; details can be found in Van de Walle *et al.* (1993) and Neugebauer and Van de Walle (1996a).

3.1.1.a Formalism for calculating concentrations of defects and impurities

Formation energies

In thermal equilibrium, the concentration is given by

$$c = N_{sites}\, exp\left(-\frac{E^f}{kT}\right). \qquad (3.1)$$

where E^f is the formation energy, and N_{sites} is the number of sites the defect or impurity can be incorporated on. For instance, for a nitrogen vacancy in GaN N_{sites} is equal to the number of nitrogen sites in the crystal, about $4.4\times10^{22}\text{cm}^{-3}$. k is the Boltzmann constant, and T the temperature. Equation (3.1) shows that defects with a *high* formation energy will occur in *low* concentrations.

Equation (3.1) assumes thermodynamic equilibrium. This assumption is expected to be satisfied at the high temperatures at which metal-organic chemical vapor deposition (MOCVD) growth of nitrides is carried out. At lower temperatures, such as those used in molecular-beam epitaxy (MBE), deviations from equilibrium may occur. The effect of growth temperature on doping characteristics will be discussed in Section 3.1.3.b.

The formation energy is not a constant, but depends on the growth conditions and the doping of the material. For example, the formation energy of a nitrogen vacancy is determined by the relative abundance of Ga and N atoms, as expressed by the chemical potentials μ_{Ga} and μ_N. In thermal equilibrium these chemical potentials are subject to the condition that $\mu_{Ga} + \mu_N = \mu_{GaN}$, i.e., their sum needs to add up to the energy of a two-atom unit of the GaN crystal. This still leaves one chemical potential variable. If the nitrogen vacancy is charged, the formation energy depends further on the Fermi level (E_F); indeed, the Fermi level (chemical potential of electrons) is the energy of the reservoir with which electrons can be exchanged to change the charge state of the defect.

Forming a nitrogen vacancy requires the removal of one N atom from the crystal; the formation energy is therefore:

$$E^f(V_N^q) = E_{tot}(\mathrm{GaN}:V_N^q) - E_{tot}(\mathrm{GaN,bulk}) + \mu_N + qE_F. \tag{3.2}$$

The quantities labeled E_{tot} are total energies for systems of atoms; these values can be obtained from state-of-the-art first-principles calculations, as described below. q is the charge state of the defect. The nitrogen vacancy is a donor, and hence the charge state can be neutral ($q = 0$) or positive ($q = +1$). (In Section 3.1.3.d we will see that the nitrogen vacancy can also occur in the +3 charge state).

In the case of impurities the formation energy is defined in a very similar fashion. For instance, for an oxygen atom residing on a nitrogen site in GaN:

$$E^f(O_N^q) = E_{tot}(\mathrm{GaN}:O_N^q) - E_{tot}(\mathrm{GaN,bulk}) + \mu_O + \mu N + qE_F. \tag{3.3}$$

The formation energy now includes a term related to the relative abundance of oxygen in the environment, i.e., the oxygen chemical potential. Similar expressions can be written down for all native defects and impurities, in all of their possible charge states.

Fermi level

The Fermi level E_F is not an independent parameter, but is determined by the condition of charge neutrality. Once equations such as (3.2) and (3.3) are formulated for every native defect and impurity in the material, the complete problem (including free-carrier concentrations in valence and conduction bands) can be solved self-consistently by imposing charge neutrality. For a given position of E_F (and fixed values of the chemical potentials) the concentration of all defects and impurities can be calculated; many of these defects and impurities are charged. Fermi-Dirac statistics can also be used to derive the concentration of electrons in the conduction band and holes in the valence band. In general, the initial choice of E_F leads to a situation where the total net charge of the system is not equal to zero. In order to satisfy charge neutrality, the problem has to be solved self-consistently, by finding the value of E_F for which the net charge is zero.

It is instructive, however, to plot the behavior of formation energies as a function of E_F in order to examine the behavior of defects and impurities when the doping level changes. Below we will present a number of plots showing the variation of formation energies with Fermi level. For clarity of presentation the atomic chemical potentials are set equal to fixed values; a general case can always be addressed by referring back to Equations (3.2) and (3.3).

Chemical potentials

We emphasized that the chemical potentials are variable parameters. However, they are subject to specific boundary conditions. To establish the range of the chemical potentials, one has to consider the various phases that can be formed, in thermodynamic equilibrium, out of the constituents. For instance, μ_{Ga} is bounded from above by the energy of a Ga atom in bulk Ga: $\mu_{Ga} \leq \mu_{Ga,bulk}$. Indeed, if one would try to raise μ_{Ga} above this level, bulk Ga metal would preferentially be formed. Similarly, μ_N has an upper bound imposed by molecular N_2.

We already pointed out that the sum of the Ga and N chemical potentials has to add up to the energy of a two-atom unit of GaN. This condition can be expressed as

$$\mu_{Ga} + \mu_N = \mu_{GaN} = \mu_{Ga,bulk} + \mu_{N_2} + \Delta H_f(\text{GaN}), \tag{3.4}$$

where ΔH_f(GaN) is the heat of formation of GaN (a negative quantity for a stable compound). Equation (3.4) shows that imposing an *upper* bound on the Ga chemical potential leads to a *lower* bound on the N chemical potential, given by

$$\mu_N^{min} = \mu_{N_2} + \Delta H_f(\text{GaN}). \tag{3.5}$$

Similarly,

$$\mu_{Ga}^{min} = \mu_{Ga,bulk} + \Delta H_f(\text{GaN}). \tag{3.6}$$

We therefore note that the Ga (as well as the N) chemical potential can vary over a range given by the heat of formation of GaN, ΔH_f(GaN).

Upper bounds on the chemical potentials of the impurities can also be established. For that purpose we must explore the various compounds that the impurity can form in its interactions with the system. For oxygen, for instance, a possible upper bound on μ_O is of course imposed by O_2 molecules. However, a more stringent constraint arises from the compound Ga_2O_3, which leads to the following upper bound:

$$2\mu_{Ga} + 3\mu_O = \mu_{Ga_2O_3} = 2\mu_{Ga,bulk} + 3\mu_{O_2} + \Delta H_f(\text{Ga}_2\text{O}_3). \tag{3.7}$$

The upper bound on the chemical potential for the impurity derived from this equation corresponds to maximum incorporation of the impurity, i.e., to the solubility limit. This limit is imposed by the condition of equilibrium with the solubility-limiting phase. For oxygen, this is Ga_2O_3; for Si it is Si_3N_4; and for Mg it is Mg_3N_2.

For clarity of presentation, we will set the chemical potentials to fixed values in the figures shown below. A general case can always be addressed by referring back to Equations (3.2) or (3.3). The fixed values we have chosen correspond to Ga-rich conditions $\mu_{Ga}=\mu_{Ga,bulk}$, and to maximum incorporation of the various impurities.

One last comment on Equation (3.1): in principle, the energy appearing in Equation (3.1) should be the *Gibbs free energy*. Use of the (zero-temperature) formation energy as defined in Equations (3.2) and (3.3) implies that contributions from vibrational entropy are neglected (contributions from *configurational* entropy, reflecting the various ways in which the defect or impurity can be incorporated in the lattice, are easily included in the prefactor). Explicit calculations of such entropy terms are computationally very demanding, and currently not feasible for the large number of defects to be addressed. However, these entropy contributions cancel to some extent, for instance when comparing relative free energies, which are important for solubilities (Qian *et al.*, 1988; Laks *et al.*, 1991). More generally, the entropy terms are small enough not to affect qualitative conclusions.

3.1.1.b First–principles calculations

The term "first principles" refers to the fact that the calculations used to derive the energies in Equations (3.2) and (3.3) do not require any adjustable parameters or any input from experiment. Most of the theoretical studies carried out for defects and impurities in GaN to date are based on density-functional theory in the local-density approximation, using pseudopotentials, a plane-wave basis set, and a supercell geometry (Neugebauer and Van de Walle, 1994a, 1996a, 1996b). This computational approach is now regarded as a standard for performing first-principles studies of defects in semiconductors.

Density functional theory (DFT) in the local density approximation (LDA) (Hohenberg and Kohn, 1964; Kohn and Sham, 1965) allows a description of the ground state of a many-body system in terms of a one-electron equation with an effective potential. The total potential consists of an ionic potential due to the atomic cores, a Hartree potential, and a so-called exchange and correlation potential that describes the many-body aspects. This approach has proven very successful for a wide variety of solid-state problems.

Pseudopotentials: Most properties of molecules and solids are determined by the *valence* electrons; the *core* electrons can usually be removed from the problem by representing the ionic core (i.e., nucleus plus inner shells of electrons) with a *pseudopotential.* State-of-the-art calculations employ nonlocal normconserving pseudopotentials (Hamann *et al.*, 1979) that are generated solely based on atomic calculations and do not include any fitting to experiment.

Calculations for nitride semiconductors pose two challenges for the use of pseudopotentials. First, nitrogen is a first-row element and hence characterized by a fairly deep potential, which requires a rather large number of terms in the plane-wave basis set (Fourier expansion) that is commonly used to expand wave functions and potentials in a variational approach. Second, the 3*d* states of Ga (and 4*d* states of In) cannot strictly be considered as core states. Indeed, the energetic position of these states is fairly close to the N 2*s* states, and the interaction between these levels may not adequately be described if the *d* states are treated as part of the core. Including the *d* electrons as valence states significantly increases the computational burden, since their wave functions are much more localized than those of the *s* and *p* states. Sometimes explicit inclusion of the *d* states as valence states is the only way to obtain accurate results (Neugebauer and Van de Walle, 1994b). In many cases, however, the effect of the *d* states can be adequately included by a correction to the exchange and correlation potential, commonly referred to as the "nonlinear core correction" (Louie *et al.*, 1982). "Soft" pseudopotentials (requiring small basis sets in the plane-wave expansions) for the various elements in the nitride system were generated using the scheme of Troullier and Martins (1991).

Supercell geometry: The last ingredient commonly used in pseudopotential-density-functional calculations for defects or impurities is the *supercell geometry*. Ideally, one would like to describe a single isolated impurity in an infinite crystal. In the supercell approach, the impurity is surrounded by a finite number of semiconductor atoms, and this structure is periodically repeated. Maintaining periodicity allows continued use of algorithms such as Fast Fourier Transforms. One can also be assured that the band structure of the host crystal is well described (which may not be the case in a cluster approach). For sufficiently large supercells, the properties of a single, isolated impurity can be derived. Convergence tests have indicated that the energetics of impurities and defects in the zinc-blende phase are usually well described by using 32-atom supercells. For the wurtzite phase (which has lower symmetry), larger supercells (containing 72 or 96 atoms) are required.

Error bars on the values derived from first-principles calculations depend on the quantity under study. When comparing energies of an impurity in different positions in the lattice, the accuracy is quite high and the error bar smaller than 0.1 eV. When looking at formation energies, or comparing energies of different charge states, the limitations of density-functional theory may play a role, and the error bar could increase to a few tenths of an eV. These error bars are still small enough not to affect any of the qualitative conclusions discussed below.

More details about the computational approach can be found in the references by Stumpf and Scheffler (1994), Neugebauer and Van de Walle (1996a, 1996b) and Bockstedte *et al.* (1997).

3.1.1.c Results for native defects in GaN

Native defects can have a profound effect on the electronic properties of semiconductors. They often act as sources of compensation; for instance, when acceptors are introduced to the material, a native defect that acts as a donor may form spontaneously, thereby counteracting the intended doping. Native defects have often been blamed for doping problems; however, they are only one of the potential reasons why doping efficiencies may be lower than expected. Other causes for doping problems include limited solubility of the dopant; large ionization energy of the dopant; incorporation of the dopant impurity in other configurations in the lattice; or compensation by foreign impurities.

Native defects have sometimes also been invoked not just as compensating centers, but also as sources of doping. The nitrogen vacancy in GaN is a prime example: for many years the nitrogen vacancy was thought to be the source of *n*-type conductivity in GaN. As early as 1983 it was pointed out that unintentional incorporation of oxygen was an alternative explanation (Seifert *et al.*, 1983). Still, it is only recently that unintentional impurities have become widely accepted as the source of *n*-type conductivity, thanks in part to contributions from first-principles theory. Direct experimental information about native defects in the nitrides is very scarce at this time; we include references where available.

Formation energies for all native point defects in GaN, in all relevant charge states, are shown in Figure 3.1. These results were obtained from first-principles calculations (Neugebauer and Van de Walle, 1994a, 1996a, 1996c). Similar results for native defects were obtained by Boguslawski *et al.* (1995). These references also contain detailed information about the atomic structure of each of the defects. For each charge state of each defect, Figure 3.1 show only the line segment that gives rise to the overall lowest energy. Thus, the change in slope of the lines represents a change in the charge state of the defect. The Fermi-level position at which the charge state changes corresponds to a transition energy, which in turn corresponds to a thermodynamic defect level that would be measured experimentally.

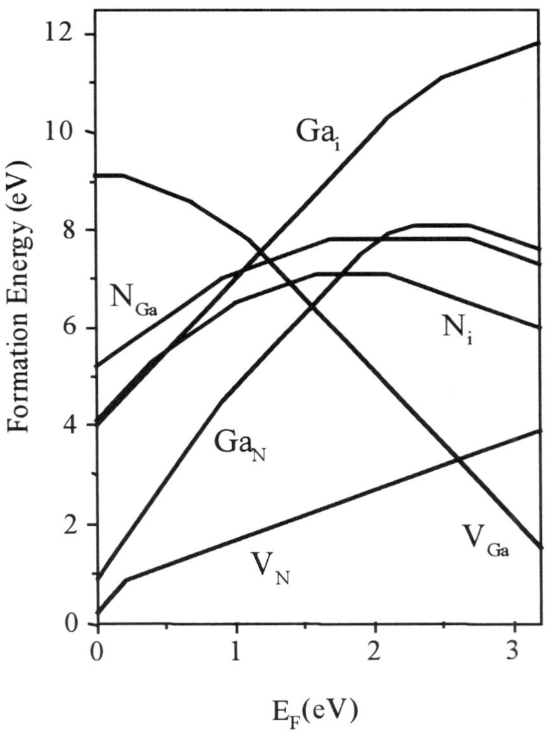

Figure. 3.1 Formation energy as a function of Fermi level for native point defects in GaN under Ga-rich conditions, obtained from first-principles calculations. E_F=0 corresponds to the top of the valence band. Based on the work of Neugebauer and Van de Walle (1994a, 1996a, 1996c).

The results shown in Figure. 3.1 are for wurtzite GaN; the formation energies in zinc-blende GaN are usually very similar (Neugebauer and Van de Walle, 1994b), although some differences may occur for self-interstitials. Figure 3.1 shows that self-interstitial and antisite defects are high-energy defects in GaN, and are thus unlikely to occur during growth. Still, these defects can be created by electron irradiation or ion implantation. One such defect, created by electron irradiation, has been identified as a complex involving interstitial Ga (Linde *et al.*, 1997).

Nitrogen vacancies

Nitrogen vacancies (V_N) do behave as donors: Figure. 3.1 shows the formation energy as a function of E_F has a positive slope, corresponding to a positive charge on the defect [see. Equation (3.2)]. When purposely created, for instance during irradiation or ion implantation, nitrogen vacancies increase the electron concentration (Look *et al.*, 1997a). However, their high formation energy makes it very unlikely that nitrogen vacancies would form spontaneously during *growth* of undoped or *n*-type GaN, and hence they cannot be responsible for *n*-type conductivity. As discussed in Section 3.1.2, unintended *n*-type conductivity in GaN must therefore be attributed to other causes, the most likely one being unintentional incorporation of donor impurities such as oxygen or silicon. Note, however, that nitrogen vacancies have a low formation energy in *p*-type GaN, making them a likely compensating center in the case of acceptor doping, as discussed in Section 3.1.3.d.

Gallium vacancies and yellow luminescence

Figure 3.1 shows that in *n*-type GaN the lowest-energy native defect is the gallium vacancy V_{Ga}, a triple acceptor. Gallium vacancies can therefore act as compensating centers in *n*-type GaN. Yi and Wessels (1996) have indeed observed evidence of compensation by a triply charged defect in Se-doped GaN.

The gallium vacancy has been proposed to be the source of the frequently observed *yellow luminescence* (YL) (Neugebauer and Van de Walle, 1996c). The YL in GaN is a broad luminescence band centered around 2.2 eV. The following features of gallium vacancies are consistent with experimental observations of the YL:

(i) The gallium vacancy is an acceptor-type defect, and hence its formation energy decreases with increasing Fermi level (see Figure. 3.1). Gallium vacancies are therefore more likely to occur in *n*-type than in *p*-type GaN. This trend is consistent with experimental observations indicating suppression of the YL in *p*-type material (Zhang *et al.*, 1995a; Götz *et al.*, 1996a; Kim *et al.*, 1996). Conversely, an increase in *n*-type doping increases the intensity of the YL (Kaneda *et al.*, 1996; Schubert *et al.*, 1997; Lee *et al.*, 1997). Schubert *et al.* (1997) also found that the defects giving rise to the YL act as compensating centers, in agreement with the gallium-vacancy model.

(ii) The concentration of gallium vacancies is lower in Ga-rich material, and higher in Ga-poor material. Experimentally, the YL has been found to be suppressed in samples grown by metal-organic chemical vapor deposition (MOCVD) with higher trimethyl-gallium (TMGa) flow rates (Zhang *et al.*, 1995a; 1996).

(iii) The Ga vacancy has a deep level (the 2-/3- transition level) about 1.1 eV above the valence band (see Figure. 3.1). Transitions between the conduction band (or shallow donors) and this deep level therefore exhibit the correct energy to explain the YL. A variety of experiments indicate that the YL is due to a transition between a shallow donor and a deep acceptor level about 1 eV above the valence band (Sánchez *et al.*, 1996; Hoffmann *et al.*, 1997; Calleja *et al.*, 1997). The calculated pressure dependence of the V_{Ga} level is also consistent with experiment (Suski *et al.*, 1995).

(iv) Gallium vacancies can form complexes with donor impurities in GaN, particularly with oxygen. The electronic structure of this complex is very similar to that of the isolated gallium vacancy, giving rise to a deep level again about 1.1 eV above the valence band. The formation energy of this complex is lower than that of the isolated vacancy; the presence of oxygen can therefore enhance the concentration of Ga vacancies and hence the YL. Complexing with oxygen is thus a likely explanation for the increase in YL intensity in the neighborhood of the interface with the sapphire substrate (Hofmann *et al.*, 1995) where the oxygen concentration is also known to be higher (see Section 3.1.2).

(v) The most direct evidence of a correlation between Ga vacancies and YL is based on positron annihilation measurements (Saarinen *et al.*, 1997). Both GaN bulk and epitaxial layers were found to contain Ga vacancies in concentrations ranging from 10^{17} to 10^{18} cm^{-3}. It was found that the concentration of V_{Ga} correlates with the intensity of the YL, providing direct evidence for the involvement of the V_{Ga} acceptor levels in the YL.

3.1.1.d Results for native defects in AlGaN and InGaN

First-principles calculations have also been performed for native defects in AlN (Mattila and Nieminen, 1996, 1997; Stampfl and Van de Walle, 1998a). The main conclusions are similar to those for GaN: self-interstitials and antisites are high in energy—with the exception of the Al interstitial in *cubic* AlN, which is a triple donor and could act as a compensating center in *p*-type material. The nitrogen vacancy is a high-energy defect in *n*-type AlN, but has a relatively low energy in *p*-type AlN. This formation energy is lower in AlN than in GaN, and therefore compensation of *p*-type $Al_xGa_{1-x}N$ by nitrogen vacancies becomes increasingly severe for increasing x (see Section 3.1.3.e). The behavior of V_{Al} is similar to V_{Ga} in GaN, but because of the larger band gap of AlN the formation energy of V_{Al} becomes significantly lower than V_{Ga} for Fermi-level positions high in the gap. This defect will therefore increasingly compensate the *n*-type conductivity in $Al_xGa_{1-x}N$ with increasing x.

Finally, in InN all native defects were found to have high formation energies (Stampfl and Van de Walle, 1998b), with the exception of the nitrogen vacancy, which has a formation energy comparable to GaN.

3.1.2 *n*-type Doping

3.1.2.a Unintentional doping

The nitrides exhibit a tendency to be *n*-type conductive if no special precautions are taken. This was long thought to be caused by the spontaneous formation of nitrogen vacancies. It is now known (see Section 3.1.1.c) that nitrogen vacancies are unlikely to form in *n*-type material. The observed *n*-type conductivity must therefore be attributed to unintentional incorporation of dopant impurities (Neugebauer and Van de Walle, 1995a).

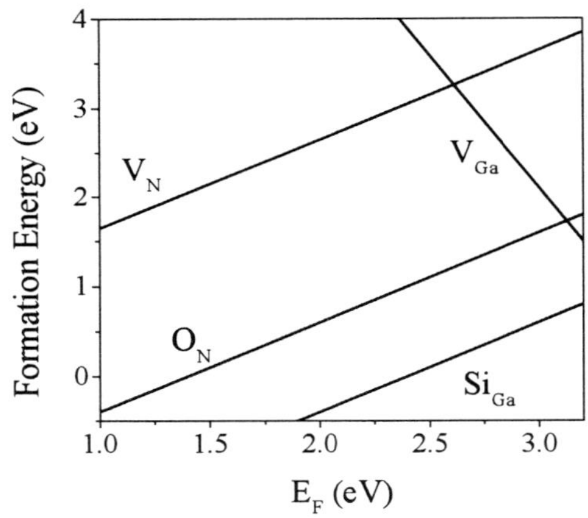

Figure. 3.2 Formation energy as a function of Fermi energy for dominant native defects (nitrogen and gallium vacancies) and donors (oxygen and silicon) in GaN. The zero of Fermi energy is located at the top of the valence band. From Van de Walle *et al.* (1998).

Figure 3.2 summarizes first-principles results for native defects and impurities relevant for *n*-type doping. It is clear that O and Si have much lower formation energies than V_N; these impurities can therefore be readily incorporated in *n*-type GaN. Both O and Si form shallow donors in GaN. The slope of the lines in Figure. 3.2 indicates the charge state of the defect or impurity [see Equation (3.3)]: Si_{Ga}, O_N, and V_N all appear with slope +1, indicating single donors.

Oxygen had been proposed as a potential source of *n*-type conductivity in GaN as early as 1983 (Seifert *et al.*, 1983). In 1992, Chung and Gershenzon observed the effect of oxygen on the electrical and optical properties of GaN grown by MOCVD. Oxygen was introduced through intentional injection of water. The carrier concentrations in their samples decreased when the growth temperature was increased, a behavior inconsistent with nitrogen vacancies as the cause of the conductivity. Rather, they proposed that less oxygen was being incorporated at the higher growth temperatures.

Still, the prevailing conventional wisdom, attributing the *n*-type behavior to nitrogen vacancies, proved hard to overcome. Recent experiments have confirmed that unintentionally doped *n*-type GaN samples contain silicon or oxygen concentrations high enough to explain the electron concentrations. Götz *et al.* (1996b, 1997) reported electrical characterization of intentionally Si-doped as well as unintentionally doped samples, and concluded that the *n*-type conductivity in the latter was due to silicon. They also found evidence of another shallow donor with a slightly higher activation energy, which was attributed to oxygen. Secondary-ion mass spectroscopy (SIMS) on hydride vapor phase epitaxy (HVPE) material also shows levels of oxygen or silicon in agreement with the electron concentration determined by electrical measurements (Götz *et al.*, 1997).

High levels of *n*-type conductivity have traditionally been found in GaN bulk crystals grown at high temperature and high pressure (Perlin *et al.*, 1997). High-pressure studies have recently established that the characteristics of these samples are very similar to epitaxial films that are intentionally doped with oxygen

(Wetzel *et al.*, 1996, 1997). The *n*-type conductivity of bulk GaN can therefore be attributed to unintentional oxygen incorporation.

Contaminating impurities may have many sources: water is the main contaminant of NH_3, the most frequently used source has in MOCVD. The fact that even high-purity (99.999%) NH_3 can be a significant source of oxygen contamination was documented in the case of InGaN growth by Piner *et al.* (1997). Fortunately, purification can remove most of the water. In MBE, oxygen may enter as a contaminant in the nitrogen source gas; or it may be due to the quartz lining of certain components, for instance plasma sources.

Contamination of GaN epitaxial films by impurities emanating from the substrate was studied by Popovici *et al.* (1997). Films grown on SiC showed large concentrations of Si; growth on ZnO produced incorporation of Zn; and growth on sapphire led to incorporation of O. The incorporation occurred in a region wider than the defective near-substrate interface layer, pointing to the possibility of impurity diffusion at the growth temperature. Siegle *et al.* (1997) also reported impurity diffusion into the near-substrate region of HVPE-grown layers, based on luminescence and Raman experiments. The impurity was likely to be oxygen, diffusing from the sapphire substrate or from a ZnO buffer.

Kim *et al.* (1997) used SIMS and Hall effect measurements to investigate the incorporation of various impurities in GaN grown by reactive MBE. They interpreted their results as favoring the nitrogen vacancy as the cause of the unintentional *n*-type conductivity. However, the authors themselves acknowledge that the particular set of samples used in their study is not well suited to rule out oxygen contamination. In fact, all of their samples contained sufficient oxygen to explain the electron concentration; therefore their results are entirely *consistent* with oxygen being the donor. Kim *et al.* emphasize, however, that the electron concentration does not increase as the oxygen concentration increases. This has been observed by others, as well (Götz *et al.*, 1998a); it is indicative of the fact that oxygen incorporates not only on nitrogen sites (where it is an electrically active donor), but also in other forms. This additional oxygen may appear, for instance, in the form of Ga_2O_3 precipitates.

Götz *et al.* (1997, 1998a) and Look and Molnar (1997) have pointed out potential pitfalls in extracting information about dopant concentrations and ionization energies from Hall-effect measurements. Experiments on unintentionally doped HVPE-grown layers indicated the presence of a thin, highly conductive layer near the interface with the sapphire substrate. This layer acts as a parasitic, parallel conduction path. Götz *et al.* (1997) suggested oxygen contamination may be responsible for the high conductivity of this layer.

We now discuss individual donor impurities in more detail.

3.1.2b Oxygen

The activation energy of oxygen has been determined from variable-temperature Hall effect measurements to be 29 meV for a donor concentration $N_D=1\times10^{18}$ cm^{-3} (Götz *et al.*, 1998b). This activation energy decreases with increasing donor concentration. Götz *et al.* also reported that the total oxygen concentration (measured by SIMS) was significantly higher than the donor concentration. In the previous section we suggested that such additional oxygen may occur in the form of Ga_2O_3 precipitates.

Intentional doping of MOCVD-grown GaN with oxygen was performed by Niebuhr *et al.* (1997), using N_2O as a source. Carrier concentrations, as measured by Hall effect, varied from $4\times10^{16}cm^{-3}$ to $4\times10^{18}cm^{-3}$ and scaled with the N_2O concentration.

DX centers

Oxygen-doped material exhibits a freezeout of carriers when it is subjected to hydrostatic pressure exceeding 20 GPa (Wetzel *et al.*, 1997; Van de Walle, 1998). This behavior is explained by a "*DX*-like" behavior of the oxygen donor. The prototype *DX* center is silicon in GaAs, which undergoes a transition from a shallow to a deep center when hydrostatic pressure is applied (Mooney, 1992). First-principles calculations for oxygen in GaN under pressure (Van de Walle, 1998) show that at sufficiently high pressure the oxygen impurity moves off the substitutional site and assumes an off-center configuration: a large outward relaxation introduces a deep level in the band gap, and the center actually becomes negatively charged (i.e., it behaves as an acceptor). Silicon donors do not exhibit this transition (see Section 3.1.2.c).

Alloying with AlN increases the band gap similar to the application of hydrostatic pressure; the behavior of impurities in AlGaN should therefore be similar to that in GaN under pressure. First-principles calculations (Park and Chadi, 1997; Van de Walle, 1998) predict a *DX* transition for oxygen in $Al_xGa_{1-x}N$ when $x>0.3$, consistent with the observed decrease in *n*-type conductivity of unintentionally doped $Al_xGa_{1-x}N$ (Koide *et al.*, 1986; Lee *et al.*, 1991; McCluskey *et al.*, 1998). Once again, Si donors do not exhibit the *DX* transition (see Section 3.1.2.b). Interestingly, the *DX* transition does *not* occur in *zinc-blende* AlGaN (Van de Walle, 1998).

One of the characteristic features of *DX* centers is the observation of persistent photoconductivity (McCluskey *et al.*, 1998a). It should be pointed out, however, that observations of persistent photoconductivity are not necessarily indicative of the presence of *DX*-like centers. Various groups (Wickenden *et al.*, 1997; Hirsch *et al.*, 1997) have reported persistent photoconductivity in *n*-type GaN. This behavior may be due to the presence of defective regions near extended defects.

The conclusion is that oxygen cannot be used as a shallow donor in $Al_xGa_{1-x}N$ with $x>0.3$. Even if another donor impurity is used (such as silicon) that does not exhibit *DX* behavior, the presence of oxygen in the layer could be detrimental to *n*-type conductivity: indeed, once oxygen undergoes the *DX* transition it behaves as an acceptor, and therefore counteracts the electrical activity of other donors. Control of oxygen incorporation in $Al_xGa_{1-x}N$ with high Al content is therefore essential.

3.1.2.c Silicon

Silicon is almost universally used for intentional *n*-type doping of GaN, AlGaN, and InGaN. Silicon doping is typically achieved by flowing SiH_4 during MOCVD growth, or using a solid Si source in MBE. The theoretical results of Figure. 3.2

show that the formation energy of silicon is quite low, and therefore it is readily incorporated in GaN. Ion implantation doping has been investigated, but is quite difficult in the nitrides. Indeed, the damage induced by the implantation process can only be annealed out at temperatures exceeding the temperature at which GaN begins to decompose. Up to 50% activation could be achieved with silicon implanted at a dose of 10^{16} cm^{-2}, and annealed at 1100 °C (Zolper *et al.*, 1997). Efficient annealing of the damage only seems to occur at temperatures higher than typical growth temperatures. Decomposition of the film can in principle be prevented by encapsulation or by working under high nitrogen pressure. Cao *et al.* (1998) reported Si implant activation efficiencies exceeding 90% by rapid thermal processing at 1400-1500°C, using encapsulation with thin AlN films.

The activation energy of silicon in GaN derived from variable-temperature Hall effect measurements is 17 meV for $N_D=3\times10^{17}$ cm^{-3} (Götz *et al.*, 1996b; 1998b). As pointed out in the case of oxygen, the ionization energy is sensitive to the concentration of the dopant.

Silicon donors do not exhibit the *DX* transition (see Section 3.1.2.b). Silicon was experimentally found to remain a shallow donor in GaN under pressure up to 25 GPa (Wetzel *et al.*, 1997) and in $Al_xGa_{1-x}N$ alloys up to x=0.44 (McCluskey *et al.*, 1998a). This is in agreement with theoretical studies predicting that Si will remain a shallow donor throughout the composition range (i.e., up to x=1.0) (Van de Walle, 1998). Another study predicts *DX*-center formation for x>0.6 (Boguslawski and Bernholc, 1997).

Silicon doping of InGaN was described by Nakamura (1994; Nakamura *et al.*, 1997; Deguchi *et al.*, 1998). Silicon doping of InGaN well and barrier layers has been reported to improve the device characteristics (threshold operating current and voltage) of laser diodes (Nakamura *et al.*, 1997). The effect may be attributed to improved structural properties (Zeng *et al.*, 1997), but also to screening of the built-in polarization fields, resulting in higher emission efficiencies (Chichibu *et al.*, 1999).

3.1.2.d Germanium

Relatively little experimental information is available for Ge. Doping is performed by flowing GeH_4 during MOCVD growth. Zhang *et al.* (1995b) reported successful *n*-type doping of $Al_xGa_{1-x}N$ with Ge. For pure GaN the carrier concentration was as high as 3×10^{19} cm^{-3}, but the doping efficiency decreased with increasing Al content of the alloy.

The activation energy of Ge derived from variable-temperature Hall effect measurements is 19 meV at a donor concentration of $N_D=3\times10^{17}$ cm^{-3} (Götz, 1998b), only slightly larger than for Si. As for oxygen, the ionization energy is sensitive to the concentration of the dopant.

Regarding possible *DX* behavior of Ge, only theoretical results are available, and they are conflicting: Boguslawski and Bernholc (1997) found Ge to exhibit a *DX*-transition for x>0.3, while Park and Chadi (1997) found Ge to be a shallow donor for all alloy compositions.

3.1.2.e Mobilities

The resistivity of a doped layer depends not only on the carrier concentration, but also on the mobility of the carriers. Analysis of mobilities is complicated by the presence of extended defects in epitaxial layers; the effect of those defects on carrier mobilities is difficult to assess. Look *et al.* (1997b) have emphasized the importance of a proper treatment of mobility when extracting data from Hall-effect measurements. Theoretical calculations of electron mobility, based on solving the Boltzmann equation, were reported by Rode and Gaskill (1995). Typical values in good-quality films at room temperature are between 500 and 1000 cm^2/Vs for electron mobility, and between 5 and 10 cm^2/Vs for hole mobility.

A few attempts have been made at modeling the carrier concentration and mobility. Fehrer *et al.* (1998) attributed the observed behavior to potential barriers caused by electron traps at grain boundaries, using a model developed for polycrystalline materials. Epitaxial growth of GaN on sapphire indeed produces columnar "grains" that are slightly tilted or twisted with respect to one another, and it is plausible that the boundaries between those regions would contain a high defect density. Band bending near the boundaries produces potential barriers that affect the electron mobility. Fehrer *et al.* (1998) obtained good agreement with their experimental data, assuming the presence of electron traps with an energy 0.31 eV below the conduction-band edge.

3.1.2.f Modulation doping

Modulation doping is a technique applied to heterostructures whereby dopants are introduced into a wider-band-gap material, with the resulting carriers transferring into an undoped, lower-band-gap material. The main advantage is that carriers moving in the lower-gap, undoped material do not suffer scattering due to ionized impurities. Nakamura *et al.* (1998a) reported modulation doping of strained-layer superlattices (SLS) for *n*-type and *p*-type cladding layers in laser diodes (see also Nakamura *et al.*, 1998b). The *n*-type layer consisted of 2.5-nm-thick undoped GaN separated by 2.5-nm-thick Si-doped $Al_{0.14}Ga_{0.86}N$ layers. The *p*-type SLS was similar, but with Mg doping of the $Al_{0.14}Ga_{0.86}N$. Improvements in device performance were reported. However, in later work Nakamura (1998b) reported similar improvements for superlattices in which only the GaN layers in the GaN/AlGaN superlattice cladding layers were doped (strictly speaking no longer a modulation-doped structure) (see also Nakamura *et al.*, 1998b). It therefore seems that the benefit of the superlattice cladding layer mostly results from structural improvements, rather than modulation doping.

3.1.2.g Deep levels

Deep levels, which can act as compensating centers, can be studied experimentally with deep-level transient spectroscopy (DLTS) and photocapacitance spectroscopy. A whole range of levels have been reported (see, e.g., Götz *et al.*, 1995a, 1996c; Hacke and Okushi, 1997; Wang *et al.*, 1998). The concentration of these levels is always below $10^{16} cm^{-3}$; the effect of these centers on compensation should therefore be quite small. The origin of these deep levels is unclear at this time. They could be due to native defects, but no correlation has been made with theoretically determined transition levels (Section 3.1.1.c). It is possible that the levels are associated with electrically active defects along dislocations or other extended defects; studies in material with lower extended defect densities would be informative. Alternatively, the levels could be due to unintentionally incorporated deep impurities, such as transition-metal impurities.

3.1.3 *p*-type Doping

The first successful *p-type* doping of GaN was accomplished by Amano *et al.* (1989), using magnesium as the dopant. They found that as-grown Mg-doped MOCVD layers were not *p*-type, but an activation treatment rendered the Mg electrically active. They used low-energy electron-beam irradiation (LEEBI) to achieve this activation.

A few years later Nakamura *et al.* (1992a, 1992b) established that the activation could also be achieved by thermal annealing, and pointed out the crucial role played by hydrogen in acceptor-doped GaN. They showed that MOCVD-grown Mg-doped films that are highly resistive after growth become *p*-type conductive after thermal annealing in an N_2 ambient. Nakamura *et al.* then annealed their samples in an NH_3 ambient at 600°C, and found that the resistivity increased significantly. They concluded that atomic hydrogen produced by NH_3 dissociation was responsible for the passivation of acceptors. This correlation between hydrogen and magnesium concentrations was further investigated by Ohba and Hatano (1994).

Given the crucial role played by hydrogen in *p*-type doping, it is important to develop a thorough understanding of the behavior of hydrogen in GaN, both as an isolated impurity, and in its interactions with dopant atoms. After a discussion of the magnesium acceptor in Section 3.1.3.a, we devote Section 3.1.3.b to the role of hydrogen in *p*-type GaN. Section 3.1.3.c describes the acceptor activation process. Compensation of *p*-type GaN is discussed in Section 3.1.3.d, and doping of AlGaN alloys in Section 3.1.3.e. Acceptors other than Mg, finally, are discussed in Section 3.1.3.f.

3.1.3.a Magnesium

Magnesium is currently the *p*-type dopant of choice. Magnesium is typically introduced by flowing bis-cyclopentadienyl magnesium (Cp_2Mg or $(C_5H_5)_2Mg$) during MOCVD growth (Amano *et al.*, 1989; Nakamura *et al.*, 1992a); or from a solid Mg source in MBE. To our knowledge, the highest room-temperature hole concentration reported for Mg doping is 3×10^{18}cm^{-3}, together with a mobility of 9 cm^2/Vs, and a resistivity of 0.2 Ωcm) (Nakamura, 1991). Eiting *et al.* (1998) have recently reported Mg-doped films with p=2×10^{18} cm^{-3} and mobility μ_p=2.2 cm^2/Vs at 300K. Obtaining a high hole concentration and also a high mobility is very challenging. One may succeed in obtaining hole concentrations above 10^{18}cm^{-3}, but at the expense of low mobility, by forming a defect band where one does not rely on acceptor ionization (Götz, 1998). The low mobility then results in higher resistivity.

The Mg activation energy has been measured with variable-temperature Hall effect, yielding 208 meV for low Mg concentrations (Götz *et al.*, 1998b). It was found to be important to take the Coulomb interaction between ionized acceptors into account. This interaction leads to lower activation energies with increasing degree of ionization.

Achieving high hole concentrations with Mg as the dopant has proved difficult: even though it is possible to increase the Mg concentration, the hole concentration levels off and even decreases past a certain point (Bour *et al.*, 1997). First-principles investigations (Neugebauer and Van de Walle, 1996d) have revealed that the magnesium *solubility* is the most important limiting factor. The solubility is determined by competition between incorporation of Mg acceptors and formation of Mg_3N_2 (see Section 3.1.1.a). Magnesium prefers the substitutional Ga site. Incorporation of Mg on substitutional nitrogen sites (Mg_N) or on interstitial sites (Mg_i) has occasionally been invoked as the cause of limited doping efficiency; however, the calculated formation energy of Mg on these sites is very high, as seen in Figure. 3.3. Incorporation of Mg on these sites is thus unfavorable, and cannot explain the observed limitations in doping.

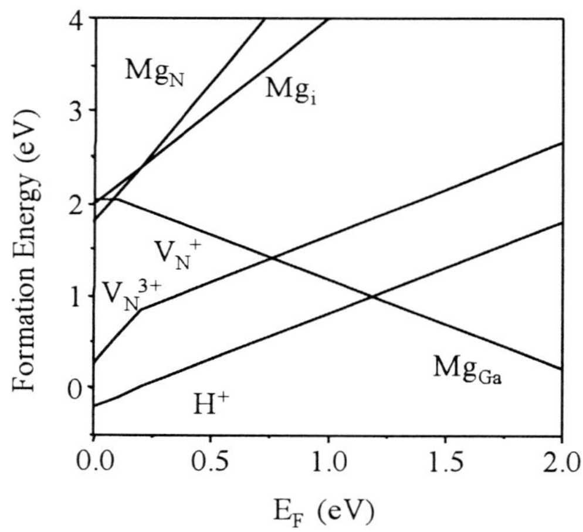

Figure. 3.3 Formation energy as a function of Fermi energy for Mg in different configurations (Ga-substitutional, N-substitutional, and interstitial). Also included are the nitrogen vacancy and interstitial hydrogen. The zero of Fermi energy is located at the top of the valence band. From Van de Walle *et al.* (1998).

3.1.3.b Role of hydrogen in doping of GaN

Hydrogen has strong effects on the properties of GaN. Many growth techniques, such as metal-organic chemical vapor deposition (MOCVD) or hydride vapor phase epitaxy (HVPE) expose the growing material to large concentrations of hydrogen. The presence of hydrogen has particularly important consequences for *p*-type doping of the material: hydrogen incorporated during growth leads to passivation of acceptors, and a post-growth processing step is required to render the acceptors electrically active. We will see, however, that the presence of hydrogen during growth is actually quite beneficial: it suppresses compensation, and enhances the solubility of acceptors.

Isolated interstitial hydrogen

The behavior of hydrogen in GaN has been analyzed in detail based on first-principles calculations (Neugebauer and Van de Walle, 1995b, 1996e; Bosin *et al.*, 1996). It was found that the formation energy of hydrogen is significantly lower in *p*-type GaN than in *n*-type GaN, corresponding to a much higher solubility in *p*-type than in *n*-type GaN (see Section 3.1.1.a).

In *p*-type GaN, H behaves as a donor (H^+); it thus compensates acceptors. The preferred location for H^+ is at the antibonding site behind a nitrogen atom. The diffusion barrier for H^+ is only 0.7 eV, which indicates a high diffusivity at moderate to high temperatures. In *n*-type GaN, H behaves as an acceptor (H^-); its most stable site is at the antibonding site behind a Ga atom. The migration barrier for H^- is 3.4 eV, indicating a very low diffusivity. For Fermi-level positions below ~2.1 eV, H^+ is favored; higher Fermi-level positions favor H^-. The neutral charge state is never stable, characteristic of a so-called "negative-*U*" center. The calculations also showed that the formation energy of H_2 molecules in GaN is high: H_2 is unstable with respect to dissociation into monatomic hydrogen.

These results are fully consistent with experimental observations showing that the ability to diffuse hydrogen into GaN at moderate temperatures depends strongly on the electrical conductivity type. Samples of *n*-type versus *p*-type GaN were exposed to monatomic deuterium at specified temperatures for one hour in a remote microwave plasma (Götz *et al.*, 1995b); depth profiles of deuterium were obtained with SIMS. The experimental results indicated that hydrogen readily diffuses and incorporates in *p*-type GaN at temperatures greater than 600°C, but not in *n*-type GaN.

Acceptor-hydrogen complexes

The behavior of isolated interstitial hydrogen provides crucial information about hydrogen interaction with impurities. Since both the solubility and the diffusivity of hydrogen in *n*-type GaN are low, hydrogen-donor complexes will rarely form. We focus here on complexes with acceptors.

In *p*-type GaN hydrogen occurs in the positive charge state and is located in an antibonding site behind a nitrogen atom; it is electrostatically attracted to negatively charged acceptors, and also assumes the antibonding configuration. In the case of Mg acceptors, this means that H is not located next to the Mg atom, but rather binds to a N atom which is a neighbor of the acceptor (see Figure. 3.4) (Neugebauer and Van de Walle, 1995b, 1996e). This is an unusual structure for an acceptor-hydrogen complex; in other semiconductors hydrogen assumes the bond-center position when binding to acceptors. As a direct consequence the vibrational frequency of the complex is *not* representative of a Mg-H bond, but rather of a N-H bond. The calculated vibrational frequency (in the harmonic approximation) is 3360 cm^{-1}. Anharmonic effects may lower this calculated frequency.

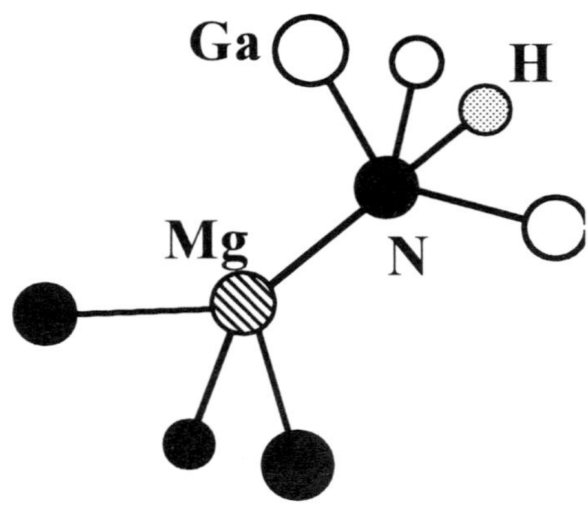

Figure. 3.4 Schematic illustration of the structure of a Mg-H complex in GaN. Hydrogen resides in an antibonding position behind a nitrogen atom.

Determination of the local vibrational modes (LVM) of the Mg-H complex in GaN has provided satisfying confirmation of the significance of hydrogen in GaN and useful information, when compared with theory, on the structure of the complex. Fourier-transform infrared absorption spectroscopy was performed on Mg-doped GaN grown by MOCVD (Götz *et al.*, 1996e). The as-grown sample was semi-insulating and displayed a LVM at 3125cm^{-1}. After thermal activation the sample became *p*-type conductive and the intensity of the LVM was reduced. Subsequent exposure to monatomic deuterium at 600°C for 2 hrs increased the resistivity and caused a new absorption line to appear at 2321cm^{-1}. This line disappeared after a thermal activation treatment. The isotopic shift clearly established the presence of hydrogen in the complex. The assignment of the LVM to the stretch mode of the Mg-H complex is in agreement with the calculated H-stretch frequency, establishing that the Mg-H complex contains a strong N-H bond. The spectroscopic evidence thus establishes that hydrogen forms a complex with a shallow acceptor and that the acceptor activation process is due to the dissociation of Mg-H complexes. Further, the decrease in the intensity of the LVM upon post-growth activation of the *p*-type conductivity is consistent with thermal dissociation of Mg-H complexes.

The calculated *binding energy* of the Mg-H complex is 0.7 eV (Neugebauer and Van de Walle, 1995c). This value implies that at the growth temperature hydrogen is not bound to the acceptors (although it is present in the Mg-doped layer); during cool-down the hydrogen atoms diffuse rapidly and eventually form hydrogen-acceptor complexes. The *dissociation energy* of the complex was calculated to be about 1.5 eV; this value is larger than the binding energy because the hydrogen needs to overcome a *barrier* in order to be removed from the Mg atom.

Effect of hydrogen on doping

Figure 3.3 compares the formation energy of hydrogen in *p*-GaN with that of the nitrogen vacancy (see Section 3.1.1.c for a discussion of nitrogen vacancies). In the absence of hydrogen, nitrogen vacancies are the dominant compensating centers, and due to charge neutrality the Fermi level will be located near the point where the formation energies of the nitrogen vacancy (V_N) and Mg_{Ga} are equal. Native defect compensation is therefore potentially an important concern for high-temperature growth techniques. The level of compensation actually increases with increasing growth temperature (Neugebauer and Van de Walle, 1996e); low-temperature growth techniques such as molecular beam epitaxy (MBE) should therefore suffer less from this problem. It is known that in MBE-grown GaN *p*-type conductivity can be achieved without any post-growth processing (Molnar *et al.*, 1993; Lin *et al.*, 1993). It should be kept in mind that the low growth temperatures employed in MBE might not be sufficient to bring the system in full thermodynamic equilibrium.

When hydrogen is present, however, compensation by nitrogen vacancies is suppressed, because the formation energy of H^+ is lower than the formation energy of V_N. Hydrogen is therefore preferentially incorporated as a compensating center, rather than nitrogen vacancies. The Mg concentration is also increased, compared to the hydrogen-free case. This follows from inspection of the formation energies in Figure 3.3: since the formation energy of hydrogen is lower than that of V_N, the Fermi level equilibration point is moved higher in the gap, leading to a lower formation energy and hence higher concentration of Mg. Incorporation of hydrogen is therefore beneficial in two respects: suppression of native defects, and enhancement of the acceptor concentration. Manipulation of the atomic hydrogen concentration during growth may therefore be used to optimize acceptor doping.

Incorporation of hydrogen does have the downside that complete compensation of the acceptors occurs. The height of the dissociation barrier (1.5 eV) should allow dissociation of the complex at temperatures of a few hundred °C. Experimentally, however, temperatures exceeding 600°C have been found necessary to activate Mg-doped MOCVD-grown GaN (Nakamura *et al.*, 1992a; Götz *et al.*, 1996d). We take this as an indication that the activation process does not merely consist of dissociating Mg-H complexes. When hydrogen leaves the Mg acceptor, it still behaves as a donor, and therefore can still compensate the acceptor as long as it remains in the *p*-type layer. The hydrogen therefore has to be removed from the *p*-type layer (e.g., into the substrate or through the surface); alternatively, the hydrogen can be neutralized, e.g., by binding to extended defects. Formation of H_2 molecules, which is common after acceptor passivation in other semiconductors, is not possible in GaN because of the high formation energy of H_2.

3.1.3.c *Activation of acceptor-doped GaN*

From the discussion in Section 3.1.3.b, it is clear that when hydrogen is present during the growth (as in MOCVD or HVPE) the acceptors will not be electrically active, due to passivation by hydrogen. This explains the need for post-growth activation.

The theoretical picture of the effect of hydrogen on acceptor doping is nicely confirmed by experimental studies. While Nakamura *et al.* (1992a) introduced hydrogen through dissociation of NH_3, Götz *et al.* (1995b) used a remote plasma to introduce the hydrogen. Götz *et al.* (1996d, 1998c) went on to investigate the activation kinetics of Mg acceptors. They annealed Mg-doped samples grown by MOCVD incrementally for 5 minutes at temperatures ranging from 500°C to 850°C. Figure 3.5(a) demonstrates the resistivity measured as a function of temperature in the as-grown sample and after each annealing step.

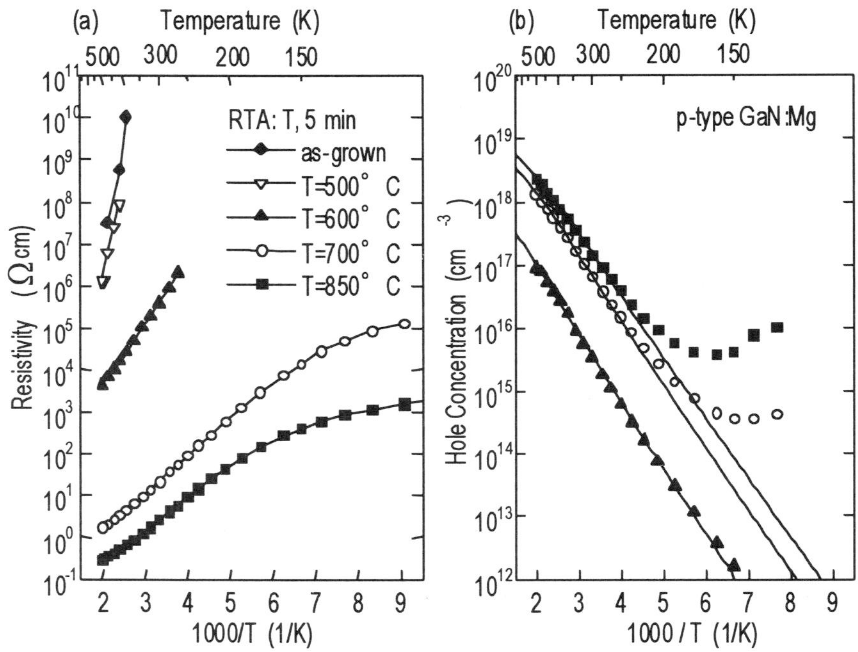

Figure. 3.5 (a) Resistivity and (b) hole concentration of a Mg-doped GaN sample as a function of reciprocal temperature. The resistivities (a) are shown for as-grown material and after annealing for 5 minutes at incremental temperatures T. The hole concentrations (b) are shown after acceptor activation at incremental temperatures T. The solid lines (b) represent calculated hole concentrations. From Götz and Johnson (1998c).

The resistivity decreased significantly after the 600°C annealing step and reached a value of about 3 Ωcm at 300 K after annealing at 850°C. In samples annealed above 700°C the saturation of the resistivity for measurement temperatures below 200 K is due to hole conduction in an acceptor impurity band. Reliable Hall measurements were possible in samples annealed at temperatures of 600°C and higher; the resulting hole concentrations are depicted in Figure. 3.5 (b). The data for the anneal at 600°C reveal acceptor activation over the entire measurement range, whereas the data for the anneals above 700°C reveal ionization of a shallow acceptor only for temperatures above 200 K, with impurity-band conduction at lower temperatures. Fitting of the Hall-effect data indicated the presence of donor compensation, at a level of about 3×10^{18} cm^{-3}.

These results are consistent with the picture of Mg-H complexes being present in as-grown, Mg-doped GaN. Hydrogen is incorporated during growth, and Mg-H complexes are formed during cooldown. After growth only a fraction of Mg atoms act as acceptors and the material is semi-insulating. The absence of impurity-band conduction after the 600°C anneal, and its appearance after the 700°C anneal clearly indicate that the activation of Mg acceptors is due to the generation of Mg-related acceptor states in the band gap of GaN, and therefore strongly supports the existence of Mg-H complexes in the as-grown material.

Pearton *et al.* (1996a) reported an activation energy for dopant activation of order 2.5 eV in bulk samples, where hydrogen retrapping is expected to be significant. Pearton *et al.* (1996b) also reported that the apparent thermal stability of hydrogen-passivated Mg acceptors depends on the ambient employed during annealing: an H_2 ambient leads to a reactivation temperature approximately 150°C higher than N_2. The authors attributed this to indiffusion of hydrogen from the H_2 ambients, causing a competition between passivation and reactivation. They also concluded that the hydrogen does not leave the sample after activation of Mg acceptors, but accumulates in regions of high defect density (in their case, the InGaN layer in a GaN/InGaN double heterostructure).

The precise mechanism by which LEEBI treatment (Amano *et al.*, 1989) activates acceptors is still controversial. The irradiation process can generate electron-hole pairs. These pairs could either directly provide the energy for releasing hydrogen from the acceptor; or the presence of minority carriers could lower the dissociation barrier. Minority-carrier injection is known to enhance the dissociation of acceptor-hydrogen complexes in other semiconductors, and thus may also be expected to promote acceptor activation in GaN. Alternatively, the irradiation process may cause local heating that effectively leads to thermal dissociation. Recent experiments by Li and Coleman (1996) monitored the LEEBI process with cathodoluminescence spectroscopy. The fact that the LEEBI process remained effective down to liquid helium temperature was interpreted as evidence for an athermal mechanism.

The influence of minority-carrier injection on the activation process was investigated by Pearton *et al.* (1996c). They found that acceptor activation takes places at a temperature as low as 175°C under minority-carrier injection (accomplished by forward bias of a p-n junction), much lower than in conventional annealing under zero-bias conditions. On the basis of these results the authors suggested that minority-carrier-enhanced dissociation of Mg-H complexes is the mechanism for activation during LEEBI treatment. An enhancement of Mg activation due to minority-carrier injection has also been reported by Miyachi *et al.* (1998).

3.1.3.d Compensation

Native defects

The nitrogen vacancy, which had a high formation energy in *n*-type GaN (see Section 3.1.1.c), has a significantly lower formation energy in *p*-type material, and acts as a compensating center. We already discussed the suppression of compensation by vacancies in the presence of hydrogen. Still, a certain concentration of nitrogen vacancies can be formed in *p*-type material, and they act as compensating centers. Figures 3.1 and 3.3 show that nitrogen vacancies can occur in a 3+ as well as a + charge state; the +/3+ transition is characterized by a large lattice relaxation (Neugebauer and Van de Walle, 1996a). The presence of nitrogen vacancies may therefore be responsible for the observed persistent photoconductivity effects (Johnson *et al.*, 1996; Li *et al.*, 1996; Van de Walle, 1997).

The nitrogen vacancy also may give rise to the blue lines (around 2.9 eV) commonly observed by photoluminescence in Mg-doped GaN (Leroux *et al.*, 1997; Kaufmann *et al.*, 1998; Park and Chadi, 1997; Van de Walle, 1997). The appearance and disappearance of photoluminescence (PL) lines during post-growth annealing of Mg-doped layers grown by MOCVD (Nakamura *et al.*, 1992a; Götz *et al.*, 1996d) may be related to the interactions of hydrogen with nitrogen vacancies. Complexes between hydrogen and nitrogen vacancies can form during growth; the calculated binding energy of the $V_N H^{2+}$ complex, expressed with respect to interstitial H in the positive charge state, is 1.56 eV. Dissociation of this complex, producing isolated nitrogen vacancies, may explain the behavior of PL lines during annealing for acceptor activation (Van de Walle, 1997).

Foreign impurities

Among potential compensating impurities, oxygen is probably the most common. The oxygen formation energy shown in Figure. 3.2 clearly extrapolates to very low values in *p*-type GaN. Any oxygen present in the growth system will therefore be readily incorporated during *p*-type growth.

3.1.3.e *p*-type doping of $Al_xGa_{1-x}N$

The maximum hole concentration achievable with Mg doping has been observed to decrease rapidly with increasing Al content x in $Al_xGa_{1-x}N$ (Tanaka *et al.*, 1994; Bremser *et al.*, 1996; Suzuki *et al.*, 1998). Magnesium has no tendency to form deep levels (so called *AX* levels, analogous to *DX* for donors) (Park and Chadi, 1997), thus ruling out a shallow-deep transition as the source of the drop in hole concentration. Calculations of the Mg acceptor indicate that its ionization energy is higher in AlN (0.4 eV) than in GaN (0.2 eV) (Stampfl and Van de Walle, 1998a). This increase in the ionization energy leads to a decrease in doping efficiency. Experimentally, based on an analysis of temperature-dependent Hall data Tanaka *et al.* (1994) found the Mg activation energy to be 35 meV deeper in $Al_{0.08}Ga_{0.92}N$ than in GaN. Assuming linearity, this would make the Mg acceptor in AlN 0.44 eV deeper than in GaN.

Compensation by native defects is another important mechanism that may contribute to the decline in hole concentrations. As discussed in Section 3.1.1.d, two defects with low formation energies may inhibit successful *p*-type doping of

$Al_xGa_{1-x}N$, namely the nitrogen vacancy and the cation interstitial. The cation interstitial has a low formation energy in zinc-blende material, but is much higher in energy in wurtzite. The nitrogen vacancy has a strikingly lower formation energy in AlN than in GaN. Compensation by nitrogen vacancies is the likely cause of the decreased doping efficiency of Mg when the Al content is raised in $Al_xGa_{1-x}N$ alloys.

3.1.3.f Alternative acceptors

Despite the successes achieved with Mg, *p*-type doping levels in GaN and AlGaN alloys are lower than desirable for low-resistance cladding layers and ohmic contacts. Achieving higher hole concentrations with Mg as the dopant has proved difficult, as discussed in Section 3.1.3.a. It would be desirable to identify acceptors that are superior to Mg.

The performance of an acceptor can be judged on the basis of three main criteria: (i) solubility; (ii) stability against compensation by other configurations of the acceptor dopant (e.g., interstitials); and (iii) depth of the acceptor level (ionization energy). Each of these aspects can be addressed on the basis of results obtained from first-principles calculations, as explained in Section 3.1. The *solubility* of a substitutional acceptor corresponds to the equilibrium concentration of the impurity in the lattice, which is determined by the formation energy. The formation energy of the impurity in configurations other than the substitutional site determines the likelihood of incorporation on those sites. Interstitial configurations tend to be favorable for elements with a small atomic radius, such as Li or Be.

Extensive theoretical investigations (Neugebauer and Van de Walle, 1996f, 1998; Fiorentini *et al.*, 1996; Bernardini *et al.*, 1997a) have not produced any candidate with characteristics exceeding those of Mg in all respects. The results show that the column-II acceptors perform noticeably better than those from column I; the latter suffer from poor solubility. Lithium also suffers from incorporation on interstitial sites, where it acts as a donor. Among column-II acceptors, Ca and heavier elements also have high formation energies and hence poor solubility. The calculated ionization energies for Zn and Ca are larger than for Mg. The smallest column-II element, Be, has a solubility comparable to Mg. Its calculated ionization energy is slightly smaller than Mg, but the difference is within the error bar of the computation and hence not conclusive (Bernardini *et al.*, 1997a; Neugebauer and Van de Walle, 1998). Because it is a small atom, Be may suffer from incorporation on interstitial sites.

Effective-mass-theory calculations of acceptor ionization energies have been performed by Mireles and Ulloa (1998). They find Be to be only marginally more shallow than Mg, and other candidates (Zn, Ca, and C) to have higher ionization energies.

Experimental work on acceptors other than Mg has been very limited. Bergman *et al.* (1987) used photoluminescence (PL) to derive an ionization energy of 0.34 eV for Zn. Doping with Zn has not resulted in *p*-type conductivity, presumably due to the large ionization energy of the Zn acceptor. Zinc was used as a dopant in InGaN for double-heterostructure blue LEDs (Nakamura, 1994). InGaN layers were grown around 800°C using diethylzinc as the Zn source. The crystal quality of the InGaN was minimally affected by the incorporation of Zn; no information was available on the amount of Zn that was incorporated. The Zn-doped films produced strong (and broad) luminescence at energies 0.5 eV lower than the emission from Si-doped InGaN; this is consistent with Zn giving rise to a rather deep acceptor level. Nakamura *et al.* (1993) also investigated Cd doping of InGaN, using dimethylcadmium and growth temperatures around 800°C. Again, a strong emission indicative of a relatively deep acceptor level was observed.

Experimental results about Be are inconclusive: Brandt *et al.* (1996) have reported high levels of *p*-type doping using Be in MBE, provided oxygen is present as a co-dopant (see also Ploog and Brandt, 1998). This result is difficult to understand based on equilibrium thermodynamics, as pointed out by Bernardini *et al.* (1997a). Other work has indicated Be indeed acts as an acceptor; results about the ionization energy are based on luminescence, with Salvador *et al.* (1996) finding a relatively large ionization energy (250 meV), while Sánchez *et al.* (1998) derived a much smaller value (90 meV).

Carbon is another case with inconclusive results: computational results indicate that the C acceptor is at least as deep as Mg (Boguslawski *et al.*, 1996; Neugebauer and Van de Walle, 1996a), and possibly much deeper (Fiorentini *et al.*, 1996). Experimentally, Fischer *et al.* (1995) found an activation energy of 230 meV.

To our knowledge, experimental studies of other acceptor impurities have only been performed using ion implantation (see, e.g., Lee *et al.*, 1996). Unfortunately, implantation damage is very difficult to remove by annealing in GaN, due to the hardness of the lattice and the low diffusivity. The presence of implantation defects then makes it difficult to reliably assess the properties of the implanted acceptors, and in our opinion no conclusive results have been obtained.

3.2 BAND OFFSETS

Heterojunctions are essential for most optoelectronic devices, to provide carrier confinement as well as optical confinement. Most nitride-based devices incorporate heterojunctions between GaN, AlN, or InN, or their alloys. The most important parameters characterizing such heterojunctions are the band discontinuities in the conduction and valence bands.

In this Chapter we focus on band offsets between GaN, AlN, InN, and their alloys. One can also form heterostructures between nitrides and other semiconductors, such as SiC, Si, or GaAs; such structures are also very interesting for device applications. However, we do not discuss them here since they are beyond the scope of this volume.

Strain plays a very important role in the band structure and the band offsets at nitride heterostructures. The lattice mismatch between AlN and GaN is about 3%; between GaN and InN, the mismatch is over 11%. The growth techniques used for fabrication of nitride devices (mainly MOCVD and MBE) allow growth of thin layers of a semiconductor on top of a layer with a different lattice constant. When the thickness of the overlayer is less than the critical layer thickness a *pseudomorphic* interface is formed and the overlayer is uniformly strained to match the in-plane lattice constant of the substrate. For growth of InGaN on GaN, this critical layer thickness has been found to be unexpectedly large (see McCluskey *et al.*, 1998b; Romano *et al.*, 1998).

Since the presence of strain in the layers has a profound influence on the band offsets, growing pseudomorphic layers with various amounts of strain can be an effective way of engineering band lineups. The amount of strain is determined by the lattice mismatch between the epitaxial layer and the substrate, or more generally, between the epitaxial layer and the layer on which it is grown. The latter could be a thick layer of a semiconductor alloy that has relaxed to its own equilibrium lattice constant; such a layer then acts in effect as a substrate with its own lattice constant. In the following we use the word "substrate" to include such configurations.

The band offsets are only well defined in the case of a pseudomorphic interface, in which the materials on either side of the junction are strained in order to match a common in-plane lattice constant, $a_{//}$. For instance, when InGaN is grown on a thick layer of GaN, the InGaN should be compressed in the plane of the interface to match the GaN lattice constant, and expanded in the perpendicular direction (by a factor determined by Poisson's ratio).

An important word of caution: The strains present in pseudomorphic layers also lead to the presence of piezoelectric fields, both in the wurtzite and in the zinc-blende structure. In addition, the wurtzite-structure semiconductors have low enough symmetry to also exhibit pyroelectricity, i.e., a spontaneous polarization in the absence of strain (Bernardini *et al.*, 1997b). These effects must be carefully taken into account when performing calculations; a lucid account is provided in the work of Bernardini *et al.* (1997b). Analysis of experimental measurements is similarly affected by the presence of such fields, as pointed out by Bernardini and Fiorentini (1998).

We are adopting the following notation: We label the valence-band offset ΔE_v, and the conduction-band offset ΔE_c. The valence-band offset at a heterojunction A/B is defined to have a positive value if the valence band maximum in B is higher in energy than the valence-band maximum in A.

3.2.1 Theory of Band Offsets at Nitride Interfaces

3.2.1.a Methodology

A variety of theoretical approaches have been developed to calculate band offsets at semiconductor interfaces; a recent review was given by Franciosi and Van de Walle (1996). The most common approach, which most of the calculations quoted here are based on, uses a superlattice geometry, in which layers of the two semiconductors under study are periodically repeated. Besides directly providing useful information about actual superlattice structures, this approach also yields results for isolated interfaces, provided the layers are sufficiently thick to ensure adequate separation between adjacent interfaces. Charge densities and potentials indeed typically converge quite rapidly to their bulk value away from the interface. It is then possible to identify a "bulk-like" region in the middle of each superlattice layer, where the value of an average potential can be determined. In addition, separate bulk calculations for the two semiconductors have to be carried out, providing information about individual band positions with respect to the average potential. This procedure for obtaining band offsets is actually similar to the one followed in x-ray photoemission spectroscopy (XPS). In XPS, typically, the separation between two representative core levels is measured across the interface; independent measurements on bulk samples are performed to obtain the energy separation between the valence-band maximum and the core levels in each material. The core level separation is then used to line up the valence bands and obtain the band offsets.

3.2.1.b Effects of strain on band offsets

In case of lattice-mismatched semiconductors, a specific in-plane lattice constant has to be chosen for the superlattice calculations; at least one, and maybe both of the materials in the heterostructure is therefore strained. The materials are strained according to their elastic constants, using the macroscopic theory of elasticity (see Van de Walle, 1989; Wright, 1997). The published elastic constants for nitrides exhibit wide variations; a critical discussion has been given by Wright (1997). This reference also contains first-principles values for these constants. We feel that these first-principles results provide reliable and consistent values for the elastic constants.

Relaxation of the atoms around the interface has to be explicitly allowed in the superlattice calculations; it was found that such relaxations have a significant effect on the band offsets (Van de Walle and Neugebauer, 1997).

One can immediately foresee a difficulty in dealing with band offsets between strained materials: each different strain combination leads to a different band offset. Instead or re-calculating the offsets for every strain situation, it is desirable to develop a formalism that allows derivation of the offsets for an arbitrary strain situation without requiring a full-fledged first-principles calculation.

In general, the effect of strain on band extrema can be expressed through *deformation potentials* (see Van de Walle, 1989). Once these deformation potentials are known (from theory or experiment), one only needs a band lineup between *unstrained* materials to derive band offsets for arbitrary strain configurations. Deformation potentials usually describe the change in the position of one band with respect to another band. The band-offset problem, however, requires knowledge of how strain affects band extrema on an absolute energy scale, since the band positions in one material will be compared with the bands in a second material. Such changes are expressed using "absolute" deformation potentials (Van de Walle, 1989). Using the band lineups at strained interfaces together with the deformation potentials allows one to derive a so-called *natural* band lineup between *unstrained* materials. This natural band lineup can be used as a starting point to calculate offsets at an arbitrary interface, by using information about the strains and the deformation potentials.

"Relative" deformation potentials for nitrides have been calculated and measured, and are available from a variety of sources (see Shan *et al.*, 1998, for a recent summary). To our knowledge, "absolute" deformation potentials have only been derived in the work of Van de Walle and Neugebauer (1997); this was for the case of zinc blende, but one expects the values for wurtzite to be very similar.

The band lineups between *unstrained* nitride semiconductors can be derived in two different ways. One approach is to line up the band structures of unstrained bulk materials using a lineup of reference levels derived from an explicit heterostructure (superlattice) calculation, designed to minimize the effect of strain on the lineup. Wei and Zunger (1996) accomplished this by using a lineup between core levels. Such a lineup may still reflect the particular strain configuration of the interface for which it was carried out. A more rigorous approach, in our opinion, is to first explicitly derive the band lineups for a strained interface, taking all the effects of strain into account, and then "subtract out" the effects of strain on the bands, using deformation potentials (Van de Walle and Neugebauer, 1997).

3.2.1c Computational results for band offsets

Most of the theoretical approaches have focused on band lineups between the binary compounds. It seems reasonable to derive results for alloys by linear interpolation. However, the band gap exhibits nonlinearity as a function of composition: the so-called *bandgap bowing*. This bowing has been found to be large in nitrides, particularly for InGaN alloys (see McCluskey *et al.*, 1998b). The band offsets should therefore also be nonlinear. Experience with other semiconductors has indicated that most of this nonlinearity occurs in the conduction band, and that linear interpolation works quite well for the valence-band offset. Whether this holds true for the nitrides has not explicitly been tested.

Tables 3.1 and 3.2 summarize a variety of computational results for nitride semiconductor band offsets. All of the calculations discussed here were based on density functional theory in the local density approximation (LDA); see the discussion in Section 3.1.1.b. It is important to acknowledge that in density functional theory the calculated eigenvalues do not necessarily correspond to quasiparticle energies. A well-known consequence of this deficiency is the failure of density-functional theory to produce the correct band gaps. Corrections beyond density-functional theory may therefore be necessary to obtain the exact band positions. Note, however, that the calculated potential-lineup term (extracted from a superlattice calculation) depends only on the charge density of the heterojunction, and as such is a ground-state property that is reliably given by density functional theory. The corrections are therefore limited to the positions of the quasiparticle energy levels that are obtained in bulk calculations for the individual materials. Quasiparticle calculations of this nature were carried out for a wide variety of semiconductors (Zhu et al., 1991); by comparing the results with LDA band structures, it was found that the average LDA error in the heterojunction valence-band offset was about 120 meV. If the error is of the same order for nitrides, this is a fairly small correction.

Table 3.1 Valence-band offsets (including strain effects) at AlN/GaN interfaces obtained from various first-principles calculations, in eV. A positive value indicates that the valence band in GaN is higher in energy than the valence band in AlN. "Method" identifies the type of calculational approach (symbols are explained in the text); WZ/ZB refers to the structure considered in the calculations (wurtzite or zinc blende); and "Orient." identifies the interface orientation. s=AlN indicates the in-plane lattice parameter of the structure is that of AlN (as if the structure is grown on a relaxed AlN "substrate"), and similarly for s=$Al_{0.5}Ga_{0.5}N$ and s=GaN.

Method	Ref.	WZ/ZB	Orient.	s=AlN	s=$Al_{0.5}Ga_{0.5}N$	s=GaN
PP+d	a	WZ	10-10			0.77
PP+d	a	WZ	0001			0.68
PP+d	b	ZB			0.74	
LMTO	c	ZB	001		0.75	
LMTO	c	ZB	110		0.81	
LMTO	c	ZB	111		0.77	
LMTO	c	WZ	0001	0.83		
PP+nlcc	d	ZB	001	0.74	0.58	0.41
PP	e	ZB	001	0.73	0.58	0.44
PP	e	WZ	0001	0.57		
PP+d	f	ZB	001	0.85		0.20

a. Satta *et al.*, 1996.
b. Chen *et al.*, 1996.
c. Ke *et al.*, 1996.
d. Majewski *et al.*, 1997.
e. Buongiorno Nardelli *et al.*, 1997.
f. Bernardini *et al.*, 1997c.

Table 3.2 Unstrained valence-band offsets between nitride semiconductors obtained from various first-principles calculations, in eV. A positive value for an interface A/B indicates that the valence band in B is higher in energy than the valence band in A. "Method" identifies the type of calculation (symbols are explained in the text); WZ/ZB refers to the structure considered in the calculations (wurtzite or zinc blende); and "Orient." Identifies the interface orientation.

Method	Ref.	WZ/ZB	Orient.	AlN/GaN	GaN/InN	AlN/InN
LMTO	a	ZB	110	0.85	0.51	1.09
LAPW	b	ZB	001	0.84	0.26	1.04
LAPW	b	ZB	111		0.50	
LAPW	b	WZ	0001	0.81	0.48	1.25
PP+nlcc	c	ZB	110	0.7	0.3	

a. Albanesi *et al.*, 1994a.
b. Wei and Zunger, 1996.
c. Van de Walle and Neugebauer, 1997.

The following symbols are used to identify the type of calculations in Tables 3.1 and 3.2:

PP: Pseudopotential calculations, in which the Ga and In *d* electrons are treated as core electrons (Buongiorno Nardelli *et al.*, 1997)

PP+d: Pseudopotential calculations explicitly including the Ga and In *d* electrons as valence electrons (Satta *et al.*, 1996; Chen *et al.*, 1996; Bernardini *et al.*, 1997c)

PP+nlcc: Pseudopotential calculations in which the Ga and In *d* states are treated using the nonlinear core correction (Majewski *et al.*, 1997; Van de Walle and Neugebauer, 1997)

LMTO: Linear muffin-tin orbital method. The LMTO calculations quoted here also use the atomic-sphere approximation (ASA). Atomic relaxations cannot be reliably calculated within this approximation, and were not included. (Albanesi *et al.*, 1994a, 1994b; Ke *et al.*, 1996)

LAPW: General potential linearized augmented plane wave calculations (Wei and Zunger, 1996)

A discussion of the physical consequences of the offset values will be given in Section 3.2.3; here we merely comment on some of the computational aspects.

Table 3.1 focuses on the AlN/GaN interface, for which the majority of calculations to date have been performed. Since the offsets are sensitive to the strain in the layers, values are listed for various values of the in-plane lattice constant. Overall, the results are fairly consistent. For the pseudopotential calculations, there seems to be no systematic dependence on the treatment of the Ga and In *d* states. Van de Walle and Neugebauer (1997) found that differences between explicit inclusion of *d* states versus use of the nlcc were small (less than 0.1 eV).

An asymmetry between growth on AlN *vs.* growth on GaN is clearly observed; the magnitude of the difference in offset seems to depend on the calculation. For instance, Bernardini *et al.* (1997c) obtained 0.85 eV for $a_{//}=a_{\mathrm{AlN}}$, and 0. 20 eV for $a_{//}=a_{\mathrm{GaN}}$. The difference between these values is somewhat larger than obtained by Majewski *et al.* (1997) or Buongiorno Nardelli *et al.* (1997), although the average of the two offsets is similar in all calculations. The difference may be due to different strain-induced splittings of the valence bands (essentially a bulk effect), probably because of details such as whether theoretical or experimental lattice constants are used, etc.

Results for band offsets between *unstrained* materials are listed in Table 3.2. Again, the various calculations have produced reasonably consistent results. We note that Buongiorno Nardelli *et al.* (1997) also calculated offsets (*including* strain) for ZB (001) GaN/InN and AlN/InN, both grown on AlN. For GaN/InN they found 0.70 eV, and for AlN/InN, 1.37 eV.

3.2.2 Experimental Results for Band Offsets

Experimental results for nitride band offsets are scarce. Martin *et al.* (1994) reported a valence-band offset of 0.8±0.3 eV for the AlN/GaN wurtzite (0001) interface using X-ray photoemission spectroscopy (XPS). This result was obtained from five different sequences, including both GaN grown on AlN, and AlN grown on GaN. In subsequent work, Martin *et al.* (1996) discussed the significant asymmetry in their measurements depending on sequence. They attributed the asymmetry to strain-induced piezoelectric fields. After approximately taking these effects into account, they reported ΔE_v values of 1.03±0.25 eV for InN/GaN, 0.67±0.24 eV for GaN/AlN, and 1.75±0.20 eV for InN/AlN. Their value for GaN/AlN is most reliable, since the lattice mismatch in this system is fairly small. However, for systems with a large lattice mismatch such as InN/GaN and InN/AlN the value of ΔE_v extracted from their measurements is ill defined. While piezoelectric effects no doubt play a role, Martin *et al.* actually neglected another contribution of strain to their measurements. Their procedure was to measure the energy difference, ΔE_{CL} between two characteristic core levels on either side of the heterojunction, and then combine this value with values for E_v-E_{CL} measured for bulk samples. As pointed out by Tersoff and Van de Walle (1987), the ΔE_{CL} value will include a contribution from strain in the heterojunction, whereas the E_v-E_{CL} reflect the positions of levels in *unstrained* bulk materials. The magnitude of the strain contribution to ΔE_{CL} is difficult to estimate, but in principle it could be sizeable. In the case of Si/Ge heterojunctions, where the lattice mismatch is about 4%, Schwartz *et al.* (1989) found that E_v-E_{CL} could be shifted by more than 0.2 eV due to the strain. For InN/GaN, where the lattice mismatch is much larger, the effects could therefore be sizeable. The effect is certainly large enough to contribute to the observed non-commutativity of the offsets reported by Martin *et al.* (1996).

Waldrop and Grant (1996) also used XPS to measure ΔE_v for AlN/GaN(0001); they obtained ΔE_v=1.36±0.07 eV, a value substantially different from all other experimental and theoretical values.

Baur *et al.* (1994), finally, estimated the AlN/GaN band offset via the 0/- acceptor level of iron. According to a heuristic approach, transition-metal impurity levels act as a common reference level to predict the band alignment in heterojunctions (Langer and Heinrich, 1985). Using this procedure they obtained an unstrained valence-band offset of 0.5 eV, a value slightly smaller than derived from other experiments and from calculations.

3.2.3 Discussion

Since the wurtzite phase differs from zinc blende only in the atomic arrangements beyond third-nearest neighbors, one does not expect major differences between the two phases. In particular, the ZB (111) orientation should be very similar to WZ (0001); Wei and Zunger (1996) indeed verified this for the GaN/InN interface. The differences noted in Tables 3.1 and 3.2 between ZB and WZ therefore should be attributed mainly to a dependence on interface orientation.

Effects of interface orientation were explicitly investigated for ZB AlN/GaN in the calculations by Ke *et al.* (1996); they actually found the differences to be less than 0.1 eV. It should be noted, however, that the atomic-sphere approximation used in the computational approach makes it difficult to properly include atomic relaxation, which may in turn affect the calculated offsets.

The results in Table 3.2 indicate that the valence-band offset between *unstrained* AlN and GaN is around 0.7-0.8 eV. This value is consistent with the more reliable experimental results. For the offset between *unstrained* GaN and

InN, theory produces values around 0.3-0.5 eV. In our opinion these theoretical numbers are more reliable than experimental results published to date.

Table 3.2 shows that the valence-band offset for AlN/GaN is larger than for GaN/InN. The difference is attributable to the role of the *d* electrons in these materials. For common-anion systems such as the III-nitrides, one might expect the valence-band maximum to be nearly aligned—this is the old "common-anion rule" which was based on the (correct) notion that the character (and thus the energetic position) of the valence-band maximum is determined by the anion *p* states. This common-anion rule was found to fail in many systems, and Wei and Zunger (1987) explained the failure in terms of the interactions between *d* states in the core and the valence-band maximum: *p-d* repulsion pushes the valence-band maximum upward, by a sizeable amount when the *d* states lie close enough to the valence-band maximum. This mechanism leads to a valence-band offset of ~0.6 eV between GaAs and AlAs: Al has no *d* states, and the Ga 3*d* states in GaAs push its valence-band maximum upward with respect to AlAs. The same effect leads to the ~0.8 eV valence-band offset between AlN and GaN. However, since both Ga and In contain *d* states at roughly the same energetic position, the valence-band offset for GaN/InN is significantly smaller (closer to the common-anion rule). The role of the *d* core states in the nitrides was also discussed by Strite (1994).

The small value of the GaN/InN valence-band offset implies a lack of hole confinement in GaN/InGaN quantum wells. The natural lineups in Table 3.2 would be modified, of course, by strain contributions. Still, Van de Walle and Neugebauer (1997) found that even after inclusion of strain the valence-band offset of GaN/InGaN is very small.

3.3 ACKNOWLEDGEMENTS

I gratefully acknowledge collaborations with Catherine Stampfl and Jörg Neugebauer, stimulating interactions with D. Bour, W. Götz, N. Johnson, and M. McCluskey, and the support of DARPA under agreement no. MDA972-96-3-014.

3.4 REFERENCES

Albanesi, E.A., Lambrecht, W.R.L. and Segall, B., 1994a. *Mat. Res. Soc. Symp. Proc.*, **339**, p. 607.

Albanesi, E.A., Lambrecht, W.R.L. and Segall, B., 1994b. *J. Vac. Sci. Technol. B*, **12**, p. 2470.

Amano, H., Kito, M, Hiramatsu, K. and Akasaki, I., 1989. *Jpn. J. Appl. Phys.*, **28**, p. L2112.

Baur, J., Maier, K., Kunzer, M., Kaufmann, U. and Schneider, J., 1994. *Appl. Phys. Lett.*, **65**, p. 2211.

Bergman, P., Ying, G., Monemar, B., and Holtz, P.O., 1987. *J. Appl. Phys.*, **61**, p. 4589.

Bernardini, F., Fiorentini, V. and Bosin, A., 1997a. *Appl. Phys. Lett.*, **70**, 2990.

Bernardini, F., Fiorentini, V. and Vanderbilt, D., 1997b. *Phys. Rev. B*, **56**, p. 10024.

Bernardini, F., Fiorentini, V. and Vanderbilt, D., 1997c. *Mat. Res. Soc. Symp. Proc.*, **449**, p. 923.

Bernardini, F. and Fiorentini, V., 1998. *Phys. Rev. B*, **57**, p. 9427.

Bockstedte, M., Kley, A., Neugebauer, J. and Scheffler, M., 1997. *Comp. Phys. Comm.*, **107**, p. 187.

Boguslawski, P., Briggs, E.L. and Bernholc, J., 1995. *Phys. Rev. B*, **51**, p. 255.

Boguslawski, P., Briggs, E.L. and Bernholc, J., 1996. *Appl. Phys. Lett.*, **69**, p. 233.

Boguslawski, P. and Bernholc, J., 1997. *Phys. Rev. B*, **56**, p. 9496.

Bosin, A., Fiorentini, V., and Vanderbilt, D., 1996. *Mat. Res. Soc. Symp. Proc.*, **395**, p. 503.

Bour, D.P., Chung, H.F., Götz, W., Romano, L., Krusor, B. S., Hofstetter, D., Rudaz, S., Kuo, C.P., Ponce, F.A., Johnson, N.M., Craford, M.G. and Bringans, R. D., 1997. *Mat. Res. Soc. Symp. Proc.*, **449**, p. 509.

Brandt, O., Yang, H., Kostial, H. and Ploog, K.H., 1996. *Appl. Phys. Lett.*, **69**, p. 2707.

Bremser, M.D., Perry, W.G., Zheleva, T., Edwards, N.V, Nam, O.H., Parikh, N., Aspnes, D.E. and Davis, R.F., 1996. *MRS Internet J. Nitride Semicond. Res.*, **1**, 8.

Buongiorno Nardelli, M., Rapcewicz, K., Briggs, E.L., Bungaro, C. and Bernholc, J., 1997. *Mat. Res. Soc. Symp. Proc.*, **449**, p. 893.

Calleja, E., Sánchez, F.J., Basak, D., Sánchez-García, M.A., Muñoz, E, Izpura, I., Calle, F., Tijero, J.M.G., Sánchez-Rojas, J.L., Beaumont, B., Lorenzini, P. and Gibart, P, 1997. *Phys. Rev. B*, **55**, p. 4689.

Cao, X.A., Abernathy, C.R., Singh, R.K., Pearton, S.J., Fu, M., Sarvepalli, V., Sekhar, J.A., Zolper, J.C., Rieger, D.J., Han, J., Drummond, T.J., Shul, R.J. and Wilson, R.G., 1998. *Appl. Phys. Lett.*, **73**, p. 229.

Chen, X., Hua, X., Hu, J., Langlois, J.-M. and Goddard, W.A., III, 1996. *Phys. Rev. B*, **53**, p. 1377.

Chichibu, S., Abare, A., Mack, M., Minsky, M., Deguchi, T., Cohen, D., Kozodoy, P., Fleischer, S., Keller, S., Speck, J., Bowers, J.E., Hu, E., Mishra, U.K., Coldren, L.A., DenBaars, S.P., Wada, K., Sota, T. and Nakamura, S., 1999. *Mat. Sci. Engin. B* 59, p. 298.

Chung, B.-C. and Gershenzon, M., 1992. *J. Appl. Phys.*. **72**, p. 651.

Deguchi, T., Shikanai, A., Torii, K., Sota, T., Chichibu, S. and Nakamura, S., 1998. *Appl. Phys. Lett.*, **72**, p. 3329.

Eiting, C.J., Grudowski, P.A. and Dupuis, R.D., 1998. *J. Electron. Mater.*, **27**, p. 206.

Fehrer, M., Einfeldt, S., Birkle, U., Gollnik, T. and Hommel, D., 1998. *J. Cryst. Growth,* **189/190**, p. 763.

Fiorentini, V., Bernardini, F., Bosin, A. and Vanderbilt, D., 1996. *Proc. 23rd Int. Conf. Physics of Semiconductors*, edited by Scheffler, M. and Zimmermann, R., (Singapore: World Scientific), p. 2877.

Fischer, S., Wetzel, C., Haller, E.E. and Meyer, B.K., 1995. *Appl. Phys. Lett.*, **67**, p. 1298.

Franciosi, A. and Van de Walle, C.G., 1996. *Surface Science Reports*, **25**, Nos. 1-4, 1-140.

Götz, W., Johnson, N.M., Street, R.A., Amano, H. and Akasaki, I., 1995a. *Appl. Phys. Lett.*, **66**, p. 1340.

Götz, W., Johnson, N.M., Walker, J., Bour, D.P., Amano, H. and Akasaki, I., 1995b. *Appl. Phys. Lett.*, **67**, p. 2666.

Götz, W., Johnson, N.M., Walker, J., Bour, D.P., Amano, H. and Akasaki, I., 1996a. *Proc. 6th Int. Conf. SiC and Related Materials*, edited by Nakashima, S., Matsunami, H., Yoshida, S. and Harima, H., *Inst. Phys. Conf. Ser.* No 142 (Bristol: IOP Publishing), p. 1031.

Götz, W., Johnson, N.M., Chen, C., Liu, H., Kuo, C. and Imler, W., 1996b. *Appl. Phys. Lett.*, **68**, p. 3114.

Götz, W., Johnson, N.M. and Bour, D.P., 1996c. *Appl. Phys. Lett.*, **68**, p. 3470.

Götz, W., Johnson, N.M., Walker, J., Bour, D.P. and Street, R.A., 1996d. *Appl. Phys. Lett.*, **68**, p. 667.

Götz, W., Johnson, N.M., Bour, D.P., McCluskey, M.D. and Haller, E.E., 1996e. *Appl. Phys. Lett.*, **69**, p. 3725.

Götz, W., Walker, J., Romano, L.T. and Johnson, N.M., 1997. *Mat. Res. Soc. Symp. Proc.*, **449**, p. 525.

Götz, W., 1998. Private communication.

Götz, W., Romano, L.T., Walker, J., Johnson, N.M. and Molnar, R.J., 1998a, *Appl. Phys. Lett.*, **72**, p. 1214.

Götz, W., Kern, R.S., Chen, C.H., Liu, H., Steigerwald, D.A. and Fletcher, R.M., 1998b. *Mater. Sci. and Engin. B* (in press).

Götz, W. and Johnson, N.M., 1998c. *Mat. Sci. Forum*, **264-268**, (Switzerland: Trans Tech Publications), p. 1375.

Hacke, P. and Okushi, H., 1997. *Appl. Phys. Lett.*, **71**, p. 524.

Hamann, D.R., Schlüter, M. and Chiang, C., 1979. *Phys. Rev. Lett.*, **43**, p. 1494.

Hirsch, M.T., Wolk, J.A., Walukiewicz, W. and Haller, E.E., 1997. *Appl. Phys. Lett.*, **71**, p. 1098.

Hohenberg, P. and Kohn, W., 1964. *Phys. Rev.*, **136**, p. B864.

Hofmann, D.M., Kovalev, D., Steude, G., Meyer, B.K., Hoffmann, A., Eckey, L., Heitz, R., Detchprom, T., Amano, H. and Akasaki, I., 1995. *Phys. Rev. B*, **52**, p. 16702.

Hoffmann, A., Eckey, L., Maxim, P., Holst, J.-Chr., Heitz, R., Hofmann, D.M., Kovalev, D., Stevde, G., Volm, D., Meyer, B.K., Detchprohm, T., Hiramatsu, K., Amano, H. and Akasaki, I., 1997. *Solid-State Electron.*, **41**, p. 275.

Johnson, C., Lin, J.Y., Jiang, H.X., Khan, M.A. and Sun, C.J., 1996. *Appl. Phys. Lett.*, **68**, p. 1808.

Kaneda, N., Detchprohm, T., Hiramatsu, K. and Sawaki, N., 1996. *Jpn. J. Appl.Phys.*, **35**, p. L468.

Kaufmann, U., Kunzer, M., Maier, M., Obloh, H., Ramakrishnan, A., Santic, B. and Schlotter, P., 1998. *Appl. Phys. Lett.*, **72**, p. 1326.

Ke, S., Zhang, K. and Xie, X., 1996. *J. Appl. Phys.*, **80**, p. 2918.

Kim, W., Salvador, A., Botchkarev, A. E., Aktas, O., Mohammad, S.N. and Morkoç, H., 1996. *Appl. Phys. Lett.*, **69**, p. 559.

Kim, W., Botchkarev, A.E., Salvador, A., Popovici, G., Tang, H. and Morkoç, H., 1997. *J. Appl. Phys.*, **82**, p. 219.

Kohn, W. and Sham, L.J., 1965. *Phys. Rev.*, **140**, p. A1133.

Koide, Y., Itoh, H., Sawaki, N., Akasaki, I. and Hashimoto, M., 1986. *J. Electrochem. Soc.*, **133**, p. 1956.

Laks, D.B., Van de Walle, C.G., Neumark, G.F. and Pantelides, S.T., 1991. *Phys. Rev. Lett.*, **66**, p. 648.

Langer, J.M. and Heinrich, H., 1985. *Phys. Rev. Lett.*, **55**, p. 1414.

Lee, H.G., Gershenzon, M. and Goldenberg, B. L., 1991. *J. Electron. Mat.*, **20**, p. 621.

Lee, J.W., Pearton, S.J., Zolper, J.C. and Stall, R.A., 1996. *Appl. Phys. Lett.*, **68**, p. 2102.

Lee, I.-H., Choi, I.-H., Lee, C.R. and Noh, S.K., 1997. *Appl. Phys. Lett.*, **71**, p. 1359.

Leroux, M., Beaumont, B., Grandjean, N., Lorenzini, P., Haffouz, S., Vennéguès, P., Massies, J. and Gibart, P., 1997. *Mat. Sci. Engin. B*, **50**, p. 97.

Li, X. and Coleman, J. J., 1996. *Appl. Phys. Lett.*, **69**, p. 1605.

Li, J.Z., Lin, J.Y., Jiang, H.X., Salvador, A., Botchkarev, A. and Morkoç, H., 1996. *Appl. Phys. Lett.*, **69**, p. 1474.

Lin, M.E., Xue, G., Zhou, G.L., Greene, J.E. and Morkoç, H., 1993. *Appl. Phys. Lett.*, **63**, p. 932.

Linde, M., Uftring, S.J., Watkins, G.D., Härle, V. and Scholz, F., 1997. *Phys. Rev. B*, **55**, p. 10177.

Look, D.C. and Molnar, R.J., 1997. *Appl. Phys. Lett.*, **70**, p. 3377.

Look, D.C., Reynolds, D.C., Hemsky, J.W., Sizelove, J.R., Jones, R.L. and Molnar, R.J., 1997a. *Phys. Rev. Lett.*, **79**, p. 2273.

Look, D.C., Sizelove, J.R., Keller, S., Wu, Y.F., Mishra, U.K. and DenBaars, S. P., 1997b. *Solid State Commun.*, **102**, p. 297.

Louie, S.G., Froyen, S. and Cohen, M.L., 1982. *Phys. Rev. B*, **26**, p. 1739.

Majewski, J.A., Städele, M. and Vogl, P., 1997. *Mat. Res. Soc. Symp. Proc.*, **449**, p. 917.

Mattila, T. and Nieminen, R.M., 1996. *Phys. Rev. B*, **54**, p. 16676.

Mattila, T. and Nieminen, R.M., 1997. *Phys. Rev. B*, **55**, p. 9571.

McCluskey, M.D.. Johnson, N.M.. Van de Walle, C.G., Bour, D.P., Kneissl, M. and Walukiewicz, W., 1998a. *Phys. Rev. Lett.*, **80**, p. 4008.

McCluskey, M.D., Van de Walle, C.G., Master, C.P., Romano, L.T. and Johnson, N.M., 1998b. *Appl. Phys. Lett.*, **72**, p. 2725.

Martin, G., Strite, S., Botchkarev, A., Agarwal, A., Rockett, A., Morkoç, H., Lambrecht, W.R.L. and Segall, B., 1994. *Appl. Phys. Lett.*, **65**, p. 610.

Martin, G., Botchkarev, A., Rockett, A. and Morkoç, H., 1996. *Appl. Phys. Lett.*, **68**, p. 2541.

Mireles, F., and Ulloa, S.E., 1998. *Phys. Rev. B*, **58**, p. 3879.

Miyachi, M., Tanaka, T., Kimura, Y., and Ota, H., 1998. *Appl. Phys. Lett.*, **72**, p. 1101.

Molnar, R.J., Lei, T. and Moustakas, T.D., 1993. *Mat. Res. Soc. Symp. Proc.*, **281**, p. 753.

Mooney, P., 1992. In *Deep Centers in Semiconductors*, ed. Pantelides, S.T., (Yverdon: Gordon and Breach), p. 643.

Nakamura, S., 1991. *Jpn. J. Appl. Phys.*, **30**, p. L1708.

Nakamura, S., Iwasa, N., Senoh, M. and Mukai, T., 1992a. *Jpn. J. Appl. Phys.*, **31**, p. 1258.

Nakamura, S., Mukai, T., Senoh, M. and Iwasa, N., 1992b. *Jpn. J. Appl. Phys.*, **31**, p. L139.

Nakamura, S., Iwasa, N. and Nagahama, S., 1993. *Jpn. J. Appl. Phys.*, **32**, p. L338.

Nakamura, S., 1994. *J. Cryst. Growth*, **145**, p. 911.

Nakamura, S., Senoh, M., Nagahama, S., Iwasa, N., Yamada, T., Matsushita, T., Sugimoto, Y. and Kiyoku, H., 1997. *Appl. Phys. Lett.*, **70**, p. 1417.

Nakamura, S., 1998a. *J. Electron. Mater.*, **27**, p. 160.

Nakamura, S. 1998b. *IEEE J. Sel. Top. Quant. Electron.*, **4**, p. 483.

Nakamura, S., Senoh, M., Nagahama, S., Iwasa, N., Yamada, T., Matsushita, T., Kiyoku, H., Sugimoto, Y., Kozaki, T., Umemoto, H., Sano, M. and Chocho, K., 1998a. *Appl. Phys. Lett.*, **72**, p. 211.

Nakamura, S., Senoh, M., Nagahama, S., Iwasa, N., Yamada, T., Matsushita, T., Kiyoku, H., Sugimoto, Y., Kozaki, T., Umemoto, H., Sano, M. and Chocho, K., 1998b. *Appl. Phys. Lett.*, **72**, p. 2014.

Neugebauer, J. and Van de Walle, C.G., 1994a. *Phys. Rev. B*, **50**, p. 8067.

Neugebauer, J. and Van de Walle, C.G., 1994ba. *Mat. Res. Soc. Symp. Proc.*, **339**, p. 687.

Neugebauer, J. and Van de Walle, C.G., 1995a. *Proc. 22nd Int. Conf. Physics of Semiconductors*, edited by Lockwood, D. J. (Singapore: World Scientific), p. 2327.

Neugebauer, J. and Van de Walle, C.G., 1995b. *Phys. Rev. Lett.*, **75**, p. 4452.

Neugebauer, J. and Van de Walle, C.G., 1995c. *Mat. Res. Soc. Symp. Proc.*, **378**, Materials Research Society, Pittsburgh, p. 503.

Neugebauer, J. and Van de Walle, C.G., 1996a. *Festkörperprobleme/Advances in Solid State Physics*, Vol. 35, edited by Helbig, R. (Braunschweig/Wiesbaden: Vieweg), p. 25.

Neugebauer, J. and Van de Walle, C.G., 1996b. *Mat. Res. Soc. Symp. Proc.*, **408**, p. 43.

Neugebauer, J. and Van de Walle, C.G., 1996c. *Appl. Phys. Lett.*, **69**, p. 503.

Neugebauer, J. and Van de Walle, C.G., 1996d. *Mat. Res. Soc. Symp. Proc.*, **395**, p. 645.

Neugebauer, J. and Van de Walle, C.G., 1996e. *Appl. Phys. Lett.*, **68**, p. 1829.

Neugebauer, J. and Van de Walle, C.G., 1996f. *Proc. 23rd Int. Conf. Physics of Semiconductors*, eds. by Scheffler, M., and Zimmermann, R., (Singapore: World Scientific), p. 2849.

Neugebauer, J. and Van de Walle, C.G., 1998 (unpublished).

Niebuhr, R., Bachem, K.H., Kaufmann, U., Maier, M., Merz, C., Santic, B., Schlotter, P. and Jürgensen, H., 1997. *J. Electron. Mater.*, **26**, p. 1127.

Ohba, Y. and Hatano, A., 1994. *Jpn. J. Appl. Phys.*, **33**, p. L1367.

Park, C.H. and Chadi, D.J., 1997. *Phys. Rev. B*, **55**, p. 12995.

Pearton, S.J., Bendi, S., Jones, K.S., Krishnamoorthy, V., Wilson, R.G., Ren, F., Karlicek, R.F., Jr. and Stall, R.A., 1996a. *Appl. Phys. Lett.*, **69**, p. 1879.

Pearton, S.J., Abernathy, C.R., Vartuli, C.B., Lee, J.W., Mackenzie, J.D., Wilson, R.G., Shul, R.J., Ren, F. and Zavada, J.M., 1996b. *J. Vac. Sci. Technol. A*, **14**, p. 831.

Pearton, S.J., Shul, R.J., Wilson, R.G., Ren, F., Zavada, J.M., Abernathy, C.R., Vartuli, C.B., Lee, J.W., Mileham, J.R. and Mackenzie, J.D., 1996c. *J. Electron. Mat.*, **25**, p. 845.

Perlin, P., Suski, T., Polian, A., Chervin, J.C., Knap, W., Camassel, J., Grzegory, I., Porowski, S. and Erickson, J. W., 1997. *Mat. Res. Soc. Symp. Proc.*, **449**, p. 519.

Piner, E.L., Behbehani, M.K., El-Masry, N.A., Roberts, J.C., McIntosh, F.G. and F.G., Bedair, S.M., 1997. *Appl. Phys. Lett.*, **71**, p. 2023.

Popovici, G., Kim, W., Botchkarev, A., Tang, H., Morkoç, H. and Solomon, J., 1997, *Appl. Phys. Lett.*, **71**, p. 3385.

Ploog, K.H. and Brandt, O., 1998. *J. Vac. Sci. Technol. A.*, 16**,** p. 1609.

Qian, G.-X., Martin, R.M. and Chadi, D.J., 1988. *Phys. Rev. B* **38**, p. 7649.

Rode, D.L. and Gaskill, D.K., 1995. *Appl. Phys. Lett.*, **66**, p. 1972.

Romano, L.T., Krusor, B.S., McCluskey, M.D., Bour, D.P. and Nauka, K., 1998. *Appl. Phys. Lett.*, **73**, p. 1757.

Saarinen, K., Laine, T., Kuisma, S., Nissilä, J., Hautojärvi, P, Dobrzynski, L., Baranowski, J. M., Pakula, K., Stepniewski, R., Wojdak, M., Wysmolek, A., Suski, T., Leszczynski, M., Grzegory, I. and Porowski, S., 1997. *Phys. Rev. Lett.*, **79**, p. 3030.

Salvador, A., Kim, W., Aktas, Ö., Botchkarve, A., Fan, Z. and Morkoç, H., 1996. *Appl. Phys. Lett.*, **69**, p. 2692.

Sánchez, F.J., Basak, D., Sánchez-García, M.A., Calleja, E., Muñoz, E., Izpura, I., Calle, F., Tijero, J.M.G., Beaumont, B., Lorenzini, P., Gibart, P., Cheng, T.S., Foxon, C.T. and Orton, J.W., 1996. *Mater. Res. Soc. Internet J. Nitride Semicond. Res.*, **1**, 7.

Sánchez, F.J., Calle, F., Sánchez-García, M.A., Calleja, E., Muñoz, E., Molloy, C. H., Somerford, D.J., Koschnick, F.K., Michael, K. and Spaeth, J.M., 1998. *Mater. Res. Soc. Internet J. Nitride Semicond. Res.*, **3**, 19.

Satta, A., Fiorentini, V., Bosin, A., Meloni, F. and Vanderbilt, D., 1996. *Mater. Res. Soc. Symp. Proc.*, **395**, p. 515.

Schubert, E.F., Goepfert, I.D. and Redwing, J.M., 1997. *Appl. Phys. Lett.*, **71**, p. 3224.

Schwartz, G.P., Hybertsen, M.S., Bevk, J., Nuzzo, R.G., Mannaerts, J.P. and Gualtieri, G.J., 1989. *Phys. Rev. B*, **39**, p. 1235.

Seifert, W., Franzheld, R., Butter, E., Sobotta, H. and Riede, V., 1983. *Cryst. Res. & Technol.*, **18,** p. 383.

Shan, W., Fischer, A.J., Hwang, S.J., Little, B.D., Hauenstein, R.J., Xie, X.C., Song, J.J., Kim, D.S., Goldenberg, B., Horning, R., Krishnankutty, S., Perry, W.G., Bremser, M.D. and Davis, R.F., 1998. *Appl. Phys. Lett.*, **83**, p. 455.

Siegle, H., Hoffmann, A., Eckey, L., Thomsen, C., Christen, J., Bertram, F., Schmidt, D., Rudloff, D. and Hiramatsu, K., 1997. *Appl. Phys. Lett.*, **71**, p. 2490.

Stampfl, C. and Van de Walle, C.G., 1998a. *Appl. Phys. Lett.*, **72**, p. 459.

Stampfl, C. and Van de Walle, C.G., 1998b. *Mater. Res. Soc. Symp. Proc.*, **482**, p. 905.

Strite, S., 1994. *Jpn. J. Appl. Phys.*, **33**, p. L699.

Stumpf, R. and Scheffler, M., 1994. *Comp. Phys. Commun.*, **79**, p. 447.

Suski, T., Perlin, P., Teisseyre, H., Leszczynski, M., Grzegory, I., Jun, J., Bockowski, M. and Porowski, S., 1995. *Appl. Phys. Lett.*, **67**, p. 2188.

Suzuki, M., Nishio, J., Onomura, M. and Hongo, C., 1998. *J. Cryst. Growth*, **189/190**, p. 511.

Tanaka, T., Watanabe, A., Amano, H., Kobayashi, Y., Akasaki, I., Yamazaki, S. and Koike, M., 1994. *Appl. Phys. Lett.*, **65**, p. 593.

Tersoff, J. and Van de Walle, C.G., 1987. *Phys. Rev. Lett.*, **59**, p. 946.

Troullier, N. and Martins, J.L., 1991. *Phys. Rev. B*, **43**, p. 1993.

Van de Walle, C.G., 1989. *Phys. Rev. B*, **39**, p. 1871.

Van de Walle, C.G., Laks, D.B., Neumark, G.F. and Pantelides, S.T., 1993. *Phys. Rev. B*, **47**, p. 9425.

Van de Walle, C.G., 1997. *Phys. Rev. B*, **56**, p. R10020.

Van de Walle, C.G. and Neugebauer, J., 1997. *Appl. Phys. Lett.*, **70**, p. 2577.

Van de Walle, C.G., 1998. *Phys. Rev. B*, **57**, p. 2033.

Van de Walle, C.G., Stampfl, C. and Neugebauer, J., 1998. *J. Cryst. Growth*, **189/190**, p. 505.

Waldrop, J.R. and Grant, R.W., 1996. *Appl. Phys. Lett.*, **68**, p. 2879.

Wang, C.D., Yu, L.S., Lau, S.S., Yu, E.T., Kim, W., Botchkarev, A.E. and Morkoç, H., 1998. *Appl. Phys. Lett.*, **72**, p. 1211.

Wei, S.-H. and Zunger, A., 1987. *J. Vac. Sci. Technol. B*, **5**, p. 1239.

Wei, S.-H. and Zunger, A., 1996. *Appl. Phys. Lett.*, **69**, p. 2719.

Wetzel, C., Suski, T., Ager, J.W., III, Walukiewicz, W., Fisher, S., Meyer, B.K., Grzegory, I. and Porowski, S., 1996. *Proc. 23rd Int. Conf. Physics of Semiconductors*, eds. Scheffler, M. and Zimmermann, R. (Singapore: World Scientific), p. 2929.

Wetzel, C., Suski, T., Ager, J.W., III, Weber, E.R., Haller, E.E., Fischer, S., Meyer, B.K., Molnar, R.J. and Perlin, P., 1997. *Phys. Rev. Lett.*, **78**, p. 3923.

Wickenden, A.E., Beadie, G., Koleske, D.D., Rabinovich, W.S. and Freitas, J.A., Jr., 1997. *Mat. Res. Soc. Symp. Proc.*, **449**, p. 531.

Wright, A.F., 1997. *J. Appl. Phys.*, **82**, p. 2833.

Yi, G.-C. and Wessels, B.W., 1996. *Appl. Phys. Lett.*, **69**, p. 3028.

Zeng, K.C., Lin, J.Y., Jiang, H.X., Salvador, A., Popovici, G., Tang, H., Kim, W. and Morkoç, H., 1997. *Appl. Phys. Lett.*, **71**, p. 1368.

Zhang, X., Kung, P., Saxler, A., Walker, D. and Razeghi, M., 1995a. *Acta Phys. Polonica* A, **88**, p. 601.

Zhang, X., Kung, P., Saxler, A., Walker, D., Wang, T. C. and Razeghi, M., 1995b. *Appl. Phys. Lett.*, **67**, p. 1745.

Zhang, X., Kung, P., Walker, D., Saxler, A. and Razeghi, M., 1996. *Mater. Res. Soc. Symp. Proc.*, **395**, p. 625.

Zhu, X. and Louie, S.G., 1991. *Phys. Rev. B*, **43**, p. 14142.

Zolper, J. C., Tan, H.H., Williams, J.S., Zou, J., Cockayne, D.J.H., Pearton, S.J., Hagerott Crawford, M. and Karlicek, R.F., Jr., 1997. *Appl., Phys. Lett.* **70**, p. 2729.

Chapter 4
Crystal Defects and Device Performance in LEDs and LDs

Fernando A. Ponce

4.1 CRYSTAL GROWTH AND MICROSTRUCTURE

It is safe to say that all semiconductors used for commercial light-emitting diodes (LEDs) and laser diodes (LDs) are produced from III-V compound thin films grown using the metalorganic chemical vapor deposition (MOCVD) technique. And while it is relative easy to mix the group III elements (Al, Ga, and In) to produce ternary and quaternary compounds, it is much harder to mix the group V elements (Sb, As, P and N). Thus, currently used III-V compounds can be grouped into the families of the antimonides, arsenides, phosphides and nitrides. The first three have a predominantly cubic lattice structure, while the nitrides tend to have a hexagonal lattice structure.

Materials with superior electronic properties and high light emission efficiencies are necessary for the manufacture of light emitting devices. In the past, these characteristics have been usually associated with crystalline perfection in the semiconductor materials. The high-optoelectronic performance of the nitride semiconductors is closely related to its microstructure in a manner that departs from traditional thinking.

Until recently, LEDs in the range from the infrared to orange were manufactured using the AlGaInP and AlGaAs systems. Lattice matched conditions were found to be necessary in order to produce high performance materials. Minimisation of extended lattice defects, such as stacking faults and dislocations, was also found to be necessary. Defect degradation in light emitting devices was linked to the motion of dislocations at the active region with the creation of the so-called dark-line defects. From this perspective, the emphasis of crystal growers had been the elimination of crystalline defects in semiconductor epitaxy.

Based on such previous experience, the logical initial approach to the growth of the nitride semiconductors was based on obtaining methods of growth of low-defect density material. However, the absence of bulk GaN substrates made this task difficult. A further level of difficulty was presented by (1) the need to grow at high temperatures (temperatures close to 1000°C are needed to thermally dissociate the stable nitrogen-containing molecules), and (2) the relatively small lattice parameter of the nitrides.

4.1.1 Lattice Structure of the Nitride Semiconductors

Like most other semiconductors, the atoms in the nitrides are tetrahedrally coordinated. The *s* and *p* orbitals in the outer electron shells combine to produce hybrid sp^3 orbitals where the probability of finding an electron in a *p*-state is 3 times as much as finding it in an *s*-state. As a result each atomic site has four nearest neighbours occupying the vertices of a regular tetrahedron, in a manner similar to the diamond-cubic structure.

Atomic layers are arranged in hexagonal closed packed planes, with alternating sub-layers containing the group III and group V elements as shown in Figure 4.1. There are various ways to stack these layers (Wyckoff, 1995). A configuration following the sequence ..*ABABAB*.., as in Figure 1a and b, results in the hexagonal wurtzite structure (space group $P6_3mc$, C^4_{6v} (#186)). A stacking sequence of the type ..*ABCABCABC*.., as in Figure 1c and d, produces the cubic zincblende structure (space group $F\overline{4}3m$, T^2_d (#216)). The indices of the stacking planes are {0001} for the hexagonal and {111} for the cubic structures. These lattices are polar in nature, since the anions and the cations occupy planes that are displaced from one another along the <0001> and the <111> directions for the hexagonal and cubic structures, respectively.

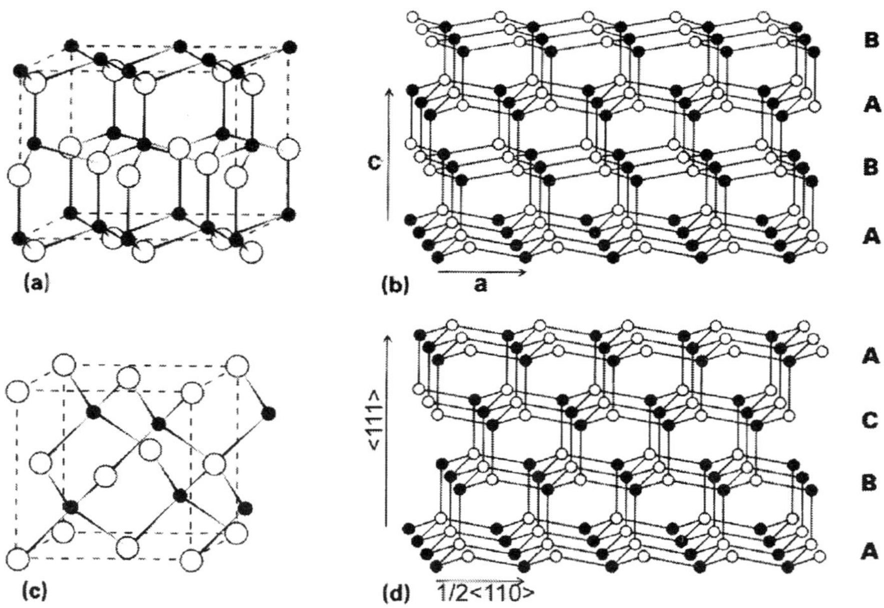

Figure. 4.1 Two main types of atomic arrangement in the nitride semiconductors. Unit cells for (a) hexagonal wurtzite (b) cubic zincblende, and (b) and (d) their respective crystalline layer sequences.

4.1.2 Thin Film Epitaxy and Substrates

The choice of substrate has important repercussions on the properties of the epilayers. As already mentioned, one critical requirement for the growth of the nitrides is stability of the substrate at high temperatures in the highly corrosive environments associated with the nitrogen sources (like ammonia). These considerations have limited the choice of substrates to materials such as SiC and sapphire. Following traditional thinking, the lattice parameter ought to play an important role as well. The values of the chemical bond lengths for the nitrides and for SiC and sapphire are depicted in Figure 4.2 and are listed in Table 4.1.

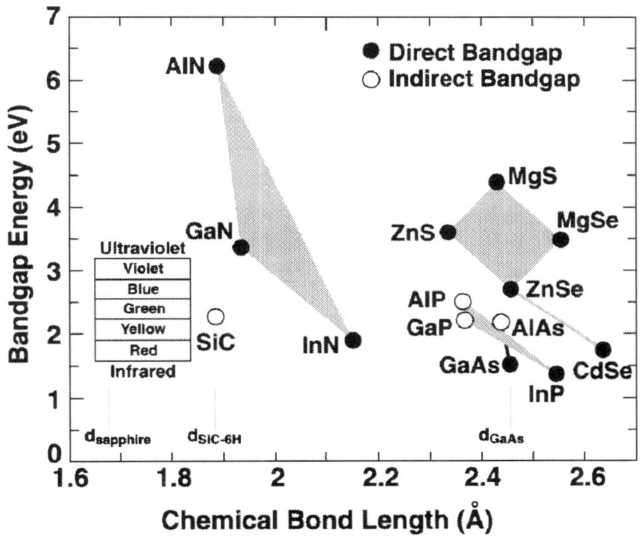

Figure. 4.2 Bandgap and chemical bond length for semiconductors used in visible light emitting devices. Bond lengths are used instead of lattice parameter in order to avoid the arbitrary values introduced by the choice of a unit cell in hexagonal vs. cubic materials. (Ponce and Bour, 1997).

Because growth happens at elevated temperatures, it is important to take into consideration the difference in thermal expansion between the thin film and the substrate. Figure 4.3 illustrates the combined effect of lattice parameter and thermal expansion compared with the coefficients of thermal expansion for sapphire, SiC and GaAs. SiC and AlN have similar thermal expansion characteristics along the basal planes, but differ from GaN and InN by more than 20%. The value for sapphire is 78% larger than AlN and 34% larger than GaN. These values are equivalent to an additional difference in lattice parameter of 0.34% and 0.20%, respectively, when cooling down from a growth temperature of 1050°C.

Table 4.1 Lattice properties of the nitride semiconductors and most commonly used substrates.

Crystalline Properties		AlN	GaN	InN	α-Al_2O_3	6H-SiC
Lattice Parameter (Å) (wurtzite structure)	a	3.1114	3.186	3.5446	4.758	3.081
	c	4.9792	5.178	5.7034	12.991	15.092
	c/a	1.600	1.625	1.609	2.730	1.633(x3)
Thermal Exp. Coefficient	a	4.2	5.59	5.7	7.5	4.2
($x10^{-6}C^{-1}$)	c	5.3	3.17	3.7	8.5	4.68
Energy gap (eV) at 300K	W	6.2 D	3.39 D	1.89 D		
(for wurtzite W, and zincblende Z lattices)	Z	5.1 D	3.2 D	2.2 D		
Lattice Parameter (Å) (zincblende structure)	a	4.38	4.52	4.98		
Bond Length (Å)		1.89	1.94	2.15		1.88
Cohesive energy per bond (eV)	E_{coh}	2.88	2.24	1.93		3.17
Density (g/cm^3)	d	3.26	6.10	6.99	3.97	3.217

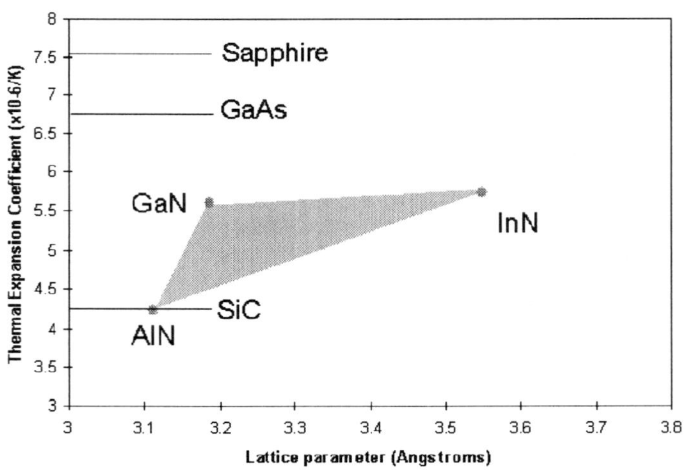

Figure 4.3 Thermal expansion along the basal plane and lattice parameter for the nitride semiconductors and some substrate materials.

4.2 EPITAXY ON SiC SUBSTRATES

The lattice of silicon carbide is also tetrahedrally coordinated. Each atomic layer is arranged in an identical manner as in the nitrides. The stacking of these layers can take place in a number of ways, leading to a high degree of polymorphism for SiC. The various lattices occur in the cubic (C), hexagonal (H), or rhombohedral (R) configurations. The structures are annotated by the symmetry (C, H, or R) following the number of layers in one period of the stacking sequence. Thus, 3C is equivalent to the zincblende structure, 2H is the wurtzite structure. The most common commercially available form is 6H-SiC.

When compared with AlN, SiC has excellent properties as a substrate for epitaxy. The lattice mismatch for [0001] epitaxy is less than 1% and the coefficient of thermal expansion for AlN and SiC are very similar. Figure 4.4 is a cross-section lattice image of the AlN/SiC interface grown by MOCVD (Ponce *et al*, 1995). The SiC lattice is characterized by the 6-layer period zig-zag characteristic of the 6H lattice structure. The AlN has a 2-layer periodicity (2H) and the lattice features (white spots) are arranged in the vertical direction. The interface is atomically planar in this figure and a high level of coherency between the two lattices is observed.

A careful analysis of the high-resolution lattice images can yield detailed information about the atomic arrangement at the AlN/SiC interface (Ponce *et al*, 1996e). It has been shown that [0001] SiC (with silicon on the top sublayer of the basal plane) produces [0001] AlN (with Al on the top sublayer of the basal plane). It is observed that the interfaces includes C-Al and Si-N bonds yielding two interface bonding configurations:

$$\cdots\text{C}\cdot\text{Si}\cdots\text{C}\cdot\text{Al}\cdots\text{N}\cdot\text{Al}\cdots, \qquad (1)$$

and

$$\cdots\text{C}\cdot\text{Si}\cdots\text{N}\cdot\text{Al}\cdots\text{N}\cdot\text{Al}\cdots. \qquad (2)$$

In this notation, long dashes represent the bonds between the widely spaced layers, and the short dash represents the bonds between the closely spaced layers. High-resolution TEM cannot distinguish between these two structures. Other possible interface bonding configurations including C-N and Si-Al have not been observed. It has been argued that the possible bonding configurations at the interface between SiC and AlN are restricted by the requirement of local charge neutrality (Ponce *et al*, 1996f). Each N atom contributes 5/4 of an electron to its bond with Si, while the Si atom contributes one electron to this bond. Therefore, there will be a local charge excess at the interface of ¼ electron per bond. A uniformly planar interface of the types (1) or (2) should be expected to be unstable due to the Coulomb energy generated by the interface charge. It has been suggested that the excess charge can be relieved by replacing one out of every four Si atoms with Al:

$$\cdots\text{C}\cdot\text{Si}\cdots\text{C}\cdot\text{Si}(3/4)\ \text{Al}(1/4)\cdots\text{N}\cdot\text{Al} \qquad (3)$$

where the Al act as acceptors residing on the Si sublattice of SiC. Alternatively, one out of four N atoms can be replaced by C:

$$\cdots\text{C}\cdot\text{Si}\cdots\text{N}(3/4)\ \text{C}(1/4)\cdot\text{Al}\cdots\text{N}\cdot\text{Al} \qquad (4)$$

where C act as acceptor residing on the N sublattice of AlN. These two stable interface atomic arrangements are depicted in Figure 4.5.

In practice, it is very difficult to produce an atomically flat surface of SiC over extended regions of the substrate. Because of the nature of the 6H structure,

In practice, it is very difficult to produce an atomically flat surface of SiC over extended regions of the substrate. Because of the nature of the 6H structure, one atomic step generates a significant lateral offset that is difficult to compensate without the generation of lattice defects. Another difficulty is that while there is a close match between SiC and AlN, no such match occurs for other compositions of the nitrides. For growth of GaN and InN on SiC, the lattice mismatch is 3.3% and 14%, respectively, as shown in Table 4.2.

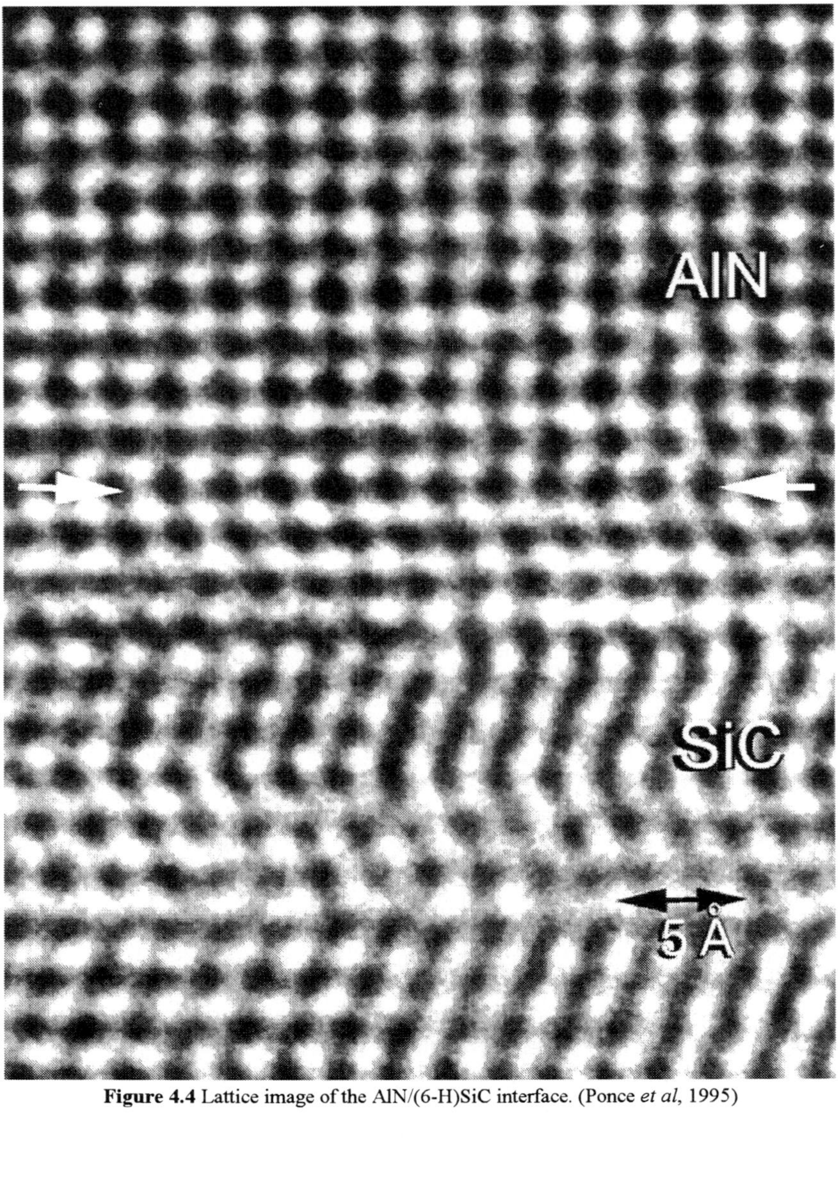

Figure 4.4 Lattice image of the AlN/(6-H)SiC interface. (Ponce *et al*, 1995)

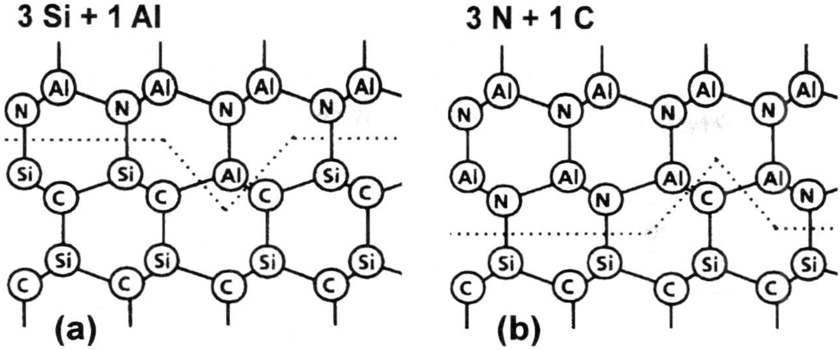

Figure 4.5 Atomic arrangement of the AlN/(6-H)SiC interface, allowing for atomic mixing at the interface. (a) ···C·Si···C·Si(3/4)Al(1/4) ···N·Al, (b) ···C·Si···N(3/4)C(1/4) ·Al···N·Al (Ponce *et al*, 1996f).

Table 4.2 Lattice mismatch (%) and resulting misfit dislocation separation (in Å) corresponding to complete misfit relaxation for basal plane interfaces.

Crystalline Properties		GaN	AlN	InN	α-Al_2O_3	6H-SiC
Lattice mismatch with	sapphire	14.8%	12.5%	25.4%	-	11.5%
	SiC	3.3%	1.0%	14.0%	-11.5%	-
	GaN	-	-2.4%	10.6%	-14.8%	-3.3%
Dislocation distance on	sapphire	17.2	20.3	10.6	-	21.9
	SiC	80.9	276.7	20.4	21.9	-
	GaN	-	114.4	27.3	17.2	80.9

4.3 EPITAXY ON SAPPHIRE SUBSTRATES

At first glance, sapphire is not an ideal substrate for the growth of the nitrides. The large lattice mismatch (14.8% with GaN and 25.4% for InN) and significant thermal expansion difference does not lead to predictions for good electronic quality materials. Nevertheless, many of the remarkable recent successes are based on thin films grown using sapphire as a substrate.

Sapphire (α-Al_2O_3) has a hexagonal structure belonging to the space group R3c (D63d, No. 167) that can be expressed both as a hexagonal as well as a rhombohedral unit cell as shown in Figure 4.6. The basic structure consists of hexagonal close-packed planes of oxygen intercalated with aluminum planes. The aluminum planes have a similar hexagonal close-packed arrangement but with 1/3 of the sites vacant, giving rise to an Al/O ratio of 2/3. Each Al atom is surrounded by 6 oxygen atoms, and each oxygen atom has 4 Al neighbours. The vacancies are aligned on the so-called r-planes, giving sapphire the ability to cleave along these rhombohedral planes. Due to these properties, the unit cell of sapphire is chosen taking into consideration the position of the Al vacancies (Kronberg, 1957). It must be noticed that the choice of unit cell is arbitrary in a periodic lattice.

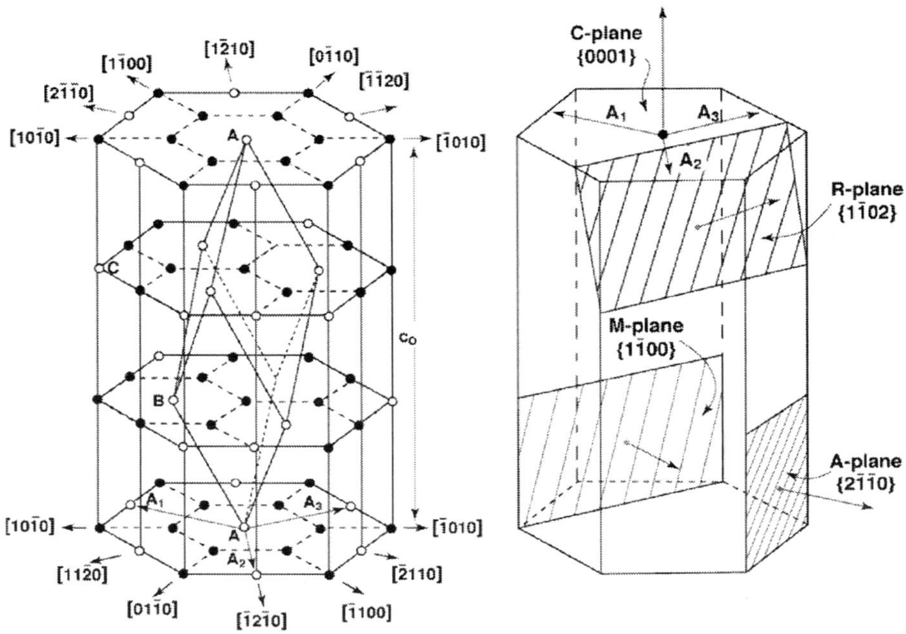

Figure 4.6 The crystal lattice of sapphire. The aluminum planes are depicted in the figure at the left. Vacancies are observed in 1/3 of the aluminum sites. Oxygen planes are intercalated with the aluminum planes, and are not shown. The morphological unit cell is drawn connecting the aluminum vacancies. Sapphire planes used in the epitaxy of the nitrides are shown at the right.

Growth of GaN by MOCVD on (0001) sapphire substrates is achieved using the two step method (Amano *et al*, 1986). A low temperature buffer layer is deposited at 550°C to overcome surface tension and to achieve uniform coverage of the substrate. Such nucleation layers are amorphous during deposition. They crystallize during heating to the growth temperatures of about 1050°C.

It is important to notice, therefore, that the nitride layer is not initially grown epitaxially on sapphire, but rather, that the crystalline structure is the result of solid phase crystallization from the amorphous phase. This process leads to a granular structure of highly oriented GaN crystallites with dimensions of the order of the thickness of the buffer layer. These crystallites have an orientation distribution of the c-axis the order of one degree of arc. The in-plane orientation has a similar distribution in orientation. During subsequent growth at high temperatures, the crystallites with c-axis orientation closest to the growth direction dominate by rapid lateral growth over neighbouring crystallites. The dominant lateral growth is a special characteristic of GaN. It is reflected first on the nucleation and growth mechanisms on sapphire substrates, as well as on the growth of heterostructures, where the interfaces are abrupt and planar. This feature is also responsible for the atomically flat characteristics of GaN surfaces grown under the conditions stated above.

4.3.1 AlN as a Buffer Layer

AlN and GaN buffer layers have been used for the growth of GaN and AlGaN layers on sapphire substrates. AlN was first used as a buffer layer by Amano and Akasaki (1986). The interface between AlN and sapphire is easily understood since it involves a transition from the oxide to the nitride forms of aluminum. The atomic arrangement at the interface, shown in Figure 4.7, gives indication of local coherence evidenced by the distortion of the atomic arrangement within a monolayer of the interface plane (Ponce *et al*, 1994). This interface is very sensitive to the surface treatment of the sapphire substrate, and determines the polarity of growth. Prolonged exposure to nitrogen sources at high temperatures leads to nitridation of the top surface of the substrate, producing crystallization along the –c direction (nitrogen on the top sub-layer of the nitride basal plane). For not so severe exposure to nitrogen, the nitride layer nucleates along the +c direction (Group III elements on the top surface of the nitride basal plane). It has been determined that the +c direction is required in order to achieve high electronic performance (Ponce *et al*, 1996c).

Figure 4.7 Lattice image of the AlN/sapphire interface. The AlN buffer layer is 6nm thick.

4.3.2 GaN as a Buffer Layer

GaN buffer layers were introduced by Nakamura (1991), and they are widely used today. The main advantage over the AlN buffer layer is in the absence of lattice mismatch between the film and the buffer layer. Figure 4.8 is a lattice image of the GaN/sapphire interface (Fertitta *et al*, 1995). Absence of coherence at the interface has been observed in several instances, although such features may be due to different initial treatment of the sapphire substrate. Like the case of AlN on sapphire, the interface is atomically abrupt and follows the surface of the sapphire. It is observed in lattice images and by X-ray diffraction that the GaN layer tends to grow parallel to the surface, rather than to the basal planes of sapphire. This again reinforces the concept of dominant lateral growth in the epitaxy of the nitrides by MOCVD, and that the growth on sapphire is not truly epitaxial.

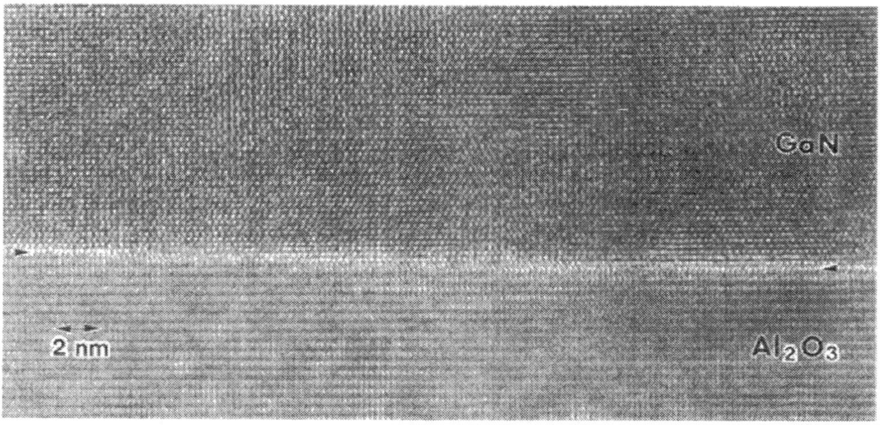

Figure 4.8 Lattice image of the GaN/sapphire interface. Here the GaN buffer layer is 8nm thick.

Figure 4.9 shows the morphology and defect structure in the buffer layer region in a Nichia light emitting diode. The buffer layer region shows defect contrast over a region of the order of 30nm thick, exhibiting a very large density of defects which terminate as the temperature of growth is increased. As indicated previously, these defects are associated with a fine array of crystallites that are slightly misoriented with each other, and coalesce with the prevalence of those with c-axis more closely aligned in the direction of growth.

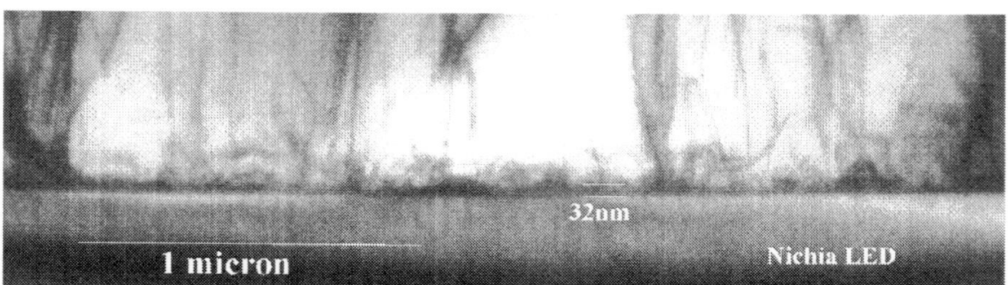

Figure 4.9 Cross-section TEM image of the buffer layer region. The buffer layer is about 30nm thick.

4.4 HOMOEPITAXIAL GROWTH OF GaN

The quality of the epilayer should be optimized with respect to the substrate when both substrate and epilayer have similar chemical and structural properties. Thus, the ideal situation for epitaxial growth occurs when bulk material is available with the same composition and structure as the desired film. Such situation is referred to as homoepitaxy.

GaN bulk crystals are difficult to produce. This is due to the high melting temperatures and high equilibrium vapor pressures at the melting temperatures of these materials (Table 4.3). In spite of these difficulties, crystals of GaN have been grown by Karpinski *et al* (1982) using high-pressure, high-temperature growth techniques. The GaN crystals grow as faceted platelets closely oriented along the (0001) lattice planes, with diagonal dimensions up to 3 to 10mm, and thickness of the order of 0.1mm. The nature of the two faces of the platelets reflect the anisotropy associated with the polar nature of the GaN structure along the c-axis. One face is typically flat with some inclusions, while the other is rough and terraced, and sometimes with a concave geometry (Liliental-Weber, 1996). The inherent chemical stability makes it very difficult to clean GaN substrates using chemical etching techniques. Furthermore, the surface of as-grown crystals present inclusions which are the result of precipitation during the cool-down step in the high pressure reactors.

Epitaxy on GaN single crystals has been successfully achieved by first carefully polishing the platelets using mechanical polishing techniques (Ponce *et al*, 1996b). The films were grown by MOCVD on surfaces with different polarity at the standard growth temperature of 1050°C. The crystalline polarity was determined by convergent beam electron diffraction, and is discussed later in this chapter (Ponce et al, 1996c). It was determined that growth in the +**c** direction resulted in the absence of dislocations visible in the transmission electron microscope, with superb electronic properties. Growth in the –**c** direction contained dislocation loops with a density of $\sim 10^8 cm^{-2}$, that originate at the substrate surface. These dislocations were observed to be associated with the intersection of stacking faults in the substrate with the substrate surface. The photoluminescence properties of films grown along the +**c** direction were significantly superior to those of GaN on sapphire (Ponce *et al*, 1996b, Pakula *et al*, 1996, Baranowski *et al*, 1997). Both Raman spectroscopy and observations by cathodoluminescence suggest that growth in the +**c** direction has a much higher compositional purity and hence sharper luminescence characteristics than growth in the –**c** direction.

Table 4.3 Melting characteristics of the nitrides (wurtzite form). These values are estimated in various ways, and have not been measured directly.

Material	Melting Temperature T_M (K)	Vapor Pressure at T_M (bar)	Melting Entropy (cal $mol^{-1}K^{-1}$)	Melting enthalpy (Kcal mol^{-1})
AlN	3487	30	16.61	57.92
GaN	2791	2000	16.01	44.68
InN	2146	$>10^5$	14.1	

4.5 DEFECT MICROSTRUCTURE IN LEDs AND LDs.

The first observations of the microstructure of bright LEDs gave surprising results. Figure 4.10 is a cross section transmission electron microscope image of a Nichia blue light emitting diode (Lester *et al*, 1995). The film was grown on basal plane sapphire substrates by MOCVD (Nakamura et al, 1994). The device layers consist of ~4μm of silicon-doped n-type GaN, followed by a GaN:Si/InGaN:(Si+Zn)/AlGaN:Mg double heterostructure, and a p-type GaN:Mg cap. The InGaN active layer was heavily doped with Zn to provide recombination centers that emit at ~450 nm. The active areas of the LEDs were ~3×10^{-4} cm^2.

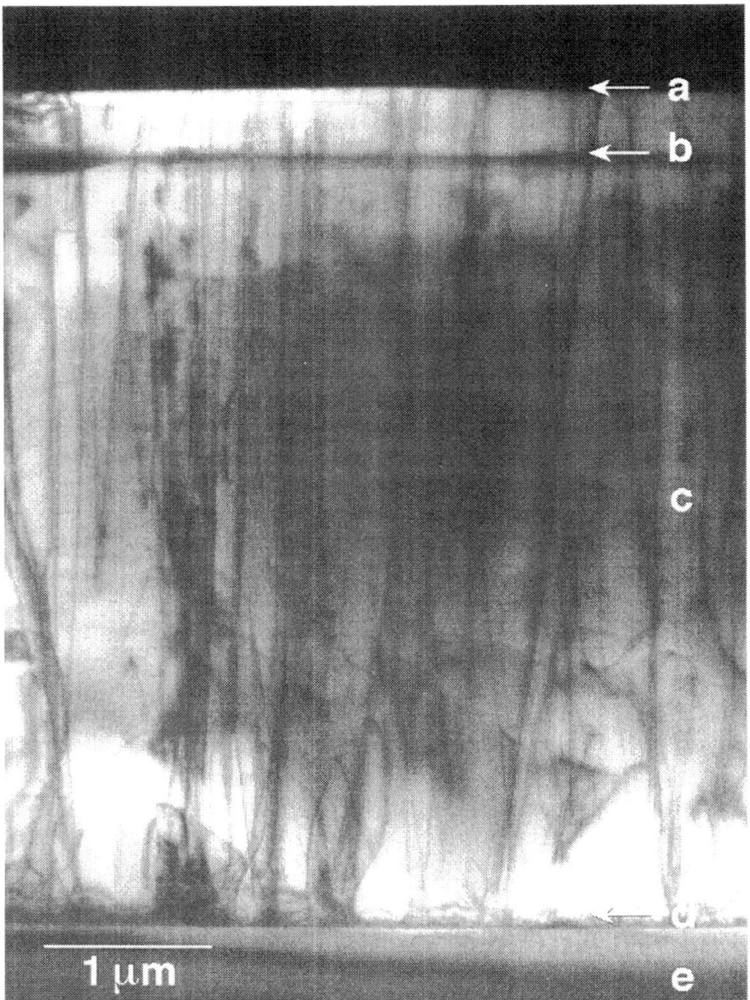

Figure 4.10 Cross-section TEM view of a high efficiency Nichia blue LED. (a) top, (b) active region, (c) main GaN layer, (d) buffer layer, and (e) sapphire substrate.

4.5.1 Large Defect Densities in High Performance Materials

The high brightness and superior electronic properties of the material were ascertained prior to the preparation of the TEM sample. The surprising result of this study was that the LEDs are highly efficient in spite of an extremely high density of structural defects. In Figure 4.10, dislocations are seen as dark lines propagating in a direction normal to the substrate. Shorter dislocations segments parallel to the substrate surface are also observed under appropriate specimen orientations (diffraction conditions). The dislocation density is of the order of $10^{10}cm^{-2}$, and does not vary substantially with distance from the sapphire interface. At this density, the average spacing between dislocations should be about 0.1μm. The TEM image is a projection over a certain thickness, and shows a certain degree of clustering of the dislocations with average separation of about 0.3μm. The dislocations cross the active region of the device.

A view from the top of the device is shown in Figure 4.11. The microstructure has a cellular form, with the dislocations located at the boundaries between defect free cells.

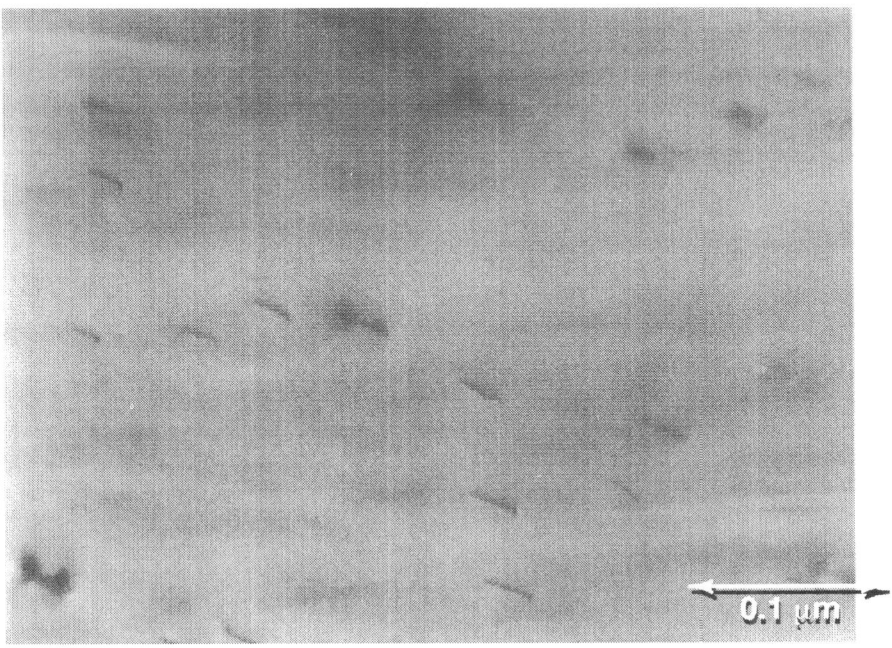

Figure 4.11 Plan-view TEM image of GaN epitaxy on sapphire. Dislocations are observed in a cellular arrangement.

The defect densities observed in these materials are far higher than those measured in LEDs produced from any other III-V system, as shown in Figure 4.12. Studies by Herzog *et al* (1972), and Roedel *et al* (1979) have shown that the efficiencies of GaAs and AlGaAs LEDs are limited by dislocation densities of only 10^4cm^{-2}. Similar observations on GaP by Brantley *et al* (1975), Stringfellow *et al* (1974), and Suzuki and Masumoto (1975), gave similar results. In all these materials, dislocations limit the efficiency of LEDs at densities several orders of magnitude smaller than in GaN-based LEDs. This indicates a critical difference between the III-V nitrides and the arsenides and phosphides, indicating that it should be possible to fabricate a wide range of minority carrier devices without the need for high degree of perfection in the epilayer.

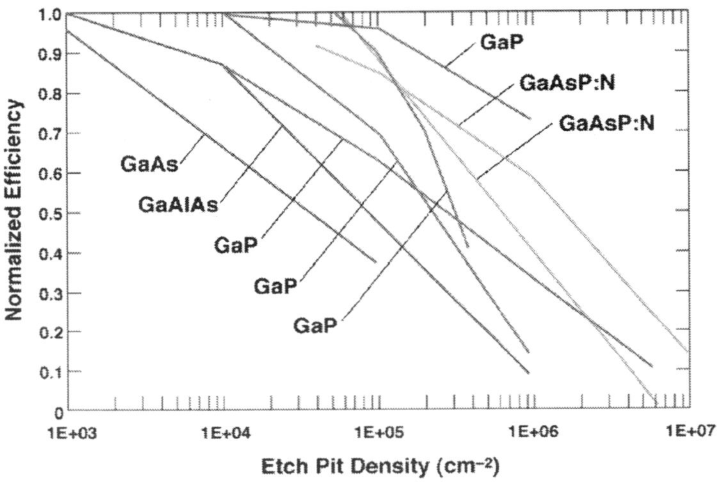

Figure 4.12 Variation of the LED efficiency with the dislocation density in various III-V compound materials (Lester *et al*, 1995).

One possible explanation of the relatively innocuous effect of dislocations is based on the ionic nature of bonding in these materials. Kurtin *et al* (1969) reported striking empirical evidence for a fundamental covalent-ionic transition in the electronic nature of solids. This means that ionic materials (such as GaN) are fundamentally different from the more covalent materials such as silicon, GaAs or InP. These differences arise from the localization of valence band excitations (holes) in ionic materials. In particular, they reported that the surfaces of ionic materials do not exhibit Fermi level pinning. Therefore, either the density of surface states is low, or the surface states lie outside the bandgap. Electrical measurements of metalized contacts by Foresi and Moustakas (1993) indicated that the surface of GaN is *not* pinned. Dislocations can be viewed as internal surfaces, and thus the preceding discussion about surfaces may also apply to dislocations. Cathodoluminescence studies reported later in this chapter show that dislocations are active in undoped materials, but not so in doped materials. It has also been determined that a good fraction of dislocations have hollow cores, i.e. the core of the dislocation is absence and instead a narrow empty tube exists. In summary, it will be seen in the remainder of this chapter that dislocations in GaN are not as simple as the dislocations observed in other semiconductors such as Si and GaAs.

The structure of GaN/sapphire can be difficult to comprehend. It must be noted that the growth of GaN/sapphire is not directly epitaxial. It is based on the controlled crystallization of the amorphous buffer layer. The conditions necessary to grow highly efficient light-emitting materials were developed by empirical methods. The growth conditions were found by optimizing the surface morphology of the films. In-situ growth monitoring techniques like reflection infrared spectroscopic methods were used to fine-tune each of the growth parameters such as temperature, gas flows, buffer layer thickness, doping levels, etc (Nakamura et al, 1992). This approach resulted in specular surfaces and a unique microstructure.

4.5.2 Columnar Structure of GaN on Sapphire

Figure 4.13 is a cross section TEM image of the region near the substrate. Dislocations appear as dark lines. The TEM sample has a wedge shape that provides some degree of robustness for measurement of the lattice orientation along the sample. This image is obtained under parallel beam illumination with two-beam conditions that enhance the contrast of the dislocations. As stated previously the typical separation of the clusters of dislocations is about 0.3μm. We have selected three regions marked by letters. A convergent electron beam focused on the sample plane is used to determine the fine orientation at each spot. The corresponding convergent beam diffraction patterns are shown in Figure 4.14. These contain the higher-order Laue zone (HOLZ) lines which are very sensitive to small variations of the crystalline orientation. In Figure 4.14 the radius of the disk is chosen to be about 5 arc.min. The position of a particular pole (intersection of HOLZ lines) is tracked along the edge of the sample. It is observed that the crystal has a well defined orientation within the domain between dislocation clusters.

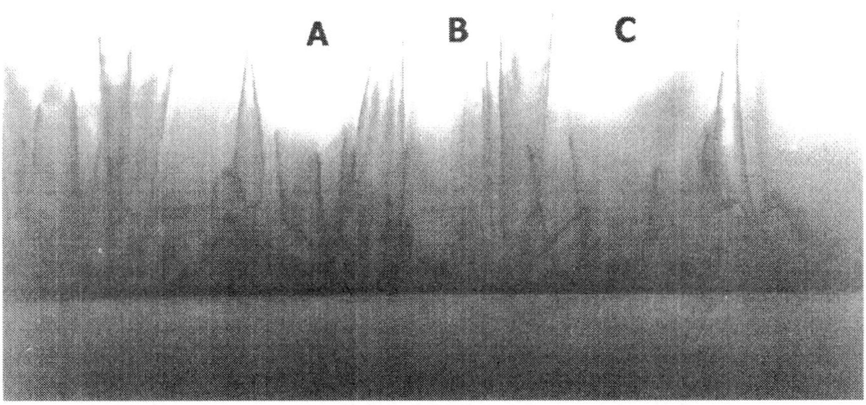

Figure 4.13 Clustering of dislocations and determination of columnar orientation.

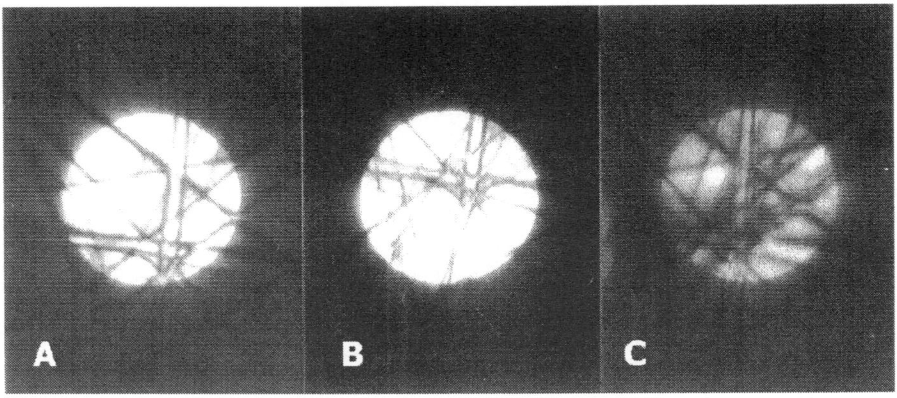

Figure 4.14 Convergent electron diffraction patterns showing variation in orientation of the lattice as the beam moves through the domains. The radius of the disk is about 5 arc min.

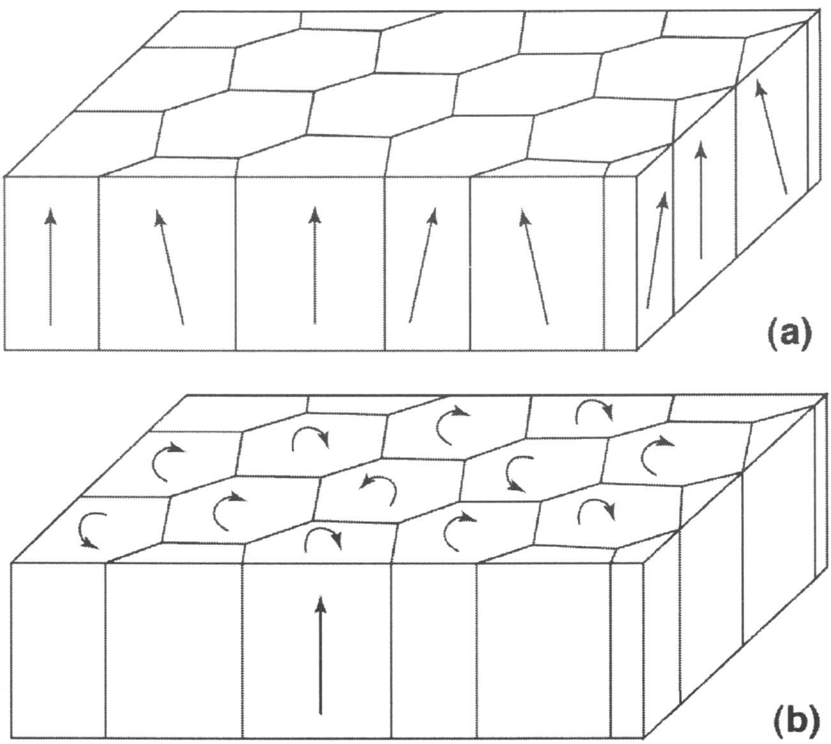

Figure 4.15 Columnar model for the III-V nitrides. The film consists of columns with a slight distribution in the crystalline orientation with two components: tilt (~5 arc min), and (b) twist (~8 arc min). Typical diameters are about 0.3 to 0.6μm.

At the dislocation boundaries, the diffraction pattern blurs and an abrupt change of orientation is observed as the beam traverses a cluster of dislocations. The relative variation of the orientation along the edge of the sample resembles the X-ray diffraction rocking curves. Thus, the cellular structure observed in Figure 4.10 and 4.11 is consistent with a columnar structure involving small misorientation between the columns. Figure 4.12 is a schematic diagram of such array, and displays the two orthogonal components of the crystal orientation of the columns (Ponce, 1997). These columns are virtually free of defects, and are slightly tilted with one another. One component is the tilt of the c-axis with respect to the normal to the film. X-ray rocking curves measure this value, which in high electronic performance materials is around 5 arc min. The other component is the rotation (or twist) of the c-axis about itself with full-width at half maximum of about 8 arc min.

4.5.3 Tilt Boundaries

Figure 4.16 illustrates the generation of tilt resulting from the introduction of extra planes. The interplanar separations in the nitrides are about 0.25nm. Thus to generate observed tilts of the order of 5 arc min, it is sufficient to have one extra plane every 172nm, which is close to the observed average diameter of the columnar structure.

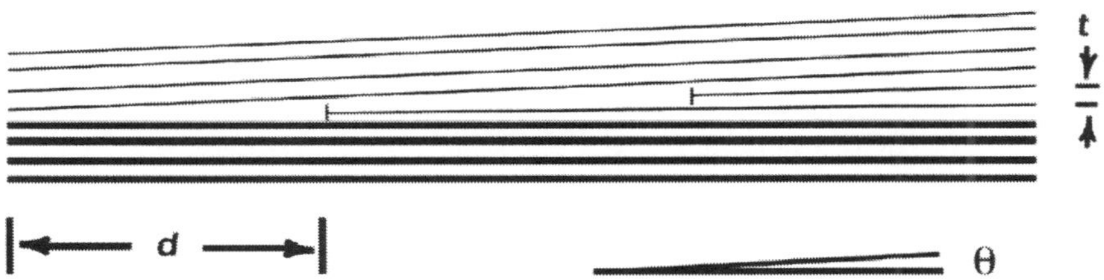

Figure 4.16 Tilt is accomplished by inserting extra planes, with edge dislocations defining the steps responsible for the tilt. A tilt of 5 arc min require steps of 0.25nm every 172nm.

A tilted domain is depicted in Figure 4.17 and a uniform matrix is considered for the sake of analysis. If the tilt is on the plane of the paper, then there should be edge dislocations in the lower boundary of the domain. In the simplest case, the Burgers vector of the edge dislocations should point normal to the basal plane. At the lateral boundaries the dislocation bends towards the surface and changes character. Since the Burgers vector cannot change within the dislocation line, the change of direction turns the edge dislocation into a screw dislocation. Thus, in a pure tilted domain, threading dislocations should be of the screw type.

4.5.4 Twist Boundaries

Figure 4.17 illustrates the case in which a domain is rotated with respect to the matrix. The lower region exhibits a "twist" boundary that consists of screw dislocations, with their Burgers vectors lying along the basal planes as shown.

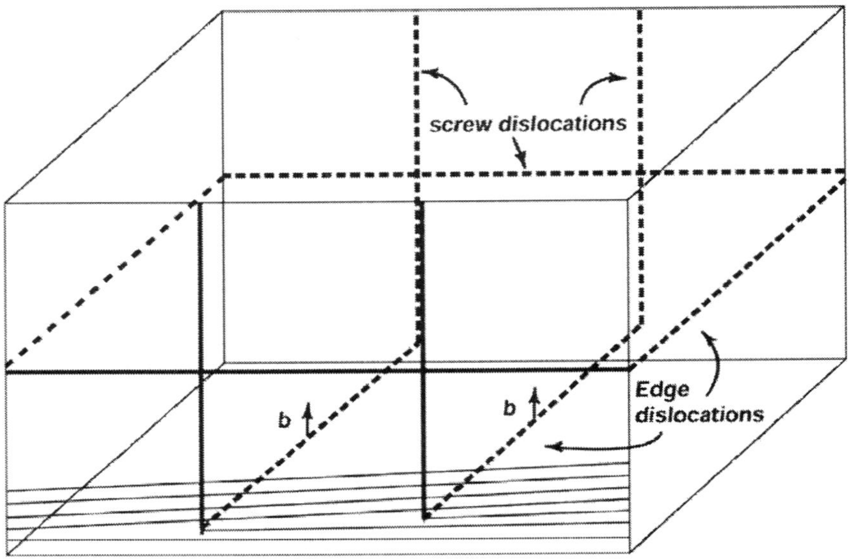

Figure 4.17 A pure tilt boundary requires the existence of edge dislocations at the base of the domain, and the threading dislocations are of the screw type.

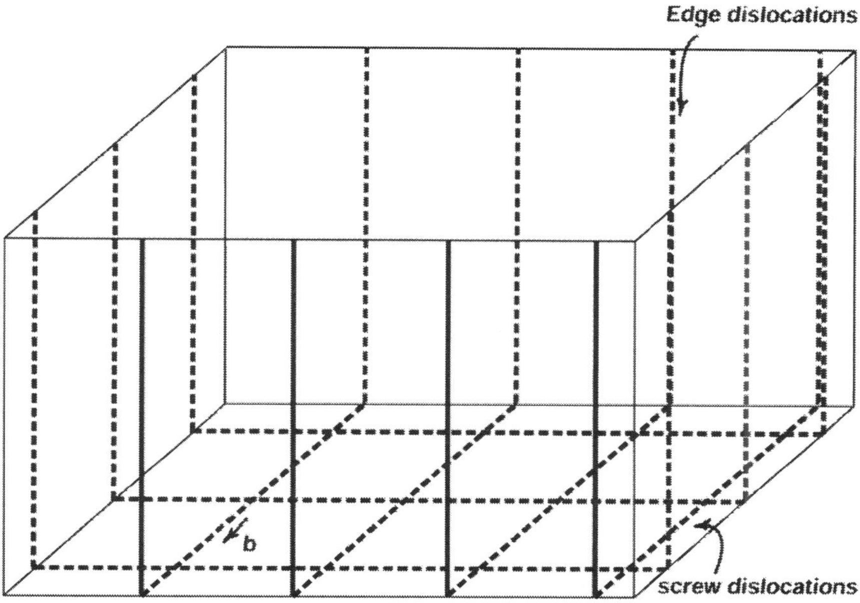

Figure 4.18 A pure twist boundaries require screw dislocations at the base of the domain, and the threading dislocations are of the edge type.

When the dislocations reach the lateral boundaries of the domain, the direction of the dislocation becomes orthogonal to the previous one, and the character of the dislocation changes into an edge dislocation. This simple analysis shows that for pure tilt and twist boundaries the threading dislocations will be either screw or edge in character, respectively. In the more general case of adjacent tilt and twist domains, dislocations can have a mixed character, i.e. have both edge as well as screw components.

4.6 POLARITY AND ELECTRONIC PROPERTIES.

Polarity plays an important role in the III-V compounds. The cubic {111} and the hexagonal [0001] directions are ordered by anion and cation sublayers. The polar structure of the wurtzite structure is shown in Figure 4.19. By convention, the +**c** direction is chosen in the direction of the bond from the cation (group III element to the anion (nitrogen).

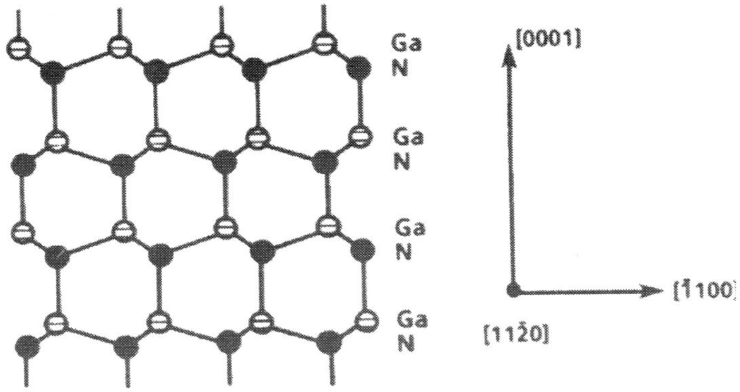

Figure 4.19 Polarity in the wurtzite structure. The [0001] direction is from Ga to N.

The determination of polarity in binary alloys was first carried out using the technique of anomalous dispersion of X-rays. More recently, convergent beam electron diffraction (CBED) has been used to determine point and space groups based on the symmetries of zone-axis diffraction patterns. The dynamical nature of electron diffraction ensures that CBED patterns are sensitive to the polarity of the material. This technique has been applied to GaN films grown on sapphire and on GaN bulk crystals. It consists of producing characteristic diffraction patterns and comparing the experimental patterns with theoretically simulated patterns and finding a match over a range of parameters.

Figure 4.20a is a TEM image showing the structure of a homoepitaxial film grown by MOCVD on the flat side of a GaN crystal (Ponce et al, 1996c). Films grown in this manner were found to have very few dislocations (rarely observed in the TEM, estimated at $10^6 cm^{-2}$) and sharp photoluminescence characteristics. The CBED pattern and calculation are shown in Figure 4.20 b and c. Coincidence is found for the experimental and theoretical results corresponding to the +**c**-direction. Figure 4.21 is from a similar study corresponding to films grown on sapphire by MOCVD. As described earlier, films grown on sapphire have a high dislocation density, and high electronic performance is achieved for growth along the +**c** direction.

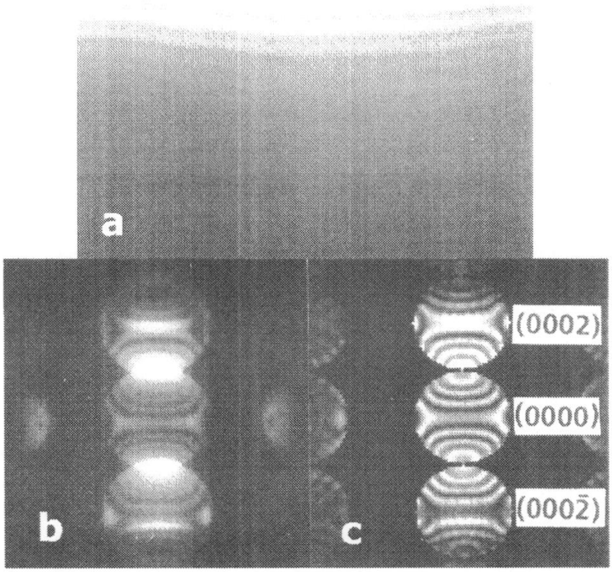

Figure 4.20 Homoepitaxial GaN: (a) TEM of epilayer, no dislocations observed. (b) experimental and (c) calculated CBED patterns. Epitaxy is in the +c direction.

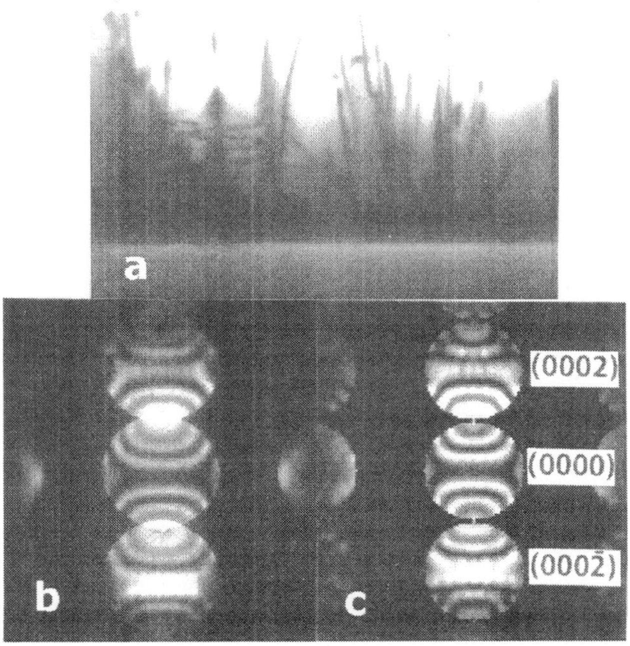

Figure 4.21 Heteroepitaxial GaN: (a) TEM of epilayer. (b) experimental and (c) calculated CBED patterns. Epitaxy is in the +c direction.

4.7 THE NATURE OF THE DISLOCATIONS.

The abundance of dislocations in high performance GaN thin films generates a number of fundamental questions regarding the nature and properties of these defects. A dislocation reflects a displacement of the lattice along the defect line. The Burgers vector, **b**, describes the lattice displacement responsible for the discontinuity at the dislocation. The properties of the dislocation are associated with the rearrangement of the interatomic array in order to maintain the atomic bonding configuration of the material while creating the lattice displacement at the core of the dislocation. The extra energy associated with the elastic distortion introduced by the dislocation is given by $E \sim \mu b^2$, where μ is the shear modulus and b is the magnitude of the Burgers vector (see, e.g., Hirth and Lothe, 1982). Direct observation indicates that dislocations propagating in the growth direction originate at the vicinity of the substrate/buffer-layer region. Short dislocation segments lying on the basal plane are observed under appropriate diffraction conditions. Oblique dislocations are also observed, and can be common under some growth conditions. By use of diffraction contrast techniques we have determined that the dislocation can have edge (**b** perpendicular to the dislocation line) and screw properties (**b** parallel to the dislocation line), as well as mixed properties. In this section, we look into more detail at the character of dislocations in GaN.

4.7.1 Determination of the Burgers Vector

The determination of the Burgers vector of a dislocation is not a straightforward operation. In the conventional diffraction contrast method, images taken under two-beam diffraction conditions are used to investigate **b** using the invisibility criterion, i.e., **g.b** = 0 when the dislocations are out of contrast (Hirsch *et al.*, 1977). Figure 4.22 is a series of TEM images taken from the same area under different diffraction conditions. It is observed that sometimes dislocations are visible and at other times they are not. Using the standard **g.b** analysis it has been shown (Ponce *et al*, 1996d) that dislocations in GaN occur with Burgers vectors parallel to **c**, **a**, and **c+a** (dislocations D1, D2, and D3 in Figure 4.22, respectively). Here, **c** and **a** are the unit cell vectors in the hexagonal wurtzite structure.

However, although this method gives the direction of **b**, its magnitude and sign remain uncertain. Cherns and Preston (1986) showed that a complete determination of **b** could be achieved using large-angle convergent beam electron diffraction (LACBED) techniques. When under LACBED conditions a Bragg line intersects a dislocation and displacement and splitting of the Bragg line can occur.

Figure 4.23 is a dark-field LACBED image taken from dislocations D1, D2 and D3 in Figure 4.22 (Ponce *et al*, 1996d). Analysis of the splitting of the Bragg lines allows us to determine the magnitude as well as the direction of the Burgers vectors for dislocations in GaN, establishing that the Burgers vectors are the full

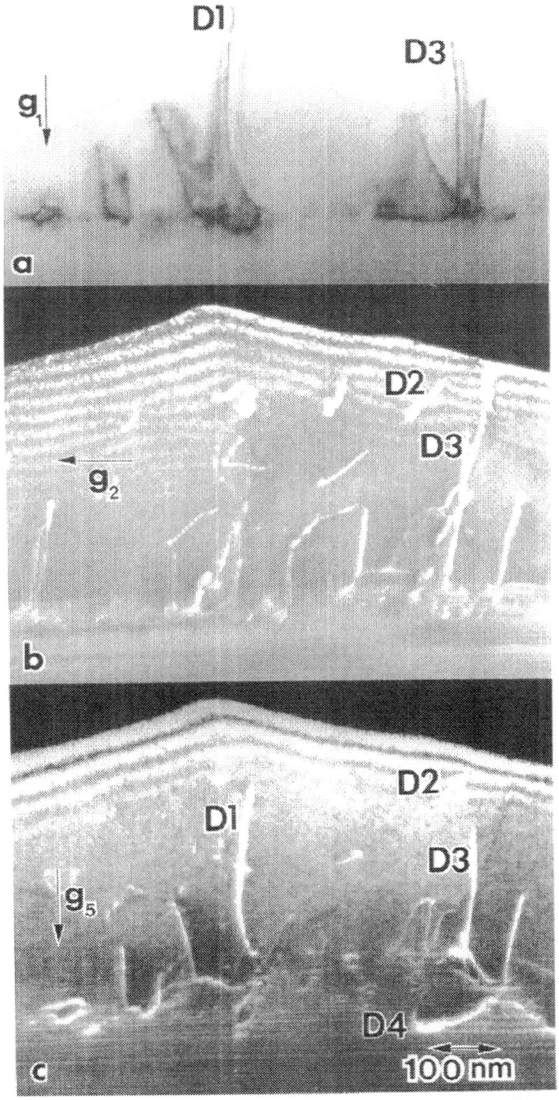

Figure 4.22 Diffraction contrast analysis of dislocations in homoepitaxial GaN along the –c-face (nitrogen on top). (a) Bright field images with g_1 = [0002] close to the [1-100] pole. (b) Dark-field image with g_4 = [01-10]. (c) Dark-field image with g_5 = [1-102] (Ponce *et al*, 1996d).

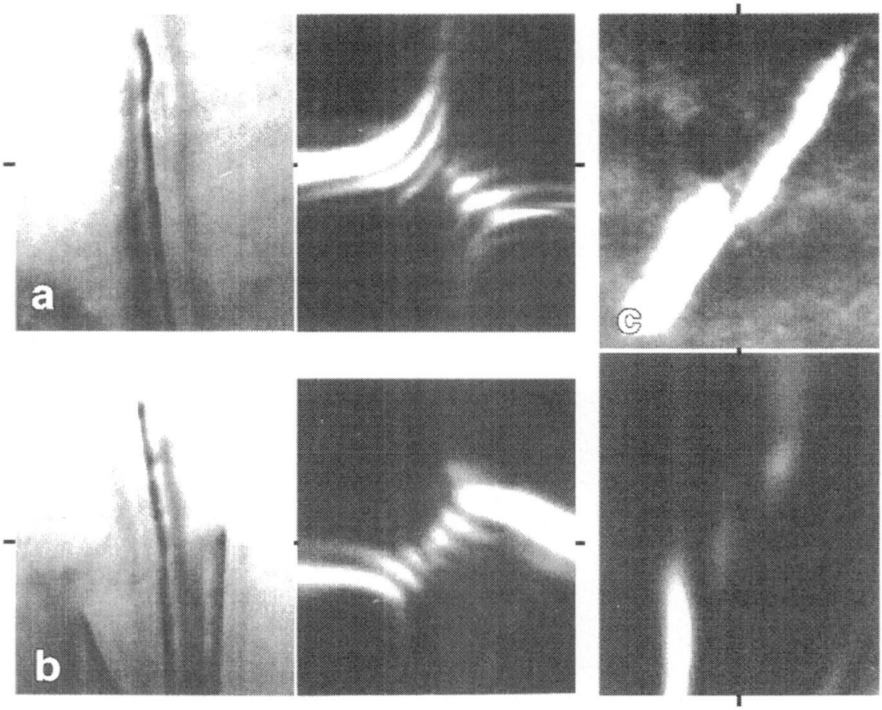

Figure 4.23 Large-angle convergent beam electron diffraction (LACBED) images to determine the magnitude of the Burgers vectors of the dislocations in Figure 4.24. (a) screw-dislocation under two-beam illumination (left) and (0006) Bragg line showing 5 extra lines at the split. The full value of b = [0001] is determined by dividing the indices of the Bragg line by the number of extra lines plus one. Similarly (b) shows a mixed dislocation c+a, and (c) an dislocation with b = a (Ponce et al, 1996d).

values of **a**, **c**, and **a+c**. This determination is significant because (a) dislocations could be in dissociated forms, as often observed in other semiconductors; and (b) the values of **c** and **a+c** are relative large when compared with the interatomic bond lengths which could lead to unstability of the core, as discussed later.

Table 4.4 Magnitude of the Burgers vectors for dislocations observed in the nitrides (in Å).

Burgers vectors	GaN	AlN	InN
a = 1/3 <1-210>	3.186	3.111	3.545
c = [0001]	5.178	4.979	5.703
c+a = 1/3 <11-23>	6.080	5.871	6.715

The combination of large Burgers vectors (see Table 4.4) and the relatively large elastic moduli for the nitrides indicate the potential instability of the dislocation core and the need for deeper understanding of the nature of the dislocation structure in the nitrides.

4.7.2 Nanopipes and Inversion Domains

Careful observation of the nature of dislocations in high performance GaN thin films provides with interesting features. Figure 4.24 is a bright-field image taken in two-beam diffracting conditions from a film grown by MOCVD on sapphire, which had a highly specular surface and superior electronic properties. In what appears at first glance to be an array of simple dislocations, using two-beam diffracting conditions yields a more complex microstructure.

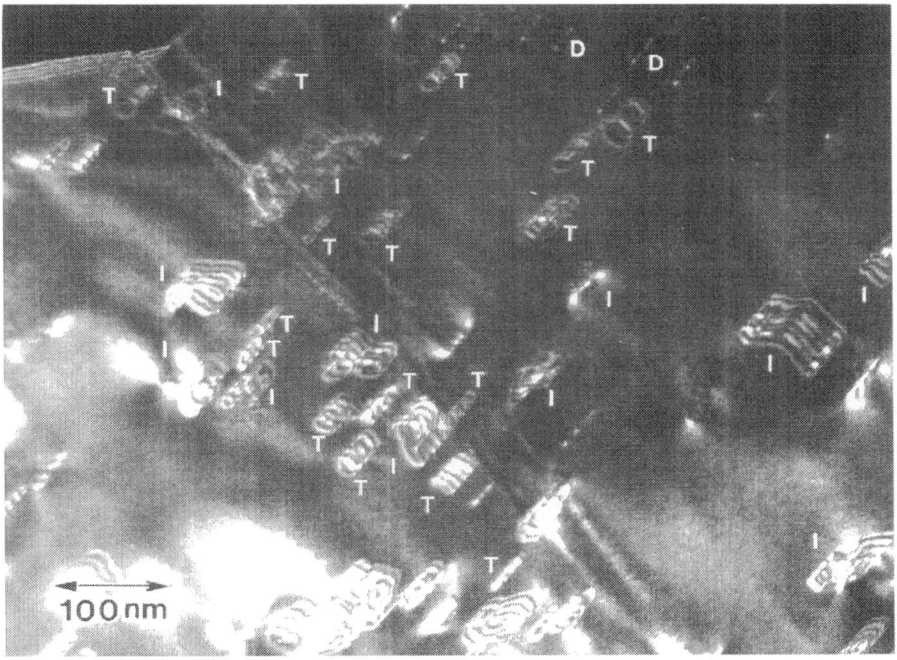

Figure 4.24 Plan-view transmission electron micrograph showing defects belong to low-angle grain boundary regions with a large density of nanopipes or empty tubes (T) and filamentary inversion domains (I). Pure dislocations (D) are also observed. This dark-field image was taken under two-beam diffracting conditions in g = 2-202. Dislocation density is ~ 1.5 x 10^{10} cm^{-2}. Only dislocations of screw-character are observed in this image.

Within the low-angle columnar structure in Figure 4.24, we observe a number of dislocation-like defects that can be identified by the diffraction effects due to the local distortion that they produce in the lattice. In addition to the standard dislocations (labelled "D" in the figure), two new types of defects are observed which are labelled "T" and "I" (Ponce *et al*, 1997b).

The defects labelled "T" are screw dislocations with empty cores. These empty tubes have a facetted configuration that is observable in plan view (Figure 4.25), with diameter ranging from 5 to 25nm. Similar observations of empty cores at dislocations in GaN have been reported by Quian *et al* (1995). The absence of a dislocation core can be explained following Frank's (1951) prediction that a dislocation whose Burgers vector exceeded a critical value should become unstable and develop a hollow tube at the core. Comparing the elastic strain energy released by the formation of a hollow core against the energy of formation of the free surfaces, Frank showed that the equilibrium core radius for the dislocation is $r = \mu b^2/8\pi^2\gamma$, where γ is the surface energy. However, using existing data it can be shown that **b** should exceed 1nm in order for hollow tubes to replace the core of the dislocations. Therefore, we feel that purely from elastic energy considerations, these nanopipes cannot be fully justified (Cherns *et al*, 1997).

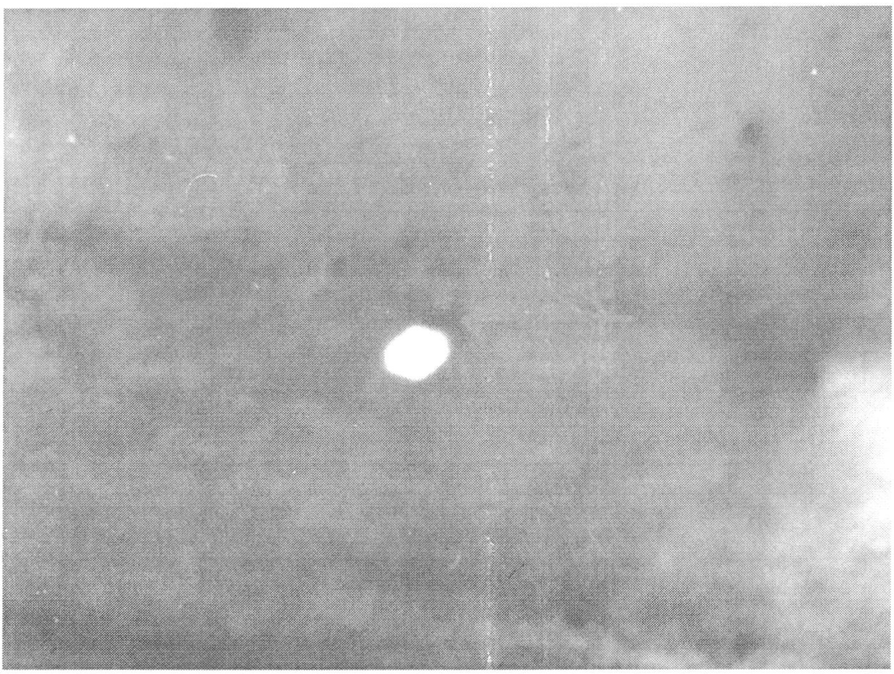

Figure 4.25 Transmission electron micrograph of nanopipe observed along its axis. It presents an irregular hexagonal cross section of about 2.4 nm in diameter.

The characteristics of the lattice displacements have been examined using LACBED. It has been demonstrated that the nanopipes are coreless screw dislocations with a Burgers vector of [0001], which for GaN has a magnitude of 0.5185nm. This screw-character of the nanopipes is independent of the tube diameter, which are too large to fit the simple form of Frank's equilibrium theory. To explain these observations, a model for the formation of nanopipes has been proposed, and is shown in Figure 4.26. It is proposed that the nanopipes grow from pinholes formed in the growing GaN epilayer as islands coalesce in the early stages of growth (Cherns et al, 1997). Trapped dislocations appear when the net displacement of the islands does not add up to zero.

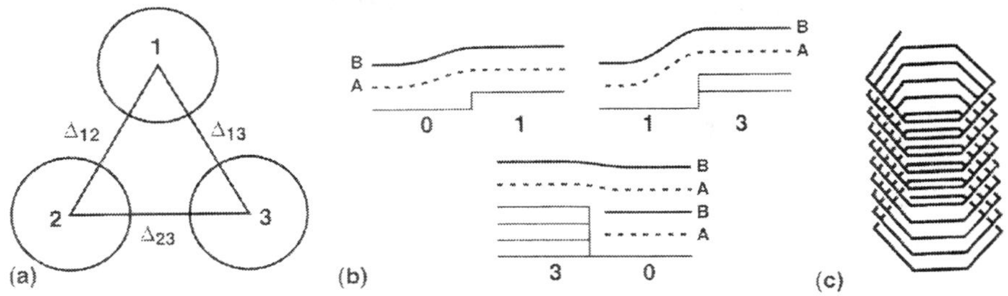

Figure 4.26 Proposed mechanism for the generation of nanopipes in epitaxial GaN/sapphire. (a) A dislocation is trapped if the island displacement Δ = Δ12+ Δ23 + Δ31 do not sum to zero. (b) Steps on the sapphire surface in units of c-sapphire/6 can join to trap a screw dislocation. (c) Addition of deposited materials near a nanopipe containing a screw dislocation preserves the helicoidal nature of the defect about its central [0001] axis (Cherns et al, 1997).

Another important observation is that neither the a nor the a+c dislocations are observed to be coreless. This is easy to understand since a component of the dislocation displacement along the basal plane would prevent the formation of a nanopipe. Coreless, screw-dislocations are stable only when propagating exactly along the c-direction. Bending of the defect will cause the nanopipe to close and to become a standard screw-dislocation.

Another important class of defects appears labelled "I" in Figure 4.24. These are filaments of GaN with inverted polarity. A detailed analysis has been carried out by CBED to ascertain the inverted direction of the lattice. LACBED has been used to characterize the nature of the inversion domain boundaries, which have been shown to be coherent (Cherns *et al*, 1998).

The existence of nanopipes and inverted domains alongside with the standard dislocations adds much complexity to the microstructure of MOCVD GaN. It may be that such complexity is necessary in order to explain the relative absence of thermal stresses in high-performance GaN thin films. There is some unpublished evidence that dislocations move at relatively low temperatures (even as low as ~250°C). Such dislocation motion would require climb and glide mechanisms that are difficult to justify given the high melting temperature of GaN. In the presence of internal surfaces (nanopipes), such mechanisms would be quite possible. The density of these defects should be closely related to the early stages of growth, including the thickness of the buffer layer. Nanopipes above a certain diameter may be subject to chemical attack, and be visible as pits after chemical etching (the pit density is typically two orders of magnitude less than the dislocation density). The existence of coreless dislocations in the proximity of the metallized top surface of a device may have negative effects on the lifetime of laser diodes, allowing metal diffusion into the p-n junction with the resulting shorting of the device.

4.8 SPATIAL VARIATION OF LUMINESCENCE.

The relationship between microstructure and light emission can be determined by using cathodoluminescence (CL) techniques in a scanning electron microscope (SEM). CL probes the optical properties of semiconductors in a similar manner as photoluminescence, the main difference being that electrons rather than photons are used to excite the material. The ability of focusing an electron beam allows the probing of the microscopic variations of the optical properties with sub-micron resolutions. The standard point resolution limit in SEM is below 10nm for accelerating voltages in the range of 1 to 15 kV. Upon entry into the material, the electrons are stopped with the production of electronic excitation within a volume that depends on the energy of the electron beam. The lower the electron beam energy the less is the volume of excitation within the semiconductor and the higher the spatial resolution of the probe. The influence of the surface is one of the limiting factors that become noticeable at voltages below 1 kV. The actual signal in CL originates from a range of depths, where the maximum depth R depends on the energy of the primary electron beam, following the relation $R \propto E^n$, with n = 1.5-1.7 (see, e.g.,Yacobi and Holt, 1990).

Using a monochromator within the CL system, images are produced of light emitted within a range of wavelengths. Thus any significant region of the spectrum can be mapped with high spatial resolution (Ponce et al, 1996a). In CL microscopy the intensity of the signal is mainly associated to the light emitted from the region excited by the probe. While the primary photons may be absorbed and may excite other emissions from volumes outside the region excited by the probe, the signal is due to excitations originating from the specific region of the material irradiated by the primary beam. Internal reflection and interference may affect the detected intensity by changing the fraction of the total emitted light that reaches the photodetector. However, these effects will not distort the shape of the CL-SEM image, which can be thought of as a map of the efficiency of luminescence excitation by the scanned electron beam at the selected range of wavelengths.

4.8.1 Undoped Material

Figure 4.27 is the room-temperature CL spectrum of an unintentionally doped layer with high specular morphology. These layers were grown under standard conditions involving a GaN buffer layer grown at 550°C followed by growth at 1050°C to a total thickness of 4μm. The spectrum consists of two peaks, one situated at the band-edge region (λ~364nm), and a broader peak centered in the yellow region (λ~559nm). The origin of the yellow luminescence band has been the subject of much study. Two models have been suggested to explain the recombination mechanism responsible for this transition. Ogino *et al* (1980) proposed a transition from a shallow donor to a deep acceptor level at $E \approx E_v + 0.86$ eV. Glaser *et al* (1995) concluded from optically detected magnetic resonance experiments that the yellow transition involves a deep donor at $E \approx E_c - 1.0$eV and therefore a shallow acceptor at $E_v = 0.2$ eV. More recently,

Van de Walle and Neugebauer (1997) have argued that gallium vacancies are sources for yellow luminescence based on first-principles calculations.

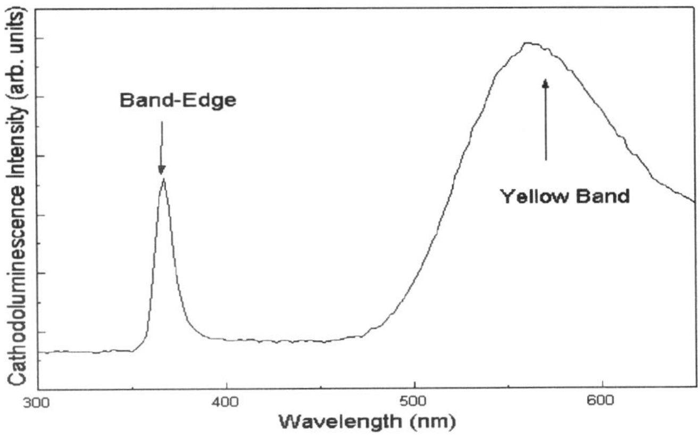

Figure 4.27 Cathodoluminescence spectrum of unintentionally-doped GaN grown by MOCVD. The CL spectrum was taken at room temperature using a 15 kV electron beam.

Figure 4.28 shows CL images taken from GaN films grown by MOCVD under conditions of exposure to ammonia at the initial thermal treatment of the sapphire surface. The resulting hexagonal crystallites have diameters ranging from 10 to 50μm. The crystallites typically have low defect densities and exhibit dislocations at the grain boundaries and in the midst of the hexagonal geometry (see e.g. Rouviere *et al*, 1996). The surface morphology is shown in Figure (a). The light emission at the band-to-band region (λ=364nm) appears in (b) and the emission in the yellow band region (λ=559nm) is shown in (c). The yellow band image coincides with the dislocation distribution observed by TEM.

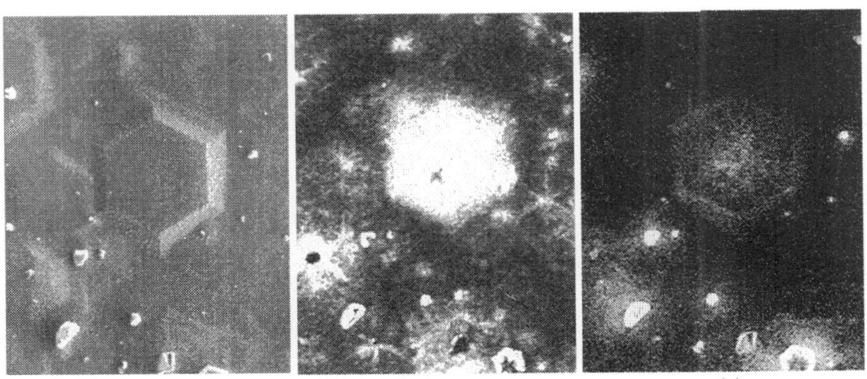

(a) (b) (c)

Figure 4.28 Undoped layer with large hexagonal crystallites. (a) secondary electron image, (b) CL at 364 nm, and (c) CL at 559nm. Strong band-to-band luminescence is observed in the large crystallites. Yellow luminescence originates from the central region of the crystallites as well as from the boundaries. The lateral width of each image is 150 μm (Ponce *et al*, 1996a).

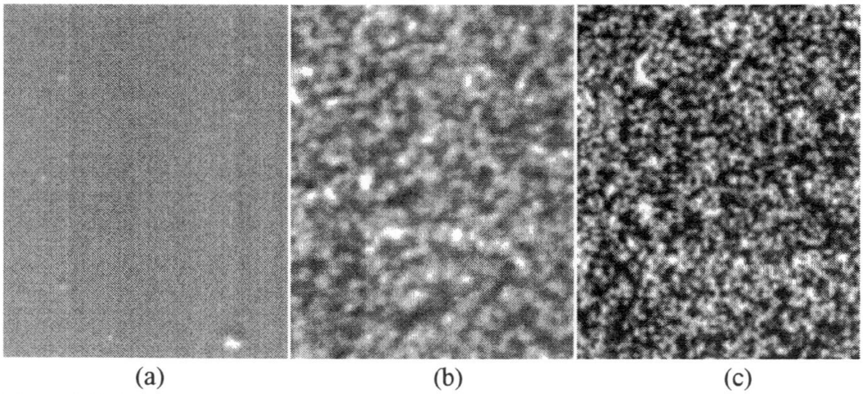

Figure 4.29 Undoped layer with specular surface morphology. (a) secondary electron image, (b) CL at GaN band-edge, λ = 364 nm, and (c) CL at yellow band, λ = 559 nm. CL images exhibit nonuniform intensity distribution, with spatial variations of the order of 0.5 μm. The lateral width of the images is ~24μm (Ponce *et al*, 1996a).

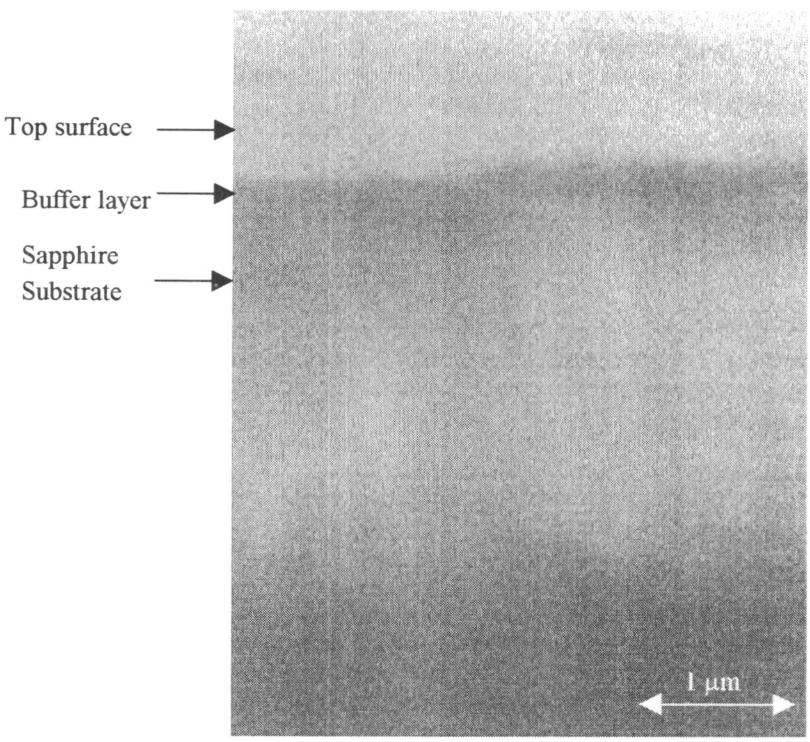

Figure 4.30 Cross-section CL image of non-intentionally doped sample shown in Figure 4.29. Dark lines resemble dislocation distribution in cross section TE.

The spatial variation of luminescence in smooth films of unintentionally doped GaN is shown in Figure 4.29. In contrast with the preceding case, the sapphire surface had not been exposed to ammonia. The rocking curve full-width at half-maximum was 6 arcmin. The secondary electron image in Figure 4.29a demonstrates the featureless planar morphology of the film. Light emission at the band-edge region (λ = 364nm) in Figure 4.29b is characterized by a granular appearance, where the bright spots represent regions of high CL emission intensity, and dark spots represent low emission intensity. The bright spots are distributed with separation ranging between 0.3 and 1.0μm. The smallest separation that is observed among features in this image is 0.3μm. Light emission in the yellow luminescence band (λ = 364nm) is shown in Figure 4.29c, where a granular structure is also observed but with smaller features than in the band-edge emission images.

The variation of luminescence along the thickness of the material can be observed by looking at a cross-section specimen, such as those prepared for transmission electron microscopy. Figure 4.30 is a CL image taken for λ = 364nm. Dark lines running longitudinally along the growth direction are observed. The morphology and density of these lines resemble the dislocation arrangement observed by TEM, for example in Figure 4.10.

These observations indicate that in undoped GaN films the yellow luminescence is spatially associated with the dislocation distribution. This leads to the hypothesis that the dislocations that are necessary to maintain local coherence in the columnar, low-angle domain boundaries, are associated with the yellow luminescence. The states in the gap that produce the deep-level (yellow) luminescence either arise directly from the atomic structure of the dislocations, or are associated with the clustering of negative point defects (vacancies, interstitials) or impurity atoms (such as oxygen and carbon) due to the strain and electrochemical lattice distortions produced by the presence of dislocations.

4.8.2 Doped Materials

The inhomogeneous spatial variation of CL in lightly doped materials (~10^{17} carriers/cm^3 at room temperature) is absent in undoped materials (~10^{16} cm^{-3}). Figure 4.31 shows the surface morphology and CL characteristics of a GaN thin film doped with silicon (~2x10^{17}cm^{-3}), grown under the same conditions as the undoped film in Figure 4.29. The secondary electron image in (a) is indicative of a specular, featureless surface. The CL band-edge emission image in (b) shows spatially-homogeneous luminescence. The yellow luminescence in (c) is significantly weaker and exhibits a high degree of uniformity down to the sub-micron range. We have also observed similar uniformity in light emission in p-type (Mg-doped) GaN films.

The uniformity of the luminescence in lightly doped samples has some interesting implications about the role of dopants in the GaN lattice. It suggests some type of interaction between dislocations and the impurities in the neighbouring lattice. The following processes could be responsible for these observations:

(a) Doping impurities such as silicon and magnesium may segregate at the dislocation core, producing an atomic coordination suitable for the elimination of electrically and optically active lattice sites such as dangling bonds. Si_3N_4 and Mg_3N_2 at the core of the dislocations could have an insulating effect. Electronic effects due to the dislocation core would disappear as a result of the chemical reaction.
(b) Compensation may occur between dopants and unwanted impurities segregated at the vicinity of the dislocation core. The mechanical and electrochemical driving forces generated by the dislocation displacement would cause such segregation.
(c) Screening of the electronic states associated with the distribution of free carriers. This is consistent with the small depletion widths that have been observed in doped materials.

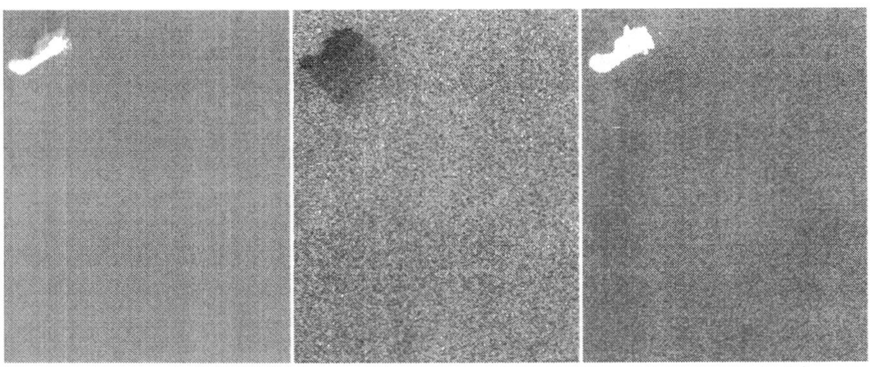

(a) (b) (c)

Figure 4.31 Silicon-doped GaN layer with specular surface morphology. (a) secondary electron image, (b) CL at GaN band-edge, $\lambda = 364$nm, and (c) CL at yellow band, $\lambda = 559$nm. The lateral width of the images is ~24μm. The upper left particle is purposely included as a reference.

4.9 MICROSCOPIC PROPERTIES OF $In_xGa_{1-x}N$ QUANTUM WELLS.

$In_xGa_{1-x}N$ thin films are used in the active region of devices. Their smaller bandgap allows light emission in the visible range (red to blue, see Figure 4.2), and increases the carrier and optical confinement with improved performance of optoelectronic devices. Particular attention has been placed recently on the properties of extremely thin $In_xGa_{1-x}N$ quantum wells.

The growth of high quality quantum wells requires the production of interfaces that are atomically planar and abrupt, with high degrees of chemical homogeneity and crystalline quality. This would be expected to be very challenging in the case of the nitride semiconductors, where the lattice of the underlying GaN film contains a high density of dislocations. In addition, the growth of $In_xGa_{1-x}N$ requires different growth conditions than those of the underlying GaN epilayer. Lower growth temperatures (700-800°C for InGaN vs. 1000-1100°C for GaN), and much higher V/III ratios are necessary in order to achieve significant indium concentration (Nakamura and Mukai, 1992). The incorporation of substantial amounts of indium is inhibited by reported miscibility gaps in the ternary composition diagram (Ho and Stringfellow, 1996). Growth is also inhibited by high indium partial pressure at growth temperature, preferential etching of indium by H_2 and NH_3, and the propensity to form indium droplets under low NH_3 cracking efficiency at reduced growth temperatures.

Studies on light emission mechanism in $In_xGa_{1-x}N$ single quantum wells suggest that electroluminescence may be due to recombination of excitons localized at certain potential minima in the quantum wells (Chichibu et al. 1996). One of the possible interpretations of those observations is that InGaN quantum wells are not uniform, and that they may behave more like an array of quantum dots (Narukawa *et al*, 1997a,b; Nakamura *et al*, 1997). Another possibility is the piezoelectric effect associated with the severely strained pseudomorphic layers (Wetzel et al, 1998).

4.9.1 The Nature of the InGaN/GaN Interface

Frank and van der Merwe (1949) predicted that in heteroepitaxy, with a difference in lattice parameter, an epilayer initially grows pseudomorphically, with the epilayer lattice elastically strained to match the size of the substrate lattice. The elastic strain energy increases with film thickness until it is comparable to the energy required to create misfit dislocations. In the formulation by Frank and van de Merwe, the energy of a relaxed interface (including the network of dislocations) is given by the relationship: $E_I \approx 9.5\ f\ G\ b/4\pi^2$, where $f = 2\ (a_f-a_s)/a_f+a_s)$ is the lattice mismatch (a_f and a_s are the in-plane lattice parameters of the film and the substrate, respectively), G is the bulk modulus, and b is the Burgers vector. The strain energy of the film (before plastic relaxation) is given by $E_h = 2\ G\ (1+\nu)\ f^2\ h\ /\ (1-\nu)$, where ν is the Poisson modulus (~ 0.25), and h is the film thickness. A critical thickness is reached when $E_I = E_h$: $h_c \approx (9.5/8\pi^2)\ (b/f)\ (1-\nu)/(1+\nu) \approx 0.072\ a_s/f$. Table 4.5 shows the variation of misfit in $In_xGa_{1-x}N$/GaN junctions as a function of indium composition, and the corresponding values for the critical thickness and misfit dislocation separation. In-depth studies of the critical thickness for the generation of misfit dislocations in specific systems have been reported (People and Bean, 1985; Hull *et al*, 1986; Bauer and van der Merwe, 1986).

Table. 4.5 Lattice mismatch between $In_xGa_{1-x}N$ and GaN. *x* is the indium composition, *a* is the lattice parameter on the basal plane. The critical layer thickness and the misfit dislocation separation for a fully relaxed interface is given in Å.

x	*a* (Å)	% misfit	Critical Thickness	Dislocation Separation
0.0	3.186	-	-	-
0.1	3.222	1.07	21.4	286
0.2	3.258	2.14	10.7	145
0.4	3.329	4.28	5.36	74
0.6	3.401	6.42	3.57	50
0.8	3.473	8.56	2.68	39
1.0	3.545	10.7	2.14	27

The growth of $In_xGa_{1-x}N$ on GaN involves a large lattice mismatch. For x>0.1, the critical layer thickness for plastic deformation is below 2nm, i.e. of the order of the thickness of the quantum well. For x>0.4, the critical layer thickness is below the value of the lattice parameter, and for x>0.8 it is below a monolayer. We should therefore expect most quantum wells with thickness greater than 1nm and x>0.2 to be plastically relaxed, and we should expect to observe misfit dislocations at the interface boundaries. In fact, however, relaxed interfaces are seldom observed in MOCVD grown InGaN thin films on GaN.

Plastic relaxation relies on the motion of dislocations from the free surface of the layer. Dislocations glide along the slip planes following the lateral elastic forces caused by the pseudomorphic growth. In the cubic zincblende structure, dislocations follow <110> directions and glide along the four {111} planes, that are configured along the faces of a regular tetrahedron. The slip planes in the wurtzite structure are shown in Figure 4.32. The known slip systems in the wurtzite structure are <11-20>(0001), [0001]{10-10} and <11-20>{10-12}; that operate on the (0001) basal planes, the prismatic planes, and the pyramidal planes, respectively. The basal plane is the most commonly observed slip system. InGaN grows pseudomorphically well beyond the critical layer thickness in Table 4.5. This indicates that the slip systems that are oblique to the basal planes are not activated under the conditions existing during epitaxial growth.

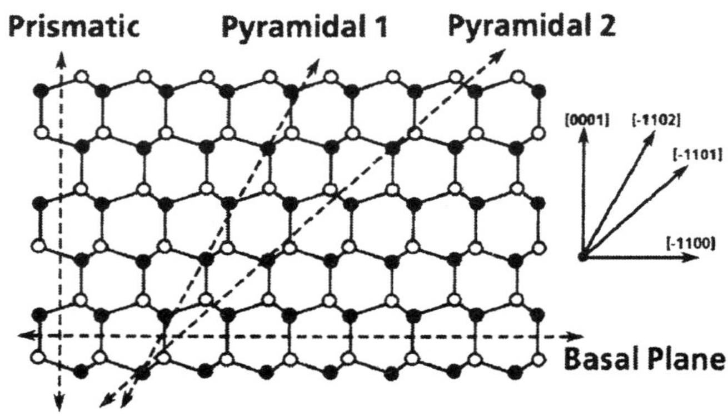

Figure 4.32 Slip systems in the wurtzite structure. The (0001) basal plane is the predominant slip system. Other slip systems are the prismatic {10-10}, and the pyramidal {10-11} and {10-12} planes.

Figure 4.33 shows the photoluminescence spectra of two single quantum wells emitting in the blue and green region of the electromagnetic spectrum (Ponce *et al*, 1998). The films were grown by metalorganic chemical vapor deposition on (0001) sapphire substrates, using conventional GaN with a 2 μm-thick Si-doped GaN film grown at 1050°C. The $In_xGa_{1-x}N$ layers were deposited at about 800°C, and capped by GaN layers 30nm thick. The top layer was kept intentionally thin in order to facilitate the optical and structural characterization of the samples. The quantum wells and the top GaN layer were not intentionally doped. The spectra were taken at a sample temperature of 2K under excitation with a HeCd laser at 325nm (with an intensity of ~ 1W/cm^{-2}). For each case, the spectrum consists of a peak at 358nm resulting from excitonic recombination in the GaN capping layer. A second dominant peak appears at longer wavelength, originating from the quantum wells with the maxima occurring at 450nm and 505nm for the respective cases. The peak position was used to estimate the indium content. Using a bowing parameter (Nakamura and Mukai, 1992) of 1, the analysis yields the values of x=0.28, 0.52, respectively. Phonon replicas are observed as shoulders on the long wavelength side.

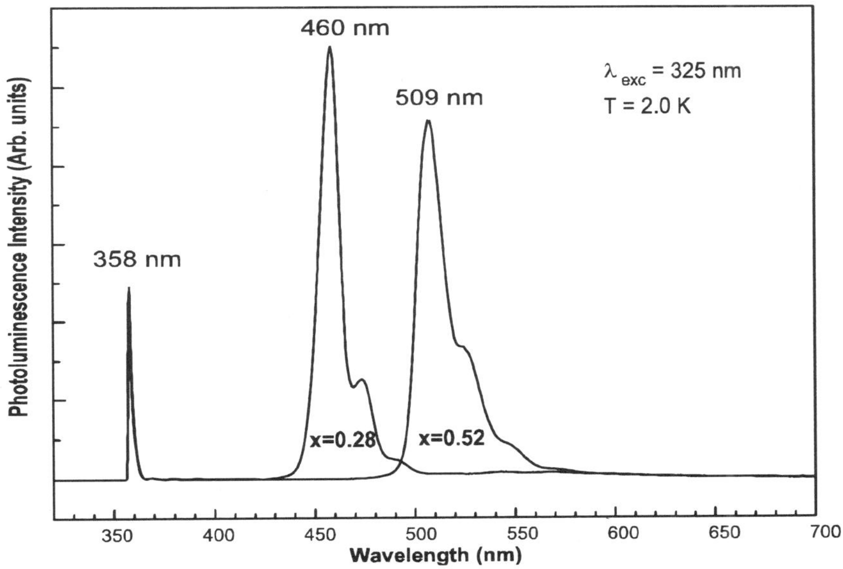

Figure 4.33 Photoluminescence spectra of $In_xGa_{1-x}N$ single quantum wells with x=0.28, and 0.52.

4.9.2 Microstructure of Quantum Wells

Figure 4.34 shows a high resolution TEM image of an $In_xGa_{1-x}N$ single quantum well with x=0.28, viewed along the [1-100] direction. The $In_xGa_{1-x}N$/GaN interfaces do not appear to be planar at a manometer scale. The average QW thickness is 1.5nm (corresponding to three basal planes of the wurtzite structure), but the quantum well presents a globular pattern with "spot" diameters of about 2.5nm, giving the impression of a connected pattern of spheroids. High-resolution lattice images such as these are meaningful only in the very thin regions of the TEM specimen, such as in Figure 2 with an estimated 6nm thickness from the position of thickness extinction contours in the wedge-shaped TEM specimen. Observations in thicker regions of this specimen give the appearance of a more homogeneous layer, consistent with thickness averaging. TEM observations of this type have been reported in the literature (Narukawa et al, 1997a). However, some questions arise about this method for determining the roughness of the quantum well on the grounds of possible artifacts due to sample preparation (possible preferential ion etching produces indium droplets in the TEM sample surface, and could be responsible for the observed contrast). Hence there is a need for further work to assess the effects of sample preparation (Ponce *et al*, 1998a)

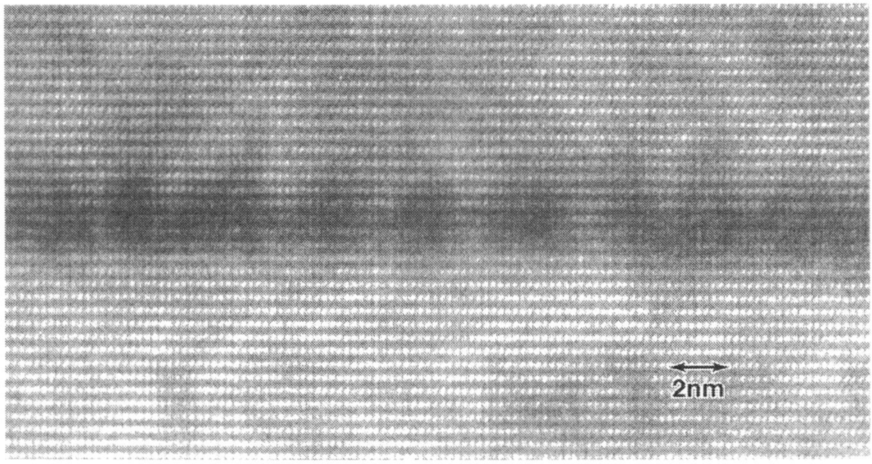

Figure 4.34 Lattice image of an $In_xGa_{1-x}N$ single quantum wells (x=0.28). Non-uniformity in the thickness could be due to variation in indium composition, or to artifacts due to TEM specimen preparation.

Figure 4.35 shows a diffraction contrast image of the same specimen, taken under conditions of high tilt (about 28 degrees towards the c-axis). Interference fringes delineate the tilted quantum well. These are the result of the net displacement of GaN planes across the quantum well due to the difference in the c-spacing of $In_xGa_{1-x}N$ and GaN. This results from an ~3% larger lattice parameter in the $In_xGa_{1-x}N$ layer plus an outward relaxation of the $In_xGa_{1-x}N$ due to pseudomorphic in-plane lattice compression. The specimen is about 170nm thick at the QW, and the fringes have a depth periodicity of about 22nm. Their parallel nature and uniformity in the figure indicates that the quantum well is uniform to within 5nm. Surface artifacts due to ion milling are seen in this figure at the thinnest portion of the specimen, with dimensions of the order of 3nm. This is unfortunate, since they coincide with the observed SQW roughness in the cross-section in Figure 4.34. Misfit dislocations at the $In_xGa_{1-x}N$/GaN interface are easily observed under these conditions, and one appears slightly to the right of the middle of the figure. Threading dislocations are observed in the lower portion of the figure. These dislocations generally propagate along the c-direction and belong to the tilted columnar structure typical of GaN epitaxy. One such dislocation crosses the QW plane at the right hand side of the figure. The number of fringes observed in the image depends on the relative tilt of the crystal lattice. In Figure 4.35, the number of fringes is 8 at the left and 7 at right hand sides of the dislocation that crosses the QW, respectively. For $s \gg 1 / \xi_g$, where s is the deviation parameter and ξ_g the extinction distance, the depth periodicity (fringe spacing) is inversely proportional to the deviation parameter (Hirsch *et al*, 1965), with 1.9 arcmin/fringe for the conditions present in this image. However, no noticeable discontinuities in the QW plane have been observed at tilted columnar boundaries. This is consistent with high-resolution images where no deviations are observed in the boundary projection.

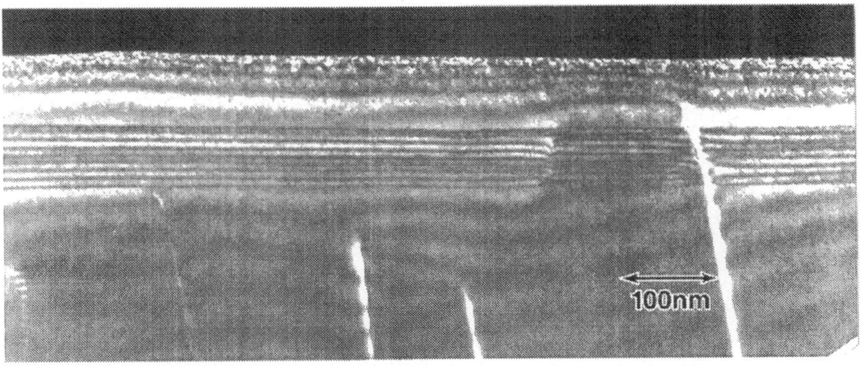

Figure 4.35 Cross-section image of SQW in Figure 4.32, tilted 28 degrees on the vertical plane perpendicular to the image. The quantum well interferes with the top surface giving rise to a series of pseudo parallel fringes. A misfit dislocation crosses the quantum well at the right, and causes an abrupt rotation of the lattice (~1.8 arc-min) that is reflected in the number of fringes in either side of the defect.

To avoid ion-milling artifacts, plan-view specimens were used without ion thinning. Larger areas are accessible in plan-view specimens versus cross-section specimens. Observations in plan-view indicate that the microstructure of the quantum wells with large nominal values of x is highly non-uniform. Figure 4.36 is a weak-beam dark-field image taken of one of these dislocated regions in a specimen with x=0.52. The wedge-shape allows the comparison of regions with and without the quantum well. At the top-right side of the figure, the GaN top layer is observed (almost featureless). Bright features are observed when the thickness (depth) of the specimen incorporates the quantum well layer. The imaging conditions are such that slight distortions of the GaN lattice planes produce a bright contrast. Arrays of dislocations are observed (e.g. at lower left of figure), which show enhanced contrast at their intersection, indicative of an edge nature, typical of misfit dislocations. The dislocation separation in the figure is about 35nm. In addition to dislocations, strained regions are observed as clusters that range from 3nm to over 10nm. The large bright regions in Figure 4.36 are in fact conglomerates of smaller clusters (that can be seen at longer exposures of the photographic plate). Small dislocation loops (diameters smaller than 5nm) have also been observed. Figure 4.37 shows a larger area, with similar features to Figure 4.36. It is observed that regions with similar diffraction contrast are typically a fraction of a micron across. A few such regions are observed in Figure 4.37. The distribution of small contrast features is however non-uniform, with some areas only showing contrast from misfit dislocations. The dislocation separations for fully relaxed InGaN/GaN interfaces are 5.8 and 11.1nm for x=0.52 and 0.28, respectively. The dislocation separation in Figure 4.36 is ≥35 nm, and varies across the specimen, indicative of incomplete and inhomogeneous plastic relaxation of lattice mismatch. Thus, these observations indicate that the QWs are not fully relaxed. The presence of dislocations only in certain regions implies that nucleation and propagation is not the same in all regions of the specimen.

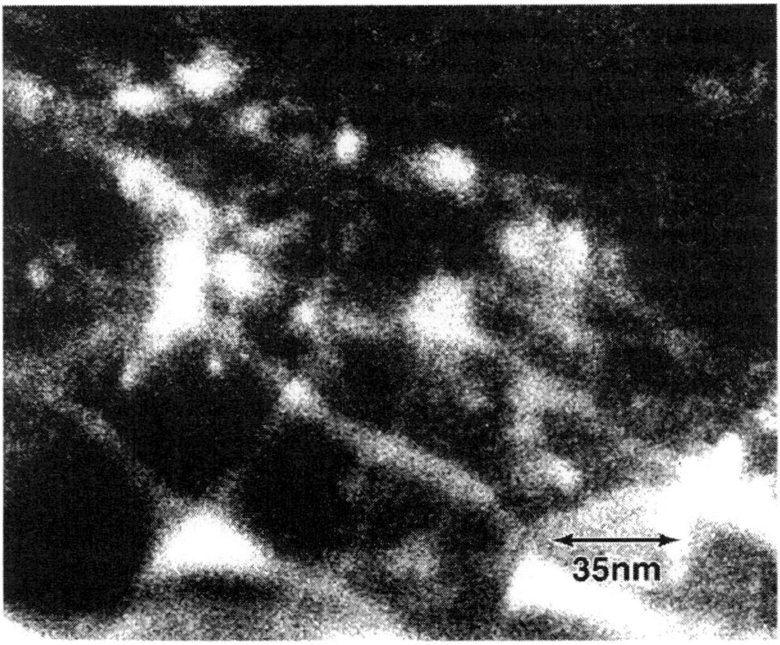

Figure 4.36 Plan-view dark-field image of $In_xGa_{1-x}N$ single quantum well with x=0.52. The sample thickness decreases towards the top-right corner. The top-right corner does not include the quantum well. Misfit dislocations and strain clusters are observed with 3 to 5 nm in diameter. Clear regions may contain clusters smaller than the resolution of the technique.

These observations indicate that: (1) the $In_xGa_{1-x}N$ quantum well follows a continuous plane implying that the MOCVD growth surface is nearly atomically smooth, in spite of the large defect density of the underlying GaN film. (2) The relative orientation of the quantum well interface and the basal planes of the wurtzite structure varies with the changes in orientation of the columnar structure of the GaN film, that we know from X-ray rocking curves to be about 5 to 7 arcmin. The $In_xGa_{1-x}N$/GaN interface must contain a non-uniform distribution of atomic steps from the wurtzite lattice. Such steps can act as inhibitors of dislocation for glide on the basal plane. Thus, the inhomogeneous plastic relaxation may be associated with the distribution of orientations in the columnar structure.

The contrast clusters that are observed in high resolution and diffraction imaging conditions could have three origins: composition fluctuations, interface roughness, and strain singularities (such as misfit dislocations loops). It is difficult to distinguish between strain caused by interface roughness and strain caused by composition fluctuation, since both may produce similar bending of the lattice planes. The specimen in Figures 4.34 and 4.35 was ion milled, and some contrast due to the ion-milled surface is visible as dark blurry spots throughout the image that perhaps could be attributed to the coalescence of indium atoms at the specimen surface. However, the contrast at the quantum well region in the high resolution lattice images is significantly higher than the variations in contrast in the field due to ion milling artifacts. Thus, the contrast is highly suggestive of inhomogeneities in the quantum well composition.

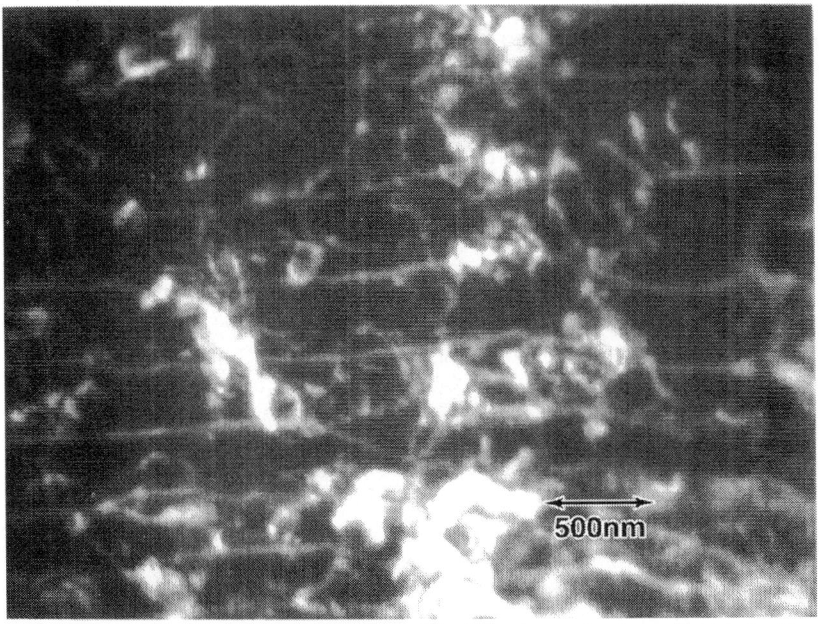

Figure 4.37 Plan-view dark-field image showing misfit dislocations and regions with high densities of strain clusters.

The microstructure of quantum wells is a fascinating problem to study. In principle, complete plastic relaxation of the lattice mismatch should be expected since critical thickness is achieved after only a few monolayers of $In_xGa_{1-x}N$ for x >0.2. With misfit dislocations expected at ~4 to 20nm separation, depending on the value of x, the dislocation density in the quantum well plane is two to three orders of magnitude higher than in the GaN epilayer. The presence of such high misfit-dislocation densities should result in variations in strain distribution that in turn should be highly sensitive to the tilt of the columnar structure in the GaN film. However, it is observed empirically that the material prefers to grow pseudomorphically, perhaps due to absence of mechanisms for plastic relaxation for growth normal to the [0001] direction in a wurtzite lattice. It can be argued that misfit dislocations can be generated at dislocations of the columnar domain boundaries, which can glide away if the strain field lies within the basal plane. If the quantum well plane is tilted with respect to the glide plane (as should be in some of the columnar domains), then the driving force for plastic relaxation will be diminished as the gliding misfit dislocation leaves the plane of the quantum well.

4.9.3 Spatial Variation of the Luminescence of $In_xGa_{1-x}N$ Quantum Wells

Given the intricate microstructure of GaN and the even more complex microstructure of $In_xGa_{1-x}N$ quantum wells, it should be interesting to probe the optical properties using cathodoluminescence. The inhomogeneous strain distribution of the preceding discussion should have a pronounced effect on the optoelectronic properties of the quantum wells.

The following CL observations (Ponce *et al*, 1998b) were performed with an electron beam accelerating voltage of 4kV and an incident beam current of <10pA. Monte Carlo simulations of the interaction of the electron beam with GaN shows that at 4kV, the maximum electron-hole pair generation was at a depth corresponding to the position of the quantum well (30nm below the surface). At this depth the lateral distribution of generated carriers was approximately 60nm. The simulations did not take into account the role of carrier diffusion to the quantum well and hence indicate the upper limit to the spatial resolution attainable under these conditions. Although the volume of excited material in the CL experiments was very small, the injection conditions were calculated to be small compared to normal LED operation.

The CL spectra in Figure 4.38 were taken under TV raster conditions (~300 μm^2) and are comparable to the photoluminescence (PL) spectra in Figure 4.33. PL data correspond to a probe area of about $1mm^2$ using the 325nm emission from a HeCd laser at 4K. The band-edge peaks for CL are weaker than from PL, due to the added depth probe selectivity of CL. The intensity and width of the quantum well peaks are similar, with 11nm and 13 nm for PL and CL, for x=0.28 respectively. The respective values for x=0.52 were 15 and 18nm. The shoulders on the red side of the spectra are phonon replica (with separation of ~91 meV).

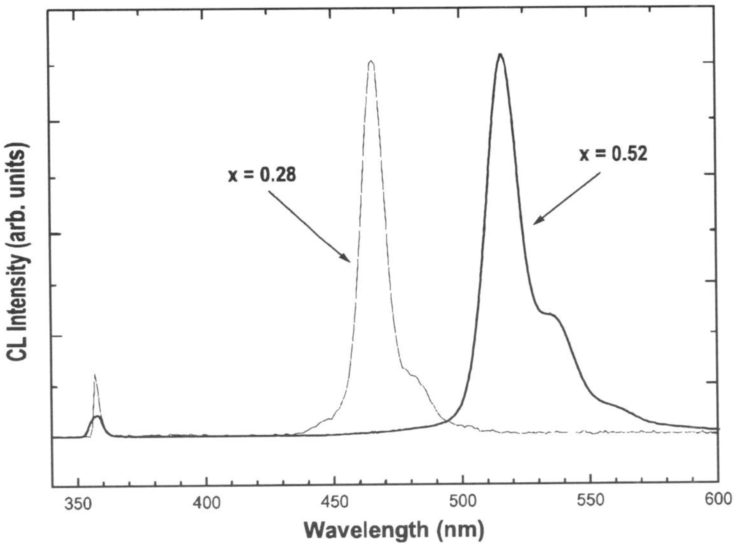

Figure 4.38 CL spectra from large scanned areas of $In_xGa_{1-x}N$ single quantum wells (x=0.28, 0.52). Data taken with an electron beam acceleration voltage of 4kV.

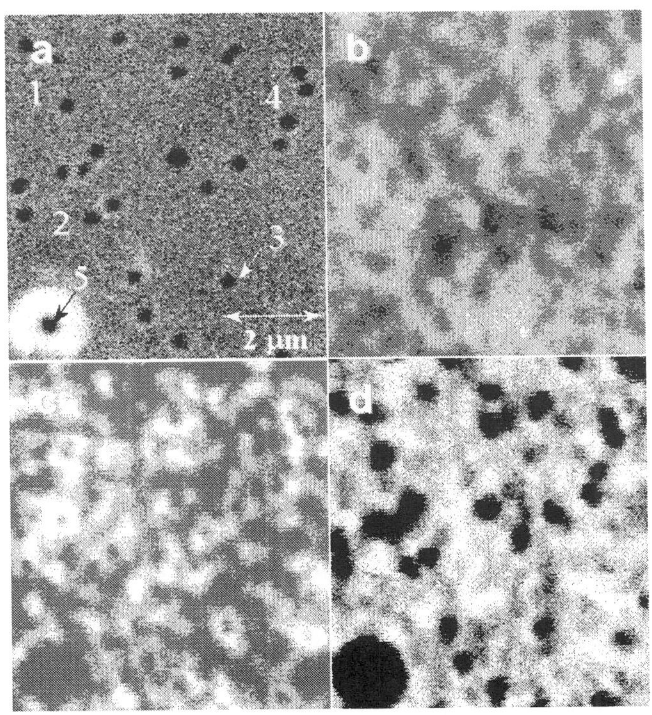

Figure 4.39 CL images of $In_xGa_{1-x}N$ single quantum wells for x=0.28. (a) Secondary electron image. (b) Band edge emission image at λ=357nm. (c) Image at λ=456nm. (d) Image at λ= 466nm.

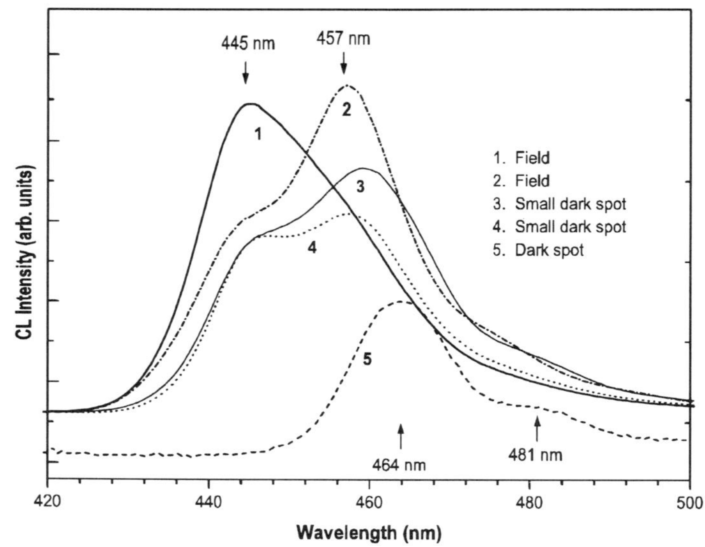

Figure 4.40 CL spectra taken in the spot mode, corresponding to spots labelled in Figure 4.39.

The spatial variation of the luminescence for x=0.28 is shown in Figure 4.36. Figure 4.36a is the secondary electron image of the region under observation. The small pinholes have diameters of the order of 40 to 80nm and are typical of epitaxy under the conditions following QW deposition. They originate from nanopipes present in the GaN epitaxy (Cherns *et al*, 1997; Ponce *et al*, 1997b; Liliental-Weber *et al*, 1998). Some of the pinholes tend to broaden into larger defects as seen in the lower left of the figure. The GaN band-edge luminescence image (with slit centered at λ=357nm), in Figure 4.26b, exhibits inhomogeneous contrast with features of the order of a fraction of a micron, and with dimensions and morphology similar to that of the columnar structure of the GaN epilayer. Figure 4.36c,d are CL images taken at λ=456 and 466nm, located at the short-λ region and the maximum of the quantum well emission, respectively. The pinholes in (a) do appear in both (c) and (d) but are not observed in (b). (c exhibits much smaller pinhole features than a and d, and pinholes in d are significantly larger than those in a.) Other luminescent features away from the topographic microstructural defects exhibit a random anisotropic pattern, with dimensions about 0.2μm across. Figure 4.40 shows the CL spectra taken in the stationary spot mode at the positions numbered in Figure 4.39a. It is interesting to note the discrepancy between the spectra recorded from selected points to that recorded from a larger area (Figure 4.38) where the peak quantum well luminescence was at 457nm. A common feature of the spectra is an additional peak, shorter in wavelength than the dominant peak.

A similar set of observations were performed on the specimen with higher indium concentration (x=0.52) specimen and these results are presented in Figure 4.41 and 4.42. The results are similar with sub-micron scale inhomogeneity in both the strength and color of the quantum well luminescence, and no direct correlation between the GaN band-edge and the quantum well luminescence (Figures 4.41b-d). The spatial variation of the luminescence peak is pronounced and more random than in the other sample. Phonon replicas are observed as in all other cases. In addition, spots with peak positions at shorter wavelengths (like positions 2 and 5 in Figures 4.41 and 4.42) exhibit extra peaks towards the shorter-λ of the spectrum. These may be emission from excited states associated with regions of the quantum well with nominal lower concentrations of indium.

The CL spectra were taken under low electron injection conditions in comparison with LED operation. No surface contamination was observed during the experiments nor any charging effects. The good match with PL indicates that the similar CL spectra reflect the same physical process observable with PL. Whereas the electron beam diameter can be smaller than 5nm, the region of excitation is estimated to have a radius of the order of 50nm at 4 keV due to the secondary scattering events (Yacobi and Holt, 1990). The lateral resolution observed in the CL images in Figures 4.39c,d are about 50nm. It is clear that emission outside the maxima of the QW spectra has sharper features, especially with regard to contrast near nanopipes. Figure 4.41 d has correspondingly broader features. It can be argued that the broadening is due to non-radiative recombination at nanopipes, and an upper limit for minority-carrier lateral-diffusion length can be determined at ~50nm from these pictures. It is interesting that the features off the peak position are sharper. We have observed that the height of the short wavelength peak at 445nm in Figure 4.42 depends on the intensity of the injected current, and thus it may be associated with an excited quantum well state.

In summary, there is a good match in the luminescence spectra between PL and CL when using a TV raster with dimensions of several microns. CL images show inhomogeneity in both the intensity and wavelength of the luminescence on a sub-micron scale. The spatial variation in the CL images has similar dimensions as the columnar microstructure observed by TEM. The peak position is small for x=0.28, and much broader for x=0.52. This indicates and increasing amount of composition inhomogeneity with average indium composition. The variations in peak wavelength at higher indium concentrations, and the appearance of peaks in the blue shoulder of the CL peaks in regions with shorter peak wavelengths, are indicative of chemical composition fluctuations along the quantum wells. Similar results were obtained by Chichibu et al (1997).

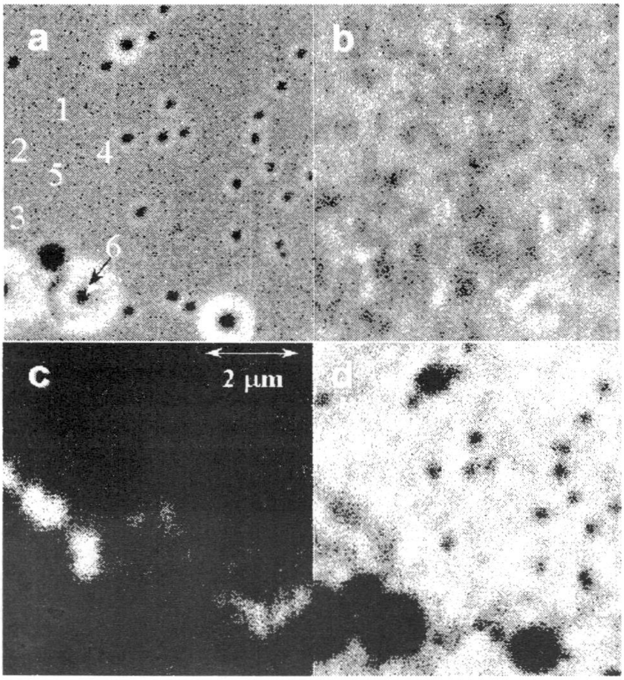

Figure 4.41 CL images of $In_xGa_{1-x}N$ single quantum wells for x=0.52. (a) Secondary electron image. (b) Band edge emission image at λ=357nm. (c) Image at λ=496nm. (d) Image at λ= 520nm.

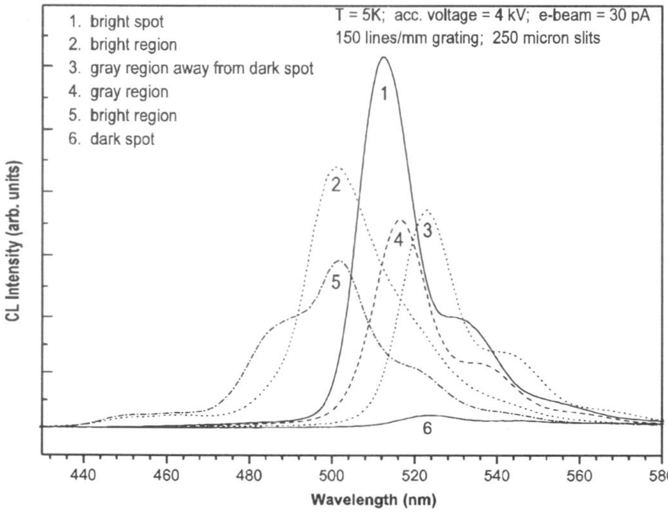

Figure 4.42 CL spectra taken in the spot mode, corresponding to spots labelled in Figure 4.41.

4.10 MICROSTRUCTURE AND DEVICE PERFORMANCE.

In view of past experience with other III-V compounds, it is surprising that GaN work so well given the complex microstructure that has been described in this chapter. It should be emphasized that the characteristics of the epitaxial layers that have had so much success in the fabrication of LEDs and LD have been achieved in an empirical manner.

There is much work to be done to fully understand the details of the extended defects in GaN and the implications on device performance. At this stage of development of understanding there are a few aspects that are becoming quite clear, these are reviewed in this final section to this chapter.

4.10.1 Stress and Point Defect Structure

Early in the development of high quality GaN thin films, Amano *et al* (1986) and Akasaki *et al* (1989) realized that there was a clear relationship between film morphology and incorporation of point defects, and introduced the concept of low temperature buffer layers. The idea developed that achieving highly planar growth morphology would optimize the semiconductor quality of the material. Control of lateral strain was considered to be significant in order to achieve control on the growth morphology.

4.10.2 Minimization of Strain by Maximizing Film Smoothness

Using ideas from Akasaki and Amano, Nakamura *et al* (1992b) derived a simple approach for *in-situ* monitoring of the growth morphology. Using *in-situ* infrared reflection techniques, the various growth parameters were optimized to produce the highest possible optical interference, i.e. a high degree of planarity of the film surface at every stage of the growth process. This straightforward approach led to optimum temperatures of deposition, buffer layer thickness, and other parameters for growth and doping of GaN, for growth of heterojunctions involving AlGaN and InGaN, etc. Only subsequent characterization determined the microscopic characteristics associated with the optimized films.

4.10.3 The role of dislocations in strain relaxation

The resulting microstructure plays the role of minimizing lateral stresses by providing a dislocation superstructure that can easily rearrange to cancel lateral stresses during growth. The tilt/twist configuration provides such dislocation superstructure. It appears that dislocation densities of 10^9 to 10^{10} are necessary for this purpose. The net result is the near absence of thermal stresses in high-quality GaN epitaxy.

4.10.4 The Role of Nanopipes and Extension to ELOG Structures

The near absence of thermal stresses implies the motion of dislocations by glide and climb during growth, and during the cooling process. Dislocation motion has been observed to below 300°C. Given the strength of the chemical bond, and the high melting temperature of GaN, it is surprising that relaxation of thermal stresses occurs at such low temperatures. The existence of nanopipes provides an explanation for ease of dislocation motion by providing internal surfaces that allow the injection of interstitials and vacancies necessary for dislocation climb. Nanopipes start at the sapphire interface, and we have proposed a model involving island coalescence during the early stages of growth. The intrinsic structure of the nanopipes would, however, allow for metal diffusion during device operation, especially in laser diodes, and the eventual shorting of the devices. There is evidence (soon to be published) that bending of nanopipes causes them to close into regular screw dislocations. Thus the net effect of epitaxial lateral overgrowth is to eliminate nanopipes from the device region without necessarily creating a rigid structure that would undermine growth morphology.

4.11 REFERENCES

Akasaki, I., Amano, H., Koide, Y., Hiramatsu, K. and Sawaki, N., 1989, *J. Cryst. Growth,* **98**, p. 209.

Amano, H., Sawaki, N., Akasaki, I. and Toyoda, Y., 1986, *Appl. Phys. Lett.* **48**, p. 353.

Baranowski, J.M., Liliental-Weber, Korona, Z.K., Pakula, K., Stepniewski, R., Wysmoiek, A., Grzegory, I., Nowak, G., Porowski, S., Monemar, B. and Bergman, P., 1997, *Mater. Res. Soc. Symp. Proc.,* **449**, p. 393.

Bauer, E. and van der Merwe, J.H., 1986, *Phys. Rev. B,* **33**, p. 3657.

Brantley, W.A., Lorimor, O.G., Dapkus, P.D., Haszko, S.E. and Saul, R.H., 1975, *J. Appl. Phys.,* **46**, p. 2629.

Cherns, D. and Preston, A.R., 1986, *Proc. 11th Int. Congress on Electron Microscopy,* (Kyoto: Japan Society of Electron Microscopy), p. 721.

Cherns, D., Young, W.T., Steeds, J.W., Ponce, F.A., and Nakamura, S., 1997, *J. Cryst. Growth*, **178**, p. 201.

Cherns, D., Young, W.T., Steeds, J.W., Ponce, F.A., and Nakamura, S., 1998, *Philosophical Magazine*, **A77**, p. 273.

Chichibu, S., Azuhata, T., Sota, T. and Nakamura, S., 1996, *Appl. Phys. Lett.,* **69**, p. 4188.

Chichibu, S., Azuhata, T., Sota, T. and Nakamura, S., 1997, *Appl. Phys. Lett.,* **70**, p. 2822.

Fertitta, K.G., Holmes, A.L., Ciuba, F.J., Dupuis, R.D. and Ponce, F.A., 1995, *J. Electron. Mater.,* **24**, p. 257.

Foresi, J.S. and Moustakas, T.D., 1993, *Appl. Phys. Lett.,* **62**, p. 2859.

Frank, F.C., 1951, *Acta Crystallo-graphica,* **4**, p. 497.

Frank, F.C. and van der Merwe, J.W., 1949, *Proc. Royal Society of London, Series A,* **198**, p. 205.

Glaser, E.R., Kennedy, T.A., Doverspike, K., Rowland, L.B., Gaskill, D.K., Freitas, J.A., Kahn, M.A., Olson, D.T., Kuznia, J.N. and Wickenden D.K., 1995, *Phys. Rev. B,* **51**, p. 13326.

Herzog, A.H., Keune, D.L. and Craford M.G., 1972, *J. Appl. Phys.,* **43**, p. 600.

Hirth, J.P. and Lothe, J., 1982, *Theory of Dislocations*, 2nd edition, (New York: Wiley Interscience), p. 159.

Hirsch, P., Howie, A., Nicholson, R.B., Pashley, D.W., Whelan, M.J., 1965, *Electron Microscopy of Thin Crystals*, Krieger, New York, Chap. 7.

Hirsch, P., Howie, A, Nicholson, R.B., Pashley, D.W. and Whelan, M.J., 1977, *Electron Microscopy of Thin Crystals,* Krieger, New York, p. 181.

Ho, I-hsiu and Stringfellow, G.B., 1996, *Appl. Phys. Lett.,* **69**, p. 2701.

Hull, R., Bean, J.C., Cerdeira, F., Fiori, A.T. and Gibson, J.M., 1986. *Appl. Phys. Lett.,* **48**, p. 56.

Karpinski, J., Porowski, S. and Miotkowska, S., 1982, *J. of Cryst. Growth,* **56**, p. 77.

Kronberg, M.L., 1957, *Acta Metallurgica*, **5**, p. 507.

Kurtin, S., McGill, T.C. and Mead, C.A., 1969, *Phys. Rev. Lett.,* **22**, p. 1433.

Lester, S.D., Ponce, F.A., Craford, M.G. and Steigerwald, D.A., 1995, *Appl. Phys. Lett.,* **66**, p. 1249.

Liliental-Weber, Z., Kisielowski, C., Ruvimov, S., Chen, Y., Washburn, J., Grzegory, I., Bockowski, M., Jun, J. and Porowski, S., 1996, *J.Electron. Mater.,* **25**, p. 1545.

Liliental-Weber, Z., Ruvimov, S., Swider, W., Kim, Y., Washburn, J., Nakamura, S., Kern, R.S., Chen, Y. and Yang, J.W., 1998, *Mater. Res. Soc. Symp. Proc.,* **482**, p. 375.

Nakamura, S., 1991, *Jap. J. Appl. Phys.,* **30**, p. L1705.

Nakamura, S. and Mukai, T., 1992, *Jap. J. Appl. Phys.,* **31**, p. L1457.

Nakamura, S., Mukai, T. and Senoh, M., 1992, *J. Appl. Phys.,* **71**, p. 5543.

Nakamura, S., Mukai, T. and Senoh, M., 1994, *Appl. Phys. Lett.,* **64**, p. 1687.

Nakamura, S., Senoh, M., Nagahama, S., Iwasa, N., Yamada, T., Matsushita, T., Sugimoto, Y. and Hiyoku, H. 1997, *Appl. Phys. Lett.,* **70**, p. 2753.

Narukawa, Y., Kawakami, Y., Funato, M., Fujita, S. and Fujita S., 1997a, *Appl. Phys. Lett.,* **70**, p. 981.

Narukawa, Y., Kawakami, Y., Fujita, S., Fujita S. and Nakamura, S., 1997b, *Phys. Rev. B,* **55**, p. R1938.

Ogino, T. and Aoki, 1980, *Jap. J. Appl. Phys.,* **19**, p. 2395.

Pakula, K., Wysmolek, A., Korona, K.P., Baranowski, J.M., Stepniewski, R., J. Grzegory, I., Bockowski, M., Jun, J., Krokowski, S., Wroblewski, M. and Porowski, S., 1996, *Solid State Commun.,* **97**, p. 919.

People, R. and Bean, J.C., 1985, *Appl. Phys. Lett.,* **47**, p. 322.

Ponce, F.A., Major, J.S. Jr., Plano, W.E. and Welch, D.F., 1994, *Appl. Phys. Lett.,* **65**, p. 2302.

Ponce, F.A., Krusor, B.S., Major, J.S., Jr., Plano, W.E. and Welch, D.F., 1995, *Appl. Phys. Lett.,* **67**, p. 410.

Ponce, F.A., Bour, D.P., Götz, W. and Wright, P.J., 1996a, *Appl. Phys. Lett.,* **68**, p. 57.

Ponce, F.A., Bour, D.P., Götz, W., Johnson, N.M., Helava, H.I., Grzegory, I., Jun, J. and Porowski, S., 1996b, *Appl. Phys. Lett.,* **68**, p. 917.

Ponce, F.A., Bour, D.P., Young, W.T., Saunders, M. and Steeds, J.W., 1996c, *Appl. Phys. Lett.,* **69**, p. 337.

Ponce, F.A., Cherns, D., Young, W.T. and Steeds, J.W., 1996d, *Appl. Phys. Lett.,* **69**, p. 770.

Ponce, F.A., O'Keefe, M.A. and Nelson, E.C., 1996e, *Philosophical Magazine* A, **74**, p. 777.

Ponce, F.A., Van de Walle, C.G. and Northrup, J.E., 1996f, *Phys. Rev. B*, **53**, p. 7473.

Ponce, F.A., 1997, *Mater. Res. Soc. Bulletin,* **22**, p. 51.

Ponce, F.A. and Bour, D.P., 1997a, *Nature,* **386**, p. 351.

Ponce, F.A., Cherns, D., Young, W.T., Steeds, J.W. and Nakamura, S., 1997b, *Mater. Res. Soc. Symp. Proc.,* **449**, p. 405.

Ponce, F.A., Cherns, D., Götz, W. and Kern, R.S., 1998a, *Mater. Res. Soc. Symp. Proc.,* **482**, p. 453.

Ponce, F.A., Galloway, S.A., Götz, W. and Kern, R.S., 1998b, *Mater. Res. Soc. Symp. Proc.,* **482**, p. 453.

Qian, W., Rohrer, G.S., Skowronski, M., Doverspike, K., Rowland, L.B. and Gaskill, D.K., 1995a, *Appl. Phys. Lett.,* **67**, p. 2284.

Qian, W., Skowronski, M., Doverspike, K., Rowland, L.B. and Gaskill, D.K., 1995b, *J. Cryst. Growth,* **151**, p. 396.

Roedel, R.J., Von Neida, A.R., Caruso, R. and Dawson, L.R., 1979, *Journal of the Electrochemical Society,* **126**, p. 637.

Rouviere, J.L., Arlery, M., Bourret, A., Niebuhr, R. and Bachem, K. H., 1996, *Mater. Res. Soc. Symp. Proc.,* **395**, p. 393.

Stringfellow, G.B., Lindquist, P.F., Cass, T.R. and Burmeister, R.A., 1974, *J. Electron. Mater.,* **3**, p. 497.

Suzuki, T. and Matsumoto, Y., 1975, *Appl. Phys. Lett.,* **26**, p. 431.

Van de Walle, C.G. and Neugebauer, J., 1997, *Mater. Res. Soc. Symp. Proc.*, **449**, p. 861.

Wyckoff, R.W.G., 1995, *Crystal Structures*, Volumes 1-4, (New York: Interscience Publishers).

Wetzel, C., Takeuchi, T., Amano, H. and Akasaki, I., 1998, *Mater. Res. Soc. Symp. Proc.,* **512**, p. 181.

Yacobi, B.G. and Holt, D.B., 1990, *Cathodoluminescence Microscopy of Inorganic Solids,* New York: Plenum.

Chapter 5
Emission Mechanisms and Excitons in GaN and InGaN Bulk and QWs

Shigefusa F. Chichibu, Yoichi Kawakami, and Takayuki Sota

5.1 INTRODUCTION

It is now established that the performance of blue/green GaN based light emitting diodes (LEDs) (Nakamura and Fasol, 1997a) is superior to conventional red AlGaAs or AlInGaP LEDs with respect to output power and reliability in spite of large threading dislocation (TD) density up to the order of 10^{10} cm^{-2} (Rosner *et al.*, 1997). The dislocation density in ZnSe based II-VI laser material is as low as less than 10^4 cm^{-2}. However, both the output power and device lifetime of GaN based purplish-blue LDs (Nakamura and Fasol, 1997a) are better than those of II-VI materials. Therefore the open question has been why InGaN LEDs exhibit highly efficient bright emissions despite of the large TD density.

This chapter treats fundamental optical processes occurring in GaN and InGaN alloys, especially the emission mechanisms of quantum well (QW) structures. Most of optical transition processes in GaN based materials rely on the excitonic feature since GaN is more excitonic than conventional III-V compound semiconductors. Basic concepts and properties of bulk excitons, perturbation effects such as strain on exciton resonance energies, and recombination dynamics of excitons and exciton molecules in GaN are introduced. Since excitons have a polarization, their properties are affected by interactions with any kinds of electromagnetic polarization such as phonons, photons, electric fields, magnetic fields, and so on, a brief description of these perturbation is given. It is also known that in quantum wells (QWs), Coulomb interaction between the electron-hole (e-h) pair is enhanced due to confinement of wavefunctions by the confinement potential between the well and the barrier. As a result, the binding energy of excitons in QWs is increased (Miller *et al.*, 1984; 1985; Bastard *et al.*, 1982) ideally as large as four times its bulk value (Shinada and Sugano, 1966). On the other hand, since the *c* [0001] axis of hexagonal nitrides corresponds to the growth axis, spontaneous and strain induced lattice polarization produces electric field across the QW plane. Hence, the energy quantization and effects of electric field on the properties of QW excitons in AlGaN/GaN and InGaN/GaN QW structures having different thicknesses and molar fractions are described. In contrast to binary GaN, the alloy compositional fluctuation induces effective bandgap inhomogeneity in InGaN

QWs. This and the QW thickness fluctuation produce energy broadening of InGaN QWs. Moreover, the growth of high quality InGaN alloys is extremely difficult due to mismatch of the thermodynamical and chemical stability between GaN and InN (Osamura *et al.*, 1975). It should be noted that the lattice mismatch between GaN and InN is as large as nearly 10%. Therefore the phase-separating nature of InGaN is also a driving force of compositional inhomogeneity in InGaN material. These characteristics are responsible for exciton localization in InGaN QWs, and the localized QW exciton is considered to play a key role to obtain bright spontaneous emissions from practical InGaN LEDs (Nakamura and Fasol, 1997a). After introducing several experimental evidences of bandgap inhomogeneity in InGaN, recombination dynamics of localized excitons and influence of the bandgap inhomogeneity on the optical gain spectra of InGaN cw laser materials are shown.

5.2 GaN BULK CRYSTALS APPROACH

5.2.1 Free and Bound Excitons

Transition structures

The elementary excitation process of semiconductors over its bandgap energy first creates free electrons in the conduction and holes in the valence bands. Next they lose energy and momentum, i.e. relax, to the lowest energy states by emitting phonons in general. Then they form a quasiparticle that is a bound state with Coulomb interaction in between the electron and the hole. The quantum of this electronic polarization is called as exciton. Due to its dipole nature with two particle and material dependent properties, excitons are of great interest in both physics and practical applications. Especially, linear and nonlinear semiconductor optics rely on the characterization and utilization of excitons (Haug, 1988). Moreover, external confinement of excitons by the potential barriers enhances their Coulomb interaction (Miller *et al.*, 1984; 1985). The introduction of the exciton was made by Frenkel (1931) in 1931 in his attempts to obtain insight into the transformation of light into the material to heat. Wannier (1937) has developed exciton energy levels that have a form of a hydrogen atomic series in this type of system in 1937. Generally in semiconductors, a weakly bound e-h pair whose wavefunction propagates more than the lattice spacing is formed and is called as Mott-Wannier exciton. Conversely in ionic crystals or molecular crystals, e-h pair is strongly bound at the matrix atom or localized at its neighbourhood. This type of exciton is called as Frenkel exciton. Excellent textbooks which treated the exciton problem in bulk (Knox, 1963; Pankove, 1971; Reynolds *et al.*, 1981) and QWs (Haug, 1988) have been published.

Wannier excitons can freely move in the crystal in a form of mobile, bound e-h pair, which is called as free exciton (FE). Thus they do not contribute to the electrical conductivity. However, they play an important role in optical absorption, reflection, and emission processes in semiconductors. The energy states of excitons resemble those of shallow impurities. The electron turns around the hole like hydrogen atom but in the material with dielectric constant $\varepsilon=\varepsilon_S\varepsilon_0$. The total energy E of the e-h pair is a sum of the energy of the center of mass motion E_K and that of relative motion E_n,

$$E = E_K + E_n. \tag{5.1}$$

Here, the center of mass motion term is

$$E_K = \hbar^2 k^2 / 2\left(m_e^* + m_h^*\right) \tag{5.2}$$

where k is the wave vector and m_e^* and m_h^* is electron and hole effective mass, respectively. On the other hand, the energy of relative motion of the exciton in the crystal can be calculated by solving the Schrödinger equation for the Coulomb interaction,

$$\left[-\frac{\hbar^2}{2m_r^*}\nabla^2 - \frac{e^2}{4\pi\varepsilon_0\varepsilon_S r}\right]F(\boldsymbol{r}) = EF(\boldsymbol{r}), \tag{5.3}$$

which gives the eigenvalue of the exciton energy as

$$E = E_n = -\frac{m_r^* e^4}{2(4\pi\varepsilon_0\varepsilon_S)^2\hbar^2}\cdot\frac{1}{n^2} = -\frac{e^2}{2(4\pi\varepsilon_0\varepsilon_S)a_B^*}\cdot\frac{1}{n^2}$$

$$= -13.6(\text{eV})\frac{m_r^*}{m_0}\left(\frac{1}{\varepsilon_S}\right)^2\frac{1}{n^2}, \tag{5.4}$$

where

$$\frac{1}{m_r^*} = \frac{1}{m_e^*} + \frac{1}{m_h^*}, \tag{5.5}$$

$$a_B^* = -\frac{(4\pi\varepsilon_0\varepsilon_S)\hbar^2}{m_r^* e^2} = a_B\frac{m_0}{m_r^*}\varepsilon_S = 0.053(\text{nm})\frac{m_0}{m_r^*}\varepsilon_S, \tag{5.6}$$

n is an integer, and m_r^* and a_B^* are reduced effective mass and effective Bohr radius of the exciton, respectively. It should be noted that these equations rely on the effective mass approximation. The ground state FE energy is thus E_b smaller than the bandgap energy E_g, where E_b is the exciton binding energy.

Studies on the exciton resonance structures in GaN began at the beginning of 1970's. Due to the crystal symmetry and relatively weak spin-orbit interaction, the valence band of the wurtzite hexagonal GaN (h-GaN) is lifted to three separate bands (Dresselhaus, 1957; Hopfield, 1960), as shown in Figure 5.1. Dingle *et al.* (1971) have assigned the energies of three (A, B, and C) excitons related to transitions from the separate valence bands (Γ_{9v}, Γ_{7v}, and Γ_{7v} in Figure 5.1) to the conduction (Γ_{7c}) band to E_{ex}(A)=3.474±0.002 eV, E_{ex}(B)=3.480±0.002 eV, and E_{ex}(C)=3.501±0.01 eV by means of the polarized optical reflection (OR) measurements at 2 K after Pankove *et at.* (1970) have observed a strong photoluminescence (PL) peak at 3.477 eV at 1.6 K. Monemar (1974) has also reported the exciton energy from the photoluminescence excitation (PLE) spectra at 1.6 K as E_{ex}(A)=3.4751 ± 0.0005 eV, E_{ex}(B)=3.481 ± 0.001 eV, and E_{ex}(C)=3.493±0.005 eV, and he deduced E_b of the ground state A-exciton as 28 meV from the E_g and E_{ex}(A) values. These data have been taken from synthesized bulk GaN crystals. Very recently, Amano *et al.* (1993) and Logothetides *et al.* (1995) have suggested a contribution of excitons in the optical absorption (OA)

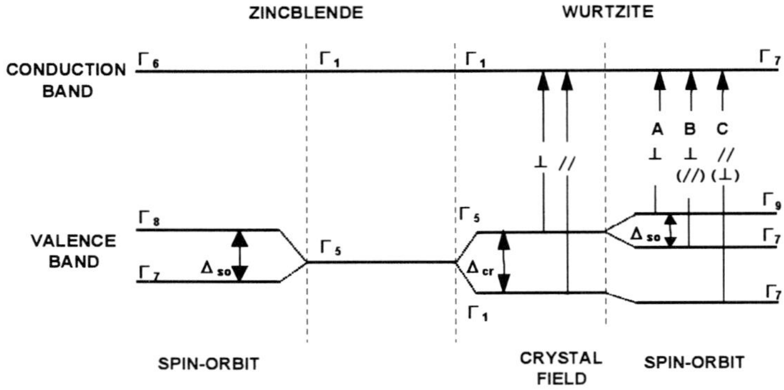

Figure 5.1 Band structures and labeling of respective transition in zincblende and wurtzite GaN. The indications ⊥ and // show that the transition is allowed for the light polarization E perpendicular and parallel to the optic (c) axis, respectively. Parentheses means that the transition is partially allowed. The value Δ_{so} and Δ_{cr} are the spin-orbit and crystal-field splittings, respectively.

spectra of zincblende cubic GaN (c-GaN) and h-GaN up to room temperature (RT). Shan *et al.* (1995;1996) have also assigned the modulation photoreflectance (PR) signals to exciton resonance structures up to RT. However, there were no reports on the observation of excitonic emissions from h-GaN at RT until 1996 (Chichibu *et al.*, 1996a), which will be shown in the following, because the growth of very high purity, high quality GaN had been quite difficult until Nakamura (Nakamura and Fasol, 1997a) grew excellent quality GaN heteroepitaxial films on sapphire (0001) substrates by two-flow metalorganic chemical vapor deposition (TF-MOCVD).

Modulation spectroscopy, particularly electromodulation, which is usually referred to as electroreflectance (ER) and electroabsorption (EA), is a useful tool to determine E_g in semiconductors (Cardona, 1969; Aspnes, 1980). The external electromodulation provides a sharp structure related to the third derivative of dielectric functions (Aspnes, 1973; 1980) for the band-to-band (B-B) transition. Moreover, exciton transition induces fine structures in OR, OA, and PLE signals and ER spectra, and in this case low-field electromodulation monitors the first derivative of the OR signal. The PR method is the contactless form of ER, and we can easily accomplish the low-field regime using PR because the electromodulation field induced by a secondary excitation light beam is small and controllable (Shirakata and Chichibu, 1996a).

Figure 5.2 shows (Chichibu *et al.*, 1996a; 1997a) represents PR and PL spectra of 4-μm-thick h-GaN on sapphire (0001) (Nakamura and Fasol, 1997a) at 10K (a) and those of c-GaN grown by molecular beam epitaxy (MBE) on 3C-SiC/Si(100) substrates (Okumura *et al.*, 1997) prepared by chemical vapor deposition (CVD) measured at 75K (b). The PR spectrum of h-GaN does not exhibit interfering fringes because of the internal multiple reflection in thin films unlike a static OR or OA spectrum, since the interfering effect is cancelled due to the nature of the signal $\Delta R/R$. However, the PR spectrum of c-GaN does. This may be due to insufficient quality of the epilayer.

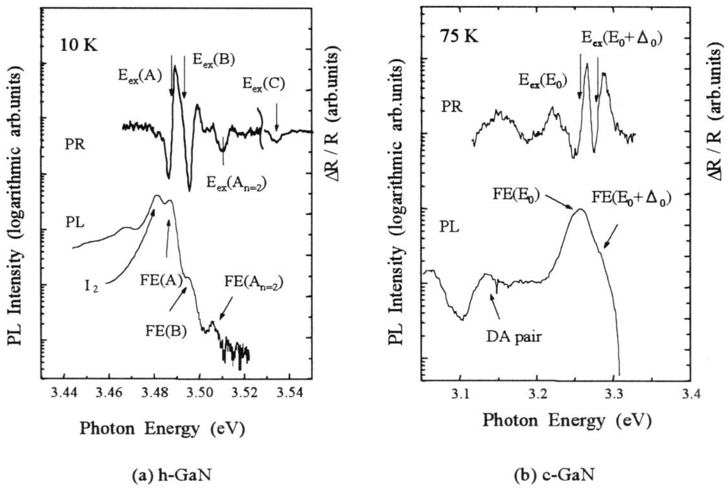

Figure 5.2 PR and PL spectra of (a) h-GaN on sapphire (0001) measured at 10 K and (b) c-GaN on 3C-SiC / Si (001) measured at 75 K.

The PR spectrum of h-GaN exhibits excitonic fine transition structures that correspond to the ground state A-, B-, C-, and the first excited state of A-exciton resonances, respectively. According to the polarization selection rules for C_{6v}^4, A- and B-transitions are allowed for the light polarization E perpendicular to the *c* axis (E⊥*c*), while C-transition is allowed for E parallel to the *c* axis (E//*c*), where E is an electric field component. Because the h-GaN film has its *c* axis parallel to the substrate normal, the polarization configuration E⊥*c* is effectively maintained in our experimental configuration. Therefore, the PR spectra shown in Figure 5.2 (a) exhibit predominant A- and B-transitions. The weak C-transition was detected by measuring quasi-polarized PR measurements (Chichibu *et al.*, 1996b; Shikanai *et al.*, 1997) where the incidence angle of the probe light was set to about 45~60° and the reflected light was nominally polarized E//*c* or E⊥*c* by a polarizer. The light was depolarized before entering the monochromator. The resonance structure is stronger for E//*c* than that for E⊥*c*. Thus, we assign the structure to a C-transition. The structure between the B- and C-transitions is assigned to the first excited state (n=2) of A-exciton transition since the structure is quenched as temperature is increased. The fact that $E_{ex}(A_{n=2})$ varies along with that of $E_{ex}(A)$ as functions of temperature and strain (Chichibu *et al.*, 1996b; Shikanai *et al.*, 1997) supports this assignment.

The PR spectrum was analyzed by the low-field ER line-shape function (Aspnes, 1973; 1980),

$$\frac{\Delta R}{R} = Re\left\{\sum_{j=1}^{p} C_j e^{i\theta_j}\left(E - E_{g,j} + i\Gamma_j\right)^{-m_j}\right\}, \tag{5.7}$$

where p is the number of the spectral function to be fitted, E is the photon energy, and C_j, θ_j, $E_{g,j}$ and Γ_j are the amplitude, phase, energy and broadening parameter of the *j*th feature, respectively. Here p=3 for h-GaN ground state excitons, as shown in Figure 5.1. For strain-free c-GaN, p is essentially 2 taking small Δ_{so} into account. m_j is a parameter that depends on the critical point type. The PR spectrum experimentally obtained was best fit to Equation (5.7) using the nonlinear least-fitting method giving an appropriate m=3, which is characteristics of an exciton resonance. Thus, the energies E_{ex}(A), E_{ex}(B), E_{ex}(C), and $E_{ex}(A_{n=2})$ and broadening parameters (Γ_A, Γ_B, Γ_C, and $\Gamma_{A_{n=2}}$) were obtained. The values of E_{ex}(A), E_{ex}(B), E_{ex}(C) and $E_{ex}(A_{n=2})$ are obtained as 3.4876, 3.4962, 3.5320, and 3.5067 eV at 10K, respectively, for the 4-μm-thick h-GaN. These values are larger by 10-26 meV than those reported for strain-free h-GaN (Dingle, 1971; Monemar, 1974). This phenomenon can be explained by the effects of biaxial compressive strain (Bir and Pikus, 1974) in h-GaN (Amano *et al.*, 1988; Naniwae *et al.*, 1990; Gil *et al.*, 1995; Chichibu *et al.*, 1996b; 1997b; Shikanai *et al.*, 1997) on sapphire, which will be discussed in detail in Section 5.2.3. From these values, E_b of A-exciton, E_b(A) is obtained to be 26 meV. This value is most reliable and is later supported by Skromme *et al.* (1997) and Yamaguchi, A. *et al.* (1998a).

In contrast to the wurtzite polytypes, data on excitons in c-GaN in the early period are not very reliable due to poor crystalline quality and mixing of the hexagonal phase. There have been some conflicting reports regarding the E_g of c-GaN (Okumura *et al.*, 1991; 1997; Paisley *et al.*, 1989; Strite *et al.*, 1991; Flores *et al.*, 1994; Menniger *et al.*, 1996a). For example, the E_g values reported scatter from 3.2 eV to 3.82 eV. However, the growth techniques of c-GaN have recently been improved to observe near-band-edge emissions from the gas-source MBE-grown epilayer at low temperature (Okumura *et al.*, 1997). Flores *et al.* (1994) have measured the PR spectra of c-GaN as a function of temperature, and have

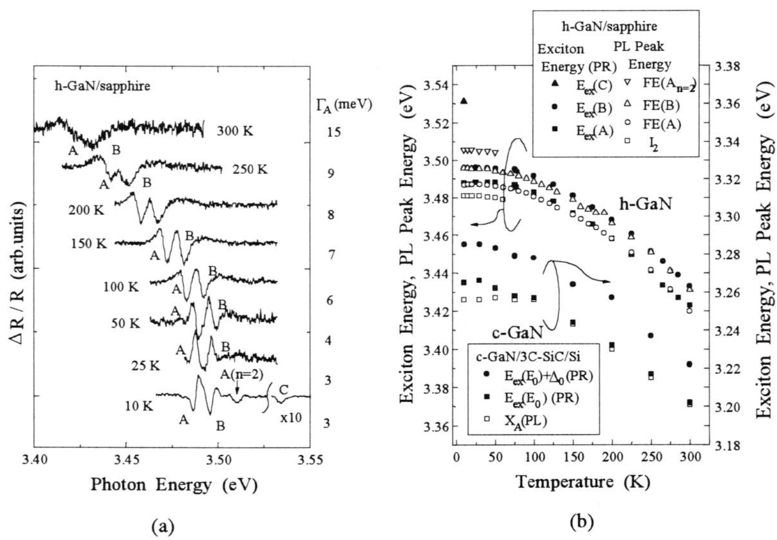

Figure 5.3 PR spectra of h-GaN as a function of temperature (a) and the exciton resonance and PL peak energy of h-GaN and c-GaN as a function of temperature (b).

determined $E_g(E_0)$ as 3.302 eV at 10K. Okumura *et al.* (1997) have determined $E_g(E_0)$ = 3.3 eV and $E_{ex}(E_0)$ = 3.27 eV by means of OR and magnetic circular dichroism (MCD) measurements. Menniger *et al.* (1996a) have reported emissions due to free and bound exciton recombinations at 3.272 and 3.263 eV, respectively at 5K. Representative PR and PL spectra of c-GaN/3C-SiC/Si(100) measured at 75 K are shown in Figure 5.2(b). The PR spectrum of c-GaN exhibits sharp two transition structures at around 3.3 eV. From the energy separation, they are assigned to the exciton transitions related to E_0 and $E_0+\Delta_0$ transitions, respectively. The values at 10 K are $E_{ex}(E_0)$ = 3.265 eV and $E_{ex}(E_0+\Delta_0)$ = 3.285 eV, leading Δ_{so} = 20 meV. This value nearly agrees with that reported by Flores *et al.* (1994), 17 meV. The E_{ex} value is consistent with those reported previously (Chichibu *et al.*, 1997a; Okumura *et al.*, 1997; Flores *et al.*, 1994; Menniger *et al.*, 1996a). For example, Menniger *et al.* (1996a) have reported the energy of FE emission in quasi-strain-free c-GaN film as 3.272 eV at 5K. A slight downshift (7 meV) of our $E_{ex}(E_0)$ value may originate from the residual biaxial tensile strain caused by the difference of the thermal expansion coefficient between c-GaN and 3C-SiC/Si structure, since the lattice parameter parallel to the substrate normal obtained from the X-ray diffraction measurements (0.4499nm) is smaller than that of strain-free c-GaN (0.4511nm) reported by Leszczynski *et al.* (1996). The E_{ex} value at RT is obtained as 3.202 eV, and it agrees well with the PL peak energy from c-GaN films of 3.2 eV recently reported by Menniger *et al.* (1996b) and Nakadaira and Tanaka (1997). From the PR spectra of c-GaN up to RT obtained, E_g of c-GaN is confirmed to be smaller by 0.21 eV than that of h-GaN.

PR spectra of the same h-GaN sample are shown as a function of temperature in Figure 5.3 (a). The PR spectra are well fitted up to RT (Chichibu *et al.*, 1996a) using Equation (5.7). The value of Γ_A increases from 3 to 15 meV with increasing temperature from 10K to RT. However, the value is still smaller than the thermal energy k_BT at RT, at which k_BT = 26 meV. The broadened PR spectrum of h-GaN at RT can be resolved into A- and B-transition. Therefore, the dominant excitation transition process is still excitonic at RT. This is also true for c-GaN (Flores *et al.*, 1994).

Luminescence properties

Typical radiative recombination processes occurring in semiconductors are schematically summarized in Figure 5.4. Luminescence spectrum of high purity, high quality semiconductors is dominated by FE emission at low temperature in general. The FE emission from h-GaN at low temperature have been observed in the early 1970s (Pankove *et al.*, 1970; Dingle *et al.*, 1971). There are various forms of exciton complexes that are composed of exciton(s) and other particles (Lampert, 1958). Neutral donor has an outer electron, which turns around the donor ion with larger orbit diameter. It binds a free hole at the position where the static dipole between the hole and the neutral donor balances and then charge neutrality of the complex is satisfied. This complex is an exciton bound to a neutral donor (I_2) and is in general called a bound exciton (BE). Neutral acceptor also produces the bound exciton (I_1). Similarly, ionized donor and acceptor can produce excitons bound to them. The transition energy of BEs associated with neutral impurities is lower than that of FEs by their localization energy, E_{loc}, which is nearly one tenth of the impurity ionization energy E_a (Haynes, 1960). The relation between E_a and

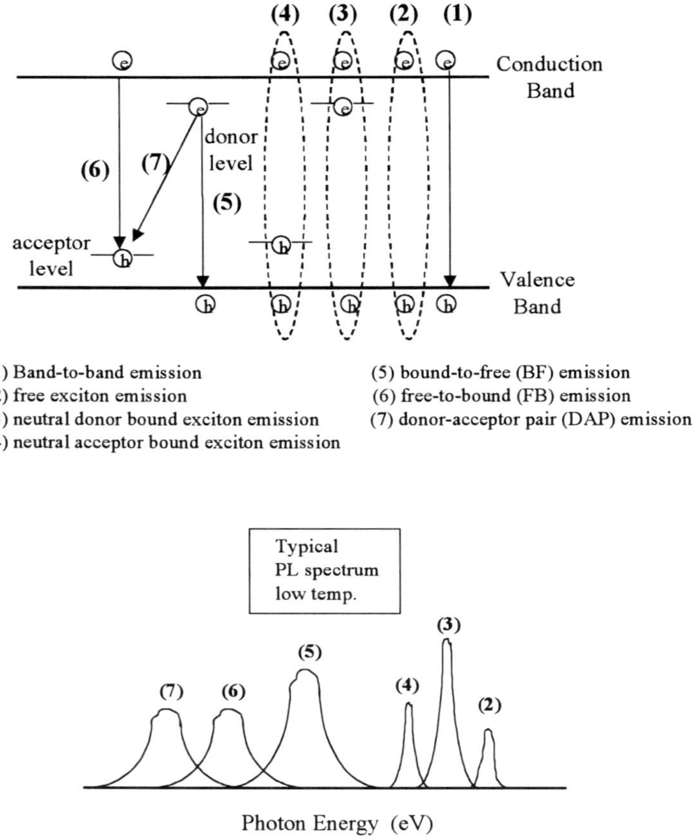

Figure 5.4 Schematic drawing of radiative recombination processes in semiconductors preferebly at low temperature where the excitonic and imputiry-related emission occur.

E_{loc} changes material to material (Bebb and Williams, 1972). Because E_{loc} is usually very small, BE recombination is dominant at very low temperatures and FE emission dominates the emission spectrum at intermediate temperatures between 20K and 100K in general depending on the stability of excitons in the matrix, that will be discussed later. Another process involving excitons at low temperature is a formation of exciton-polariton, which is a complex between the electromagnetic field (photons) and excitons (or oscillators) that have the same resonance frequency as the photons at finite value of very small wavevector $\boldsymbol{k}$ (Toyozawa, 1959; Hopfield, 1967; Tait *et al.*, 1967). The contribution of the exciton-polariton in optical spectra of h-GaN has recently been reported (Gil *et al.*, 1997; Stepniewski *et al.*, 1997). Excitons couple with several kinds of phonons. A remarkable feature of the PL spectrum of GaN at low temperature is the appearance of longitudinal optical (LO) phonon replicas of FE and BE lines (Dingle, 1971). In impurity doped semiconductors, emissions due to recombination of free carriers (electrons or holes) and the impurity levels (acceptors or donors) are dominant at intermediate temperature or RT depending on E_a of the impurity. They are referred to as free-to-bound (FB) or bound-to-free (BF) emissions. The simultaneous existence of donor and acceptor impurities introduce a pair type emission between them, and is called donor-acceptor pair (DAP) recombination. The B-B emission occurs in direct bandgap semiconductors with relatively high temperature, high carrier density, small E_b, or high number of active phonons which produces free carriers.

The PL spectrum of the 4-μm-thick h-GaN/sapphire exhibits fine structures near the band edge labelled I_2, FE(A), FE(B), and FE($A_{n=2}$) in order of increasing the peak energy from 3.4811, 3.4867, 3.4958, and 3.5057 eV at 10K, respectively, as shown in Figure 5.2 (a). The full-width at half maximum (FWHM) of each peak is smaller than 4 meV. Because the peak energies agree with those of the E_{ex} values obtained from the PR spectrum, FE(A), FE(B), and FE($A_{n=2}$) for h-GaN are assigned to a recombination of the ground states A- and B-free excitons and the first excited state of the A-exciton, respectively. Similarly, the PL spectrum of c-GaN at 75K exhibits a peak labelled FE(E_0) at 3.257 eV and a shoulder labelled FE(E_0+Δ_0) at 3.277 eV in addition to the broad band labelled DA pair at 3.137 eV, as shown Figure 5.2 (b). They are assigned to the ground state FE recombinations.

In Figure 5.5 plotted is PL spectra of h-GaN and c-GaN as a function of temperature. Relative intensity of I_2 compared to FE(A) decreases with increasing temperature, while that of FE(B) increases. Because the energy separation between FE(A) and FE(B) is as small as 8 meV and broadening of the peaks is apparent with the increase of temperature, PL spectra are numerically deconvoluted, assuming two to five peaks having Lorentzian distributions. It is found that PL spectra above 70K are dominated by FE(A) and FE(B). The broadened PL peak at RT is an envelope of FE(A) and FE(B) structures, which is revealed by the temperature dependence of the PL measurements. Both FE(A) and FE(B)

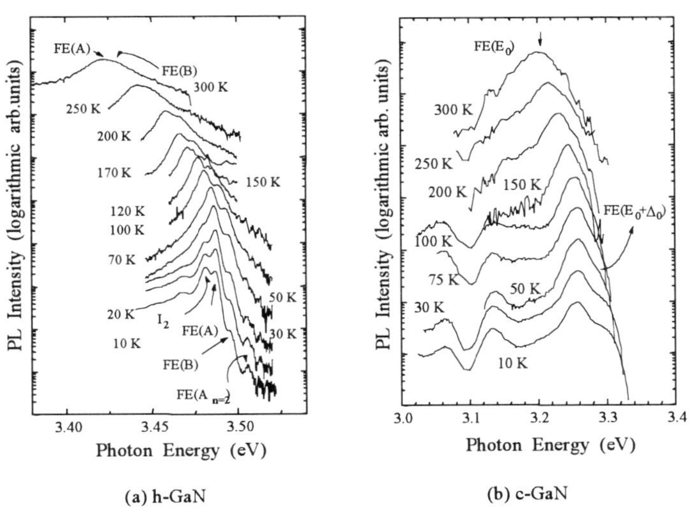

Figure 5.5 PL spectra of (a) h-GaN and (b) c-GaN as a function of temperature. The PL spectrum of h-GaN at RT is an envelope of FE(A) and FE(B). The PL peaks of both samples at RT are assigned to free excitonic emissions.

emissions have Lorentzian profiles. Peak energies of FE(A) and FE(B) at RT are obtained as 3.421 and 3.429 eV, respectively. The peak energies remain unchanged when the excitation intensity changes, and the peak intensities increase linearly with increasing excitation intensity (the slope of the logarithmic plot is n = 1.00±0.01) even at RT. In Figure 5.3 (b) are summarized the energies of FE resonances and the PL peaks of h-GaN and c-GaN as a function of temperature. The PL peak energies of FE(A) and FE(B) agree well with exciton resonance energies E_{ex}(A) and E_{ex}(B), respectively between 10K and RT. The FWHM value of FE(A) increases from 5 meV to 21 meV with increasing temperature from 10K to RT, as is the case with Γ_A. However, the value is still smaller than k_BT up to RT.

In popular III-V semiconductors like GaAs or InP, FEs dissociate easily due to internal or external perturbations since the e-h pair is weakly bound by the Coulomb force; (i) the presence of high number of charges (carriers, impurities ions, or defects) with excitons that screens the Coulomb interaction between the e-h pair, (ii) the electric field which pulls the e-h pair to the opposite directions, (iii) thermal energy k_BT more than E_b which releases FEs from the bound state, and (iv) coupling with LO phonons which can give FEs the energy more than E_b. In particular, E_b of InGaAlAsP materials is smaller than k_BT of RT and LO phonon energy is close to k_BT of RT. Therefore FEs can no longer contribute to their emission mechanisms at RT. However, for column III nitride semiconductors, Chichibu *et al.* (1996a; 1996c) first published a paper which reported the observation of PL peaks due to the recombination of FEs even at RT from h-GaN in 1996 as explained below. Soon after Monemar *et al.* (1996a) and Kai *et al.* (1996) also assigned the PL peak at RT to excitonic recombination, although they did not measure the exact E_{ex} at RT.

In Figure 5.6 are summarized PL, PR and OA spectra of h-GaN/sapphire measured at RT. The PL and PR spectra were obtained from the 4-μm-thick

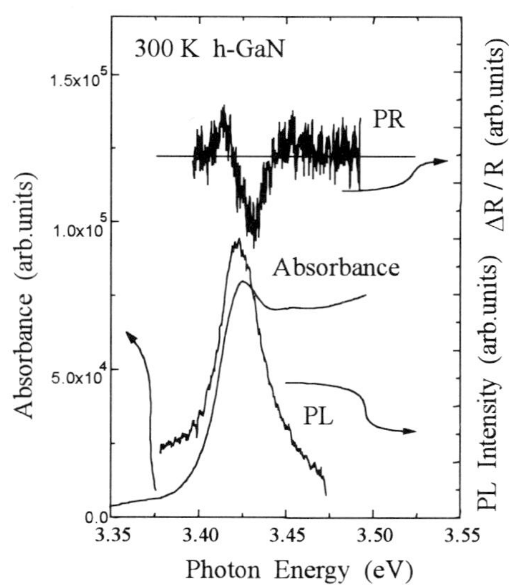

Figure 5.6 Room-temperature PL, PR and OA spectra of h-GaN/sapphire.

undoped h-GaN. The OA spectrum was obtained from a 1-μm-thick h-GaN/sapphire (Chichibu *et al.*, 1997a) because the transmitted light intensity from the 4-μm-thick sample was so weak that we could not measure the spectrum. Since the degree of biaxial compressive strain differs in each, the spectral position (horizontal axis) of the OA spectrum was appropriately adjusted by referring to the E_{ex}(A) values of both samples at 77K (Chichibu *et al.*, 1996b; 1997b; Shikanai *et al.*, 1997). The OA spectrum exhibits a distinct excitonic absorption peak similar to the result reported by Amano *et al.* (1993). The energy of A (B) excitonic absorption, which is obtained from the fitting (Deguchi *et al.*, 1997a), agrees with E_{ex}(A) [E_{ex}(B)] and FA(A) [FE(B)]. This result also confirms that the PL peak is due to FE recombinations at RT.

It is natural to observe FE emission from three-dimensional (3D) bulk GaN at RT, since this material satisfies several restrictions for excitons to survive at RT;

(i) exciton effective Bohr radius a_B^* is smaller than the screening length, D,

$$a_B^* \leq D = \sqrt{\varepsilon_S \varepsilon_0 k_B T / n_0 e^2}, \tag{5.8}$$

which is satisfied when the total charge density n_0 in the system is smaller than the critical charge density N_{crit} (Haug and Koch, 1990),

(ii) the electric field, F, across the FE is weaker than the critical ionization field,

$$F_{crit} \leq (E_b / a_B^*), \tag{5.9}$$

(iii) $E_b \geq k_B T = 26 meV,$ (5.10)

and

(iv) small numbers of active LO phonons at RT, which is satisfied if the LO phonon energy $\hbar\omega$ is far larger than k_BT of RT (26 meV).

These restrictions strongly depend on the material parameter. For the 3D GaN, the latter two are easily satisfied. The E_b value of the A-exciton is 26 meV, which is comparable to k_BT at RT (iii). The LO phonon energies in GaN are very large [91 meV for A_1(LO) mode and 93 meV for E_1(LO) mode (Azuhata *et al.*, 1995; 1996)] and few LO phonons are thermally activated even at RT (iv). The former two restrictions depend on the external parameters; injected (excited) carrier density, impurity concentration, and the electric field. Due to the small dielectric constant (ε_S=9) (Chichibu *et al.*, 1996b; Shikanai *et al.*, 1997; Azuhata *et al.*, 1995; 1996) and large effective masses (m_e^*=0.2m_0 and m_h^*=1.1m_0) (Suzuki *et al.*, 1995) of h-GaN, a_B^* of the ground state A-exciton is estimated using Equation (5.6) to be as small as 3.4nm, which is 3.5 times smaller than that of GaAs (12nm). Therefore the charge density to screen the Coulomb interaction (N_{crit}) at RT is as high as 1.1x10^{18}cm^{-3} according to Debye-Hückel screening (Haug and Koch, 1990) (i). Furthermore, F_{crit} is as high as 7.6x10^4 V/cm due to the large E_b and small a_B^*, which means that excitons in GaN are field-resistant particles. Important material parameters of GaAs (popular III-V), ZnSe, ZnS, ZnO (II-VI), $CuGaS_2$, $CuAlS_2$ (I-III-VI_2), and wurtzite and zincblende GaN are summarized in Table 5.1. These excitonic material properties, which come from the strong Coulomb interaction in GaN, make it possible to observe FE emissions at RT (Chichibu *et al.*, 1997a).

Table 5.1 Exciton-related material parameters in popular III-V, II-VI, I-III-VI$_2$, and GaN.

material	E_g at 300K (eV)	effective masses (in units of m_0) m_e^*	m_h^*	ε_S	exciton binding energy E_b (meV)	exciton Bohr radius a_B^* (nm)	$\hbar\omega$ (LO) (meV)	N_{crit} at 300 K (cm^{-3})
GaAs	1.43	0.067	0.45	13.1	4.6	12	36	1.3x10^{17}
ZnSe	2.67	0.16	0.75	8.7	18	3.8	31	8.4x10^{17}
ZnS(W)	3.56	0.34	0.6(W)	8.2	36	2.0	45	2.9x10^{18}
ZnO	3.37	0.28	1.8	8.5	46	1.9	75	3.8x10^{18}
$CuGaS_2$	2.49	0.13	0.69(av.)	7.6	29	3.6	50	8.4x10^{17}
$CuAlS_2$	3.49	0.3(?)	1.0(?)	7.0	70	1.5	62	4.0x10^{18}
c-GaN	3.23	0.17	0.86	8.2	29	3.1	93	1.2x10^{18}
h-GaN	3.43	0.2	1.1(av.)	9(av.)	26	3.4	93	1.1x10^{18}

Taking the degree of screening in weakly excited (n=10^{16}–10^{17}cm^{-3}) high-quality h-GaN into account, we conclude that the PL spectra of h-GaN are dominated by excitonic emissions up to RT. The contribution of excitons on the PL spectra at RT has also been found in $CuAlS_2$ (Chichibu *et al.*, 1995) and $CuGaS_2$ (Shirakata *et at.*, 1996b) both of which have large E_{ex} values. The I_2 peak in Figures 5.2(a) and 5.5(a) is assigned to the recombination of excitons bound to a neutral donor because the peak shows a strong thermal quenching property and the energy separation from FE(A) (7 meV) agrees with that reported previously (Dingle, 1971; Naniwae *et al.*, 1990).

For c-GaN, the energy of the PL peak close to $E_{ex}(E_0)$, which is labelled FE(E_0) in Figure 5.5 (b) and X_A(PL) in Figure 5.3 (b), at 10K is 3.256 eV, and is nearly 9 meV smaller than $E_{ex}(E_0)$ = 3.265 eV. This energy agrees well with the difference between the peak energy of FE (3.272 eV) and that of BE (3.263 eV) reported for the strain-free c-GaN (Menniger *et al.*, 1996a). As shown in Figure 5.3(b), X_A shifts to higher energy with increasing temperature from 10K to 75K and then shifts to lower energy with increasing temperature up to RT. It is noted that the X_A peak energy agrees with $E_{ex}(E_0)$ above 75 K. Therefore, X_A is dominated by FE recombinations FE(E_0) and FE($E_0+\Delta_0$) at those temperatures. The temperature coefficients of E_{ex} between 180 K and RT for both polytypes are close to each other (3.6–3.7x10^{-4} eV/K). Therefore, we assign that the X_A peak is an envelope of I_2 and two FEs, and I_2 is dominant at low temperature and FEs are dominant above 75K. The transition structure in the PR spectra of c-GaN is also dominated by FE resonance up to RT similar to h-GaN, which is evidence that exciton transition is dominant in both polytypes. Menniger *et al.* (1996b) have first claimed that the PL peak at 3.2 eV at RT was excitonic emission, although they have not measured E_{ex}.

It is known that in QWs, Coulomb interaction of excitons is enhanced due to confinement of wavefunctions in the wells. As a result, E_b in QWs is increased

(Miller *et al.*, 1984; 1985; Bastard *et al.*, 1982) ideally as large as four times its bulk value (Shinada and Sugano, 1966), which will be treated in Section 5.3.1. One of other important properties of excitons is their decay dynamics (Harris *et al.*, 1995; Kawakami *et al.*, 1996), which will be shown in Section 5.2.2. Moreover, high excitation of semiconductors preferably at low temperature produces dense exciton gas in which exciton molecules are produced. Hvam and Ejder (1976) have explained new PL peaks at 3.453 and 3.448 eV under high excitation intensities as inelastic exciton-exciton scattering interactions. The simplest form is a biexciton that is composed of two excitons interacting with each other. Detailed explanation of biexcitons in GaN (Okada *et al.*, 1996; Kawakami *et al.*, 1996) will be given in Section 5.2.2.

5.2.2 Biexcitons in GaN

5.2.2.1 Introduction

Free excitons in hexagonal-GaN (*h*-GaN) are composed of three bands labelled E_{XA}, E_{XB} and E_{XC}, which are transitions from the conduction (Γ_{7c}) to the A (Γ_{9v}), B (Γ_{7uv}) and C (Γ_{7lv}) valence bands, respectively (Dingle, 1971). The binding energy (E_{ex}) and the effective Bohr radius (r_B) of the A exciton (E_{XA}) in *h*-GaN have been reported to be about 28 meV (Monemar, 1974) and about 2.9nm, respectively. Since the E_{ex} value is larger than the thermal energy of RT, excitonic emissions have been observed up to RT in high-quality *h*-GaN epilayers (Chichibu *et al.*, 1996a). It is interesting to compare the values of E_{ex} and r_B in GaN with those of ZnSe, which are 18 meV and 3.5nm, respectively. The Mott's screening density of excitons in GaN is estimated to be about 2×10^{18}cm^{-3} which is about two times as large as that in ZnSe. Consequently, a variety of interesting phenomena based on many body effects of excitons such as exciton-exciton collision, biexciton (exciton molecule), electron-exciton interaction, etc., can be expected in GaN. Among them, biexciton is a molecular-like quasiparticle, which is formed by two interacting excitons with opposite spins. Emission due to biexcitons may lead to high optical gain and low-lasing threshold (Shaklee *et al.*, 1971; Holst *et al.*, 1998), in contrast to the direct recombination of free excitons, where no laser action is possible. However, it was not until recently that the biexciton in *h*-GaN was experimentally identified (Okada *et al.*, 1996; Kawakami *et al.*, 1996). This is not only because recombinations in many samples have been dominated by the non-radiative recombination channel, but also because the appearance of biexcitonic luminescence was obscured by the impurity-bound-excitons. In this chapter, dynamical behavior of free excitons, impurity-bound excitons and biexcitons in *h*-GaN epitaxial layer is assessed by employing time-resolved photoluminescence (TRPL) spectroscopy.

5.2.2.2 Experimental procedure

The GaN epitaxial layer assessed here was grown on a (0001) oriented sapphire (Al_2O_3) substrate with a thickness of 4μm by a two-flow metalorganic chemical vapor deposition (TF-MOCVD) technique. The source gases used were trimethylgallium [$(CH_3)_3Ga$] and ammonia (NH_3). The detailed growth condition has been reported elsewhere (Nakamura, 1991).

In order to make assignment of each emissions under low excitation conditions, the cw PL was measured by the excitation of He-Cd laser (325nm) and it was compared with the reflection spectrum. Such PL detections were carried out using a cooled-charge coupled device (CCD) and a 50-cm monochromator with a 600-lines/mm grating. TRPL measurements were performed with a synchroscan streak camera in conjunction with a 25cm monochromator using a 1200 lines/mm grating. The spectral resolution of both measurements are as small as 0.2nm. Pulsed excitation was provided by the frequency-tripled beam of a mode-locked Al_2O_3:Ti laser, which was pumped by Ar^+-laser.[1] The wavelength, the pulse width and the repetition rate were 266nm, 1.5 ps and 80.00000 MHz, respectively. All measurements have been done at 18K.

5.2.2.3 Results and discussion

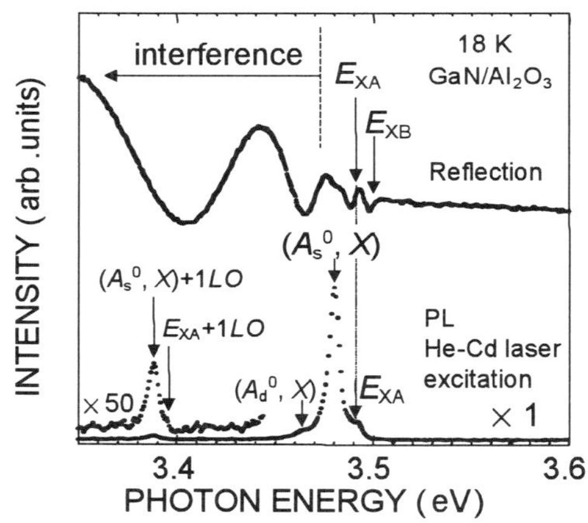

Figure 5.7 PL spectrum under cw excitation (He-Cd laser: 325nm) condition. Reflection spectrum is also shown in the figure.

PL and reflection spectra from a sample are shown in Figure 5.7. By the comparison between the two, the PL peak at 3.4921 eV is attributed to the A free exciton (E_{XA}). This energy position is shifted towards the higher photon energy side by about 14 meV compared to that of the nearly strain-free homoepitaxial *h*-GaN layer (Pakula *et al.*, 1996). This is because the epilayer suffers biaxial compression of the order of 10 kbar (Gil *et al.*, 1995) at the face perpendicular to the (0001) direction, which is mainly induced by the difference of thermal expansion coefficient between the substrate and the epilayer. The main PL peak is located at 3.4805 eV. The energy difference between this peak and the E_{XA} is as

[1] Instrumentation of TRPL spectroscopy is described in Section 5.2.4.4.1.

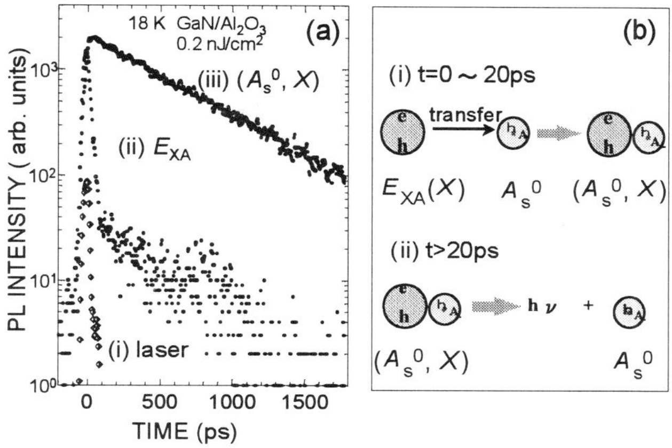

Figure 5.8 (a) PL decay spectra of the E_{XA} and the (A_s^0, X) under fluence of 0.2 nJ/cm^2. (b) The model of dominant optical processes under low photo-excitation.

large as 11.6 meV, which corresponds to the localization energy of excitons to the impurity centers. According to the reported value (Pakula *et al.*, 1996), this emission is attributed to the recombination of exciton bound to a shallow acceptor $[(A_s^0, X)]$. In this particular sample, no donor bound exciton $[(D^0, X)]$, which should be observed at about 6 meV lower energy side from the E_{XA} (Naniwae *et al.*, 1990), was detected. The emission at 3.4633 eV is probably ascribed to the exciton bound to a deep acceptor $[(A_d^0, X)]$.

Time integrated PL spectra taken under pulsed excitation condition are very similar to the cw PL spectrum shown in Figure 5.7 if the fluence (I_{ex}) which

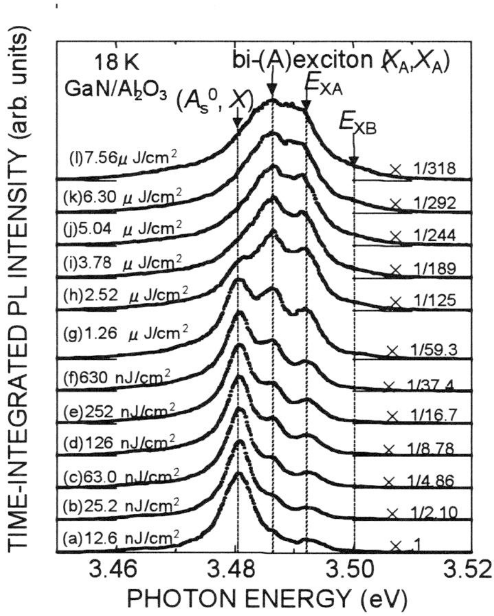

Figure 5.9 Time-integrated PL spectra taken under pulsed excitation for different fluence.

corresponds to the excitation energy density per pulse is smaller than 1 nJ/cm^2. The transient response of spectrally integrated PL is shown in Figure 5.8 for both the E_{XA} and the (A_s^0, X) bands under I_{ex} =0.2 nJ/cm^2. The lifetime of PL (τ_{PL}) is expressed as

$$\frac{1}{\tau_{PL}} = \frac{1}{\tau_{rad}} + \frac{1}{\tau_{non\text{-}rad}} + \frac{1}{\tau_{trans}}, \qquad (5.11)$$

where τ_{rad} and $\tau_{non\text{-}rad}$ represent lifetimes of radiative and nonradiative recombinations, while τ_{trans} is the term due to the transfer process to lower energy levels. The E_{XA} decay [I(t)] was fitted by the double exponential characteristics as

$$I(t) = A_1 \exp(-\frac{t}{\tau_1}) + A_2 \exp(-\frac{t}{\tau_2}), \qquad (5.12)$$

where the best fitting was attained with τ_1 = 14 ps, τ_2 = 457 ps and A_1/A_2 = 98.5. The intensity of (A_s^0, X) becomes maximum at t = 22ps after the peak of the E_{XA} is reached. The (A_s^0, X) decay was well fitted with single exponential curve whose lifetime is 585 ps. The fast component of 14 ps in the E_{XA} suggests that the photo-generated free A exciton is effectively trapped to the shallow acceptor level. This is confirmed by the fact that A_1/A_2 decreases with increasing I_{ex} which can be understood as a result of the saturation of trap centers. Long decay times of both the (A_s^0, X) emission and the slow component of the E_{XA} emission indicate the good optical quality of the sample.

Figure 5.9 shows the time-integrated PL spectra taken at various excitation intensity. The I_{ex} value ranged from 12.6 nJ/cm^2 to 7.56 μJ/cm^2. The (A_s^0, X)

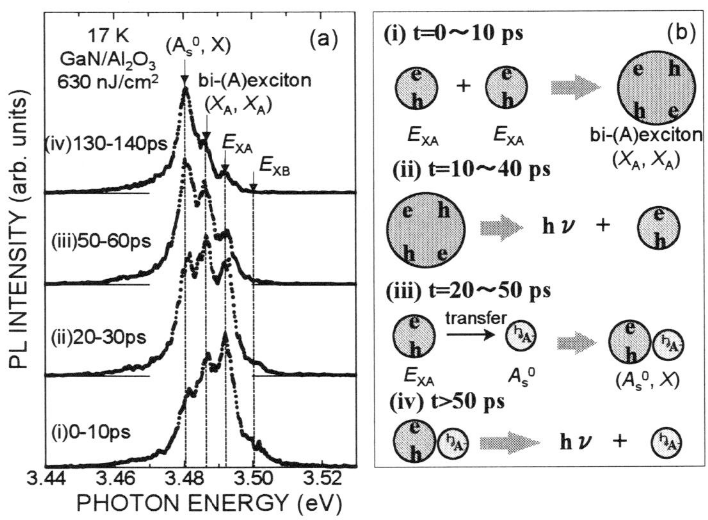

Figure 5.10 (a) Time-resolve PL spectra taken under fluence of 630 nJ/cm^2 at time delays of (i) 0-10 ps, (ii) 20-30 ps, (iii) 50-60 ps and (iv) 130-140 ps. (b) The model of dominant optical processes under moderately high photo-excitation.

intensity itself tends to saturate at higher excitation due to the filling of shallow acceptor centers. Accordingly, the E_{XA} and the E_{XB} bands increase in intensity gradually. A new peak appears at 3.4864 eV if the I_{ex} exceeds the value of about 10 nJ/cm^2. This band grows superlinearly with the I_{ex}.

In order to assess the recombination dynamics under moderately high excitation condition, time-resolved PL spectra under I_{ex}= 630 nJ/cm^2 were plotted, as shown in Figure 5.10 at time delays of (i) 0-10 ps, (ii) 20-30 ps, (iii) 50-60 ps and (iv) 130-140ps after the pulsed excitation. Just after the excitation (i) the E_{XA} is the predominant emission, while the main peak changes to the new emission band in the time-window of (ii). Since the lifetime of this band is short, it almost disappears at the window of (iv). Judging from these excitation dependence and temporal behavior, the biexcitonic emission, where a biexciton (exciton molecule) emits a photon leaving an exciton (Fig. 5.10 (b) (ii)), is the most plausible model for the new peak. If it is the case, the energy separation of this peak from the E_{XA} (5.7 meV) corresponds to the binding energy of biexciton (E_b) (Kawakami *et al.*, 1996). This value agrees very well with the recently published value of (5.7$\pm$0.3) meV, which was obtained from four wave mixing (FWM) experiments by tuning the excitation lasers to the low-energy tail of the E_{XA} resonance with two beams having perpendicular linear polarization (Zimmermann *et al.*, 1997). As shown in Table 5.2, the E_b/E_{ex} value for GaN is calculated to be 0.20 which is comparable to those in II-VI widegap semiconductors such as ZnSe (0.19) (Nozue *et al.*, 1981), CdS (0.20) (Shionoya *et al.*, 1973), ZnS (0.25) (Yamada *et al.*, 1996) and ZnO (0.25) (Koch *et al.*, 1978), as well as to that in CuCl (0.13) (Mysyrowicz *et al.*, 1968). Considering the Hayne's rule for the evaluation of the E_b value (Huang, 1973), the E_b value of 5.7 meV is reasonable for GaN. It should be noted that the transition energy of the biexciton is by chance very close to the (D^0, X) in the case of GaN. Therefore, the absence of the (D^0, X) emission is preferable for the study of biexcitons in GaN.

The transient curve of this biexciton emission under I_{ex}= 630 nJ/cm^2 is shown in Figure 5.11. The decay was fitted by double exponential curve with τ_1 = 39 ps, τ_2 = 225 ps and A_1/A_2 = 2.40. It is likely that this fast decay component represents the radiative lifetime of biexciton because it is almost constant under the excitation condition in the range from 63.0 nJ/cm^2 to 2.52 μJ/cm^2. The slow decay

Table 5.2 Effective Bohr radius, binding energies of excitons and biexcitons in various widegap semiconductors.

Material	Effective Bohr radius (nm)	Binding energy of exciton (meV)	Binding energy of biexciton (meV)
ZnSe	3.5	18	3.5
CdS	2.7	28	6.4
ZnS	2.5	36	9
ZnO	1.4	59	15
CuCl	0.7	213	27.8
GaN	2.9	28	5.7

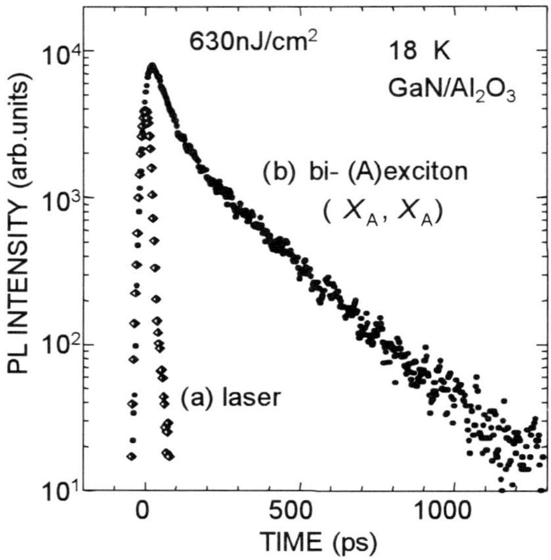

Figure 5.11 PL decay spectrum of the biexcitonic emission band under fluence of 630 nJ/cm^2. The decay shows two exponential processes whose fast and slow decay components are 39.5 ps and 225 ps, respectively.

component of 225 ps probably represents the recurring process between excitons and biexcitons.

5.2.2.4 Conclusion

The recombination dynamics of the *h*-GaN epitaxial layer has been characterized by means of TRPL spectroscopy. The biexcitonic emission has been observed under high excitation conditions. The binding energy of biexciton is estimated to be about 5.7 meV. Future subject is an identification of biexcitons in low-dimensional structures based on nitride semiconductors, in which the enhancement of E_b is attained (Kleinman, 1983; Nair and Takagahara, 1997) possibly leading to the stabilization of biexcitons even up to RT.

Acknowledgement

Experimental results reported here is a collaborative work between Fujita laboratory at Kyoto University and Dr. Shuji Nakamura's group at Nichia Chemical Industries Ltd. Thanks are due to Prof. Sg. Fujita for valuable discussion and comments. The author is grateful to Z.G. Peng and Y. Narukawa for their contribution to the optical characterization. A part of this measurement was done using the facility at the "Venture Business Laboratory (VBL)" in Kyoto University.

5.2.3 Strain effects

All samples investigated in the above two subsections were grown on sapphire substrates using the TF-MOCVD method. Lattice mismatch between GaN and sapphire leads to compressive strains ε_{xx} ($=\varepsilon_{yy}$) in the lateral plane of the GaN epilayers. As the stress component ε_{zz} vanishes, there exists tensile strain ε_{zz} along the Z-axis, i.e., the normal direction of the sample surface. It is known that the thicker the epilayer thickness is, the smaller the magnitude of such residual biaxial strains becomes. It is also known that the residual biaxial strains change the valence band structure and the change is reflected in the optical spectra near the band gap region. Therefore, from a physical view point, a systematic optical study on the near bandgap region using samples with various thickness and/or those grown on different substrates is expected to clarify not only the effect of strains on the valence band but also important physical parameters of the valence band. The systematic study was carried out by Chichibu *et al.* (1996b; 1997b), Shikanai *et al.* (1997), and Azuhata *et al.* (1997). Independently, Gil *et al.* (1995) and Gil and Briot (1997b) made a similar study. Later, many researchers studied the strain effect on the valence band at each stage of achieving sample quality refinement. On the other hand, from an engineering viewpoint, the strain effect attracts attention as a tool to reduce the density-of-states (DOS) of the topmost valence bands because the low DOS is believed to be favorable to the laser action (Kamiyama *et al.*, 1995; Nido 1995). Uniaxial strain effect was recently reported by Yamaguchi A. *et al.* (1997; 1998b).

The biaxial strain dependence of the valence band in bulk materials with the wurtzite structure is reviewed in this subsection. Simple calculations needed for analyzing the experimental data are presented. Then, the experimental data of Chichibu *et al.* (1996b; 1997b) and Shikanai *et al.* (1997) are introduced, which summarize the ε_{zz} dependence of the A, B, and C exciton resonance energies. Discussions are given on physical parameters obtained analyzing the data with the equations derived theoretically. A brief discussion is also given on the effect of the uniaxial strain.

Let us consider the valence band energy spectra near the Γ point in the Brillouin zone under the biaxial strain. In this case, the following relations hold:

$$\varepsilon_{xx} = \varepsilon_{yy} = -\frac{C_{33}}{2C_{13}}\varepsilon_{zz} \qquad \text{and} \qquad \varepsilon_{yz} = \varepsilon_{zx} = \varepsilon_{xy} = 0, \tag{5.13}$$

where C_{ij} denote the components of the elastic stiffness constants. The starting Hamiltonian suitable for the space group $C_{6v}{}^4$ is given in the following matrix form (Bir and Pikus, 1974),

$$\boldsymbol{H} = \left\| \begin{matrix} F & 0 & -H^* & 0 & K^* & 0 \\ 0 & G & \Delta & -H^* & 0 & K^* \\ -H & \Delta & \lambda & 0 & I^* & 0 \\ 0 & -H & 0 & \lambda & \Delta & I^* \\ K & 0 & I & \Delta & G & 0 \\ 0 & K & 0 & I & 0 & F \end{matrix} \right\|, \tag{5.14}$$

where

$$
\begin{aligned}
&\Delta=\sqrt{2}\Delta_3, \quad H=i\left(A_6k_zk_+ + D_6\varepsilon_{z+} + A_7k_+\right), \quad I=i\left(A_6k_zk_+ + D_6\varepsilon_{z+} - A_7k_+\right),\\
&F=\Delta_1+\Delta_2+\lambda+\theta, \quad G=\Delta_1-\Delta_2+\lambda+\theta, \quad K=A_5k_+^2+D_5\varepsilon_+,\\
&\lambda=A_1k_z^2+A_2k_\perp^2+D_3\varepsilon_{zz}+D_4\varepsilon_\perp, \quad \theta=A_3k_z^2+A_4k_\perp^2+D_3\varepsilon_{zz}+D_4\varepsilon_\perp,
\end{aligned} \tag{5.15}
$$

with

$$
\begin{aligned}
&k_+=k_x+ik_y, \quad k_\perp^2=k_x^2+k_y^2,\\
&\varepsilon_{z+}=\varepsilon_{xz}+i\varepsilon_{yz}, \quad \varepsilon_\perp=\varepsilon_{xx}+\varepsilon_{yy}, \quad \varepsilon_+=\varepsilon_{xx}-\varepsilon_{yy}+2i\varepsilon_{xy}.
\end{aligned}
$$

Here $\boldsymbol{k}$, ε, A_i (i=1-7) and D_i (i=1-6) denote, respectively, the wave vector, the strain, the valence band parameters, and the deformation potential constants. The quantity Δ_1 is called the crystal-field splitting and the quantity $3\Delta_2$ is usually put equal to $3\Delta_3$, and called the spin-orbit splitting. For simplicity, we use the cubic approximation throughout this subsection. In this approximation the constants Δ_i, A_i, and D_i satisfy the relations (Bir and Pikus 1974),

$$
\begin{aligned}
&\Delta_2=\Delta_3, \quad 4A_5-\sqrt{2}A_6=-A_3, \quad 2A_4=-A_3=A_1-A_2, \quad A_7=0,\\
&4D_5-\sqrt{2}D_6=-D_3, \quad 2D_4=-D_3=D_1-D_2.
\end{aligned} \tag{5.16}
$$

It is straightforward to solve the eigenvalue problem with Equation (5.14). The energies at the Γ point, i.e., $\boldsymbol{k}=0$, are

$$
E_A^0=\Delta_1+\Delta_2, \tag{5.17a}
$$

$$
E_{B,C}^0=\frac{\Delta_1-\Delta_2}{2}\pm\left[\left(\frac{\Delta_1-\Delta_2}{2}\right)^2+2\Delta_2^2\right]^{1/2}, \tag{5.17b}
$$

for the strain-free crystal and

$$
E_A^1=E_A^0+\left(D_1-\frac{C_{33}}{C_{13}}D_2\right)\varepsilon_{zz}+\left(D_3-\frac{C_{33}}{C_{13}}D_4\right)\varepsilon_{zz}, \tag{5.18a}
$$

$$
\begin{aligned}
E_{B,C}^1=&\frac{1}{2}\left[\Delta_1-\Delta_2+\left(D_3-\frac{C_{33}}{C_{13}}D_4\right)\varepsilon_{zz}\right]+\left(D_1-\frac{C_{33}}{C_{13}}D_2\right)\varepsilon_{zz}\\
&\pm\left\{\frac{1}{4}\left[\Delta_1-\Delta_2+\left(D_3-\frac{C_{33}}{C_{13}}D_4\right)\varepsilon_{zz}\right]^2+2\Delta_2^2\right\}^{1/2},
\end{aligned} \tag{5.18b}
$$

for the crystal under the biaxial strain. Here the subscripts B and C correspond, respectively, to plus and minus signs in the right hand side of Equations (5.17b) and (5.18b). In Equations (5.17) and (5.18), the relations given by Equation (5.13) are used. $E_A^1(E_A^0)$, $E_B^1(E_B^0)$, $E_C^1(E_C^0)$ denote the bands belonging to the irreducible representations Γ_9, Γ_7, and Γ_7, respectively. It is obvious that Equation (5.18) reduces to Equation (5.17) for $\varepsilon_{zz} = 0$. Note that Equations (5.17a) and (5.17b) are nothing but the equations derived by Hopfield (1960). Expressions for the strain dependence of the effective masses at near Γ point are also obtained straightforwardly, though they are not shown here to save space. Note that the effective mass of the A-valence band does not depend on the strain.

To interpret experimental data, the expressions for the resonance energies are needed. To derive these, let us define the conduction band minimum energy at Γ point, E_{CB}, by

$$E_{CB} = E_{gA} + E_A^0 + \Xi_D \varepsilon_{zz}, \tag{5.19}$$

where E_{gA} is the energy gap between the conduction and the A-valence bands in the strain-free crystal and Ξ_D the combined dilational component of the deformation potential acting on the conduction band. Here the conduction band is assumed as isotropic. This is not a bad approximation as shown below. The resonance energies observed experimentally are given by

$$E_\alpha = E_{CB} - E_{ex,\alpha} - E_\alpha^1, \quad \alpha = A,\ \ B, \text{ and } C. \tag{5.20}$$

Here $E_{ex,\alpha}$ is the exciton binding energy of the band α. The latest experiment (Yamaguchi, A. *et al.*, 1998b) could determine those of the A- and B-excitons, but not of the C-exciton. Thus, ambiguity enters into the calculations at this stage because the estimation of $E_{ex,\alpha}$ depends on the valence band parameters and the electron effective masses, and the back ground dielectric constants. In this subsection the estimation of $E_{ex,\alpha}$ is made according to Shikanai *et al.* (1997) as shown below.

Figure 5.12 shows the A-, B-, and C-exciton ground state resonance energies and the first excited resonance energy of the A-exciton as a function of ε_{zz}, which are determined at T = 10K by the PR method for a number of GaN epilayers with various thickness grown on sapphire, Si, and 6H-SiC substrates. In Figure 5.12 are included the data measured by Monemar *et al.* (1996b) for the strain-free crystal. The magnitude of ε_{zz} was determined measuring the c-axis length by X-ray diffraction method for each sample and the relaxed c-axis length was taken as c_0 = 0.5185nm according to Leszczynski *et al.* (1996). The following are found from Figure 5.12; the resonance energies, E_A, E_B, E_C , and $E_{A,n=2}$ increase with increasing ε_{zz}. E_A is a linear function of ε_{zz} and the ground state binding energy of the A-exciton, $E_{ex,A}$, is not affected by strain, as expected from the theory mentioned above. The least-square fit to the data of E_A and $E_{A,n=2}$ yields

$$E_A = 3.478 + 15.4\varepsilon_{zz} \tag{5.21a}$$

and

$$E_{A,n=2} = 3.498 + 15.4\varepsilon_{zz}. \tag{5.21b}$$

From the difference between E_A and $E_{A,n=2}$, $E_{ex,A}$ is estimated to be 26 meV, which is simply set to four-thirds of the energy difference. This value is comparable to $E_{ex,A}$ of 24 meV reported by Gil *et al.* (1995). The linear slope of 15.4 eV for the A-exciton is comparable to that of 12 eV reported by Amano *et al.* (1988). From

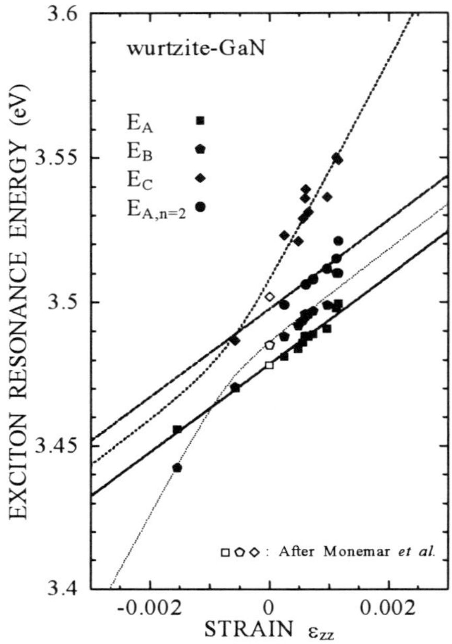

Figure 5.12 Strain dependence of the free exciton resonance energies in wurtzite GaN. $\varepsilon_{zz} = (c-c_0)/c_0$ denotes strain along the Z-axis, where c_0 is the lattice constant, c, for the unstrained crystal. Closed plots are the experimental data given by Chichibu *et al.* (1996b; 1997b) and Shikanai *et al.* (1997). The open symbols are after Monemar *et al.* (1996b). Lines are the free-exciton resonance energies theoretically calculated. E_A, E_B, E_C, and $E_{A,n=2}$ represent the resonance energies of the A-, B-, and C-free excitons, and the first excited states of the A-exciton, respectively.

Equations (5.19), (5.20), (5.21a), and $E_{ex,A}$= 26 meV, E_{gA} is determined as 3.504 eV. Since the slope of E_A with respect to ε_{zz}, 15.4 eV, is expressed from Equations (5.18a), (5.19), and (5.20) as $\Xi_D-(D_1-D_2D_{33}/C_{13})-(D_3-D_4C_{33}/C_{13})$, parameters to be solved, i.e., Δ_1, Δ_2=Δ_3, and $\Xi_D-(D_1-D_2C_{33}/C_{13})$, can be treated as adjustable parameters. Note that Ξ_D and $(D_1-D_2C_{33}/C_{13})$ cannot be determined separately at this stage. Using the nonlinear least-square method, the values for those parameters are obtained as Δ_1= 22meV, Δ_2=Δ_3= 5 meV, $\Xi_D-(D_1-D_2C_{33}/C_{13})$ = 38.9 eV, and,

$$(D_3-D_4C_{33}/C_{13}) = 23.6 \text{ eV}. \tag{5.22}$$

The strain dependence of $E_{ex,B}$ and $E_{ex,C}$ has to be known to reproduce the experimental data. To estimate them based on the obtained $E_{ex,A}$ of 26 meV, the values is needed for the electron effective masses, the valence band parameters, and the dielectric constants. Since Suzuki *et al.* (1995) reported simultaneously the values for the former two parameters, their values are used herein. According to them, effective masses of the electron are $m_{//}^*/m_0$ = 0.2 and $m_\perp^*/m_0$ = 0.18 and the valence band parameters are A_1 = –6.56, A_2 = –0.91, A_3 = 5.65, A_4 = –2.83, A_5 = –3.13 and A_6 = –4.86. Here $m_{//}^*$ ($m_\perp^*$) denotes the electron effective mass parallel (perpendicular) to the c-axis. The values of the electron effective masses show that the conduction band is nearly isotropic. This guarantees Equation (5.19). When $E_{ex,A}$ is calculated taking account of the anisotropy of the effective reduced masses with the above-mentioned parameters values and with typical values for the background dielectric constants $\varepsilon_{\perp,0} = 9.28$ and $\varepsilon_{//,0} = 10.1$ (Azuhata *et al.*, 1995; 1996), $E_{ex,A}$ is underestimated as 19 meV. To reproduce the value of 26 meV experimentally obtained, slightly smaller values of $\varepsilon_{\perp,0} = 7.87$ and $\varepsilon_{//,0} = 8.57$ have to be used, where the ratio of $\varepsilon_{\perp,0}/\varepsilon_{//,0}$ is fixed to 0.92 as obtained using the above typical values. The resonance energies, E_A, E_B, and E_C, calculated with the use of the obtained values for the physical parameters discussed above are shown in Figure 5.12, where the binding energy of the first excited state for the A-exciton is simply set to be a quarter of $E_{ex,A}$. Good agreement is achieved between the experimental data and the theoretical curves.

It is found from Figure 5.12 that controlling the magnitude of residual strain can modify the energy gaps by a considerable amount. For large in-plane tensile strain, the valence band structure may be considered as one simple band. Before this situation is seen, the band anticrossing occurs between the B- and C-valence bands. At the anticrossing point, characters of them are interchanged. Strictly speaking, the reliability of the results derived herein depends on the valence band parameters used herein. Therefore, further work should be done to eliminate the ambiguity. Bearing this in mind, brief discussions will be given on the values for several physical parameters obtained up to date in the following.

The values for Δ_1, Δ_2, and Δ_3 reported so far are summarized in Table 5.3. Generally, ab-initio calculations overestimate the value of Δ_1. The value of Δ_1

Table 5.3. The values for the band-splitting parameters, Δ_1, Δ_2, and Δ_3, and the deformation potential constants, D_3 and D_4.

Reference	Δ_1 (meV)	Δ_2 (meV)	Δ_3 (meV)	D_3 (meV)	D_4 (meV)
Experimental					
Dingle *et al.* (1971)	22	3.7	3.7	-	-
Gil *et al.* (1995)	10.0	6.2	6.2	-	-
Volm *et al.* (1996)	11	5.7	5.7	-	-
Korona *et al.* (1996)	9.3	6.6	6.6	-	-
Reynolds *et al.* (1996)	24.7	5.8	5.8	-	-
Shikani *et al.* (1997)	22	5	5	8.82	-4.41
Edwards *et al.* (1997)	9.8	5.7	5.7	-	-
Yamaguchi *et al.* (1997)	-	-	-	-	-3.4
Stepniewski *et al.* (1997)	8.8	5.9	5.9	-	-
Yamaguchi *et al.* (1998a)	10	5.5	6.0	7.2	-3.6
Theory					
Suzuki *et al.* (1995)	72.9	5.4	5.4	3.03	-1.52
Volm *et al.* (1996)	35	6	6	-	-
Ohtoshi *et al.* (1996)	-	-	-	6.6	-3.3
Majewski *et al.*(1996)	35.3	5.3	6.7	6.61	-3.55
Wei and Zunger (1996)	43	4.3	4.3	-	-
Chen *et al.* (1996)	37.5	4	4	-	-
Kim *et al.* (1997)	36	5.0	5.9	5.7	-2.85
Shimada *et al.* (1998)	50.4	-	-	5.80	-3.25

obtained herein is close to that obtained for zincblende GaN (Flores *et al.*, 1994). The data given by Yamaguchi, A. *et al.* (1998a) are for the sample grown using the lateral epitaxial overgrowth (LEO) technique. The values of Δ_1 obtained experimentally appear to have sample dependence, i.e., the residual strain dependence. Volm *et al.* (1996) pointed out that $\Delta_1 = 0.035 + a\varepsilon_{xx}$ eV with a = –9 eV. Further studies will be needed on the value of Δ_1 for strain-free GaN.

The shear deformation potential constants are estimated from Equation (5.22) as $D_3 = 8.82$ and $D_4 = -4.41$ eV for $C_{13} = 11.4\times10^{11}$ and $C_{33} = 38.1\times10^{11}$ dyn/cm^2 (Yamaguchi, M. *et al.*, 1997). The values for D_3 and D_4 reported so far are also tabulated in Table 5.3. Both the variations of the reported values are large. When they are determined experimentally, this is in part attributed to uncertainty of the values for the elastic stiffness constants as will be seen in the Section 5.2.4. Ab-initio calculations of the values for the deformation potential constants were carried out by Suzuki and Uenoyama (1996) and Shimada *et al.* (1998). The former (latter) reported the following values: $D_1 = -13.9$ (–20.0), $D_2 = -13.7$ (–14.2), $D_3 =$ 2.99 (5.80), $D_4 = -1.63$ (–3.25), and $D_5 = -2.04$ (–2.85) eV. The differences are attributed to the fact that the latter takes account of the internal strain correction but the former does not. Except for Table 5.3, experimental results (Perlin *et al.*, 1992; Camphausen *et al.*, 1988; Naniwae *et al.*, 1990; Shan *et al.*, 1995) and theoretical calculations (Perlin *et al.*, 1992; Christensen and Gorczyca 1994) were reported on the pressure coefficient and the deformation potential constants for the band gap of GaN and related nitrides.

Yamaguchi, A. *et al.* (1997; 1998a) investigated the change of the valence band structure of GaN grown on sapphire substrate under uniaxial stress along the normal direction of the (11-20) surface. It was observed that the energy separation between the A and the B valence bands increased as increasing the magnitude of the stress. This indicates that the uniaxial stress is effective to reduce the DOS at the valence band top. They also determined the deformation potential constant D_5 as –3.3 eV from analysis of their data based on Equation (5.14), which is close to the value reported by Shimada *et al.* (1998).

As having been already mentioned many times, there is ambiguity on the valence band parameters (Suzuki *et al.*, 1995; Kim *et al.*, 1997; Majewski *et al.*, 1997; Yeo *et al.*, 1998) and the electron effective masses (Knap *et al.*, 1997; Perlin *et al.*, 1996; Drechsler *et al.*, 1995; Wang *et al.*, 1996). Further studies are indispensable to remove the ambiguity.

5.2.4 Phonons in nitrides

Information on phonon structures in semiconductor is important in considering the electronic transport, the non-radiative electron relaxation process, and so on. From the experimental point of view, optical phonons at near the Γ point have been usually studied by Raman and infrared spectroscopy and acoustic phonons by Brillouin, resonance ultrasound, and surface acoustic-wave spectroscopy. Phonons, with which carriers interact strongly, can be studied using the phonon replica peaks of PL and low energy tail portions of OA spectra. However, full information on phonon structure over entire Brillouin zone (BZ) is only given by inelastic neutron scattering method. On the other hand, from the theoretical point of view, recent first-principles calculations can make it possible to calculate not only phonon frequency at the Γ point but also the phonon dispersion over entire BZ. In this subsection phonon modes measured and/or calculated to date is reviewed mainly for hexagonal GaN with wurtzite structure. A brief description is also given on carrier-phonon interaction.

Space group of wurtzite group III nitrides, MN (M=B, Al, Ga, and In) is C_{6v}^{4}. The unit cell contains two molecules. Therefore, there are twelve phonon modes for a given wave vector. According to the factor group analysis at the Γ point, optical phonons belong to the irreducible representations:

$$\Gamma_{opt} = A_1(Z) + 2B + E_1(X,Y) + 2E_2, \tag{5.23}$$

and acoustic phonons to $\Gamma_{ac} = A_1 + E_1$ where letters in parentheses describe the polarization direction. Optical phonons belonging to $A_1(Z)$, $E_1(X,Y)$ and E_2 modes are Raman active, those to $A_1(Z)$ and $E_1(X,Y)$ modes are infrared active, and those to B modes are silent. $A_1(Z)$ and $E_1(X,Y)$ mode phonons are polar phonons and E_2 mode phonons are non-polar ones. For the polar modes, there exist transverse optical (TO) and longitudinal optical (LO) modes.

Optical phonon frequencies of GaN measured to date are summarized in Table 5.4, where results using Raman and IR spectroscopy are given. Except for these, a number of groups (Reynolds *et al.*, 1998; for an excellent review of studies before 1992, see Strite and Morkoç, 1992) have reported optical phonon frequencies deduced from phonon replicas in PL spectra. Most of the measurements have observed similar phonon frequencies, in spite of the different growth technique of the sample and the difference of the sample quality varying from sample to sample. In experiments using GaN epitaxial layers grown on sapphire substrates, one should pay special attention to distinguish the phonon modes of GaN from those of sapphire. Quite recently Deguchi *et al.* (1998b) carried out Raman scattering measurements using a "*pure* GaN substrate" grown using LEO technique and identified all symmetry allowed optical phonon frequencies at the Γ point, to find that they are lower by 2-3cm^{-1} than those in GaN on sapphire (Azuhata *et al.*, 1995) except for the low frequency E_2 mode.

One of basic features of optical phonons in wurtzite-type crystals (Argüello *et al.*, 1969) involving GaN, is that the electrostatic forces dominate over the anisotropy in the short-range forces, in other words, the LO-TO splitting is much greater than the A_1-E_1 splitting. Thus A_1 and E_1 modes are mixed to become quasi-LO, quasi-TO and pure TO modes except when phonons propagate parallel to the XY-plane or to the Z-axis. Neglecting this fact, missassignment of phonon frequencies for the A_1(LO) and/or E_1(LO) modes occurs as seen in several literatures.

Note that optical phonon frequencies at Γ point except for LO modes have also been calculated using a frozen phonon approach in the framework of first-principle calculations by several groups (Miwa and Fukumoto, 1993; Gorczyca *et al.*, 1995; Kim *et al.*, 1996; Karch *et al.*, 1998; Shimada *et al.*, 1998). They are tabulated in Table 5.5. Most calculated frequencies for silent B mode phonons (Miwa and Fukumoto, 1993; Gorczyca *et al.*, 1995; Kim *et al.*, 1996; Karch *et al.*, 1998) lie in the range of 330-337cm^{-1} for low frequency B mode and of 677-720

Table 5.4 Optical phonon frequencies for GaN at the center of the Brillouin zone obtained experimentally using Raman (R) and/or infrared (IR) spectroscopy. TRR indicates time-resolved Raman spectroscopy. Question marks represent that the values are not specified in literatures.

Reference	Substrate	Film thickness (μm)	Temperature (K)	Phonon frequency (cm^{-1})						Exp.
				E_2		E_1		A_1		
				low	high	TO	LO	TO	LO	
h-GaN										
Wetzel *et al* .(1996a)		bulk	RT		565			529		R,IR
Kisielowski *et al.* (1996)		bulk	RT		566.2					R
Perlin *et al.* (1992)		needle,platelet	RT	144	568	560		531		R
Manchon *et al.* (1970)		needle	RT	145	568	559		533		R,IR
Cingolani *et al.* (1986)		needle	RT	143	568	559		533	710	R
Manchon *et al.* (1970)		needle	20	144	569	560		534		R,IR
Chtchekine *et al.* (1997)	sapphire	8	RT		(564)	(564)		532		R
Chen *et al.* (1997)	sapphire	2	RT		567				736	R
Li *et al.* (1998)	sapphire	1	RT		572					R
Shu *et al.* (1998)	sapphire	?	RT		564			534	731	R
Kuball *et al.* (1998)	sapphire	2	RT		567	559		534	734	R
Lee *et al.* (1998)	sapphire	3.5	RT		570.4	560.8	742.5	533.0		R
Demangeot *et al.* (1997)	sapphire	3	RT		568	559	740	532	734	R,IR
Davydov *et al.* (1997a)	sapphire	300	RT	144	568	559	?	531.8	?	R
Siegle *et al.* (1995)	sapphire	2	RT		570	561	742	533	735	R
Zhang *et al.* (1996)	sapphire	6	RT		578	569	750	545	735	R
Cingolani *et al.* (1986)	sapphire	up to 200	RT				741			R
Lemos *et al.* (1972)	sapphire	20	RT		568	559		533		R
Kubota *et al.* (1989)	sapphire	?	RT		568					R

Table 5.4 (continued)

Humphreys *et al.* (1990)	sapphire	0.36	RT		568			534		R
Burns *et al.* (1973)	sapphire	10-50	RT	144	569	560	749	532		R
Filippidis *et al.* (1996)	sapphire	200	RT	145						R
Kozawa *et al.* (1994)	sapphire	2	RT		572	563	745	534	736	R
Behr *et al.* (1996)	sapphire	10	RT				741	530		R
Murugkar *et al.* (1995)	sapphire	2.7 to 3.2	RT		570				738	R
Azuhata *et al.* (1995)	sapphire	2	RT	144	569	561	743	533	735	R
Tsen *et al.* (1996)	sapphire	2	25						741	TRR
Sheih *et al.* (1995)	sapphire	2	25		574			538	741	R
Zhang *et al.* (1997)	sapphire	1.3	10	144.2	569.2	560.0	746.6	533.5	739.3	R
Siegle *et al.* (1996)	sapphire	400	4.2		567			532		R
					569			534		
Siegle *et al.* (1997a)	sapphire	220	4.2		567	560	742	533		R
Ponce *et al.* (1996)	sapphire	?	RT?	143	569	559	742	532	733.4	R
Kuo *et al.* (1995)	sapphire	2	RT?		567	558	742	534	736	R
Popovici *et al.* (1997)	sapphire	1-3,Mg-doped	RT?		569	557			736	R
Wetzel *et al.* (1996b)	sapphire	2	RT?						736	IR
Kirillov *et al.* (1996)	sapphire	10	RT?	141.75	566.5	557				R
Deguchi *et al.* (1998)	pure 80 μm GaN substrate		RT	144	567	558	740	531	733	R
Tabata *et al.* (1996)	GaAs(001)	?	RT		571	556		537	737	R
Giehler *et al.* (1995)	GaAs(001)	0.2	80		569					R,IR
Ramsteiner *et al.* (1996)	GaAs(001)	?	10		569					R
Ramsteiner *et al.* (1998)	GaAs(001)	0.2	10?		569					R
Kirillov *et al.* (1996)	GaAs(001)	2.2	RT?		566.5				734	R

Table 5.4 (continued)

c-GaN				TO	LO	R
Ramsteiner *et al.* (1996)	GaAs(001)	?	10	555		R
Miyoshi *et al.* (1992)	GaAs(001)	0.4	RT	550	730	R
Ramsteiner *et al.* (1998)	GaAs(001)	0.2	?	555	740	R
Tabata *et al.* (1996)	GaAs(001)	?	RT	555	741	R
Mirjalili *et al.* (1998)	GaAs(001)	0.6	RT	552	738	R,IR
Siegle *et al.* (1995)	GaAs(001)	0.6	RT	555	740	R
Humphreys *et al.* (1990)	sapphire	0.81	RT		740	R
Giehler *et al.* (1995)	GaAs(001)	0.1	80	554	740	R,IR

Table 5.5 Calculated results for optical phonon frequencies of h-GaN at the center of the Brillouin zone.

	Phonon frequency (cm^{-1})					
	A_1(TO)	E_1(TO)	E_2	E_2	B	B
Reference			low	high	low	high
Miwa and Fukumoto (1993)	534	556	146	560	335	697
Gorczyca *et al.* (1995)	537	555	150	558	330	677
Kim *et al.* (1996)	570	585				
Karch *et al.* (1998)	541	568	143	579	337	720
Shimada *et al.* (1998)	544	566	185	557	526	584
Wei *et al.* (1997)	534	557	144	569	335	697

cm^{-1} for high frequency one. Shimada *et al.* (1998) reported somewhat different values of 526 cm^{-1} and 584 cm^{-1} for low and high frequency B modes, respectively. Wei *et al.* (1997) calculated the phonon frequencies at Γ point using a classical rigid ion model. They reported A_1(LO) and E_1(LO) mode phonon frequencies as 730 and 747cm^{-1}, respectively.

Optical phonon frequencies are known to change due to the hydrostatic pressure (Perlin *et al.*, 1992) and the residual biaxial strain (Kozawa *et al.*, 1995; Demangeot *et al.*, 1996; Davydov *et al.*, 1997). Among them, the low frequency E_2 mode is insensitive to the pressure and strain compared with other modes. Careful measurements of the optical phonon frequency shift can make it possible to estimate the residual biaxial strain in epitaxial layers.

LO mode phonons can couple to plasma oscillation of carriers via the Frölich interaction and this elementary excitation mode is called a LO phonon-plasmon coupling mode. In this case it follows that the LO phonon frequencies observed shift toward the higher frequency region and their line width becomes broader as the carrier concentration increases. This phenomena is used as an indicator of the residual carrier concentration of GaN (Perlin *et al.*, 1995; Zhang *et al.*, 1996; Wieser *et al.*, 1997; Harima *et al.*, 1997; Demangeot 1998) as in the other cubic semiconductors.

Optical phonon frequencies measured and calculated for h-InGaN and h-AlGaN are summarized in Table 5.6. For details, see each reference.

The dispersion relation of acoustic phonon is described as $\omega = qv_s$ within the Debye approximation, where ω is the angular frequency, q is the absolute value of the wave vector, and v_s is the sound velocity. ω can be measured using various experimental methods. v_s can be expressed using the elastic stiffness constants, C_{ij}, for example, see Vacher and Boyer (1972). Therefore, measurements of acoustic phonon frequencies give information on elastic stiffness constants.

In wurtzite structure, nonvanishing C_{ij} are C_{11}, C_{33}, C_{12}, C_{13}, C_{44}, and C_{66} with $C_{66} = (C_{11}-C_{12})/2$. Table 5.7 tabulates the values of C_{ij} for wurtzite GaN obtained experimentally and calculated to date. Depending on samples and experimental methods, the variation of the values for C_{ij} is large in comparison with optical phonon frequencies at the Γ point. According to Wright (1997), the results by Savestenko and Shele (1978) should be excluded, and the value of C_{33} given by Schwartz *et al.* (1997) is also too small. Although the value of C_{13} given by

Table 5.6 Optical phonon frequencies for AlGaN and InGaN at the center of the Brillouin zone obtained experimentally using Raman (R) and/or infrared (IR) spectroscopy.

Reference	Substrate	Film thickness (μm)	Temperature (K)	Phonon frequency (cm^{-1})						Exp.
				E_2		E_1		A_1		
				low	high	TO	LO	TO	LO	
AlGaN										
Brafman *et al.*(1968)		whisker	RT		665	667	910	667	910	R
Filippidis *et al.*(1996)		whisker	RT	249	657	670	913	610		R
McNeil *et al.*(1993)		bulk&poly	RT	252	660	673	916	614	893	R
Perlin *et al.*(1993)		needle	RT	241	660		924?	607		R
Carlone *et al.*(1984)	sapphire	1-10 amorphous	RT	298	746	788	825	514	650	R,IR
Sanjurjo *et al.*(1983)	sapphire	?	RT			671.6	895	659.3	888	R
Wetzel *et al.*(1996)	GaN/sap.	1							752	IR
Hayashi *et al.* (1991)										R
Carlone *et al.*(1984)	Si(111)	1-10	RT	303	426	614	821		663	R,IR
InGaN										
Kwon *et al.*(1996)	sapphire	?	RT		495				596	R

Yamaguchi, M. *et al.* (1998) is too small compared with other results, it is acceptable for of the following reasons. The values of C_{ij} in the GaN substrate are similar to the lower bound values of bulk GaN given by Polian *et al.* (1996) taking into account their experimental errors. Furthermore the ratio of $2C_{13}/C_{33}$ for the GaN substrate and the GaN on sapphire (Yamaguchi, M. *et al.* 1997) are 0.419 and 0.598, respectively, and, thus, the former is closer to the ratio of 0.38 obtained lattice constants measurements (Detchprohm *et al.* 1992) in relaxed GaN epilayer on GaN.

Here we would like to give a comment on the electron-phonon interaction. It is known that the interaction between LO phonons and carrier is dominated by the Frölich interaction in polar semiconductors. The strength is expected to be strong in GaN because of the small dielectric constant. Sheih *et al.* (1995) was the first to demonstrate that by using picosecond Raman spectroscopy, the Frölich interaction is much stronger than the deformation potential interaction in GaN. After this work a femtosecond Raman spectroscopic study was carried out by Tsen *et al.* (1997). They found that the electron-LO phonon scattering rate and the corresponding mode phonon lifetime are 50fs and 5ps, respectively. Tsen *et al.* (1998) also measured the nonequilibrium phonon decay time and its temperature dependence, to light on the anharmonic decay process of LO phonons.

There is a little information on phonon dispersion relations over the first Brillouin zone. For wurtzite GaN, phonon-density-of-states (PDOS) of a bulk GaN powder was measured by Nipko *et al.* (1998) using time-of-flight neutron spectroscopy at T = 8K. PDOS consist of a low-energy portion (0-45meV), a high-energy portion (65-95meV), and a gap in between. The two broad bands at about 23 and 39 meV contain the acoustic phonons and the first group of optical phonons, whereas, the two sharp bands at about 75 and 86 meV arise from the upper optical phonon branches. They have also calculated such phonon dispersion curves as to reproduce the measured PDOS with a rigid-ion model, to obtain lattice specific heat and Debye temperature of 570K at T = 0K. The feature of the dispersion curves is that the optical phonon branches are rather flat than those calculated by Azuhata *et al.* (1996). The calculated Debye temperature is close to that estimated by Slak *et al.* (1987). Davydov *et al.* (1997) calculated the phonon dispersion relations for some special points and lines using a rigid ion model. They also

Table 5.7 Elastic stiffness constants of wurtzite GaN in units of GPa.

Reference		C_{11}	C_{12}	C_{13}	C_{33}	C_{44}	C_{66}
Exp.							
Savastenko and Sheleg (1978)	bulk	296	120	158	267	24	88
Polian *et al.* (1996)	bulk	390	145	106	398	105	123
Schwarz *et al.* (1997)	bulk	377	160	114	209	81.4	109
Deger *et al.* (1998)	GaN/sapphire	370	145	106	398	105	123
Yamaguchi, M. *et al.* (1997)	GaN/sapphire	365	135	114	381	109	115
Yamaguchi, M. *et al.* (1998)	GaNsubstrate	373	137	80.4	381	93.6	118
Calc.							
Kim *et al.* (1997)		396	144	100	392	91	126
Wright (1997)		367	135	103	405	95	116
Shimada *et al.* (1998)		350	140	104	376	105	115

discussed the second-order Raman spectra. Siegle *et al.* (1997b) also studied the zone-boundary phonons.

For cubic GaN, phonon dispersion curves was intensively calculated by Karch *et al.* (1998) using first-principles linear response theory within local density approximation. The feature of calculated dispersion curves is the rather flat optical branches. Their feature of cubic GaN is similar to that of AlAs. Azuhata *et al.* (1998) succeeded in reproducing the dispersion curves using a modified bond charge model, to point out that the two phonon density-of-states for cubic GaN is similar to the two phonon absorption spectra observed experimentally for wurtzite GaN as in the case of ZnS (Detsch, 1962; Marshall and Mitra, 1964).

In summary, more data are needed to understand the deeply mechanical and thermal properties of GaN and related III-nitrides with both wurtzite and cubic structures.

5.3 InGaN BULK AND QWs FOR PRACTICAL DEVICES

As described in the former Sections, most of optical transition processes in GaN based materials rely on the excitonic feature even at RT provided that the restrictions for excitons to survive are maintained. Major developments of III-nitride semiconductors (Nakamura and Fasol, 1997a; Akasaki and Amano, 1997) have led to the commercial production of blue, green (Nakamura and Fasol, 1997a), UV (Mukai *et al.*, 1998a), and amber (Mukai *et al.*, 1998b) InGaN single quantum well (SQW) LEDs and to the demonstration of InGaN multiquantum well (MQW) purplish-blue LDs (Nakamura *et al.*, 1996a; 1996b; Nakamura and Fasol, 1997a) with lifetimes exceeding 10,000 hours for cw operation at RT using nearly dislocation free overgrown GaN on sapphire (Nakamura *et al.*, 1997b; 1998a) and pure GaN substrates (Nakamura *et al.*, 1998b) prepared by the lateral epitaxial overgrowth (LEO) technique. Although all practical devices contain the InGaN active region, the emission mechanisms have been not fully understood due to complex material physics and engineering of the InGaN alloy, such as phase separation or compositional inhomogeneity (Osamura *et al.*, 1975; Koukitsu *et al.*, 1996; Ho and Stringfellow, 1996; Singh *et al.*, 1997; Matsuoka, 1997; Kisielowski, 1997; 1998) due to large mismatches in the lattice constants, vapor pressure, and optimum growth temperature, large effective masses (Suzuki *et al.*, 1995) and polarization due to the wurtzite crystal lattice.

In contrast to binary GaN, the alloy compositional fluctuation induces effective bandgap inhomogeneity in InGaN QWs. This and the QW thickness fluctuation produce broadening of the InGaN QW energy states. Chichibu *et al.* (1996d), Narukawa *et al.* (1997a), and several researchers (Scholz *et al.*, 1997; Jeon *et al.*, 1996) have assigned the spontaneous emission from InGaN QWs to the recombination of excitons localized at certain potential minima (Chichibu *et al.*, 1996d; 1997c; 1997d; 1998a; 1998b; 1998c; 1999a; Narukawa *et al.*, 1997a; 1997b; Jeon *et al.*, 1996). However, since the c [0001] axis of hexagonal nitrides corresponds to the growth axis, spontaneous and strain induced lattice polarization produces the pyro- and piezoelectric field, F, across the QW plane or double-heterostructure (DH) active region. Thus several groups have explained the importance of the quantum confined Stark effect (QCSE) (Miller *et al.*, 1984;

1985) due to the piezoelectric field, F_{PZ}, in strained wurtzite InGaN QWs (Akasaki and Amano, 1997; Takeuchi *et al.*, 1997a; 1997b; Chichibu *et al.*, 1996d; 1998b; 1998c; 1999a; Bergman *et al.*, 1998; Im *et al.*, 1998; Wetzel *et al.*, 1998). In particular, the blueshift of the electroluminescence (EL) peak in InGaN SQW LEDs with increasing drive current (Nakamura and Fasol, 1997a) has been explained (Chichibu *et al.*, 1996d) by the combined effects of a reduction of QCSE due to Coulomb screening of F_{PZ} (Akasaki and Amano, 1997; Takeuchi *et al.*, 1997a; Deguchi *et al.*, 1998a) and band-filling of the energy tail states (Chichibu *et al.*, 1996d). Moreover, characteristic optical gain mechanisms were reported (Kuball *et al.*, 1997; Deguchi *et al.*, 1997b; 1998a; Mohs *et al.*, 1997; Song *et al.*, 1998; Chichibu *et al.*, 1998a; 1998b; 1998d) for InGaN LD wafers although gain spectra which can be explained by the well-known electron-hole plasma (EHP) (Haug and Koch, 1990; Chow *et al.*, 1994) lasing model were reported (Akasaki and Amano, 1997; Deguchi *et al.*, 1997b; 1998a; Chichibu *et al.*, 1998a; Frankowsky *et al.*, 1996; Wiesmann *et al.*, 1996; Kuball *et al.*, 1997).

In order to obtain an insight into what dominates the emission properties of InGaN QWs and 3D layers for further optimization of InGaN MQW LDs, it is necessary to investigate the effects of effective bandgap inhomogeneity and electric field (F) across the QW on the QW exciton physics separately and consistently.

In this Section, fundamental electronic modulations and decay dynamics in strained wurtzite III-nitride, in particular $In_xGa_{1-x}N$, QWs were treated to explore the reason why practical InGaN devices emit bright luminescences in spite of the large TD density. Energy quantization and effects of F, which is a sum of the fields due to spontaneous and piezoelectric polarizations and the pn junction field, on the energy levels are first described to show how F changes the energy states in the QWs. Indeed F causes the redshift of the ground state resonance energy through QCSE (Miller *et al.*, 1984; 1985). The absorption spectrum is modulated by QCSE, quantum-confined Franz-Keldysh effect (QCFK) (Miller *et al.*, 1986), and Franz-Keldysh (FK) effects from the barriers when, for the first approximation, potential drop across the well ($F\times L$) exceeds the valence band discontinuity, ΔE_V. Under large $F\times L$, holes are first confined in the triangular potential well formed at one side of the well. This produces an apparent Stokes-like shift in addition to the in-plane net Stokes-like shift on the absorption spectrum. The QCFK and FK further modulate the electronic structure of the wells with L greater than 3D FE effective Bohr radius, a_B. When $F\times L$ exceeds ΔE_C, both electron (e) and hole (h) confined levels drop into the triangular potential wells at opposite sides of the wells, which reduces the wavefunction overlap.

Next the existence of excitons under very high F is shown experimentally using binary GaN QW sandwiched by AlGaN cladding layers. Then the effects of in-plane carrier localization in InGaN are described using samples having various well thickness, L, InN molar fraction, x, and Si-doping density, n, in the GaN or InGaN barriers and wells. After introducing several experimental evidences of bandgap inhomogeneity in InGaN, recombination dynamics of localized excitons and influence of the bandgap inhomogeneity on the optical gain spectra of InGaN cw laser materials are shown. While strong F_{PZ} naturally modulates the QW energy states, effective in-plane (lateral) localization of quantized excitons was found to play an important role for the QWs especially for L smaller than a_B^*. Doping of Si in the barriers partially screens the F resulting in a smaller Stokes-like shift, shorter recombination decay time, and higher emission efficiency.

Finally, the use of InGaN was found to overcome the field-induced oscillator strength lowering due to the spontaneous and piezoelectric polarization. Effective in-plane localization of the QW excitons (confined excitons, or quantized excitons) in quantum disk (Q-disk) size potential minima, which are produced by non-random alloy potential fluctuation enhanced by the large bowing parameter and F, produces confined e-h pairs whose wavefunctions are still overlapped when $L<a_B$. Their Coulomb interaction is more pronounced for $F \times L < \Delta E_V$.

5.3.1 Quantized Energy Levels

In recent years, advances in crystal growth techniques made it possible to realize LEDs and LDs using wurtzite III-nitride compounds. They are fabricated on sapphire, 6H-SiC, and, quite recently, pure GaN substrates and, thus, the device surface is the (0001) plane. The heart of those devices consists of SQW or MQW structures to control electronic and, consequently, optical properties. The electronic structures in SQWs and MQWs deviate substantially from those of bulk materials, because electrons are confined in the one space direction, i.e., the [0001] direction. QW structures consisting of the III-nitride compounds are basically the type-I structure, i.e., both the lowest electron and hole states are at least confined in the well material. However, attention should be paid to the facts that the III-nitride compounds are piezo-active materials and, furthermore, those with wurtzite structure have spontaneous polarization. Therefore, the piezoelectric field due to biaxial strains arising from the lattice mismatch between the well and the barrier materials, may modify the electronic band structure and, consequently, optical properties in QWs through the quantum confined Stark effect (QCSE) (Miller *et al.*, 1984; 1985) or quantum confined Franz-Keldysh effect (QCFK) (Miller *et al.*, 1986) depending on both the well width and the strength of the piezoelectric field (Chichibu *et al.*, 1998b; 1998c; 1999a).

In this subsection, general discussions of some basic quantum confined effects are given. Enhancement of the exciton binding energy in SQW is shown based on numerical calculations in a simple model GaN SQW. In the later subsection, effects of the piezoelectric field on the quantum confined electronic states will be discussed using concrete numerical calculations.

Elementary properties of electrons are discussed in such a model SQW that its barrier regions extend semi-infinitely on both sides of the well having the width of L_w in the Z direction. The Z direction is again defined as along the [0001] direction. Energy gap difference ΔE_g between the larger band gap of the barrier material and the small band gap of the well material causes confinement potentials ΔV_e for electrons in the conduction band (CB) and ΔV_h for holes in the valence band (VB), i.e., $\Delta E_g = \Delta V_e + \Delta V_h$. The potential discontinuity occurs within one atomic layer between the well and the barrier and, thus, SQW is modelled by a one-dimensional finite potential well with infinitely steep walls.

Before considering the finite potential well, it is instructive to consider the symmetric infinite potential well defined by

$$V(z) = \begin{cases} 0 & for \quad |z| < L_w/2, \\ \infty & for \quad |z| > L_w/2. \end{cases} \tag{5.24}$$

For simplicity of calculations, let us consider the electrons in the nondegenerate and energetically isotropic CB. Schrödinger equation describing the motion of the electrons in the SQW is given by,

$$\{H_e(\boldsymbol{r}) + V(z)\}\psi(\boldsymbol{r}) = E\psi(\boldsymbol{r}) \tag{5.25}$$

where H_e is the effective mass Hamiltonian for the electrons in CB of a bulk material used in the well and is assumed to have an eigenvalue of

$$E_e(\boldsymbol{k}) = E_g + \frac{\hbar^2}{2m_e^*}\left(k_\perp^2 + k_z^2\right) \tag{5.26}$$

Here $\boldsymbol{k}_{|} = (k_x, k_y)$ is the electron wave vector, $\boldsymbol{k}$, component in the X-Y plane, k_z is the Z component of $\boldsymbol{k}$, m^*_e is the effective mass of the electrons, and E_g is the energy gap in the bulk material.

From Equation (5.24), while there is no quantum confinement in the k_x-k_y plane, the electrons are completely confined inside the well in the Z-direction. Therefore, the wave function of the electrons have the plane wave component $exp\{i(k_xx+k_yy)\}$ in the X-Y plane and the envelope function describing quantized standing waves in the Z direction. The latter function must vanish at the interface of the well and the barriers. With these arguments, we can write,

$$\psi(\boldsymbol{r}) = \phi_e(z)e^{i(k_xx+k_yy)}u_c(\boldsymbol{k},\boldsymbol{r}), \tag{5.27}$$

where $\phi_e(z)$ is the slowly varying envelop function and $u_c(\boldsymbol{k}, \boldsymbol{r})$ is the Bloch function of CB.

Replacing $\hbar k_z \to p_z = (\hbar/i)\partial/\partial z$ in Equation (5.26), we obtain the following one-dimensional Schrödinger equation,

$$H_e(z)\phi_e(z) = \left\{-\frac{\hbar^2}{2m_e^*}\frac{\partial^2}{\partial z^2} + V(z)\right\}\phi_e(z) = E_z^e\phi_e(z), \tag{5.28}$$

with,

$$E_z^e = E - E_g - \frac{\hbar^2 k_\perp^2}{2m_e^*}. \tag{5.29}$$

Equation (5.28) can be easily solved analytically using the boundary conditions $\phi_e(-L_w/2) = \phi_e(L_w/2) = 0$ as mentioned before and the following normarized envelop functions and the eigenvalues are derived,

$$\phi_{e,even}(z) = \sqrt{\frac{2}{L_w}}\,cos\left(\frac{\pi mz}{L_w}\right), \tag{5.30a}$$

$$\phi_{e,odd}(z) = \sqrt{\frac{2}{L_w}}\,sin\left(\frac{\pi mz}{L_w}\right), \tag{5.30b}$$

$$E_{z,m}^e = \frac{\pi^2\hbar^2}{2m_e^*L_w^2}m^2, \tag{5.31}$$

with $m = 2n-1$ for even states, $m = 2n$ for odd states, where $n=1,2,3......$. Combining Equations (5.29) and (5.31), the total energy of the electrons subject to one-dimensional confinement is given by,

$$E^{e}_{z,n} = E_g + \frac{\hbar^2}{2m_e^*}\left(\frac{\pi^2 n^2}{L_w^2} + k_\perp^2\right), \qquad n = 1,2,3,....... \tag{5.32}$$

Equation (5.32) represents a succession of energy subbands consisting of energy parabola $\hbar^2 k^2_\perp/2m_e^*$ separated by $\pi^2\hbar^2(2n+1)/2m_e^* L^2_w$ (n=1,2,3,.....). The ground state energy $\pi^2\hbar^2/2m_e^* L^2_w$ and the energetically separation between two adjacent quantized states become larger proportional to $1/L^2_w$ as decreasing L^2_w. They also depend on m_e^* and vary from material to material.

What happens for the finite potential well? In this case, the finite barrier height leads to the penetration of wave functions for the electrons in the well into the barrier regions due to the tunnelling effect. Thus the quantum-confined energy levels depend on the barrier height and are no longer obtained analytically. Furthermore, strictly speaking, the difference in effective masses and in the background dielectric constants in the well and in the barrier materials makes the boundary conditions somewhat difficult. However the basic features of the wave functions and the energy spectra are unchanged. Note that as long as the barrier height is positive, there always exists, at least, the ground state.

The Schrödinger equation given by Equation (5.26) can be solved numerically using, for example, the Noumerov method, in which Equation (5.26) is directly integrated numerically by achieving convergence of each eigenvalue under the restriction that the logarithmic derivative of the corresponding wave function must be continuous. This method will be applied to a concrete example in the later subsection.

Before mentioning the band structure of VB in the well consisting of wurtzite III-nitride, the quantum confinement effects on degenerate bands should be addressed. The best studied and the most instructive example to demonstrate those effects is found in SQWs consisting of cubic III-V or II-VI compound semiconductors (for example, see, Chow *et al.*, 1994). The one-dimensional confinement leads to the symmetry reduction. The degeneracy of heavy and light hole bands at the Γ point in bulk is removed due to the symmetry reduction and the band mixing or, the band anticrossing, is found in the dispersion relations along the k_x (k_y) direction. However, for wurtzite III-nitrides, symmetry of bulk is originally low enough and the valence bands consist of nondegenerate so called the A-, B-, and C-valence bands as mentioned in Section 5.2.3. Note that each band is doubly degenerate with respect to spins. Thus no further removal of the band degeneracy occurs, while complex band mixing between valence subbands is expected under the one-dimensional confinement.

Procedures to calculate the valence band structure for the holes in the well with wurtzite structure is as follows (Uenoyama and Suzuki (1995)): Let the confinement direction be along the Z-axis. Starting point is again the 6×6 $\boldsymbol{k}\cdot\boldsymbol{p}$ Hamiltonian (Bir and Pikus (1974)) suitable for the valence band in the wurtzite structure, Equation (5.14). It is described using the total angular momentum eigenstates as the basis functions. Introducing the slowly varying envelop functions associated with the diagonal matrix elements of the starting Hamiltonian, i.e., the envelop function labelled by the eigenvalues of the orbital angular momentum, m_ℓ, the generalized envelop functions of the hole subbands can be written down by the linear combination of the product of the envelop function labelled by m_ℓ and the associated basis function. Replacing $\hbar k_z \rightarrow p_z = (\hbar/i)\ \partial/\partial z$ as done in the electrons case, the Schrödinger equation like differential equation for the envelop function labelled by m_ℓ is derived from the diagonal matrix elements of the starting Hamiltonian. The independent three differential equations determine quantum confinement energies corresponding to the A-, B-, and C-valence bands and the envelop function labelled by m_ℓ. Using those, the original 6×6 differential eigenvalue problem is transformed into the 6×6 algebraic eigenvalue problem for coefficients introduced to write down the generalized envelop function. Applying the cubic approximation, the *k*-linear terms are eliminated. Using a unitary transformation, the 6×6 Hamiltonian matrix can be block diagonalized as two 3×3 Hamiltonian matrices. Note that the two Hamiltonians are independent of the angle in the k_x-k_y plane and the subband states are isotropic in $\boldsymbol{k}$ space. Diagonalizing one of the two matrices, the hole subbands and the corresponding envelop functions are obtained. The information of the valence subbands is indispensable to estimate the joint density-of-states, the materials gain, and so on.(Chow *et al.*, 1994)

So far as optical properties such as OA and PL under low excitation condition are concerned in the well, information on the energetically low-lying quantum confined energy states near the Γ point becomes important. Exciton is also expected to play an important role in such optical properties in the well. Here we briefly review the exciton in bulk before considering the exciton confined in the well. We restrict ourselves to the Wannier excitons (Wannier, 1937). The exciton is such an elementary excitation that the electron and the hole are attracted via the Coulomb interaction to form a bound state. Note that the exciton itself can move freely in a crystal. Thus it is easily understood that the extension of the wave functions of the electron and the hole and the degree of the overlap of them determine the degree of binding the electron to the hole, i.e., the binding energy of the exciton. Therefore, the larger the Coulomb attraction is, the larger the binding energy is. The distance between the electron and the hole is usually known as the effective Bohr radius, a_B^* of the exciton. The exciton binding energy, E_x, and a_B^* are reported to be about 26 meV and 3.5nm, respectively, in bulk GaN. E_x is unexpectedly larger than that in other typical cubic III-V compound semiconductors. This is partly because GaN is a wide gap material and, consequently, has small background dielectric constant. Note that the background dielectric constant describes the screening of the Coulomb interaction in an unexcited crystal. The large effective masses of the electrons and the holes also lead to the large E_x. The following should be also pointed out: when the density of photoexcited and/or background carriers becomes higher, the Coulomb interaction is screened and the binding energy becomes smaller, as described in Section 5.2.1.

Let us consider the exciton in the well. When the well width is much larger than a_B^* in bulk, no drastic change of E_x is expected because the excitons feel nearly the same environment as in bulk. As decreasing the well width down to the same order of or less than a_B^*, both the electron and hole wave functions are confined in the well and the overlap of them becomes large. This leads to enhancement of the exciton binding energy. It is shown theoretically that the exciton binding energy in the two-dimensional limit is four times as large as that in bulk (Bastard *et al.*, 1982; Haug and Koch, 1990). However, in actual systems, the extremely narrower well width leads to the leakage of a major part of the wave functions out of the barriers and results in the decrease of the exciton binding energy. Therefore, the exciton binding energy monotonically increases up to its maximum values and, then, decreases again as decreasing the well width. This feature has been already shown experimentally and theoretically for the exciton confined in the well consisting of cubic III-V semiconductors.(Moore *et al.*, 1990; Hou *et al.*, 1990; Reithmaier *et al.*, 1991; Joyce, 1992; Atanasov *et al.*, 1994)

Here a brief description is given on calculational methods of the exciton binding energy in the well. In the three-, two-, and one-dimensional limits, exciton problems can be solved analytically, for example, from Wannier equation (for example, see, Haug and Koch, 1990) assuming isotropic systems. However, since QW has a finite well width, the exciton problems in such a quasi-two-dimensional case cannot be solved analytically. Bastard *et al.* (1982) first solved them performing variational calculations for the exciton Hamiltonian within the effective mass approximation. Let us introduce the variational method. Isotropic effective masses and background dielectric constant are assumed according to them. The starting Hamiltonian describing the relative motion of the exciton is

$$H = H_e(z_e) + H_h(z_h) + H_{e-h}(\boldsymbol{r}), \tag{5.33}$$

where H_e (H_h) is the Hamiltonian describing the electron (hole) motion confined in the well just as Equation (5.28) and the corresponding eigenvalues are $H^e_{z.n}$ ($H^h_{z.n}$). $H_{e\text{-}h}$ is the Hamiltonian consists of the kinetic energy part of the relative motion of the exciton in the X-Y plane and the Coulomb potential attracting the electron and the hole in the three-dimensional form. z_e (z_h) and $\boldsymbol{r}$ denote the z-coordinate of the electron (hole) and the relative position vector between the electron and the hole. For the exciton associated with the ground states of both the confined electron and hole states, the following product wave function is assumed as a variational function,

$$\Psi = A\xi(\boldsymbol{r})\phi_{e,1}(z_e)\phi_{h,1}(z_h), \tag{5.34}$$

where A is the normalization constant and $\xi(\boldsymbol{r})$ is the exciton envelop function as a trial function. Various forms are proposed for the trial function (Basterd *et al.*, 1982; Greene *et al.*, 1984; Iotti and Andreani, 1995; Antonelli *et al.*, 1996; and references cited therein). Variational parameter(s) is included in the trial function. Let us denote it as α here. The exciton ground state under consideration is obtained as

$$E_{ex} = \min_{\alpha}\langle \Psi | H | \Psi \rangle, \tag{5.35}$$

and, thus, E_x defined as positive value is given by

$$E_x = E^e_{z,1} + E^h_{z,1} - E_{ex}. \tag{5.36}$$

The exciton envelop function with the obtained variational parameter(s) is used to calculate the oscillator strength of the exciton. Note that the functional forms of ϕ_e and ϕ_h may vary according to situations.

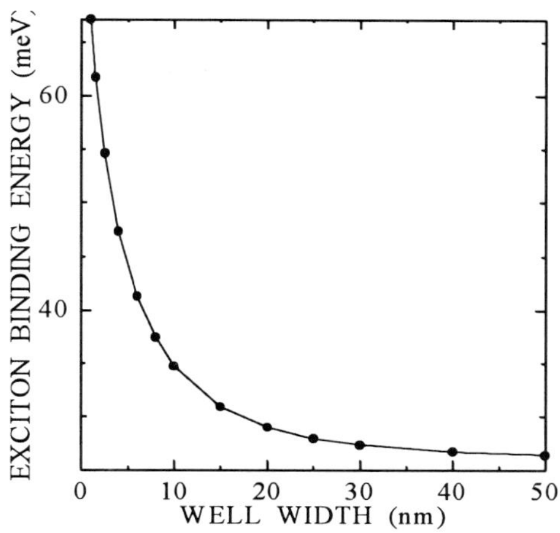

Figure 5.13 Well width dependence of the A-exciton binding energy of GaN SQW with infinite potential well. Calculation was carried out using variational method taking account of both the effective mass anisotropy and the anisotropy of the background dielectric constant.

Figure 5.13 illustrates the upper bound of the A-exciton binding energy as a function of the well width. Calculations were carried out assuming the perfect confinement of the electron and the hole in the GaN well. However, taken account of were the anisotropic effective masses of the electrons and the holes and the anisotropy of the background dielectric constants.

Thus, a trial function with two variational parameters was used. The values of physical parameters used in the calculations were after Suzuki *et al.* (1995), Suzuki and Uenoyama (1996), Chichibu *et al.* (1996, 1997), and Shikanai *et al.* (1997). As expected, while E_x decreases down to the bulk value of 26 meV as increasing the well width, it increases up to 104 meV as decreasing the well width . For the well width of 3.0nm, the upper bound of the binding energy reaches up to over 50 meV. The actual value is lower than this value because of the finite confinement potentials. Calculations performed varying the values of the confinement potentials for the electrons and the holes within reasonable values showed that the maximum binding energy occurred around at the well width of 2-2.5nm.

With the above arguments, the lowest exciton resonance energy associated with the ground states in the well is written as,

$$E_{res} = E_g + E^e_{z,1} + E^h_{z,1} - E_x, \tag{5.37}$$

where $E^e{}_{z,1}$ ($E^h{}_{z,1}$) is the ground state energy of the quantum confined states for the electrons (holes) in the well. The exciton resonance was clearly observed in the PL spectra of GaAs/AlAs superlattices (Goldstein *et al.*, 1983) at temperature $k_BT < E_x$. The resonance energy was also found to be sensitive to monatomic well size fluctuation in the narrower well. In absorption spectra of GaAs/AlAs superlattices, the heavy-hole and the light-hole exciton resonance were clearly seen even at room temperature, which were followed by the absorption due to the two-dimensional continuum states of the ground state and absorption structures arising from higher states (Ishibashi *et al.*, 1981). The room temperature exciton resonance in the absorption spectra were also observed even under the strength up to F=7.3×10^4 V/cm of electric field perpendicular to the wells in GaAs/AlAs MQW with the well width of 9.5 nm (Miller *et al.*, 1984; 1985). This is due to just the confinement effect. The well width dependence of the absorption spectra were also investigated in AlGaAs/GaAs superlattices at low temperature (Masselink *et al.*, 1985). It follows that the well width dependence of the absorption spectra is determined by competition between that of quantum confinement energy level and that of exciton binding energy.

5.3.2 Piezoelectric Field

Perfectly lattice matched conditions between the well and the barrier materials are implicitly assumed in the theory mentioned in Section 5.3.1. Generally speaking, lattice constants in the lateral plane of the well and of the barrier materials are different and there exist strains in one or both of the components. For thin films including superlattices, biaxial strains remain in materials. As having already been seen in Section 5.2.3, the biaxial strains change the energy gaps, the curvatures of the valence bands, and consequently, the effective masses of the holes. Therefore, the quantum confined valence subbands are modified by the biaxial strains. For details, see, for example, Kamiyama *et al.* (1995), Suzuki and Uenoyama (1996) and Niwa *et al.* (1997). The biaxial strains also lead to the piezoelectric field along the Z direction in wurtzite materials. It inclines the one-dimensional confinement potential. The quantum confinement energy levels and the corresponding envelop functions in such potential are different from those in the original confinement potential. This is easily understood that imaging the field pushes the electrons and the holes toward the opposite walls in the well because of opposite charges. Thus the overlap of the electron and hole wave functions is reduced and this reduction is expected to result in the decrease of the exciton binding energy. Therefore, optical properties under the presence of the piezoelectric field are expected to be considerably modified from those in the absence it. Takeuchi *et al.* (1997a; 1997b; 1998a; 1998b) first pointed out that the piezoelectric field plays an important role in InGaN/GaN quantum wells. Sakai *et al.* (1998) confirmed this. Similarity was found in GaN/AlGaN quantum wells (Im *et al.*, 1998). Effects of piezoelectric field on the quantum-confined states is discussed in the present subsection.

In crystals with wurtzite structure, the piezoelectric tensor is written as,

$$\begin{pmatrix} 0 & 0 & 0 & 0 & e_{15} & 0 \\ 0 & 0 & 0 & e_{15} & 0 & 0 \\ e_{31} & e_{31} & e_{33} & 0 & 0 & 0 \end{pmatrix}, \tag{5.38}$$

where e_{ij} are the piezoelectric constants. Therefore, under the presence of biaxial strain given by Equation (5.13), the piezoelectric polarization along the Z-axis, $P_{z,\text{piz}}$, arises as,

$$P_{z,piz} = 2e_{31}\varepsilon_{xx} + e_{33}\varepsilon_{zz}. \tag{5.39}$$

Charge neutrality condition leads to the electric filed along the Z-axis, F_z, as,

$$F_z = -P_{z,piz} / \varepsilon_{r//}\varepsilon_0, \tag{5.40}$$

where ε_0 and $\varepsilon_{r//}$ are the dielectric constant in vacuum and the relative dielectric constant along the Z-axis. The crystals with wurtzite structure has also non-vanishing spontaneous polarization along the Z-axis, $P_z^{eq.}$, from symmetry. Thus, the total polarization along the Z-axis is, in principle,

$$P_z = P_z^{eq.} + P_{z,piz}. \tag{5.41}$$

Depending on the sign of ε_{xx}, the direction of $P_{z,piz}$ is parallel or antiparallel to that of $P_z^{eq.}$, see Table 5.8. In taking account of the spontaneous polarization, $P_{z,piz}$ should be replaced by P_z in Equation (5.40). In calculating F_z with the use of Equation (5.40) (or the modified one), one should bear it in mind that the estimation of ε_{xx} has to be made taking account of the coherent growth of binary and ternary III-nitrides compounds (Takeuchi *et al.*, 1997b).

There are several groups reporting the values of e_{ij}. Table 5.8 summarizes reported the values of e_{ij} for wurtzite III-nitrides and e_{14} for zincblende, where the calculated spontaneous polarization values are also given. Note that Bykovski *et al.* (1996) obtained the values of e_{ij} for the wurtzite GaN transforming the value of e_{14} extracted from the unpublished data on GaN mobility. Using the values given by them, Takeuchi *et al.* (1997b) calculated F_z in InGaN strained layer on GaN along (0001) as a function of the InN mole fraction. They found that F_z was 1.08 MV/cm

Table 5.8 Piezoelectric constants and spontaneous polarization for III-nitrides in units of C/m^2. ZB and WZ denote zinc blende and wurtzite structures, respectively.

Reference		ZB	WZ			
		e_{14}	$P_z^{eq.}$	e_{33}	e_{31}	e_{15}
BN						
Shimada *et al.* (1998)	Calc	-0.64	-	-0.85	0.27	-
AlN						
Tsubouchi and Mikoshiba (1985)	Exp.	-	-	1.55	-0.58	-
Davydov and Tikhonov (1996)	Calc	0.67	-	-	-	-
Bernardini *et al.* (1997)	Calc	-	-0.081	1.46	-0.60	-
Shimada *et al.* (1998)	Calc	0.59	-	1.29	-0.38	-
GaN						
O'clock and Duffy (1973)	Exp.	0.6	-	-	-	-
Bykhovski *et al.* (1996)	Exp.	0.375	-	0.43	-0.22	-0.22
Davydov and Tikhonov (1996)	Calc	0.68	-	-	-	-
Bernardini *et al.* (1997)	Calc	-	-0.029	0.73	-0.49	-
Shimada *et al.* (1998)	Calc	0.50	-	0.63	-0.32	-
InN						
Bernardini *et al.* (1997)	Calc	-	-0.032	0.97	-0.57	-

when the InN mole fraction was 0.13. Except for Table 5.8, the piezoelectric coefficient, d_{ij} (m/V), of GaN was reported by Muensit and Guy (1998) and Im *et al.* (1998). Those groups do not take account of the spontaneous polarization.

Here and hereafter, taking account of the facts that the variation of e_{ij} is large and the treatment of spontaneous polarization is not known, changes of the quantum confined states are considered as a function of F_z not of ε_{xx} (ε_{zz}). Schrödinger equation describing the electron motion in the SQW under F_z (z) is given by

$$\left\{-\frac{\hbar^2}{2m_e^*}\frac{\partial^2}{\partial z^2}+V(z)+ezF_z(z)\right\}\phi_e(z)=E_z^e\phi_e(z), \tag{5.42}$$

where e is the proton charge. When the exciton effect is neglected, information on the quantum confined states in the confinement potential $V(z)$ under $F_z(z)$ is derived solving Equation (5.42).

To show the effect of $F_z(z)$, let us consider the well width dependence of the ground state resonance energies in a model InGaN/GaN SQW neglecting the exciton effect. The barriers are assumed to extend semi-infinitely on both sides of the well. The energy gaps of the barrier and the well are set to be 3.4 eV and 3.0 eV, respectively. The latter value is close to that of strained $In_{0.1}Ga_{0.9}N$ (McCluskey *et al.*, 1998). The other parameter values of the well material are assumed to be the same as those in GaN. The band offset ratio is set to be $\Delta V_e/\Delta V_h$ = 4 (Van de Walle and Neugebauer, 1997). Only F_z in the well is taken into account and its direction is from the left barrier to the right one across the well.

Calculations were carried out using the Noumerov method mentioned in Section 5.3.1 and the calculated results are shown in Figure 5.14 by closed circles, squares, and diamonds for F_z = 0, 500 and 800 kV/cm, respectively. Figure 5.14 shows the following. The larger F_z leads to the larger red shift of the resonance

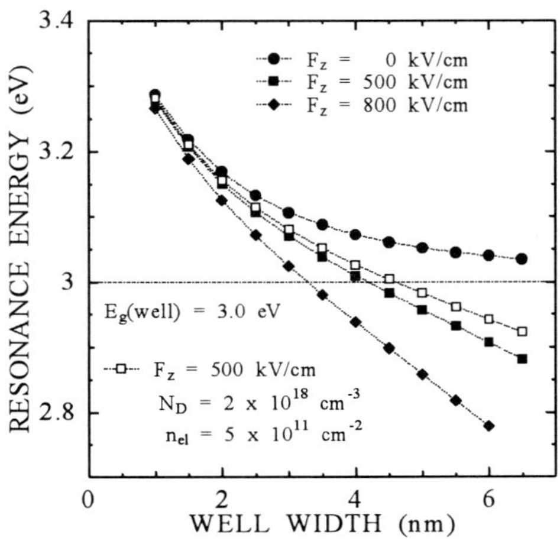

Figure 5.14 Well width dependence of the ground state resonance energies in the model SQW under the electric filed of 0 (closed circles), 500 (closed squares), and 800 kV/cm (closed diamonds). Open squares represent the resonance energies in the modulation doped SQW.

energy. The effect of F_z becomes remarkable for the wider well width. At the well width of 3.0 nm, typical well width used in LED and LD devices (Nakamura and Fasol, 1997), the amount of red shift is 32 meV and 82 meV for F_z = 500 and 800 kV/cm, respectively. Note that these values depend on the amount of the band discontinuity and the band offset ratio. Note that the band-offset ratio reported so far is scattered. Since the band-offset ratio directly influences the magnitude of the confinement potentials for the electrons and the holes, calculated results change. Therefore, in comparing experimental data with theoretical calculations, special attention should be paid.

Figure 5.15 shows the electron and hole wave functions in the ground states for the cases of F_z =0 (a), 500 (b) and 800 kV/cm (c) at the well width of 3.0nm, where the energy gap is set to be zero to save the space. Figure 5.15 clearly illustrates that F_z pushes the electrons and the holes toward the opposite walls in the well. It is also easily understood that the overlap of the electron and hole wave functions reduces as increasing F_z. Under the values of parameters used herein, the hole feels the triangular potential for F_z = 500 kV/cm and both the electron and the hole for F_z=800 kV/cm. In the case of F_z= 800 kV/cm, the hole bound state is very close to the continuum states in the barrier material and the confinement of holes seems to be weak. Figure 5.15 also shows that changes of the confined hole energy levels are large. This is attributed to the small confinement potential for the holes, i.e., the band-offset ratio, in the present model.

In LED and LD devices, the barriers or both the barriers and the wells are heavily doped to make p-n junction. Impurity doping introduces the space charges in the devices and they may in part cancel out or enhance the electric field strength in the well. Imaging a partial cancellation of the electric field, let us consider a model SQW with the modulation doping, in which the right hand side barrier is doped with n-type impurity. Let the impurity concentration of N_D= 2×10^{18}cm^{-3} and

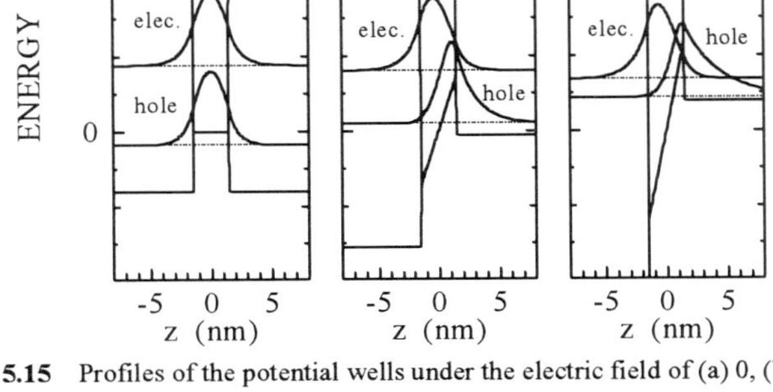

Figure 5.15 Profiles of the potential wells under the electric field of (a) 0, (b) 500, and 800 kV/cm (c) and those of the electron and hole wave functions in the SQW with the well width of 3.0nm. Energy gap is set to be zero to save space. Horizontal lines in each panel represent the ground state energy levels for the electron and the hole.

the sheet carrier density of electrons supplied in the well be $n_{el} = 5\times10^{11}$ cm^{-2}. In this case, both Equation (5.42) and the Poisson equation have to be solved simultaneously. For simplicity, calculations were carried out within Hartree approximation.

Figure 5.16 compares the electron and hole wave functions in the model SQW without (a) and with (b) the modulation doping. The modulation doping changes the potential profile at the right interface between the well and the barrier. This leads to the substantial reduction of F_z in the well. Therefore, the overlap of the electron and hole wave functions becomes larger. The well width dependence of the resonance energy in the modulation doped SQW under F_z = 500 kV/cm is shown in Figure 5.14 by open squares. The reduced F_z results in the blue shift of the resonance energies in comparison with those in the SQW without the modulation doping. Thus it follows that the heavily doping may weaken the piezoelectric field effect.

The upper bound of the zero-field exciton binding energy in the well with the width of 3.0nm is over 50 meV as shown in Section 5.3.1. The corresponding effective Bohr radius is about 2.5nm or less. Thus, the field strength needed for the exciton ionization is roughly estimated as 200 kV/cm. However, Figures 5.15 and 5.16 show that the electron and hole wave functions are still confined in the well up to, at least, 500 kV/cm. Thus, the excitonic effects, would be found in the optical spectra in SQWs and MQWs with well width of 3.0nm or less even under considerably large field strength (Chichibu *et al.*, 1999a). Combining the behaviour of the confined one-particle energy levels and the exciton binding energy under the electric field, the following scenario is drawn (Deguchi *et al.*, 1998a; Chichibu *et al.*, 1998b; 1998c; 1999a). Let us start from the zero-field lowest resonance energy

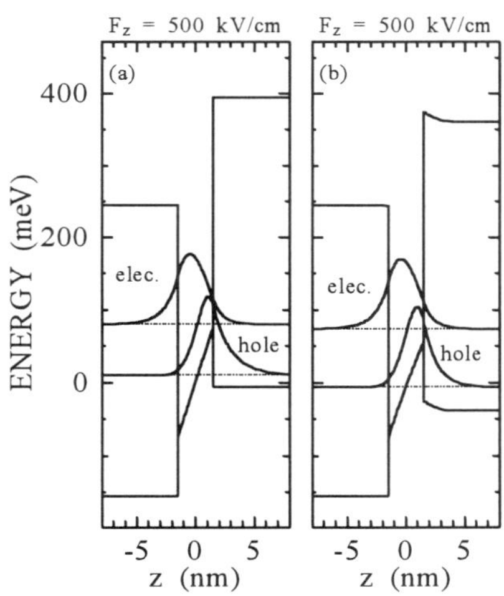

Figure 5.16 Effects of the modulation doping for profiles of the potential wells and the electron and hole wave functions in the SQW with well width of 3.0nm. (a) undoped SQW and (b) modulation doped SQW, where the impurity concentration in the right hand side barrier is 5×10^{18} cm^{-3} and the sheet carrier density of electrons supplied in the well is 2×10^{11} cm^{-2}. For both cases the electric field in the well is 500 kV/cm. Energy gap is set to be zero to save space. Horizontal lines in each panel represent the ground state energy levels for the electron and the hole.

found in the absorption spectra at the well width of 3.0nm, i.e., slightly less than a_B^*. At first, the large excitonic absorption is found because of strong confinement. Applying the electric field across the well, the Stark shift of the resonance energy is observed as shown in Figure 5.14. Note that Figure 5.14 neglects the exciton binding energy correction. While the exciton binding energy decreases as increasing the field strength, the confinement of the electron and hole wave functions in the well result in the exciton binding energy larger than its bulk value probably by ~500 kV/cm. This is just the quantum confined Stark effect (QCSE) first pointed out by Miller *et al.* (1984; 1985).

For QWs with well width larger than a_B^*, where the exciton ionization occurs under relatively weak electric field, another features may be found in the optical absorption spectra. That is, the quantum confined Franz-Keldysh effect (QCFK) (Miller *et al.*, 1986) may be observed (Bulutay *et al.*, 1999). As shown in Figure 5.15, the electric field destroys the symmetry of the wave functions because it pushes the electron and the hole toward the opposite walls in the well. This results in the violation of the selection rule for the optical transition. Thus, the original step-like absorption spectra reflecting the confined energy subbands are smeared. For large electric field strength under which the lower subbands fall into the triangular potentials shown in Figure 5.15(c), a part of the electron and hole wave functions penetrates into the energy gap region. In this case, the photon-assisted tunnelling of electrons from VB to CB leads to an electric-field-dependent absorption tail in the photon energy region lower than the lowest zero-field resonance energy.

So far, the internal electric field effects were studied using emission spectra alone in GaN/AlGaN and InGaN/GaN SQWs and MQWs. There is no experimental data of absorption spectra, to the best of the author's knowledge, except for our recent report (Chichibu *et al.*, 1999a). To get further insight into the internal electric field effects, experiments on the absorption spectra seems to be indispensable because a direct comparison of the experimental data is allowed with the theoretical calculations based on QCSE and/or QCFK. Furthermore, there is no information on the spontaneous polarization from the experimental side. This is partly because experimentally deduced internal electric field strength has both contributions from the spontaneous polarization and the strain induced piezoelectric polarization. Further experimental and theoretical studies are needed to clarify the piezoelectric field effects in SQWs and MQWs consisting of III-nitride.

5.3.3 Spontaneous Emission of Localized Excitons

5.3.3.1 Introduction

InGaN alloys are attracting special interest because they are adopted as active regions of ultra-violet (UV) (Mukai *et al.*, 1998a), blue (Nakamura and Fasol, 1997a), green (Nakamura and Fasol, 1997a), and amber (Mukai *et al.*, 1998b) SQW LEDs and all MQW LDs (Nakamura *et al.*, 1996a; 1996b; 1997a; 1997b; 1998a; 1998b; Akasaki *et al*, 1996; Itaya *et al.*, 1996; Bulman *et al.*, 1997; Kuramata *et al.*, 1997; 1998; Mack *et al.*, 1997; 1998; Nakamura, F. *et al.*, 1998; Kneissl *et al.*, 1998; Yamada *et al.*, 1997; Kimura *et al.*, 1998; Kobayashi *et al.*,

1998), and they emit bright luminescences in spite of the large TD density up to $10^{10}cm^{-3}$ (Ponce and Bour, 1997).

In this subsection, several important physics which affect the emission properties of InGaN QWs are discussed. First, the influence of F discussed in Section 5.3.2 was re-examined experimentally on AlGaN/GaN and GaN/InGaN strained QWs to conclude that there exists QW excitons against polarization-induced and junction high electric fields across the strained QWs provided that $L<a_B$. Next the presence of effective bandgap inhomogeneity, which produces certain quantum disk (Q-disk)-size local potential minima in InGaN practical LEDs is shown. Finally the origin of the QW exciton localization was investigated by comparing optical properties of InGaN QWs grown on dislocated GaN-base on sapphire and nearly TD-free GaN overgrown laterally on the SiO_2 mask.

5.3.3.2 Characteristic optical spectra of InGaN QW devices

It has first been shown that optical spectra of blue and green SQW LEDs and purple MQW LD (Nakamura and Fasol, 1997a) were characterized by the presence of the Stokes-like shift, which is a redshift of the spontaneous emission peak compared to its absorption (excitation) band (Chichibu *et al.*, 1996d), as shown in Figure 5.17. Both the photovoltage (PV) and electroabsorption (EA) spectra show

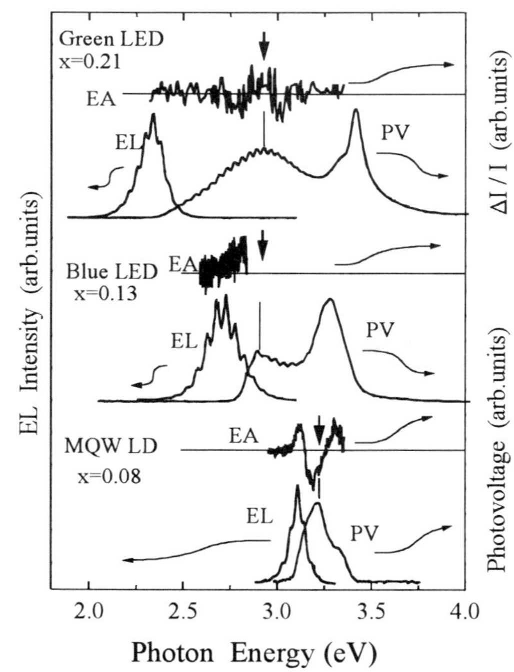

Figure 5.17 Electroluminescence (EL), photovoltage (PV), and electroabsorption (EA) spectra of InGaN quantum well devices measured at room temperature. The InN molar fraction is derived assuming that the QW is fully strained.

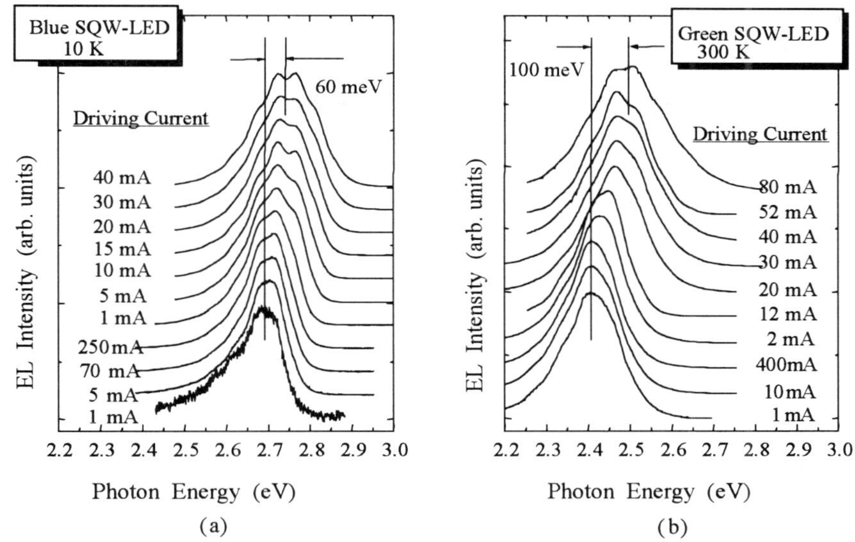

Figure 5.18 Electroluminescence (EL) spectra of (a) InGaN SQW blue LED at 10 K and (b) green LED at 300 K as a function of driving current. The amount of the blue shifts are 60 to 100 meV.

simultaneous redshift and spectral broadening with increasing x, and the electroluminescence (EL) peak appeared at the low energy tail of those absorption-type transition structures.

Another characteristic property of InGaN SQW LEDs is a blueshift of the EL (PL) peak with increasing driving current (excitation power density), as shown in

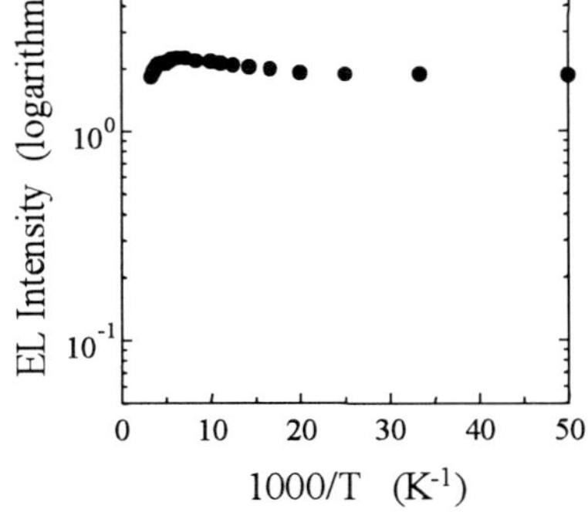

Figure 5.19 EL intensity of InGaN SQW blue LED as a function of temperature. The forward current was kept constant at I=20 mA.

Figure 5.18. Note that in Figure 5.18 periodic modulation in the spectra is due to multiple internal reflection fringes. In addition to this, the internal quantum efficiency of blue/green SQW LEDs is very high, as shown in Figure 5.19. The EL intensity does not show a distinct quenching property up to 250K.

5.3.3.3 Framework

Optical properties peculiar to wurtzite InGaN and III-nitride materials in comparison to familiar III-V semiconductors such as GaAs or InP known to date, are as follows:
(i) Nitrides show an excitonic character in their optical properties as described in Section 5.2.1. Indeed, dominant resonance structures due to A and B FEs were found in both absorption and photoreflectance spectra of 3D GaN layers, and FE emission has been found to dominate the RT PL spectrum of GaN (Chichibu *et al.*, 1996a; 1997a; Monemar *et al.*, 1996a). It is also shown in Section 5.3.1 that E_b increases in QWs due to confinement of wavefunctions.
(ii) As described in Section 5.3.2, wurtzite structure has the highest symmetry compatible with the existence of spontaneous polarization. Moreover, strain-induced piezoelectric tensor of wurtzite has three non-vanishing independent components. In the absence of the external fields, the total macroscopic polarization, P, is the sum of the spontaneous polarization in the equilibrium structure, P_0, and of the strain induced piezoelectric polarization, P_{PZ}.
(iii) InGaN alloys have an immisible gap (Koukitsu *et al.*, 1996; Ho and Stringfellow, 1996; Singh *et al.*, 1997; Matsuoka, 1997) and they usually show broad luminescence band (Nakamura and Fasol, 1997a), which reflects effective bandgap inhomogeneity in InGaN.

5.3.3.4 Polarization in wurtzite lattice

As described in Section 5.3.2, since III-nitride epilayers grown on sapphire (0001) or SiC $(0001)_{Si}$ substrates are along the *c*-axis, we shall consider only the polarization along the *c*-axis, $P_3=P_{0,3}+P_{PZ,3}$. Bernardini *et al.* (1997) calculated $P_{0,3}$ in AlN, GaN, and InN as –0.081, –0.029, and –0.032 C/m^2, respectively. $P_{0,3}$ induces the electric field according to $F_{0,3}= -P_{0,3}/\varepsilon_0\varepsilon_r$, and is estimated to be 3.4 MV/cm for GaN, which is nearly 50 times the ionization field, F_i, of FEs in 3D GaN (7.6×10^4 V/cm) (Chichibu *et al.*, 1996a: 1996c; 1996d; 1997a; 1998a; 1999a). From the fact that FE emissions can be observed from 3D GaN even at RT (Chichibu *et al.*, 1996a; 1997a; Monemar *et al.*, 1996a), $F_{0,3}$ would be smaller than F_i. This may be due to internal screening of the polarization-induced field, because the layer is usually too thick and any kinds of internal charges contribute to Coulomb screening. An expression for $P_{PZ,3}$ under the biaxial strain $(a-a_0)/a_0=\varepsilon_{xx}=\varepsilon_{yy}=-(C_{33}/2C_{31})\varepsilon_{zz}$ is given by $P_{PZ,3}=2(e_{31}-e_{33}C_{31}/C_{33})\varepsilon_{xx}$, where a is the lattice constant, a_0 is that of strain-free material, C_{ij} are the elastic stiffness constants, and e_{ij} are the piezoelectric constants. Bernardini *et al.* (1997) reported the following values in units of C/m^2: $e_{33}=1.46$ and $e_{31}=-0.60$ for AlN; $e_{33}=0.73$ and $e_{31}=-0.49$ for GaN; $e_{33}=0.97$ and $e_{31}=-0.57$ for InN. For in-plane compressive strains usually found in InGaN, the direction of F_{PZ} is from Ga face to N face (surface to substrate) according to Hellman (1998) if we consider the Ga surface growth (+C).

The direction of $F_{PZ,3}$ is opposite to $F_{0,3}$ and the pn junction field F_{pn}. This means that total electric field F exists in the QW regions of practical devices is a sum of polarization-induced fields and the pn junction field, $F_3 = F_{0,3}+F_{pn}-F_{PZ,3}$. Here we simply model the effect of resultant F on the band structures of strained QWs.

5.3.3.5 Schematic band diagrams of InGaN QWs under an electric field

From the framework (i) and (ii), physical scenarios of the optical transitions in InGaN QWs are drawn as follows. Since the critical thickness of $In_xGa_{1-x}N$ ($0<x<0.15$) is reported to be greater than 40nm (Akasaki and Amano, 1997; Takeuchi *et al.*, 1997), coherent growth of InGaN is assumed. This strain causes F_{PZ} but excitons should also be significant due to the confinement. Therefore the problem treated here is the behavior of confined energy levels in QWs under F, as discussed in Sections 5.3.1 and 5.3.2. In QWs, FE absorption is observed even at RT under high F across the QW (Miller *et al.*, 1984; 1985) due to quantum confinement of the wavefunctions which enhances the Coulomb interaction between the e-h pair to increase E_b (Bastard *et al.*, 1982). Miller *et al.* (1984; 1985) have observed an excitonic absorption in GaAs QWs, which was redshifted by 2.5 times the zero-field E_b for $F = 10^5$ V/cm (50 times the F_i of 3D excitons). They have explained the redshift in terms of field modulation of quantized energy levels by QCSE. We estimate in Section 5.3.1 that E_b in $GaN/Al_{0.1}Ga_{0.9}N$ QWs under $F = 0$ as a function of L introducing finite barrier height. The obtained E_b value for $L = 3$ nm is 47 meV and F_i is estimated to be 6.0×10^5 V/cm taking into account the shrinkage of e-h distance in z-direction down to 0.78nm. E_b is 37 meV for $L = 6$ nm.

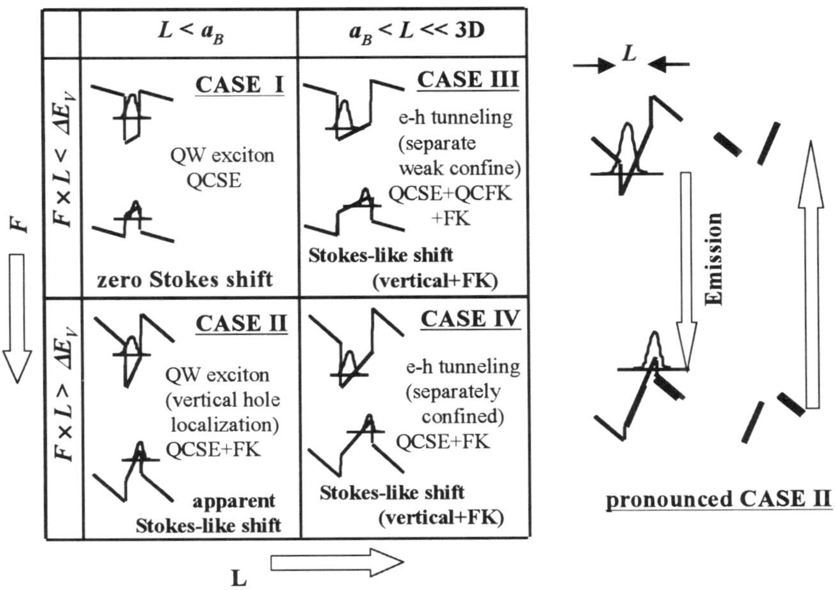

Figure 5.20 Schematic band diagrams of GaN/InGaN QWs under the electric field F. Each case represents the restrictions determined among F, L, ΔE_V, and a_B.

Taking the small fraction (1/4) of ΔE_V against the conduction band discontinuity ΔE_C into account (Van de Walle and Neugebauer, 1997), schematic band diagrams of InGaN/GaN QWs are drawn in Figure 5.20 for several cases; i.e. restrictions among $F\times L$, ΔE_V, and a_B. Since the restriction $F\times L>\Delta E_V$ breaks before breaking $F\times L>\Delta E_C$ with increasing F or L, restrictions between $F\times L$ and ΔE_V and between L and a_B (quantum size effect) only are drawn in Figure 5.20. Note that in-plane (lateral) bandgap inhomogeneity is omitted.

In CASE I, both electron and hole wavefunctions are confined in the well and have unique quantized energy levels where zero Stokes-like shift is expected though the resonance energy slightly shifts to lower energy due to QCSE. In CASE II, at least the hole wavefunction drops into the triangular shape potential well at one side of the well, and continuum states are formed at the rest of the potential slope inside the QW region. In addition the penetration of the barrier continuum wavefunction due to FK effect cannot be neglected. Therefore vertical component Stokes-like shift is produced. For the extreme situation of CASE II (pronounced CASE II), both the e-h wavefunctions drop into the triangular potential wells formed at opposite sides of the well resulting in z-directional separation. Note that the energy position of the barrier absorption may be shifted to lower energy due to FK effect between the barriers. In CASE III and IV, the absorption tail would be modulated due primarily to QCSE (Miller *et al.*, 1984; 1985) and QCFK (Miller *et al.*, 1986). It is obvious that in CASES III and IV the e-h wavefunction overlap is greatly reduced and the oscillator strength decreases, which may increase the radiative recombination lifetime τ_r.

5.3.3.6 Quantum well excitons in AlGaN/GaN structure

In order to verify the QCSE model shown in Figure 5.20, first we compared binary GaN SQW and In-doped GaN (InGaN) SQW with AlGaN barriers. GaN SQW was prepared in order to eliminate effective bandgap inhomogeneity due to InN molar fraction inhomogeneity peculiar to InGaN alloys. Figure 5.21 shows the optical transmittance at 10K and RT, and PL spectra at RT of a 5-nm-thick GaN/$Al_{0.15}Ga_{0.85}N$ DH structure grown on thick $Al_{0.3}Ga_{0.7}N$ base layer and those of 5.5-nm-thick $Al_{0.15}Ga_{0.85}N$/InGaN/$Al_{0.1}Ga_{0.9}N$ DH structure (SQW UV-LED) (Mukai *et al.*, 1998c) grown on thick GaN base. Since the former sample was grown on thick $Al_{0.3}Ga_{0.7}N$ to observe QW absorption spectrum, both GaN/AlGaN/AlGaN and InGaN/AlGaN/GaN SQWs suffer from compressive biaxial strain. Accordingly, F_{PZ} is opposite to the pn junction field. Against F_{PZ} up to the order of 500 kV/cm, the GaN/AlGaN SQW exhibits an excitonic absorption peak from 10 K to up to RT, as predicted form the arguments in increased E_b in QWs (Chichibu *et al.*, 1996d; 1998a; 1998b; 1999a). Observation of the QW excitonic absorption in GaN/AlGaN SQW under the strong F has also been predicted by Bulutay *et al.* (1999). Therefore InGaN SQW would also show the excitonic character. Although the bandgap of GaN is larger than that of InGaN, the PL peak energy of InGaN/AlGaN SQW is higher than that of GaN/AlGaN SQW and the QW resonance energy of GaN/AlGaN should be smaller than that of InGaN/AlGaN structure, which means that QCSE due to F dominates the optical transitions in the binary GaN wells. Accordingly, the PL intensity of GaN/AlGaN SQW was far lower than that of InGaN/AlGaN, although the self-absorption due to the GaN base

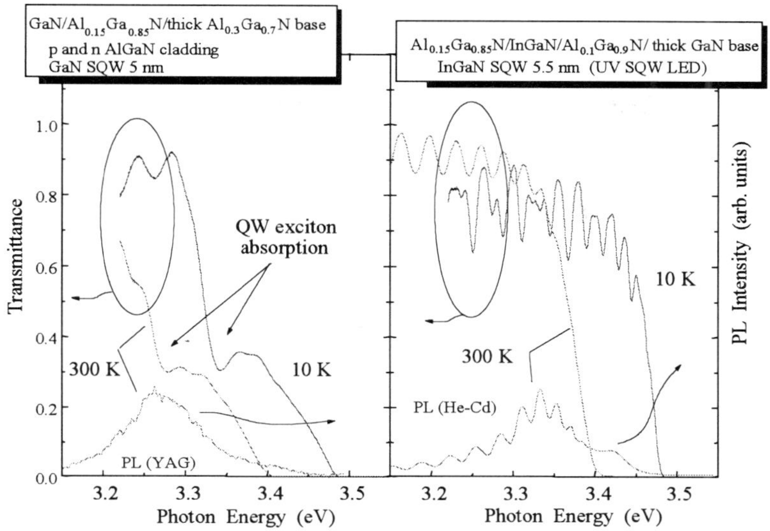

Figure 5.21 Optical transmittance and high excitation PL spectra at 10K and RT of a 5-nm-thick GaN/$Al_{0.15}Ga_{0.85}N$ DH structure grown on thick $Al_{0.3}Ga_{0.7}N$ base layer and those of 5.5-nm-thick InGaN/$Al_{0.1}Ga_{0.9}N$ DH structure (SQW UV-LED) grown on thick GaN base.

is remarkable for the particular UV-LED structure. Note that the PL peak energy agrees with the excitonic absorption peak at RT and excitons are delocalized for the GaN/AlGaN SQW, which is different from the results obtained from InGaN QWs. Note that in Figure 5.21, absorption due to the InGaN SQW cannot be seen clearly. Strong absorption due to the thick GaN base is considered to hide broadened QW excitonic absorption due to the InGaN SQW. However, the result also supports the idea that the joint density of states (JDOS) of InGaN tail states is smaller than that of GaN.

5.3.3.7 CASE studies of $In_xGa_{1-x}N$ QWs

In order to study systematically QW exciton localization due to the in-plane effective bandgap inhomogeneity, effects of F_{PZ}, and Coulomb screening of F_{PZ}, several series of samples were prepared; (i) a series which belongs to CASE I, II, and IV (the same x with L variable); MQWs with 14 periods of undoped $In_{0.1}Ga_{0.9}N$ wells (L was varied from 1.2 to 6nm) with 4.3-nm-thick GaN:Si (n = 2×10^{18} cm^{-3}) barriers capped by a 100-nm-thick $Al_{0.06}Ga_{0.94}N$, (ii) a series which belongs to CASE I, II, and extreme CASE II (the same L with x variable); 3-nm-thick $In_xGa_{1-x}N$/GaN:Si SQW structures capped by a 6-nm-thick GaN with x = 0.03, 0.1, 0.15, and 0.22; and (iii) partly Coulomb screened CASE II (the same L and x with variable n) 3-nm-thick $In_{0.13}Ga_{0.87}N$/GaN MQWs with different Si-doping concentrations in the barriers. To investigate the effect of the built-in field F_{bi}, we also prepared several actual LED and LD wafers (Nakamura and Fasol, 1997a) having InGaN QWs with GaN or InGaN barriers. All samples were grown on a 2-3 μm-thick GaN template on sapphire (0001) substrate by MOCVD.

Conventional optical measurements such as low excitation PLE, EL, low and high excitation static and time-resolved (TR) PL, and cathodoluminescence (CL) measurements were carried out between 4 K and RT. Monochromatic spatially-resolved CL measurements were carried out at 10 K.

5.3.3.8 Optical properties as a function of QW thickness L

Low excitation PLE and PL spectra at RT of $In_{0.1}Ga_{0.9}N$/GaN MQWs with various L are shown in Figure 5.22. The PL spectra exhibited a broad luminescence peak due to InGaN QWs, whose FWHM was nearly 120 meV. The PL peak showed a redshift by 360 meV with increasing L from 1.2 to 6.2nm, and the intensity decreased for $L>3.6$ nm ($L>a_B$). The PLE spectra exhibited broadened absorption tail except for the $L=1.2$nm case. The broadening was pronounced for wells with $L>a_B$. These results agree with the scenario that the QW configuration changes from CASE I, II, to IV due to increasing L assuming constant x and F.

The apparent bandgap energy determined as the energy where the PLE signal intensity drops to half the maximum, PL peak energies, and Stokes-like shifts are plotted as a function of L in Figure 5.23. As expected, the apparent Stokes-like shift increased from nearly 50 meV to 220 meV. Since in CASE I, zero Stokes-shift is expected if the well had a homogeneous bandgap energy, the observed Stokes-like shift for CASE I ($L<a_B$) directly shows the presence of in-plane net Stokes-like shift due to localized energy states in the QW. Similar results were obtained at 4K where the Stokes-shift was nearly 100 meV ($L<2.5$nm). The observed blueshift of the high excitation PL peak indicates presence of high F across the wells.

In order to estimate F, confined energy levels and wavefunctions in $In_{0.1}Ga_{0.9}N$ QWs are calculated as functions of F, L, and n in the barrier by

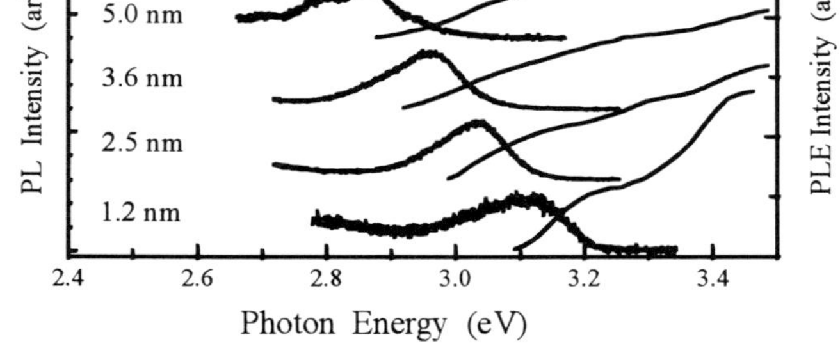

Figure 5.22 Low excitation PL and PLE spectra at RT of $In_{0.1}Ga_{0.9}N$ / GaN:Si MQWs for different well thickness, L. The apparent bandgap is defined as the energy where the PLE signal intensity drops to half the maximum.

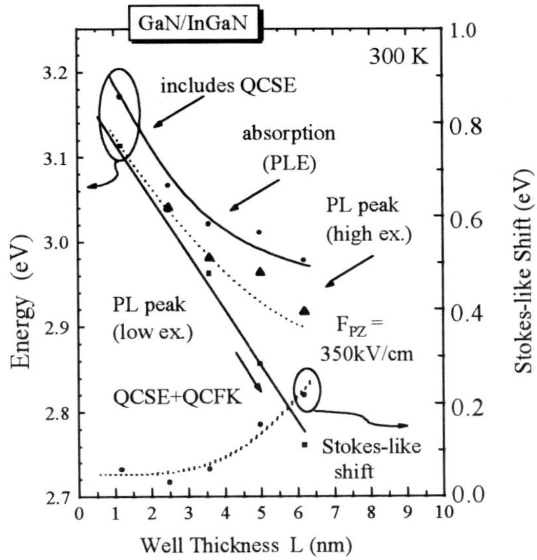

Figure 5.23 High and low excitation PL peak energies, apparent bandgap energy estimated from PLE spectra, and apparent Stokes-like shift of $In_{0.1}Ga_{0.9}N$ wells at RT as a function of L.

variational method neglecting E_b within the Hartree approximation by solving the Schrödinger equation and Poisson equation simultaneously and self-consistently, as described in Sections 5.3.1 and 5.3.2. We did not fit the data but simply calculated and compared the relation between the low excitation PL peak energy and L (L<3.6nm) in Figure 5.22. As a result, the zero-field bandgap of the 3D InGaN well is obtained to be 2.92 eV and ΔE_C and ΔE_V are estimated to be about 400 and 100 meV, respectively. F is estimated to be nearly 3.5×10^5 V/cm which

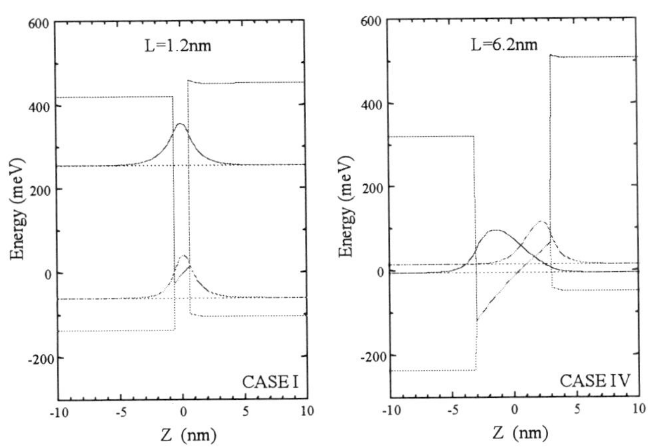

Figure 5.24 Quantized energy levels and wavefunctions of e-h pair in $In_{0.1}Ga_{0.9}N$ SQW under the condition of F = 400 kV/cm.

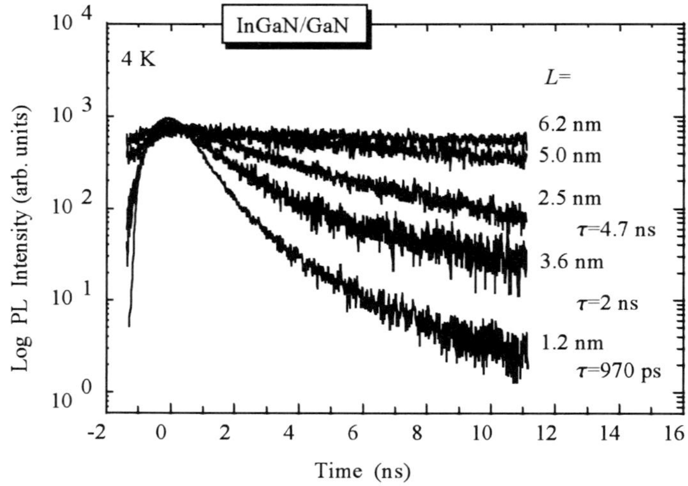

Figure 5.25 TR-PL signal from $In_{0.1}Ga_{0.9}N$ / GaN MQWs for various well thickness, *L*.

gives Stark shift of nearly 45 meV in the 3-nm-thick QW. Examples of the calculated quantized energy levels and e-h wavefunctions for InGaN under F = 400 kV/cm are shown in Figure 5.24. Under $F = 3.5\times10^5$ V/cm, $F\times L$ exceeds ΔE_V for L>3.4 nm. The hole confined level would already be formed in the triangular potential well (CASE II) between 2.5 and 3.6nm. Beyond this, the system belongs to CASE IV where the e-h pair is confined in opposite sides of the well, as shown in Figure 5.24. Figure 5.14 is drawn in the same format as Figure 5.15. This may explain the extremely long decay time τ in terms of reduction of the oscillator strength for large L at low temperature where nonradiative recombination is suppressed. At 4 K, τ was nearly 35 ns for L = 5nm (>a_B), as shown in Figure 5.25. This may cause serious degradation of the emission intensity at RT if nonradiative centers at interfaces or bulk are activated. Relatively shorter τ (0.97–4 ns) for L<2.5nm at 4K indicates that the overlap of the e-h wavefunction is still large because L is smaller than a_B. Note that the estimated F_i (6.0×10^5 V/cm) is larger than F (3.5×10^5 V/cm), which implies that Coulomb interaction between the e-h pair still remains. This kind of particles can be regarded as QW excitons.

5.3.3.9 Optical properties as a function of InN molar fraction x

Next sample series have a 3-nm-thick InGaN SQW with various x, which correspond to CASE I, II, and pronounced CASE II. Note that change in x changes in ΔE_C, ΔE_V, and F simultaneously. The PLE spectra broaden and the PL peak energy shows pronounced redshift with increasing x, as shown in Figure 5.26. The InGaN SQW with small amount of In (x = 0.03) exhibited a sharp onset of the PLE spectra at 3.3 eV, as is the case with L = 1.2nm in Figure 5.22. If we assume a linear relation between F and x, which is reasonable since F_{PZ} is a linear function of the residual strain, F_{PZ} is estimated to be nearly 1.1×10^5 V/cm. The PLE spectra were taken at 10 K, and the sample is considered to belong to CASE I, as shown in

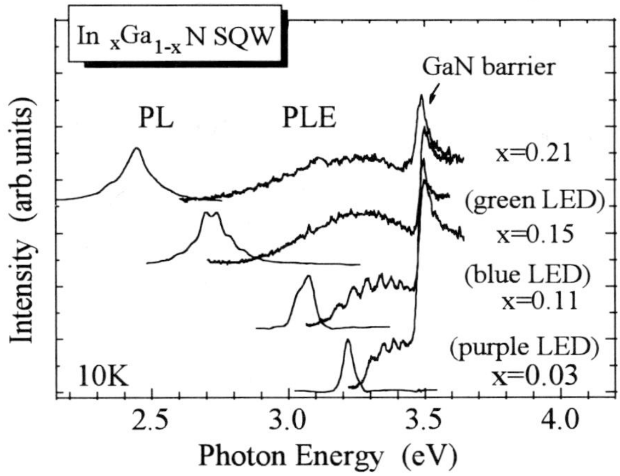

Figure 5.26 Low excitation PL and PLE spectra at 10K of 3-nm-thick $In_xGa_{1-x}N$ / GaN:Si SQW structures with different InN molar fraction, *x*. The *x* values are estimated assuming coherent growth of InGaN QWs.

Figure 5.27. The appearance of the Stokes shift in the CASE I sample again indicates the presence of localized energy states. With increasing *x*, first the hole confined level drops into the triangular potential, as shown in Figure 5.27 (400 kV/cm), and that of electron will also drop into the triangular potential at opposite side for higher *x* (pronounced CASE II in Figure 5.20). This scenario explains the broadness of the PLE spectra with increasing *x*. In addition, the spatial separation of the e-h pair normal to the QW plane may reduce the wavefunction overlap and increase the emission decay time with increasing *x* as shown in Figure 5.28. However, as shown in Figure 5.27, the overlap of e-h wavefunction is still large.

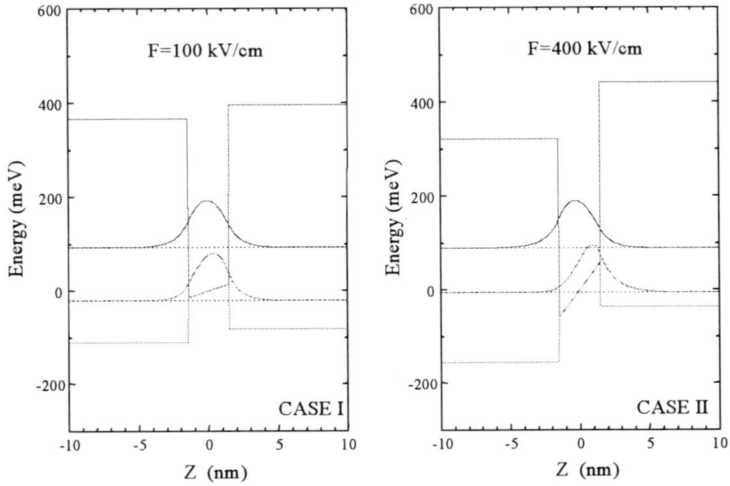

Figure 5.27 Quantized energy levels and wavefunctions of e-h pair in 3-nm-thick InGaN SQW under the condition of F = 100kV/cm (x = 0.03) and 400kV/cm (x = 0.1).

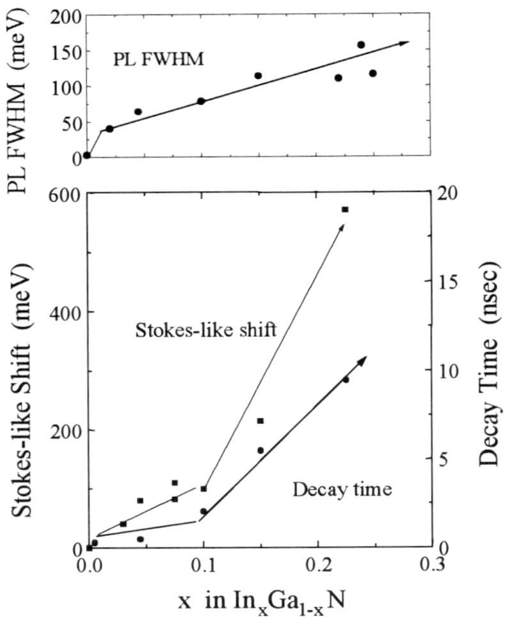

Figure 5.28 PL FWHM, apparent Stokes-like shift, and TR-PL decay time of 3-nm-thick $In_xGa_{1-x}N$ QWs as a function of x.

Both the Stokes-like shift and the decay time had a critical x of nearly 0.1, as shown in Figure 5.28. This composition is considered to correspond to the point that the band diagram changes from CASE I to CASE II or pronounced CASE II where the absorption due to continuum states from barriers becomes remarkable. On the other hand, the FWHM of the PL peak suddenly increases by alloying (or doping) very small amount of In, which enhances the degree of effective bandgap inhomogeneity, and increases monotonically with increasing x (Chichibu *et al.*, 1998a; 1998b ;1999a; Shan *et al.*, 1998; O'Donnell *et al.*, 1999; Martin *et al.*, 1999) as shown in Figure 5.28.

5.3.3.10 In-plane localization of QW excitons in InGaN QW structures

As described, InGaN QWs and bulk layer have an in-plane effective bandgap inhomogeneity, which is a source of the effective QW exciton or carrier localization. Blue and green InGaN SQW LEDs show blueshift of the EL peak energy with increasing driving current, as shown in Figure 5.18. This has been explained by combined effects of Coulomb screening of F_{PZ} and band filling of the localized energy states by QW excitons (Chichibu *et al.*, 1996d). Really recently, Mukai *et al.* (1998a; 1998c) reported that current and temperature dependences of the EL peak energy of 5.5-nm-thick $In_xGa_{1-x}N$ SQW LEDs change drastically at a critical emission peak wavelength of 375nm. When x is very small, LEDs do not show any blueshifts but show temperature dependence. Conversely, when x is larger than corresponding emission peak of 375nm, the LEDs shows nearly temperature-independent but current density-dependent EL peak energy. Moreover,

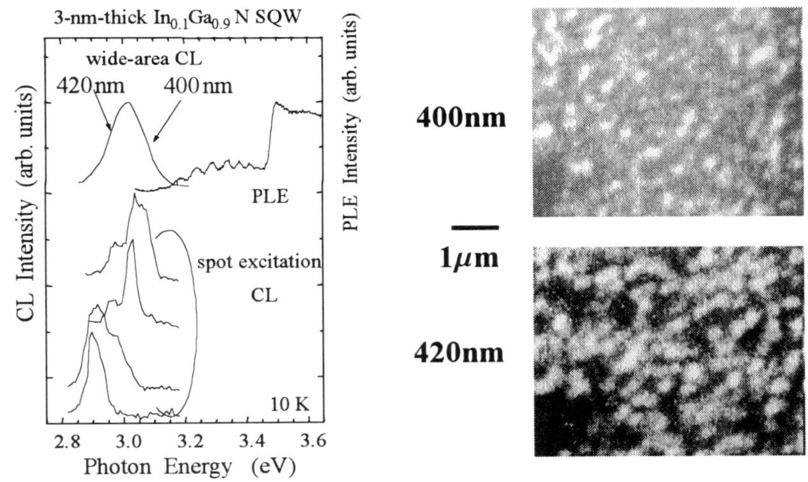

Figure 5.29 Wide-area (10 μm×10 μm) scanning and spot excitation CL spectra of 3-nm-thick $In_{0.1}Ga_{0.9}N$ SQW capped by a 6-nm-thick GaN. The broad CL peak (FWHM = 100 meV) consists of several sharp emission peaks (FWHM = 20 meV) having different peak energies. Monochromatic CL images taken at 400 and 420nm showed primarily the complimentary relation. However, some parts exhibit both 400 and 420nm emissions. The resolution of the CL mapping is limited by the diffusion length, which was nearly 60nm in lateral direction. Both measurements were carried out at 10K.

the output power of UV LEDs increase with increasing wavelength in UV region longer than 370nm where the self-absorption due to GaN base layer is already negligible. These results indicate that the emission mechanisms change with x, and those of GaN QW and InGaN QW are totally different, as introduced in Figure 5.21.

Typical wide-area and spot-excitation CL spectra at 10K of a 3-nm-thick $In_{0.1}Ga_{0.9}N$ SQW capped by a 6-nm-thick GaN are shown in Figure 5.29. Although the effect of the multiple interference fringes on the CL spectra is not negligible, wide area (10 μm×10 μm) broad CL band (FWHM nearly 100 meV) consists of many sharp peaks (FWHM less than 20 meV) (Chichibu *et al.*, 1997d; 1999a) at 10K. This clearly shows that the broad CL band consists of sharper emission peaks having various peak energies. Therefore, there exist several *structures* composed of an InN-rich part surrounded by a GaN-rich part in each spot area. This *structure* may be due to mesoscopic compositional undulation and it can act as quantum dot (QD), quantum disk (Q-disk), or segmented QW depending on the lateral size of the potential confinement (Sugawara, 1995). Note that the energy states are confined at least along the *c*-axis.

Monochromatic scanning CL images of the same GaN-capped $In_{0.1}Ga_{0.9}N$ SQW taken at wavelengths of 400nm and 420nm are also shown in Figure 5.29. A careful comparison between the two images using a triangular-shaped dark marker on the bottom-left corner indicated following results; (a) each bright (white in the figure) area consists of emissions from real spaces of about 60-400nm in lateral size, (b) some dark areas in one figure correspond complementary to bright ones in the other figure, and (c) some areas exhibit both 400 and 420nm CL emissions. These results can also be explained by the existence of emission bandgap inhomogeneity whose lateral interval is smaller than 60nm, which value is the spatial resolution of the system (diffusion length). In monochromatic CL imaging system we used, the diffusion length is determined as the narrowest dark region width between bright areas. Supporting evidences of the short diffusion length in InGaN QWs were reported by Rosner *et al.* (1997) and Speck *et al.* (1998). They showed less pronounced CL contrast with small dark region for InGaN SQW compared to that of binary GaN. Sato *et al.* (1998) estimated the PL intensity as a function of TD density and also supported the short diffusion length model in InGaN QWs (Chichibu *et al.*, 1997d; Vertikov *et al.*, 1998). The result (c) indicates that the real area that emits CL is smaller than the diffusion length (60nm). Ponce *et al.* (1998a) and Kisielowski (1997; 1998) estimated the structural size of InN-rich clusters to be less than 10nm and 20nm, respectively. Such CL nanostructure was also found for compositions as low as $x = 0.03$. These results mean that large FWHM of the PL peak and broadness of the PLE signal edge for the CASE I QWs represent the in-plane effective emission bandgap inhomogeneity. If we believe these lateral size of the *structure*, it can act as Q-disk, since the lateral size is in between a_B and exciton resonance wavelength (Sugawara, 1995), which is the emission wavelength divided by the refractive index and is nearly 160nm in this case. Note that e-h wavefunctions in the Q-disks are confined with respect to the z-direction and Q-disks or Q-dots can improve the emission efficiency of QW excitons due to lateral accumulation of carriers that prevent them from being trapped into nonradiative pathways. This may be the reason why InGaN exhibits bright spontaneous emissions. Naturally, it overcomes disadvantages of using wurtzite materials due to the polarization-induced electric fields. The e-h wavefunctions are still overlapped even in the pronounced CASE II. The emission lifetime is affected by both the e-h pair separation due to F and in-plane localization for CASE II and pronounced CASE II. In Q-disks, the spontaneous emission lifetime becomes long since the emission is prohibited when the wave vector of exciton center-of-mass motion is above the critical energy (Sugawara, 1995).

5.3.3.11 Origin of exciton localization

One of the possible origins of the in-plane bandgap inhomogeneity is an inhomogeneous distribution of F due to strain fluctuation. If we attribute the in-plane net Stokes-like shift to this only, the potential fluctuation can be levelled by filling carriers to screen the F. However, the FWHM of PL or EL in InGaN QWs does not change or even increase with increasing excitation level (Chichibu *et al.*, 1996d; Nakamura and Fasol, 1997a). Therefore, an intrinsic effective bandgap inhomogeneity is the most probable.

There are two major possible origins for the formation of effective bandgap inhomogeneity or localized energy tail states. One is the nonrandom alloy compositional disorder due to large differences in thermal stability and lattice parameter between GaN and InN and the other is QW thickness fluctuation. The latter can theoretically be observed even for GaAs, InP, or GaN QWs without any compositional inhomogeneity. As described, the structural size is less than 60nm to up to 300nm at 10K.

Keller *et al.* (1998a) found a spiral growth of InGaN initiated by the mixed character TDs. If such a growth mode accumulates excess In to the growth step edge, compositional inhomogeneity would be produced in addition to the QW thickness fluctuation due to the spiral growth. To investigate this, optical properties of InGaN QWs were investigated as a function of TD density (Chichibu *et al.*, 1998e; 1999a; 1999b) using LEO technique. A series of InGaN QWs were grown on dislocated GaN grown from the SiO_2 mask opening (window) and nearly TD-free GaN laterally overgrown on SiO_2 from the opening (wing) at the same time in order to minimize run-to-run or place-to-place unstable reproducibility.

Figure 5.30 shows PL and PLE spectra at 10K of InGaN SQW and MQW structures grown on LEO window and wing. As shown, they showed exactly the same properties. The luminescence decay time τ measured by the time-resolved PL measurements was 6 ns at RT for the MQW structure, and was independent of the TD density, as shown in Figure 5.31. In Figure 5.31, TR-PL signals of InGaN SQWs grown on window and wing under unoptimized growth conditions are also shown. The decay time τ_{RT} is as short as 260 ps and is independent of the TD density. In these particular samples, nonradiative recombination process determines the RT lifetime. The low temperature PL lifetime was also independent of the TD density. In addition, the energy resolved τ of nearly TD-free InGaN MQW increased with decreasing the photon energy, as shown in Figure 5.32. This is characteristics of the localized electronic system. The relation between τ and E was fitted using

$$\tau(E)=\frac{\tau_r}{[1+exp\{(E-E_{me})/E_0\}]}, \tag{5.43}$$

where E_0=60 meV represents the depth in the tail states, E_{me} = 2.88 eV is the energy similar to the mobility edge, and τ_r = 12 ns is the radiative lifetime. These values are

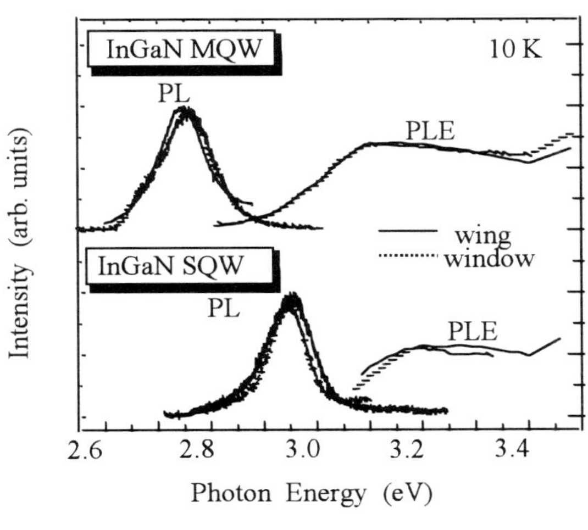

Figure 5.30 PL and PLE spectra at 10 K obtained from InGaN SQW and MQW structures grown on LEO window and LEO wing regions.

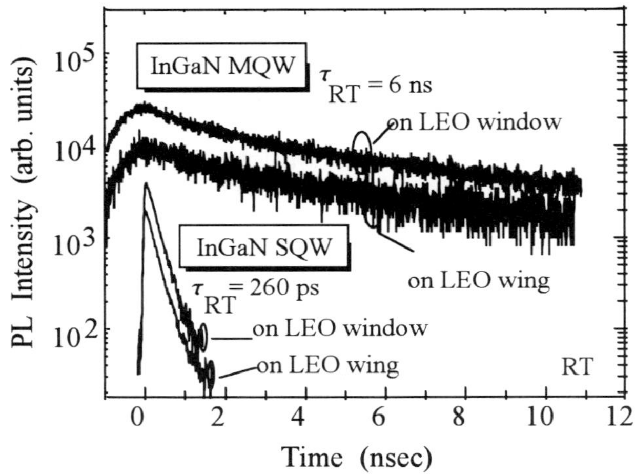

Figure 5.31 Time-resolved PL signals from InGaN SQWs and MQWs grown on LEO wing and window regions measured at RT.

reasonable for the device-quality InGaN QW systems (Nakamura and Fasol, 1997a) with highly lattice mismatched nonrandom alloy broadening.

It should be noted that the data shown in Figures 5.30 and 5.31 show that major limiting factor which determines the emission mechanisms in InGaN QWs is not related to TDs, since all the optical properties do not depend on TD density, as Mukai *et al.* have reported (1998d). Conversely, the in-plane net Stokes-like shift is affected by changing the barrier growth rate (Keller *et al.*, 1998b) and InN molar fraction of the QWs (Chichibu *et al.*, 1996d; 1997c; 1998a; 1998b; 1999a).

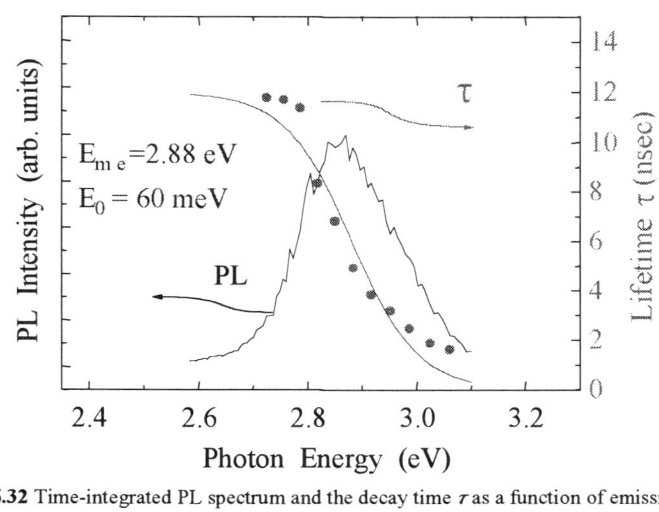

Figure 5.32 Time-integrated PL spectrum and the decay time τ as a function of emission energy measured at RT of InGaN MQW grown on the LEO wing.

	II-VI Laser	**III-Nitride Laser**
substrate	**GaAs**	**sapphire**
base material	**ZnSe**	**GaN**
active region	**ZnCdSe**	**InGaN**
lasing λ (nm)	**480∼520**	**380∼430**
Stokes-like shift	**~30 meV (4K)**	**~110 meV (4K)**
	~ 0 (300K)	**~ 55 meV (300K)**

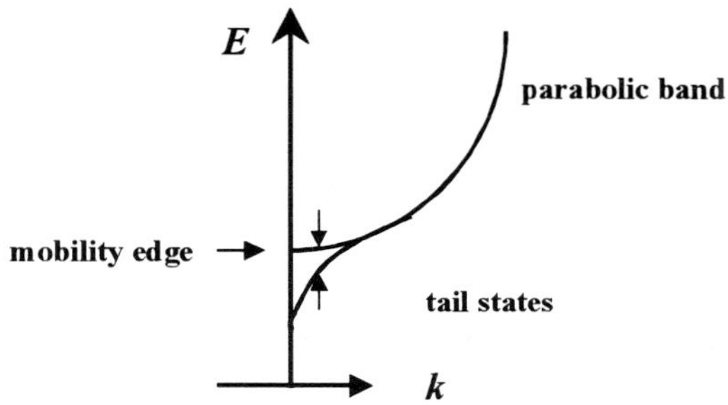

Figure 5.33 Schematic illustration for the joint density of states (JDOS) of the electronic system with localized energy tails states.

Therefore the in-plane effective bandgap inhomogeneity is initiated by the growth parameters, point defects, or thermodynamics rather than phase separation initiated by TDs. Anomalous temperature independence of the luminescence peak energy (Chichibu *et al.*, 1997c; Nakamura and Fasol, 1997a) cannot be explained only by an inhomogeneous F across the QW plane. Since the change in x changes F_{PZ} simultaneously, the Q-disk size/segmented QW potential may, at least partly come from the nonrandom alloy potential fluctuation emphasized by the large bowing parameter (Takeuchi *et al.*, 1997b; Akasaki and Amano, 1997; McCluskey *et al.*, 1998) in InGaN.

Figure 5.33 schematically represents JDOS of a localized system having energy tails. A remarkable difference between the JDOS of CdZnSe QW lasers and InGaN QW lasers is that the room-temperature Stokes-like shift is nearly zero for CdZnSe while it is still nearly 50 meV for InGaN QWs (100 meV at LT). For LDs, effective bandgap inhomogeneity more than 50 meV is too large to obtain a uniform EHP in the well. Indeed, some InGaN MQW LDs showed EHP lasing in the tail states (Jeon *et al.*, 1996; Deguchi *et al.*, 1997a; 1997b; 1998a; 1998c; Mohs *et al.*, 1997; Chichibu *et al.*, 1998a; 1998b; 1998d; Vertikov *et al.*, 1998), that is discussed in Section 5.3.5. This may cause the increase of threshold current density of InGaN LDs in terms of reduction in the differential gain (Domen *et al.*, 1998). However, if the gross gain produced in the tail states exceeds total internal loss in the LD cavity, stimulated emission will be coming out from the structure since the smaller the JDOS, the lower the gross carrier density to produce population inversion. A schematic model of the emission mechanisms in InGaN QWs is shown in Figure 5.34.

As described above, importance of QW exciton localization into the energy

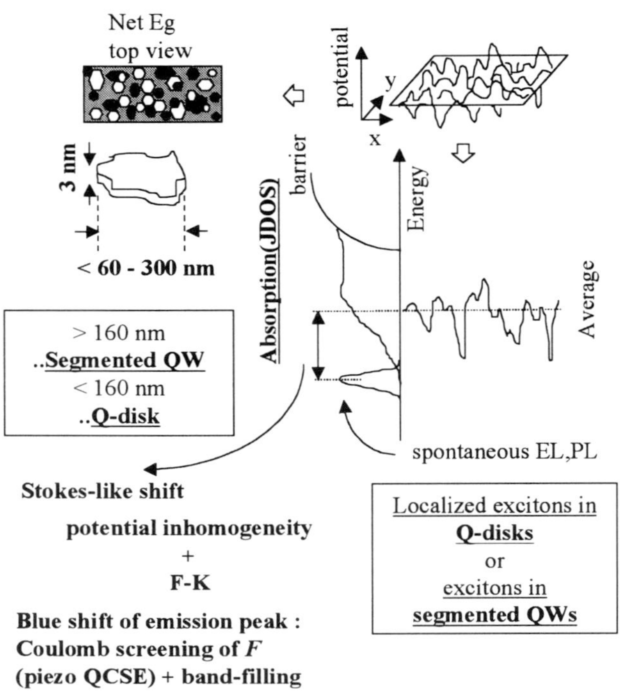

Figure 5.34 Schematic illustration for the joint density of states (JDOS) and potential inhomogeneity in InGaN QWs with localized energy tails states.

tail states was explained in addition to high-field effects in InGaN QWs. This Q-disk size potential minima make the diffusion length short and effectively keep carriers away from nonradiative pathways, which is a distinct difference between GaN and InGaN QWs. Strong F_{PZ} separates e-h pair into triangular potential wells formed at opposite sides of the well and apparent Stokes-like shift is dominated by F_{PZ} for CASE II, III, and IV. The oscillator strength of the separate e-h pair is very small for CASE III and IV. State-of-the-art InGaN LEDs and LDs generally have QWs with L = 2.5-3.5nm and most of blue, green, and amber LEDs may belong to pronounced CASE II. However, the overlap of e-h wavefunctions is recovered by screening F_{PZ} by injecting carriers or doping Si, resulting in huge blueshift of the emission peak and a decrease of the apparent Stokes-like shift. Note that vertical e-h pair distance is still smaller than a_B, resulting in large overlap of their wavefunctions. Therefore the localized QW excitons have strong Coulomb interaction provided that $L<a_B$. In order to understand precise mechanisms of the bandgap inhomogeneity and carrier localization, and to obtain long wavelength LDs expanding from pure green to red wavelengths, fabrication and investigation of cubic zincblende InGaN QWs are mandatory to eliminate the modulation due to strain-induced piezoelectricity.

5.3.3.12 Effects of Si-doping in InGaN QW structures

Doping of Si in GaN or InGaN has been reported to result in improved structural and optical properties (Nakamura *et al.*, 1992; 1993; 1997a; Ruvimov *et al.*, 1996; Grudowski *et al.*, 1997; Chichibu *et al.*, 1998b; 1998d; 1999a; Deguchi *et al.*, 1998a; Keller *et al.*, 1998b; Minsky *et al.*, 1998). In particular, the PL intensity at RT of Si-doped InGaN was found to be nearly 36 times stringer than that of undoped InGaN films. In addition, a reduction of TD density from $5\times10^9 cm^{-2}$ to $7\times10^8 cm^{-2}$ was measured (Ruvimov et al., 1996) for GaN films doped with Si. InGaN QWs have also been evaluated by RT PL and a higher PL intensity has been observed for QWs with an intentional *n*-type Si-doping concentration (Grudowski et al., 1997). Here we summarize the effect of Si-doping in GaN or InGaN barriers of InGaN QW structures.

Si-doping has two major advantages, namely structural improvement and electronic improvement of QWs. The former relies on particular GaN growth process having large TD density. The latter is simply due to Coulomb screening of the internal electric field in the QW due to polarization.

Generally, InGaN/GaN QW structure is grown at relatively low temperature around 800°C in order to adjust the growth temperature to optimized one for the InGaN well. This temperature is too low to grow good quality GaN barriers. Keller *et al.* (1998b) has found that growth of undoped GaN at low temperature forms island-like spiral structure initiated by TDs exist in the underlying template GaN layer, but formation of the spiral is effectively suppressed by doping with Si during the GaN barrier growth even at low temperature. This allows to grow smoother GaN/InGaN interfaces resulting in narrower FWHM of satellite peaks of super lattice (SL) reflection in the XRD spectra (Cho *et al.*, 1998). For the case of InGaN wells with InGaN barriers, doping of Si in the InGaN barrier again improves the

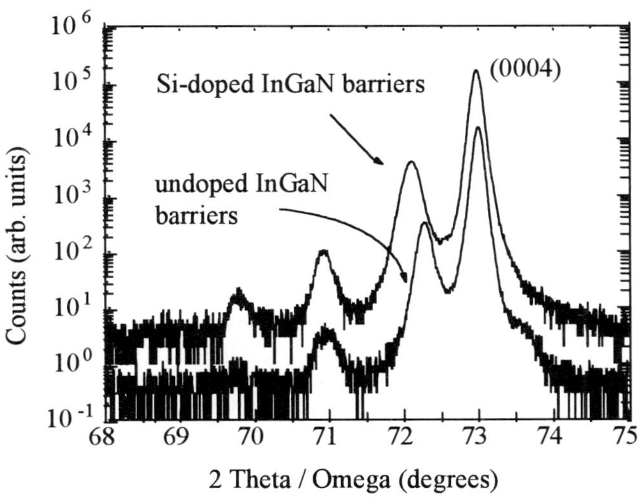

Figure 5.35 2θ-ω scan of 12 periods of $In_{0.13}Ga_{0.87}N/In_{0.04}Ga_{0.96}N$ MQW structure grown on a 2μm-thick GaN template.

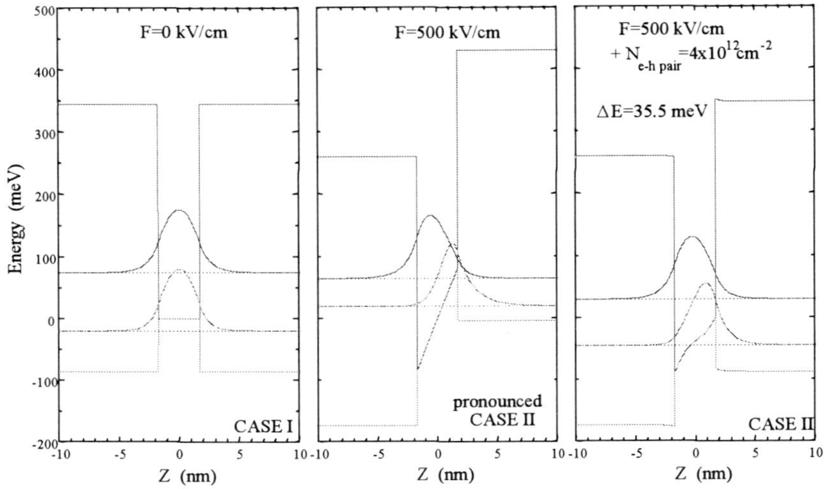

Figure 5.36 Quantized energy levels and wavefunctions of e-h pair in 3-nm-thick InGaN SQW under the condition of $F = 0$kV/cm, $F = 500$kV/cm, and $F = 500$kV/cm and $N_{e\text{-}h\,pair} = 4\times10^{12}$cm^{-2}.

interface structure, as shown in Figure 5.35. Higher order peak is found in the Si-doped MQW structure (Chichibu *et al.*, 1998d). Such flattening of the interface *would* improve nm to μm-scale effective bandgap inhomogeneity in the QW *if* the QW thickness fluctuation dominates the in-plane Stokes-like shift of InGaN QWs or *if* InN-clustering is initiated by TDs. However, the results in this Section and recent observations (Mukai *et al.*, 1998d; Chichibu *et al.*, 1998e; 1999b) show that TDs are not the major source of the effective bandgap inhomogeneity in InGaN QWs. It is obvious that the InN molar fraction of the well is affected by the well growth rate (Nakamura and Fasol, 1997a). The finding that change in the barrier growth rate also changes the InN molar fraction of the InGaN well (Keller *et al.*, 1998b) indicate that another surface activity or In-related phenomenon may be the source of QW exciton localization.

One of the effective methods to screen polarization-induced F is by doping the wells or barriers. If the injected or doped charge density is enough to screen F_{PZ}, the pronounced CASE II QWs may recover wavefunction overlap (Deguchi *et al.*, 1998a; Chichibu *et al.*, 1998b; 1998c; 1999a;). Quantized energy levels and wavefunctions of a 3-nm-thick $In_{0.13}Ga_{0.87}N$ SQWs for $F = 0$, $F = 500$ kV/cm, and $F = 500$ kV/cm with injected e-h pair density of 4×10^{12}cm^{-2} are shown in Figure 5.36. As shown, the wavefunction overlap and transition energy were recovered by the Si-doping. To confirm this experimentally, 3-nm-thick $In_{0.13}Ga_{0.87}N$/GaN MQWs with different Si-doping concentrations in the GaN barriers were examined at RT (Chichibu *et al.*, 1998b; 1998d; 1999a). F_{PZ} in the undoped MQW is estimated to be 5-6$\times10^5$ V/cm. The PL peak shifted to higher energy and the absorption tail in the PLE spectrum decreased by doping Si up to 10^{19}cm^{-3}, as shown in Figure 5.37. The apparent resonance energy shifted to higher energy by 50 meV, which nearly agrees with Figure 5.36, and the apparent Stokes-like shift

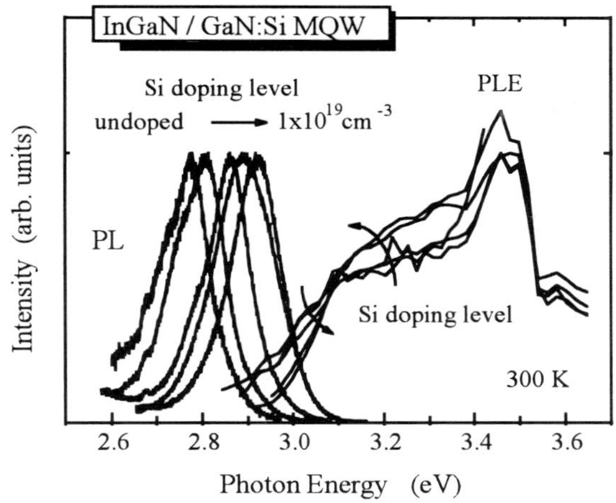

Figure 5.37 Low-excitation PL and PLE spectra at 300 K of $In_{0.13}Ga_{0.87}N$/GaN MQWs for different Si-doping concentrations (n). PLE edge was sharpened and the PL peak blueshifted with increasing Si-doping concentration.

decreased from 220 meV to 120 meV. At the same time, the emission decay time τ decreased from 14 ns to 850 ps with increasing Si-doping, as shown in Figure 5.38. These results indicate that F_{PZ} was effectively screened, and the overlap of the e-h wavefunction was recovered. Note that screening of F_{PZ} reduces the slope of the conduction and valence bands within the QW to result in sharp onset of the absorption tail. This reduces the "vertical" component of the apparent Stokes-like shift, too. Of course, increased electron density due to the modulation doping may improve the radiative recombination rate, and the electrons will fill the energy tail states at the same time.

The sample with the highest doping level showed the shortest $\tau = 850$ ps and the brightest emission. This value seems to be close to the intrinsic radiative

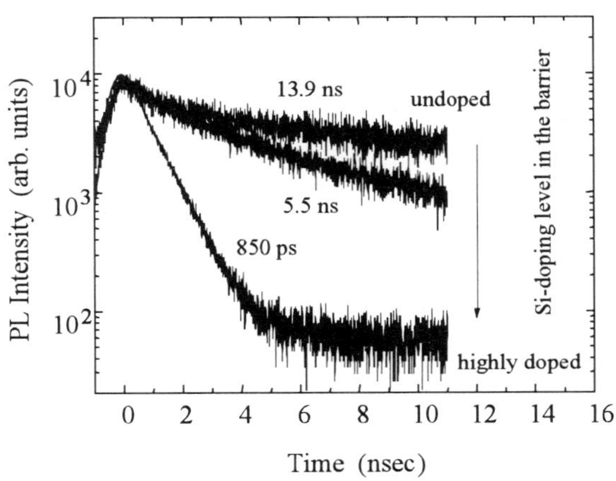

Figure 5.38 Low-excitation PL and PLE spectra at 300 K of $In_{0.13}Ga_{0.87}N$/GaN MQWs for different Si-doping concentrations (n). PLE edge was sharpened and the PL peak blueshifted with increasing Si-doping concentration.

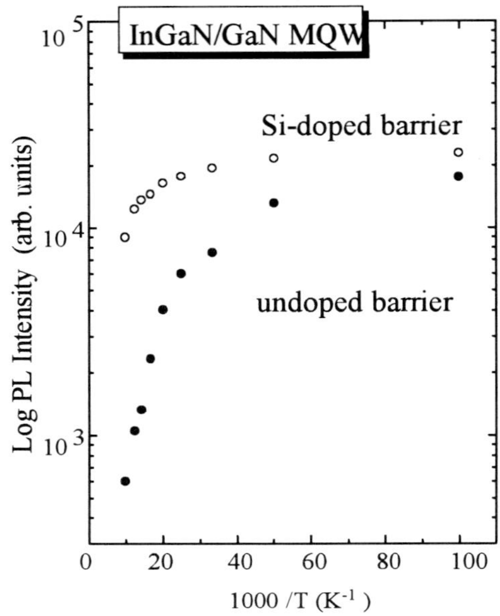

Figure 5.39 PL intensity as a function of temperature for $In_{0.1}Ga_{0.9}N$/GaN MQW structure having 3 InGaN wells and undoped or Si-doped (10^{18} cm^{-3}) GaN barriers.

recombination lifetime at RT. The longer τ usually observed for spontaneous emissions in practical SQW LED and LD wafers having InN mole fraction less than 10% may be due to combined effects of e-h pair separation due to F and in-plane localization of QW excitons for CASE II and pronounced CASE II. In Q-disks potential, the spontaneous emission lifetime becomes long since the emission is prohibited when the wave vector of exciton center-of-mass motion is above the critical energy (Sugawara, 1995).

Another important effect by Si-doping is reduction of nonradiative centers in the GaN:Si barriers. Figure 5.39 shows PL intensity as a function of temperature for $In_{0.1}Ga_{0.9}N$/GaN MQW structure having 3 InGaN wells and undoped or Si-doped (10^{18} cm^{-3}) GaN barriers. As shown, thermal quenching behavior of the PL intensity due to activation of nonradiative channels with increasing temperature is less pronounced for the MQW with Si-doped GaN barriers. This result shows that Si-doping in the GaN barrier reduces the number of nonradiative recombination centers in the MQW region including the interfaces.

5.3.3.13 Summary of subsection

The electronic states responsible for highly efficient spontaneous emission from InGaN bulk and QWs seem to suffer from every electric field effect as shown above. Striking finding is that red and amber InGaN SQW LEDs emit the light having photon energies less than the InN bandgap. Therefore it is not odd to consider that the energy states do not have unique discrete energy states; they have quasi two dimensional electronic states. This picture agrees with the idea that spontaneous emission from InGaN QWs is due to recombination of confined QW excitons localized at Q-disk size potential minima. Further experimental and theoretical studies are needed to clarify the spontaneous emission mechanisms in InGaN bulk and QWs.

Acknowledgments

The authors are grateful to Professors U.K. Mishra, L.A. Coldren, J.E. Bowers, E. Hu, J.S. Speck, and K. Suzuki, and Drs. S. Keller, S. B.Fleischer, D. Cohen, J. Ibbetson, C. Bulutay, T. Azuhata, and T. Uenoyama for helpful discussions. The authors would like to thank A. Abare, M. Mack, M. Minsky, T. Deguchi, A. Shikanai, K. Torii, P. Kozodoy, M. Hansen, P. Fini, and H. Marchand for help in the experiments. S.F. Chichibu and T. Sota are thankful to Professor H. Nakanishi and Professor Y. Horikoshi for continuous encouragement.

5.3.4 Localized Exciton Dynamics

5.3.4.1 Introduction

Time-resolved photoluminescence (TRPL) spectroscopy is a useful technique to evaluate the optical properties of semiconductors. As for GaN-based semiconductors, targets as photonic devices can be classified in three subjects.

The first subject is to achieve the efficient LEDs. Typical blue or green LEDs commercially available show the external quantum efficiency (η_{ext}) of about 10% (Nakamura *et al.*, 1995a; 1995b). However, further improvement of the efficiency is desired to extend the application area of LEDs, such as the replacement of light

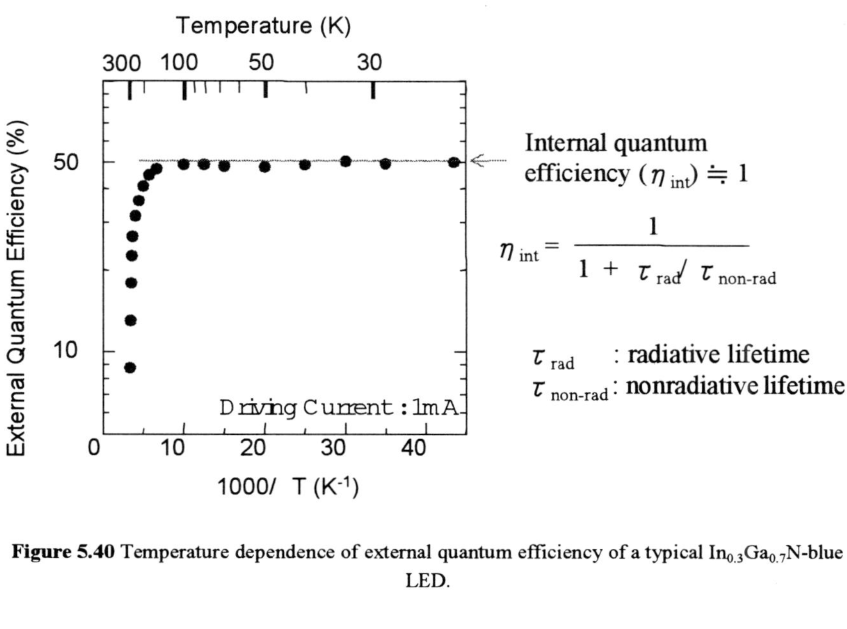

Figure 5.40 Temperature dependence of external quantum efficiency of a typical $In_{0.3}Ga_{0.7}N$-blue LED.

bulb or fluorescent lamp (vacuum tubes) by LEDs (solid state devices). Figure 5.40 shows the temperature dependence of η_{ext} value of a typical $In_xGa_{1-x}N$-based blue LED operated under constant current of 1 mA. The ηext value rapidly decreases with increasing temperature above about 200K. This result indicates that the device performance at RT is still limited by the non-radiative recombination processes, and that it should be possible to achieve the η_{ext} more that 40% if the pathway to the non-radiative recombination centers are eliminated sufficiently. Therefore, it is very important to assess the mechanism of both radiative and non-radiative processes in order to clarify the key to get higher efficiency.

The second subject is related to the LDs. Since the first operation of LD has been demonstrated at 400nm under a continuous wave (CW) mode at RT (Nakamura *et al.*, 1996b), the device lifetime has been made good progress, and at current stage the maximum operation time is estimated to be 10,000 hours (Nakamura *et al.*, 1997b). However, accelerated tests under elevated temperature and higher optical-power operation shows that the device life is still less than the level of commercialization. So far, the lowest threshold current density (I_{th}) achieved in the InGaN-based LD is about 2 kA/cm^2 which is one order of magnitude higher than that in LDs based on other semiconductors. Consequently, it is necessary to reduce the value of Ith to achieve the stable operation. Moreover, it should be noted that the tuning wavelength for the stable CW-operation of LDs is currently in the range between 380nm and 430nm, which is much narrower than that of LEDs. The reason why the operation wavelength is limited at about 430nm at the long wavelength side may be because the optical gain cannot be generated sufficiently in the devices with In-rich active layers due to the inhomogeneous broadening effect, as well as due to the limited number of density-of-states (DOS). Further breakthrough is therefore required to realize pure blue and green LDs using InGaN-based semiconductors. Such targets can be facilitated by the well-understanding of emission mechanism, as well as by the well-designing of the LD structures.

The third subject is related to the development of new photonic devices besides LEDs or LDs utilizing ultrafast phenomena or optical nonlinearity. For such purposes, it is essential to understand the recombination mechanism by assessing dynamical behavior of excitons and/or carriers based on radiative, non-radiative, localization and many body processes, and then to make a positive feedback to the fabrication of photonic devices.

5.3.4.2 Radiative and non-radiative processes

Recombination processes of carriers (electrons and holes), which are generated by either photo-excitation or forward-biasing across p-i-n structures, can be classified to two competing pathway; radiative and non-radiative processes. Various types of radiative and non-radiative processes are depicted in Figure 5.41.

If the semiconductors are excited by the short pulse, the transient luminescence intensity is generally expressed by the following equation as a function of time after excitation,

$$I(t) = I_0 \exp\left(-\frac{t}{\tau_L(T)}\right), \tag{5.44}$$

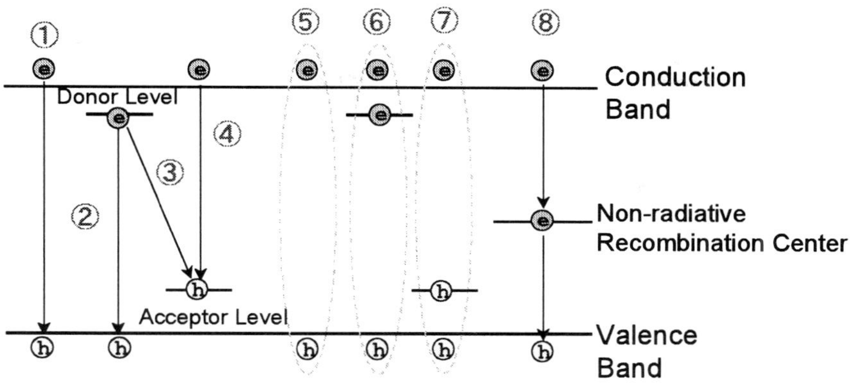

Figure 5.41 Recombination processes in semiconductor. ① band to band, ② bound to free, ③ donor to acceptor (DAP), ④ free to bound, ⑤ free exciton, ⑥ exciton bound to neutral donor, ⑦ exciton bound to neutral acceptor, ⑧ non-radiative recombination.

where $\tau_L(T)$ is the decay time of luminescence at a given temperature (T K). The inverse of $\tau_L(T)$ is the sum of three different types of transition probability,

$$\frac{1}{\tau_L(T)} = \frac{1}{\tau_{rad}(T)} + \frac{1}{\tau_{non-rad}(T)} + \frac{1}{\tau_{trans}(T)}, \tag{5.45}$$

where $\tau_{rad}(T)$ and $\tau_{non\text{-}rad}(T)$ denote the radiative and non-radiative lifetimes, and $\tau_{trans}(T)$ is the transfer time toward lower-lying energy levels. Details of the transfer time is discussed in the following section. If the radiative recombination occurs at the bottom of energy levels, the term of the transfer time can be neglected so that the equation is simplified as shown below.

$$\frac{1}{\tau_L(T)} = \frac{1}{\tau_{rad}(T)} + \frac{1}{\tau_{non-rad}(T)}. \tag{5.46}$$

The internal quantum efficiency [$\eta_{int}(T)$] of the emission can also be written in terms of $\tau_{rad}(T)$ and $\tau_{non\text{-}rad}(T)$, where

$$\eta_{int}(T) = \frac{\tau_{non-rad}(T)}{\tau_{rad}(T) + \tau_{non-rad}(T)}. \tag{5.47}$$

Therefore, it is possible to estimate the temperature dependence of radiative and non-radiative lifetimes, by monitoring both $\tau_L(T)$ and $\eta_{int}(T)$. However, it is generally difficult to measure the $\eta_{int}(T)$ value directly by the experiment.

$\tau_{non\text{-}rad}(T)$ can be expressed by the following equation,

$$\frac{1}{\tau_{non-rad}} = v_{th} \times \sigma \times N_t, \tag{5.48}$$

where v_{th}, σ, and N_t denote thermal velocity, cross section captured to non-radiative recombination center (NRC) and the density of NRC, respectively. There is a case where non-radiative recombination processes can be ignored, if the crystal quality is relatively good (low N_t) and the measurement is done at sufficiently low temperatures. In fact, we sometimes see the temperature dependence of luminescence intensity $I(T)$ where, $I(T)$ is constant $[I(T) = I_C]$ in a low temperature region ($T<T_C$), and then decreases gradually with increasing temperature ($T>T_C$). In such a case, it can be assumed that internal quantum efficiency is nearly equal to unity at temperature below T_C as shown in Figure 5.40, so that the luminescence decay time in this temperature range corresponds to the radiative lifetime. Then, the internal quantum efficiency can be expressed by $\eta_{int}(T) = I(T) / I_C$, and both radiative and non-radiative lifetimes at temperature above T_C can be given by the following equation (Miller *et al.*, 1980),

$$\tau_{rad}(T) = \tau_L(T) \times \frac{I_C}{I(T)} \tag{5.49}$$

and

$$\tau_{non-rad}(T) = \tau_L(T) \times \frac{I_C}{I_C - I(T)}. \tag{5.50}$$

5.3.4.2.1 Radiative lifetimes correlated with the dimensionality of excitons

Currently, active layers of many light emitting devices are composed of QWs whose property is considered as two-dimensional (2D) nature. According to the Hanamura's theory (Hanamura, 1988), radiative lifetime of 2D-exciton ($\tau_{2D,K=0}$), whose center-of-mass momentum is zero ($\boldsymbol{K} = 0$), is given by,

$$\tau_{2D,K=0}^{-1} = \frac{8\pi\omega_{ex}}{\hbar c}|\phi(0)|^2 \mu_{cv}^2 = 24\pi\left(\frac{\lambda_{ex}}{a_B}\right)^2 \gamma_S, \tag{5.51}$$

where ω_{ex} is the transition frequency of exciton, $\phi(0)$ is the wave function of the exciton-internal-motion, μ_{cv} is the transition dipole moment between the relevant conduction band and valence band, λ_{ex} is the transition wavelength of exciton, a_B is the Bohr radius of exciton, and the $\gamma_S=4\mu_{cv}^2/3\hbar\lambda_{ex}^3$ is the contribution per unit cell to the band-to-band radiative decay rate. This equation predicts that a radiative lifetime of 2D is as fast as a few ps. Such a fast spontaneous emission has been observed in the high-quality GaAs/AlAs QW at 2K in which no shift was observed between luminescence and luminescence excitation (Deveaud *et al.*, 1991). However, in most cases, various types of QWs show the radiative lifetime of a few hundred ps at low temperature. This is owing to the effect of exciton localization. Since there exists interfacial disorder, compositional fluctuation, and impurity centers in the actual QWs, 2D-excitons are trapped to local potential minima. Such localized states can be treated as a quasi zero-dimensional (0D) system like a quantum dot (QD). As can be understood from Heisenberg's uncertainty principle, such a localization of exciton-wave-function in real space results in the spreading of center-of-mass momentum in $\boldsymbol{K}$-space. Hence, the recombination probability of excitons is considerably lowered compared with the case of no localization because the number of excitons, which satisfies the momentum conservation rule with photon ($\boldsymbol{K} = \boldsymbol{q}_c$, $\boldsymbol{q}_c = \omega_{ex} n_r / c$), becomes very small. Sugawara (1995) derived the radiative lifetime of localized excitons at low temperature as shown below,

$$\tau_{rad,Q0D}^{-1} = \frac{e^2 n_r \omega_{ex} \mu_{cv}^2 F^2 |\phi(0)|^2}{\pi m_0^2 \varepsilon_0 \hbar c^3} \times 2\pi\beta^2, \tag{5.52}$$

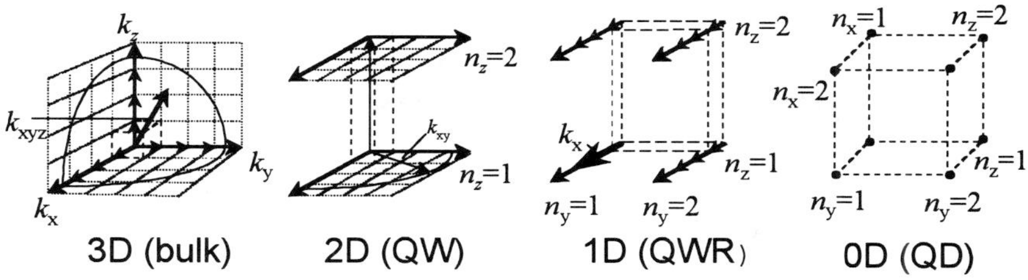

Figure 5.42 Distribution of DOS in K-space in 3D, 2D, 1D and OD.

where n_r denotes the refractive index, F denotes the overlap-integral of wave functions between electron and hole in the z-direction, and β denotes the degree representing the spread of exciton-wave-function where $\sqrt{2}\ \beta$ corresponds to the radius of the localization center in the plain of the QW. It is noted that Equation (5.52) is valid if $\beta \gg \lambda_{ex}/n_r$, and that we sometimes call the system weak localization, quantum disk (QD), mesoscopic QD and QD depending on the value of β. It is therefore theoretically predicted that radiative lifetime of localized exciton decreases with increasing the zero-dimensionality. It is also noted that Equation (5.52) is not valid above a certain temperature where localized excitons are thermalized to higher localized levels, or delocalized to the extended states.

Before discussing how radiative lifetime of localized excitons in QWs ($\tau_{rad, Q0D}$) changes with temperature in, temperature dependence of radiative lifetime is derived for 3D (bulk), 2D (QW) , 1D (quantum wire (QWR)) and 0D (QD) without the effect of localization originating from the disorder. The effect of thermalization can be estimated by the following parameter ;$\zeta(T)$ which express the fraction of excitons with $\boldsymbol{K}<\boldsymbol{q}_c$. If the Maxwell-Boltzman distribution is assumed for excitons,

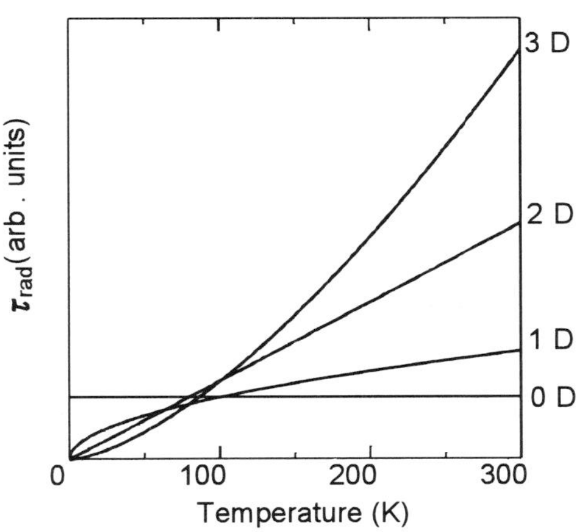

Figure 5.43 Temperature dependence of radiative lifetime in various dimensionality.

$$\tau_{rad}(T) = \tau_0 \zeta(T)^{-1}, \tag{5.53}$$

and

$$\zeta(T) = \frac{\int_0^{\Delta} D(E)\exp(-E/k_B T)dE}{\int_0^{\infty} D(E)\exp(-E/k_B T)dE}, \tag{5.54}$$

where $D(E)$ denotes the density of states of exciton, $\Delta = \hbar^2 q_c^2/2M$ ($M = m_e + m_h$) is the maximum momentum energy of exciton which satisfies the momentum-conservation rule with photon. Equivalent temperature of Δ (Δ/k_B) is typically about 1K. Considering the E-dependence of DOS ($D_{3D}(E) \propto \sqrt{E}$, $D_{2D}(E)$ = const., $D_{1D}(E) \propto 1/\sqrt{E}$, $D_{0D}(E)$: form of delta function), $\zeta(T)$ for each dimension can be derived as shown below if $T \gg \Delta/k_B$ (Feldman *et al.*, 1987; Akiyama *et al.*, 1994),

$$\tau_{rad}(T) = \begin{array}{ll} \tau_{0,3D}(k_B T/\Delta)^{1.5} & (3D) \\ \tau_{0,2D}(k_B T/\Delta) & (2D) \\ \tau_{0,1D}(k_B T/\Delta)^{0.5} & (1D) \\ \tau_{0,0D} & (0D) \end{array} \tag{5.55}$$

These characteristics are schematically shown in Figure 5.43. Although the radiative lifetime of 0D-exciton is the slowest at low-temperature among four different dimension, it could be the fastest at RT due to superior coherency of exciton.

In the actual QWs, the depth and the spatial dimension of localization centers differ according to the structures. In the case of weak localization, excitons are easily thermalized to higher sub-bands within localization centers due to the small energy difference between the lowest quantized-level and upper sub-bands. In this case, radiative lifetime increases with temperature because the oscillator strength is averaged by the contribution from other upper-transitions. If temperature is increased further, excitons may be delocalized to extended states with 2D-DOS. In such a case, radiative lifetime increases in proportion to temperature as mentioned

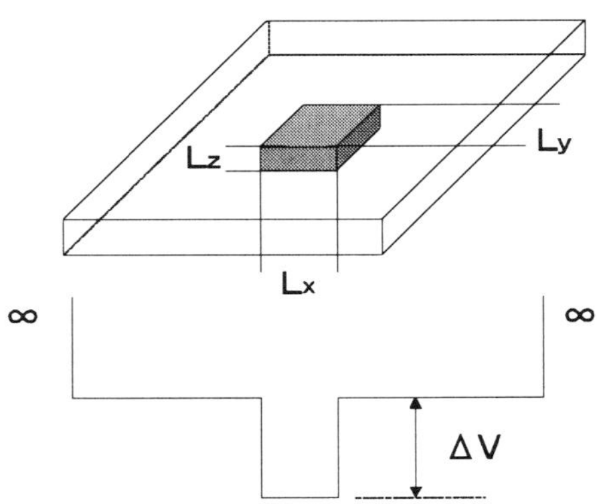

Figure 5.44 Schematic of band structure used for the theoretical model.

above. If the degree of localization is large enough, temperature dependence of radiative becomes less dominant. Therefore, despite the macroscopic characterization, temperature dependence of radiative lifetime is one of the most effective tool for the evaluation of dimensionality (Gotoh *et al.*, 1997). Here, we derive the conventional formula to analyse dimensionality of excitons in thin quantum boxes of intermediate regime between 0D and 2D. As shown in Figure 5.44, the model used for the theoretical calculation is the quantum box (L_x=L_y) within the well. The electron-hole wave function are given by he following Shrödinger equation:

$$\left(-\frac{\hbar^2}{2m_e}\nabla_e^2-\frac{\hbar^2}{2m_h}\nabla_h^2+\Delta V_e(\boldsymbol{r}_e)+\Delta V_h(\boldsymbol{r}_h)\right)\Psi=E\Psi, \tag{5.56}$$

where $\Delta V_e(\boldsymbol{r}_e)$ and $\Delta V_h(\boldsymbol{r}_h)$ are spatial distributions of the potential for electrons and holes. For simplicity, the term of electron-hole Coulomb interaction is omitted in this equation. In this case, wave function of electron and hole can be expressed as shown below.

$$\Psi_i \cong \xi_i^j(x)\xi_i^k(y)\xi_i^l(z) \qquad (i = \text{electron, hole}) \tag{5.57}$$

The overlap integral between ψ_e and ψ_h is given by,

$$F=\int_{-\infty}^{\infty}\psi_e\psi_h dxdydz. \tag{5.58}$$

Since the oscillator strength is proportional to the square of F, radiative lifetime at T = 0K can be given by the following equation:

$$\tau_{rad}(T=0K)=\tau_C\times F^2. \tag{5.59}$$

Thermal effect on τ_{rad} can be obtained by calculating the averaged oscillator strength as shown below.

$$\tau_{rad}(T)=\tau_C\times F_{av}^2, \tag{5.60}$$

$$F_{av}^2=\frac{\sum_i F^2\exp(-E_i/k_BT)}{\sum_i\exp(-E_i/k_BT)}. \tag{5.61}$$

In order to apply this model to the case of $In_xGa_{1-x}N$ QW, $\tau_{rad}(T)$ was calculated for the band structures shown in Table 5.9. These structures are selected to compare the experimental results shown in the following section with the theory. In this calculation, effective mass of electron and hole are assumed that m_e= 0.2m_0, m_h= 1.1m_0 for all structures, bandgap energy of $In_xGa_{1-x}N$ ($E_g(x)$) is given by $E_g(x)$ = 3.4(*1-x*)+1.9*x*-3.2*x*(*1-x*), and the ratio of offset between conduction-band-offset and valence-band-offset is 6:4.

Figures 5.45 and 5.46 show the calculated results for $In_xGa_{1-x}N$ QW structures (x = 0.1 and 0.2, respectively) with L_z = 3 nm assuming that the localization centers are composed of quantum boxes with various parameters.

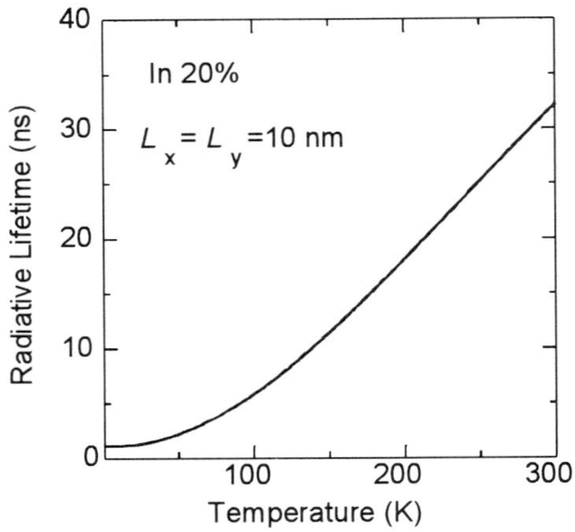

Figure 5.45 Theoretical calculation for the temperature dependence of radiative lifetime in $In_{0.10}Ga_{0.90}N$ QW (L_z = 3 nm) whose localization center is composed of quantum box with $L_x = L_y$ = 10 nm, and with In composition of 20nm.

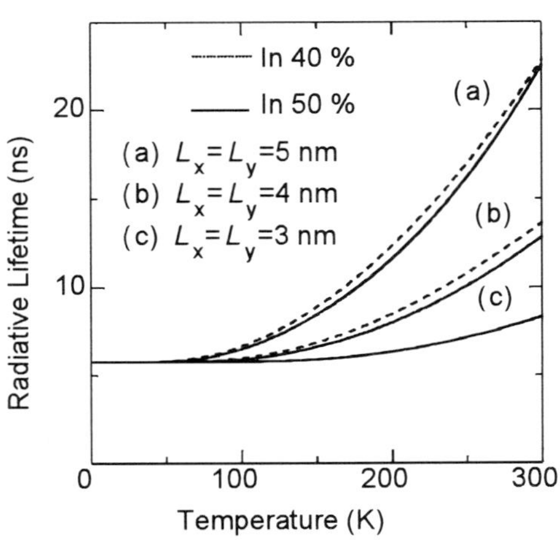

Figure 5.46 Theoretical calculation for temperature dependence of radiative lifetime in $In_{0.20}Ga_{0.80}N$ QW (L_z = 3nm) whose localizaion centers are composed of (a) $L_x = L_y$ = 5 nm, (b) $L_x = L_y$ = 4 nm and (c) $L_x = L_y$ = 3nm, and with In composition of 40 % or 50%.

In this calculation, radiative lifetimes at 0K are normalized to the experimental data taken at 20K.

5.3.4.2.2 Non-radiative recombination process

In spite of high dislocation densities (10^8-10^9cm^{-3}) of GaN-based semiconductors grown on sapphire, the external quantum efficiency of blue-and green LEDs is substantially high. In fact, Mukai *et al.* (1998d) reported that the efficiency of blue LED whose defect density is lowered by the epitaxially laterally overgrowth (ELO) technique is almost the same as that of conventional LEDs grown directly on sapphire. On the other hand, Sugahara *et al.* (1998) showed by means of the comparison between TEM image and CL-mapping that dislocations, whose types are both pure edge and mixed (screw and edge), do act as NRC. These results suggest that the diffusion length of carriers and/or excitons are so small that such dislocations do not affect the emission efficiency, and that the device performance is limited by other types of non-radiative recombination centers.

It was found by means of electrically detected magnetic resonance (EDMR) (Carlos and Nakamura, 1997) or Photo-calorimetric study (PCS) (Sasaki *et al.*, 1998) that energy levels of NRC are distributed at midgap state. The EDMR revealed a broad resonance at g = 2.01 acting as NRC whose origin is probably point-defect-like centers. However, detailed atomic nature of NRC has not been clarified yet.

5.3.4.3 Models of the exciton/carrier localization

In this section, feasible models are discussed in order to account for the localization phenomena in $In_xGa_{1-x}N$-based semiconductors. It has been reported that there exists a large Stokes shift between absorption edge and the emission peaks in the spectra obtained from $In_xGa_{1-x}N$ active layers. This indicates that excitons or carriers involved in the radiative recombination processes in $In_xGa_{1-x}N$ ternary alloys are largely localized. Since the emission spectra observed are generally much broader than the theoretical value of the alloy broadening effect (Zimmerman, 1990), it follows that the energy levels within the $In_xGa_{1-x}N$ layers are distributed by disorder such as the fluctuation of well width and (or) the separation of In composition. A large blue shift of peak energy is observed in the photoluminescence (PL) spectra as a function of excitation energy density. Such a

Table 5.9 Band structures calculated for the temperature dependence of radiative lifetime. Well width (L_z) is assumed to be 3nm for all structures.

In composition in the well	In-plain side length of quantum box (L_x=L_y)	In composition in the quantum box
10 %	10 nm	20 %
20 %	3 nm	50 %
20 %	4 nm	40 %
20 %	4 nm	50 %
20 %	5 nm	40 %
20 %	5 nm	50 %

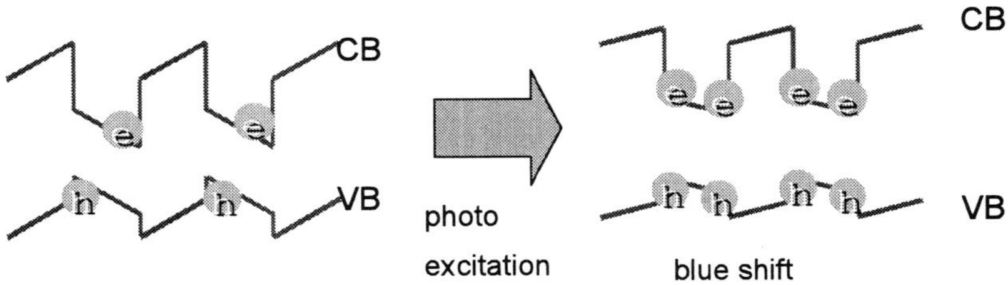

Figure 5.47 Schematic band structure of QW under piezo-electric field.

feature is also reported in the LED structure as rapid current-induced blue shift of electroluminescence (EL) peak energy. The model accounting for these phenomena can be classified into four models as shown below.

(M1) A large piezoelectric field is induced in the (0001)-oriented $In_xGa_{1-x}N$ active layers under biaxial strain (Takeuchi *et al.*, 1997b). Such an internal electric field causes QCSE, where the transition energies between electrons and holes are red-shifted compared to the zero-field condition. Since the electric fields are screened by the photogenerated (or injected) carriers, the blue-shift of the emission peak with increasing photo-excitation (or current) may be understood as the screening effect, as described in Figure 5.47. Takeuchi *et al.* (1998b) estimated the magnitude of the piezo-electric field to be 1.3 MV/cm in the $In_{0.16}Ga_{0.84}N$ MQWs, by analyzing a shift of the PL peak as a function of applied voltage. The QCSE induces the reduction of oscillator strength (probability of radiative recombination in another word) with increasing well width. This may be the reason why the well width in actual $In_xGa_{1-x}N$-based LED or LD structures is typically as thin as 2-3nm. If the effect of QCSE plays a dominant role in the recombination process, temporal behaviour of luminescence is expressed by the multi-exponential decay curve because recombination probability changes with carrier density in the well. Such characteristics have been observed in the structures with thicker well width (Sohmer *et al.*, 1997).

(M2) If the In compositions within the QW layer are fluctuated, excitons generated by current injection are localized to local potential minima and

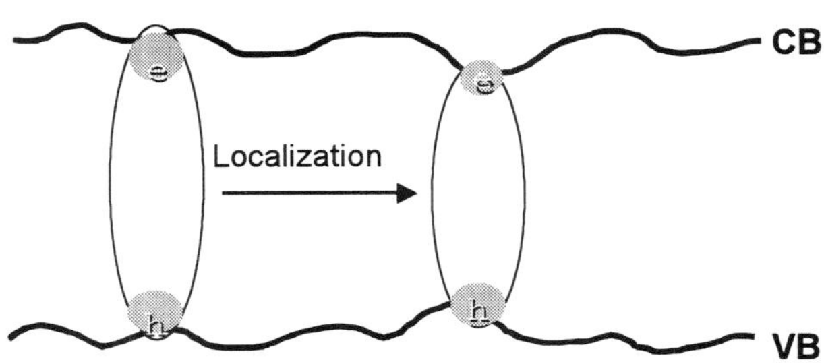

Figure 5.48 Schematic of band structure under random fluctuation of potential.

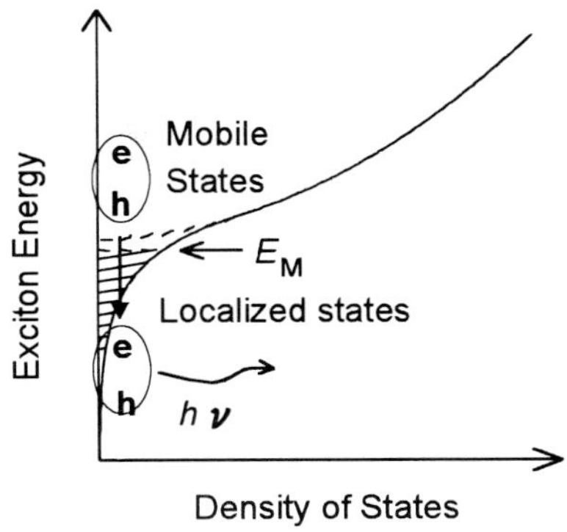

Figure 5.49 Density of states of exciton energy induced by the potential fluctuation. Localization process of excitons is also described.

recombine radiatively (Chichibu *et al.*, 1996d; Taguchi *et al.*, 1996; Narukawa *et al.*, 1997a; Satake *et al.*, 1998). Therefore, such a dynamical process causes the exciton localization, and may account for large band filling effect because the density of tail state is very low, as schematically shown in Figures 5.48 and 5.49. The universal behavior has been reported in this weak localization model, where the ratio between the Stokes shift and the linewidth of exciton absorption is about 0.6 that is independent on the system (Yang *et al.*, 1993). In this case, luminescence decay time depends on the monitored photon energy. This is because exciton transition is not only due to radiative recombination but also transfer process to the lower-lying energy levels as described in Equation (5.45). If the density of tail states is approximated as exponential function as $g(E) = \exp(-E/E_0)$ in which E_0 represents the depth of localization, and if radiative lifetime is constant within the emission band, it is possible to estimate both E_0 and τ_{rad} by fitting the experimental data of $\tau_L(E)$ with theoretical equation as shown below (Gourdon and Lavallard, 1989).

Suppose that density of exciton with energy E at time t after pulsed excitation is expressed as $n(E,t)$. If non-radiative process can be neglected, is given by the following equation,

$$\frac{dn(E,t)}{dt} = -\frac{n(E,t)}{\tau_{rad}} - \sum_{B'<B} W(E \to E')n(E,t) + \sum_{B'>B} W(E' \to E)n(E',t), \tag{5.62}$$

$$\begin{aligned} W_{tr}(E) &\propto \int_B^{\infty} g(E')dE' \\ &\propto \exp(-E/E_0), \end{aligned} \tag{5.63}$$

where $W(E \to E')$ denotes the transfer probability from E to E'. If the temperature is low enough, all transfer processes take place from high energy`to low energy. In this case, the transfer probability ($W_{tr}(E)$) can be give as shown below. Consequently, $n(E,t)$ and $\tau_L(E)$ can be obtained as follows,

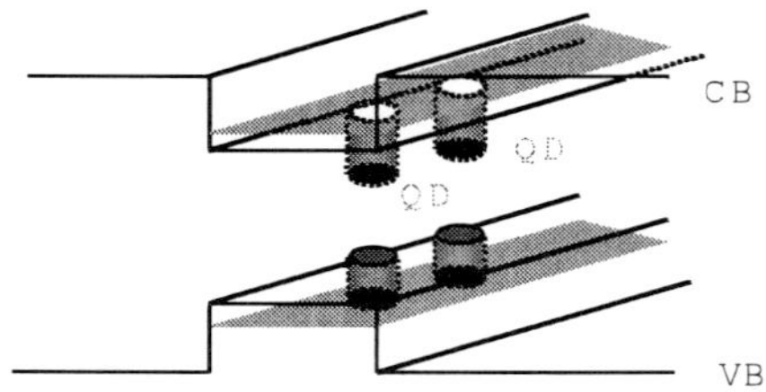

Figure 5.50 Schematic of energy levels formed in QW.

$$n(E,t) = n_0 \exp(-t/\tau(E)), \tag{5.64}$$

$$\tau(E) = \frac{\tau_{rad}}{1+\exp\{(E-E_{me})/E_0\}}, \tag{5.65}$$

where E_{me} represents the characteristic energy analogous to the mobility edge.

(M3) It has been found that the degree of exciton localization is enhanced with increasing In composition (x value) in the $In_xGa_{1-x}N$ active layers, and that the model of weak localization collapses at a certain composition (x>0.15-0.2). In fact, structural analysis using cross sectional TEM revealed the formation of In-rich clusters with estimated diameter of a few nm in $In_xGa_{1-x}N$ layers (Narukawa *et al.*, 1997b; Kisielowski *et al.*, 1997). Moreover, cathodoluminescence (CL)-mapping technique showed the spatially-resolved CL images (Chichibu *et al.*, 1997d), indicating the exciton localization into In-rich regions originating from compositional undulation whose lateral size was estimated to be smaller than 60 nm. Therefore, in the case of large localization, it is reasonable to presume that In-rich QD-like region rather than the small fluctuation of In composition is self-formed within the QW due to the phase separation. Figure 5.50 shows schematically energy levels of QDs formed within the QW. In this case, the broadness of the emission spectra are due to the size and/or compositional fluctuation of QDs, as well as due to the component of higher quantized levels within QDs as depicted in Figure 5.51. Therefore, macroscopic optical data do not reflect the property of singular QD, but should be interpreted as the ensemble of many QDs whose energy levels are distributed inhomogeneously. The transient PL in this case may be attributed to the transfer process from upper quantized levels

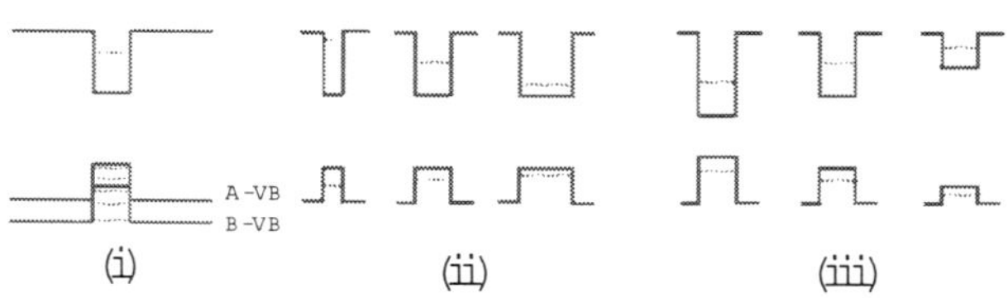

Figure 5.51 Feasible mechanism for the broadness of the luminescence linewidth in QDs. (i) formation of sub-bands, (ii) fluctuation of the size, and (iii) fluctuation of the potential depth.

toward the lowest level, and also to the size and/or compositional dependence on radiative lifetime. The difference between (M2) and (M3) is noted. In the case of (M2), excitons are mobile within the local potential minima under elevated temperature or high injection levels and energy levels of localized states are interconnected with the extended states. However, in the case of (M3), the transfer probability between localized centers is very low, and such centers are separated energetically from extended states.

If the In-rich active region (QD) is surrounded by In-poor region isotropically, QD is almost under hydrostatic stress as a result of the reduction of biaxial term. Accordingly, it is expected that the piezoelectric field is lowered than the case of uniform QW. Recently, PL measurement has been done for highly localized samples as a function of applied voltage across *p-i-n* junction. Preliminary results show that the PL peak shift was much smaller than the value calculated theoretically, assuming the uniform QW.

(M4) Theoretical calculation predicts that an anti-bonding *s*-like state produced by nitrogen (N) vacancy appears at 0.3 eV below the conduction-band (CB) edge for GaN, which gets shallower with increasing In composition in $In_xGa_{1-x}N$ and then resonant with the CB for $In_{0.18}Ga_{0.82}N$ (Yamaguchi, E. and Junnarkar, 1998). The energy difference between the resonant states and the CB edge increases with x value. This result can provide a consistent model for explaining localization behavior. Although no experimental evidence is obtained whether or not the N vacancy plays an important role for determining optical properties in $In_xGa_{1-x}N$, it is interesting to note that this model also predicts the zero-dimensional behaviour of excitons because the localization occurs at N vacancies in the real space.

5.3.4.4 Experimental results and discussion

5.3.4.4.1 Instrumentation of time-resolved luminescence spectroscopy

Figure 5.52 shows the experimental apparatus for the TRPL. Pulsed photo-excitation for the TRPL is provided by the frequency doubled (2ω) or frequency tripled (3ω) beams of a mode-locked Al_2O_3:Ti laser (ω) which was pumped by Ar^+ laser. It is possible to make a selective excitation to $In_xGa_{1-x}N$ active layers by using frequency doubled beam whose tuning range is from 350nm to 530nm. Two types of pulse-width, 1.5 ps and 100 fs can be selected by adjusting the optics in Al_2O_3:Ti laser. The pulse width of 1.5 ps is suitable for the TRPL measurement if the monochromacy is needed for the photo-excitation. Repetition rate of the laser is 80.00000 MHz whose time-interval is 12.5 ns. It can be reduced to 4 MHz by an acoustic optic (AO) modulator in order to avoid the multi-excitation depending on the decay times. The detection is performed by means of a fast scan streak camera in conjunction with monochromator using gratings whose grooves are 100 lines/mm, 300 lines/mm and 1200 lines/grooves. Time-resolution for the detection is about 2 ps.

Even if there is no short-pulse laser system, luminescence dynamics can be assessed by applying pulsed-voltage across p-i-n junction. Figure 5.53 shows the time-resolved electroluminescence (TREL) system (Kawakami *et al.*, 1997). The pulse width and repetition rate of pulsed voltage applied to the LED are 8 ns and 4 MHz, respectively. The condition of impedance matching can be attained with a

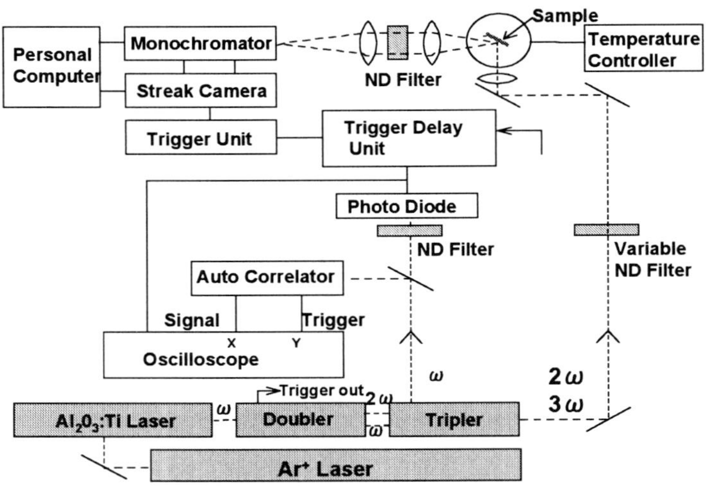

Figure 5.52 Experimental apparatus for the TRPL measurement.

variable resistor by minimizing the reflected current-pulse monitored by the fast digital oscilloscope. Although the pulse width of applied voltage is rather broad, the time resolution of about 500 ps can be achieved by deconvoluting the current-pulse-signal with the transient EL signal monitored by the streak camera.

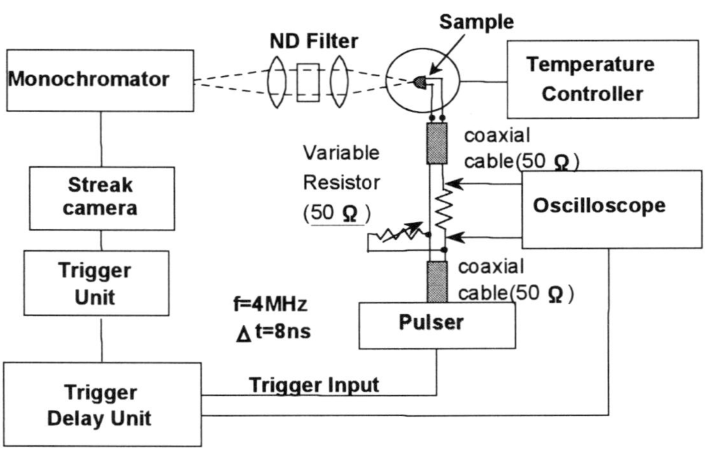

Figure 5.53 Experimental apparatus for the TREL measurement.

5.3.4.4.2 Recombination dynamics in $In_xGa_{1-x}N$-based light emitting devices

Thermodynamic calculation shows that the miscibility gap takes place in $In_xGa_{1-x}N$ ternary alloys between x = 0.06 to x = 0.93 at the typical growth temperature of 800°C under the thermal equilibrium condition (Ho and Stringfellow, 1996; Matsuoka, 1997). Therefore, it is likely that localization of excitons arises from the intrinsic nature of $In_xGa_{1-x}N$ where perfect crystal is very difficult to be grown. In this study, optical assessment has been performed for $In_xGa_{1-x}N$-based LEDs or LDs whose In composition in active layers ranges from $x = 0$ to $x = 0.3$. All the samples assessed here were grown at the group of NICHIA by the two-flow metalorganic chemical vapor deposition (MOCVD) technique (Nakamura, 1991).

Ultraviolet light emitting diodes

The development of ultra-violet (UV) LEDs (Akasaki and Amano, 1997; Mukai *et al.*, 1998a) is very important for the application such as the replacement of Hg lamp and the new excitation light-source of fluorescent lamp. Nevertheless, it is generally difficult to fabricate efficient UV-LEDs compared to visible LEDs not only due to the reduced potential confinement but also because non-radiative recombination centers are easily generated in active layers with wider bandgap. Recently, Mukai *et al.* (1998a) reported that the addition of small amount of In (less that a few % in composition) to undoped GaN active layers resulted in the improvement of η_{ext}. They achieved the η_{ext} value as large as 7.5%.

Figure 5.54 schematically shows the LED structures. In order to assess the effect of small amount of In in the active layer, two types of UV-LEDs were investigated (Narukawa *et al.*, 1999a). The layers are composed of n-GaN:Si layer (4 μm), n-$Al_{0.10}Ga_{0.90}N$:Si cladding layer (30nm), undoped- $In_{0.02}Ga_{0.98}N$ active layer (40nm) or undoped-GaN active layer (40nm), p-$Al_{0.15}Ga_{0.85}N$:Mg cladding layer (60nm) and p-GaN:Mg layers (120nm). The whole layer is coherently grown onto n-GaN:Si layer. Therefore, there is almost no strain in the GaN active layer, and the compressive strain in the undoped-$In_{0.02}Ga_{0.98}N$ active layer is estimated to

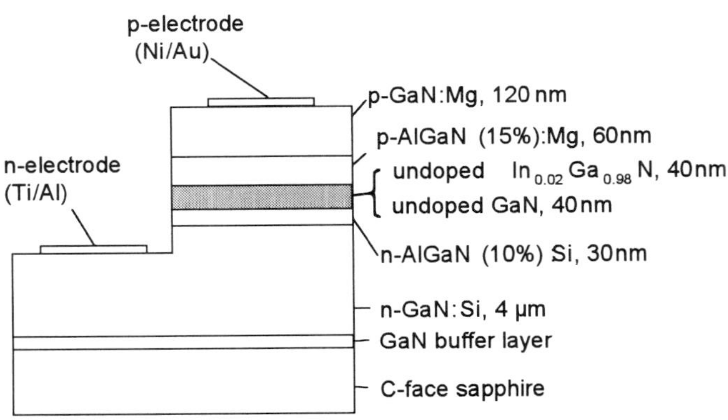

Figure 5.54 Schematic structure of UV-LEDs whose active layer is composed of undoped-GaN, or undoped-$In_{0.02}Ga_{0.98}N$ whose thickness are 40nm.

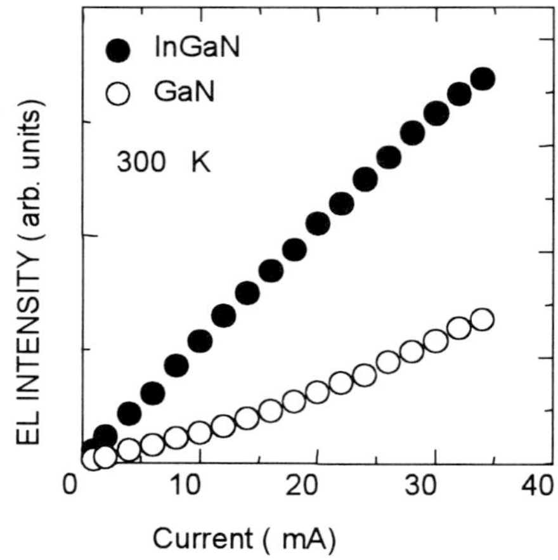

Figure 5.55 *I-L* characteristics UV-LEDs. The η_{ext} value of $In_{0.02}Ga_{0.98}N$-UV-LED is about 4%.

be about 0.2%.

Figure 5.55 shows the EL intensity as a function of current taken at GaN- and $In_{0.02}Ga_{0.98}N$-UV-LEDs. It was found that the $In_{0.02}Ga_{0.98}N$-UV-LED is three times more efficient that the GaN-UV-LED.

Figure 5.56 shows electroluminescence (EL) and photo-induced-voltage (PV) spectra of GaN- and $In_{0.02}Ga_{0.98}N$-UV-LED taken at 20K. In the EL spectrum from GaN-LED, besides donor bound exciton ((D^0,X)), donor-acceptor pair (DAP) related emissions from both GaN active and AlGaN cladding layers were observed at 20K. The EL peak of $In_{0.02}Ga_{0.98}N$ LED is located at 3.412 eV which is on the low-energy side of the (D^0,X) in GaN by about 80 meV. No impurity related emission was observed from the $In_{0.02}Ga_{0.98}N$ LED. PV spectrum shows that the Stokes shift of this emission is about 15 meV.

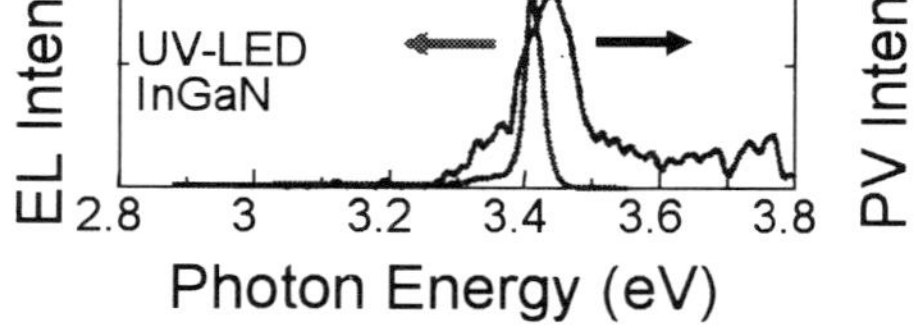

Figure 5.56 EL and PV spectra of GaN- and $In_{0.02}Ga_{0.98}N$-UV-LED taken at 20K.

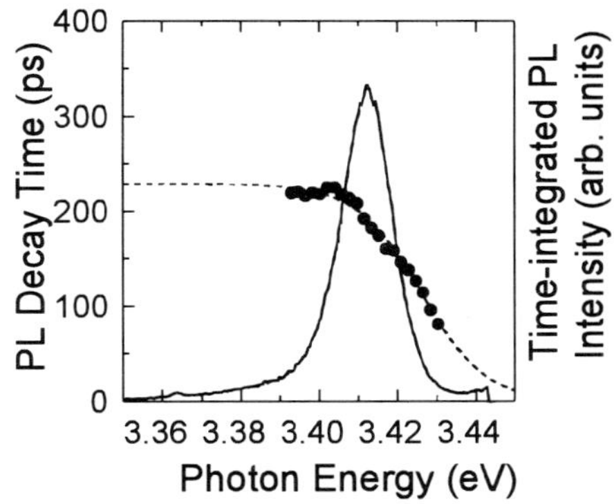

Figure 5.57 Time-integrated photoluminescence (TIPL) spectrum and the PL decay time as a function of monitored photon energy at 20K taken at the $In_{0.02}Ga_{0.98}N$-UV LED.

TRPL measurement was taken at $In_{0.02}Ga_{0.98}N$ at 20K. The energy of the photo-excitation was 3.457 eV at this temperature. As shown in Figure 5.57, PL decay time decreases with increasing monitored photon energy. This is because the decay of localized exciton is not only due to radiative recombination but also due to the transfer process to the tail state. If the density of tail state is approximated as $\exp(-E/E_0)$, and if the radiative recombination lifetime (τ_{rad}) does not change with emission energy, observed lifetime ($\tau(E)$) can be expressed by the Equations (5.43) or (5.65). The best fit could be obtained using τ_{rad}= 230 ps, E_0= 8.2 meV and Eme=3.426 eV.

Figure 5.58 shows the temperature dependence of TIPL peak energy of the $In_{0.02}Ga_{0.98}N$-UV LED and the $In_{0.3}Ga_{0.7}N$-blue LED. The peak energy of the UV LED increases between 20K and 70K, and then decreases according to the

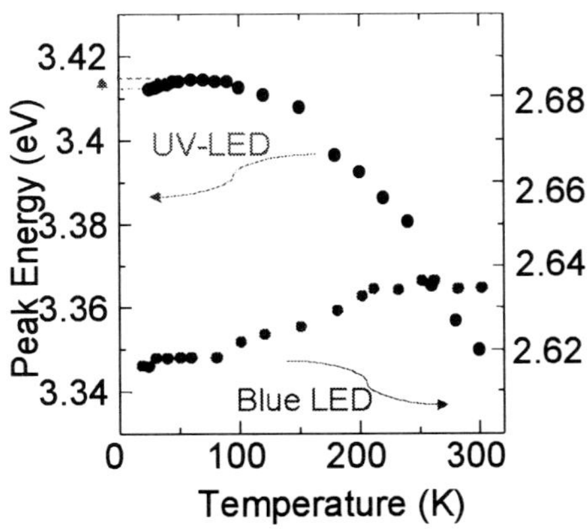

Figure 5.58 Temperature dependence of PL peak energy in the $In_{0.02}Ga_{0.98}N$-UV LED and the $In_{0.3}Ga_{0.7}N$-blue LED.

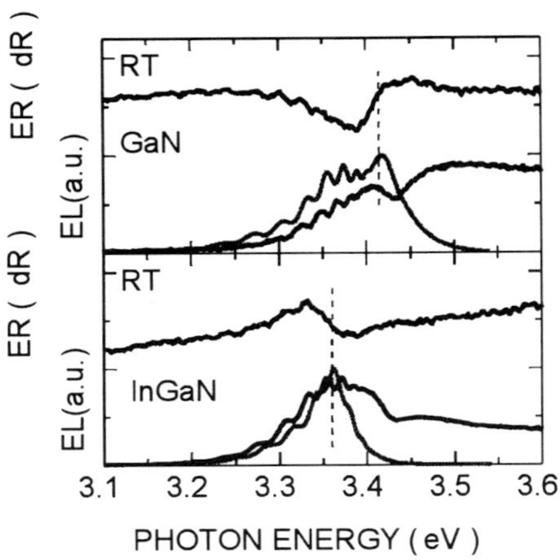

Figure 5.59 EL, PV and ER spectra at RT taken at GaN- and $In_{0.02}Ga_{0.98}N$-UV LEDs.

temperature dependence of bandgap energy. This indicates that excitons are trapped at shallow potential minima at low temperature whose origin may be either impurities or compositional fluctuation of In, and are delocalized above 70K. This is in contrast to the *T*-dependence of blue LED in which peak energy increases with temperature even up to RT. This is probably because of the deep localization of carriers or excitons in the blue LED.

EL, PV and electro-reflectance (ER) spectra at RT are depicted in Figure 5.59 for GaN- and $In_{0.02}Ga_{0.98}N$-UV LEDs. There was almost no energy difference between emission energy and absorption edge (Stokes Shift) in both LEDs. Therefore, in these LEDs, it was confirmed that the mechanism of spontaneous emission at RT is contributed from the delocalized carriers probably forming free excitons.

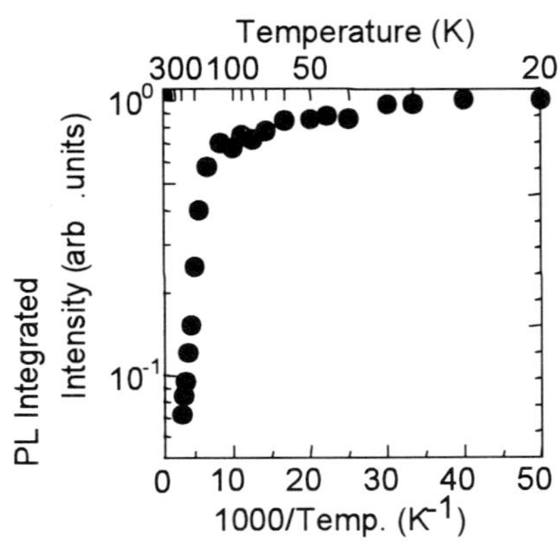

Figure 5.60 Temperature dependence of integrated PL intensity taken at the $In_{0.02}Ga_{0.98}N$-UV LED.

In order to estimate how the internal quantum efficiency changes with temperature, PL integrated intensity was plotted as a function of temperature as shown in Figure 5.60. In this measurement, the energy of photo-excitation was changed according to temperature in order that the selective photo-excitation to the $In_{0.02}Ga_{0.98}N$ active layer was made. At low temperature, PL intensity does not strongly depend on temperature. If the internal quantum efficiency at 20K (η_{int}(20 K)) is assumed to be unity, η_{int} (T K) can be estimated by the equation of $\eta_{int}(T) = I(T)/I_C$ as discussed in the section of 5.3.4.2, using the ratio between PL intensity at TK ($I(T)$) and that at 20K (I_C). By this way, internal quantum efficiency of $In_{0.02}Ga_{0.98}N$-UV LED is estimated to be about 7% at RT.

As shown in Equations (5.49) and (5.50), it is possible to estimate how radiative and non-radiative lifetimes change with temperature, by monitoring both PL decay time and internal quantum efficiency. Figure 5.61 shows temperature dependence of τ_{rad} and $\tau_{non\text{-}rad}$ in the $In_{0.02}Ga_{0.98}N$ –UV LED obtained by this method. The τ_{rad} value increases gradually at low temperature below 50K because of the localization of excitons, and increases proportionally to $T^{1.5}$ between 50K and 200K. It is probable that excitons are delocalized and become nearly free three dimensional (3D)-excitons above 50K because theoretical calculation for the oscillator strength predicts the relation of $\tau_{rad} \propto T^{1.5}$ for 3D-excitons (Equation (5.55)), in which the wave-vector selection rule is taken into account between the radiation field and the exciton center-of-mass motion. This is also consistent with the fact that the thickness of the $In_{0.02}Ga_{0.98}N$ active layer is as large as 40nm. On the other hand, $\tau_{no\text{-}nad}$ is almost constant at 2 ns. This is different from the temperature dependence of GaN single epilayer, where the $\tau_{no\text{-}nad}$ value decreases with increasing temperature reaching at 200 ps at RT. It is suggested that the density of NRC (N_t in Equation (5.49)) is greatly suppressed by the addition of a small amount of In to the active layer. Moreover, almost no temperature dependence of $\tau_{no\text{-}nad}$ value may indicate that the origin of NRC is altered by the addition of In.

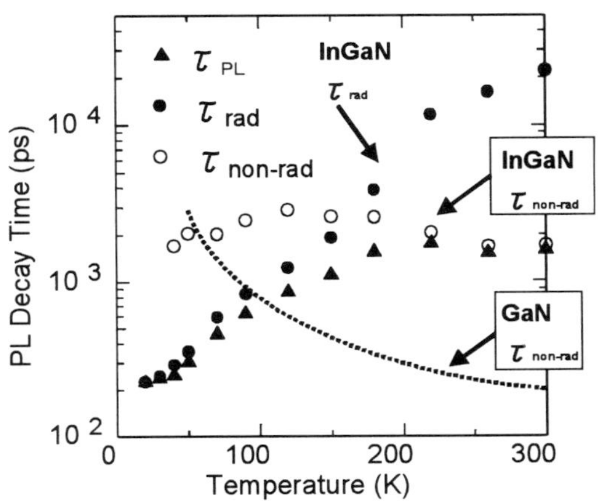

Figure 5.61 Temperature dependence of τ_{PL}, τ_{rad} and $\tau_{non\text{-}rad}$ taken at the $In_{0.02}Ga_{0.98}N$ –UV LED. For comparison, T-dependence of $\tau_{non\text{-}rad}$ of GaN-UV LED is depicted.

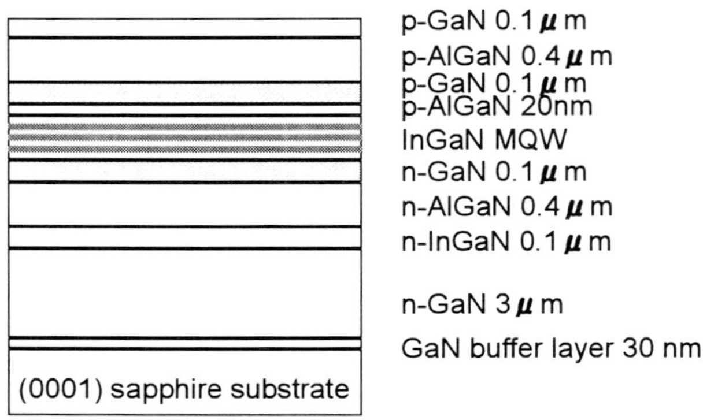

Figure 5.62 Schematic of LD structures whose active layers are composed of (a) the Si doped $In_{0.10}Ga_{0.90}N$ (3 nm)/$In_{0.02}Ga_{0.98}N$ (6 nm) MQW with 3 periods (near-UV LD), (b) the undoped $In_{0.20}Ga_{0.80}N$ (3 nm)/$In_{0.05}Ga_{0.95}N$ (6 nm) MQW with 7 periods (violet LD).

In this section, radiative and non-radiative recombination processes have been assessed in GaN- and $In_{0.02}Ga_{0.98}N$-UV LEDs by means of ER, PV, and TRPL spectroscopy. It was found that the effect of small In-addition to the GaN active layer plays an important role for suppressing non-radiative recombination process.

Near-ultraviolet and violet laser diodes

Two-types of LD structures, whose In content in the active layers are roughly (a) 10% and (b) 20%, are assessed in this study (Narukawa *et al.*, 1997a; 1997b; 1999b). As illustrated in Figure 5.62, both samples consist of the separate confinement heterostructure (SCH) where (a) the Si-doped $In_{0.10}Ga_{0.90}N$ (3nm)/ $In_{0.02}Ga_{0.98}N$ (6nm) MQW with 3 periods, (b) the undoped $In_{0.20}Ga_{0.80}N$ (3nm)/ $In_{0.05}Ga_{0.95}N$ (6nm) MQW with 7 periods are sandwiched between GaN waveguide

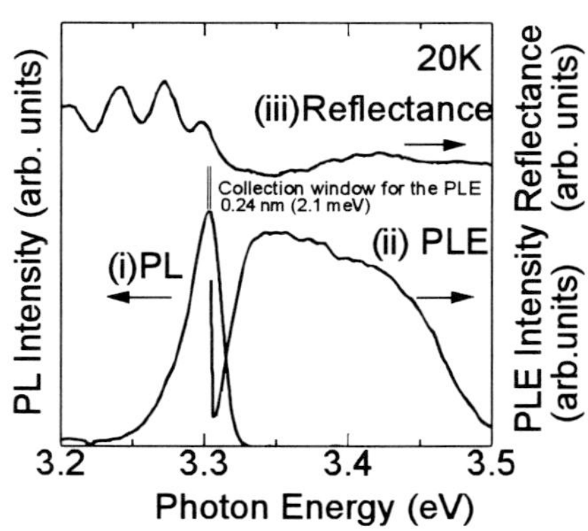

Figure 5.63 PL, PLE and reflectance spectra at 20K taken at the near-UV LD structure.

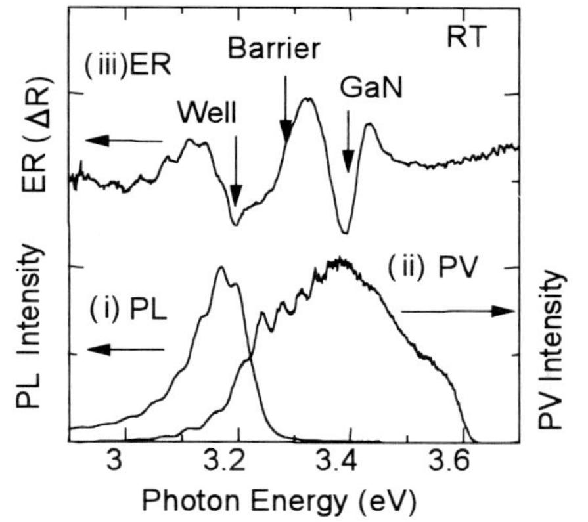

Figure 5.64 PL, PV and ER spectra at RT taken at the near-UV LD structure.

layers (0.1 μm in each) and (a) $Al_{0.15}Ga_{0.85}N$, (b)$Al_{0.07}Ga_{0.93}N$ cladding layers (0.4 μm in each). The top of the $Al_xGa_{1-x}N$ clad and the GaN waveguide are Mg-doped p-type layers, while the bottom of the clad and the waveguide are Si-doped n-type layers. The LD-operation at RT was performed at 390nm under CW mode for the sample (a) (near-UV LD), and at 420nm under pulsed mode for the sample (b) (violet LD).

Figure 5.63 (i) shows the CW PL spectrum at 20K taken at the near-UV LD structure (sample (a)). PL spectrum from the active layer consists of single peak located at 3.303 eV with the linewidth of 28 meV. It was difficult for this particular sample to apply proper electric field to the pn junction at this cryogenic temperature probably because of the freezing out of carriers. Therefore, photoluminescence-excitation (PLE) and reflectance measurements were performed instead of PV or ER spectroscopy as depicted in Figure 5.63 (ii) and (iii). Although it is difficult to observe a clear absorption edge in the reflectance spectrum due to interference fringes, both absorption spectra (PLE and reflectance) suggests that the Stokes-shift is approximately a few tens meV at 20K.

Figure 5.64 shows the PL, PV and ER spectra at RT taken at the near UV-LD structure. The absorption edge of $In_{0.10}Ga_{0.90}N$ well layer, $In_{0.02}Ga_{0.98}N$ barrier layer and GaN cladding layer were roughly located at 3.23eV, 3.32V and 3.43eV, respectively. Therefore, almost no stokes shift is observed in the main PL peak.

Since the degree of localization is so small for this sample that the excitons involved in the radiative recombination become free (2D) at RT due to the thermalization. Such a weak localization at low temperature is probably attributed to the model (ii), which originates from small potential fluctuation of either In composition or well width.

Figure 5.65 shows PL decay times of the near-UV LD at 20K as a function of monitored emission energies. Time-integrated PL is also inserted in the figure. Decay times increase with decreasing emission energies, and tend to saturate on the low energy side of the band. As discussed before, this can be understood as the effect of exciton localization. Theoretical analysis of the PL decay time as a function of emission energy was done assuming the density of tail state as the form

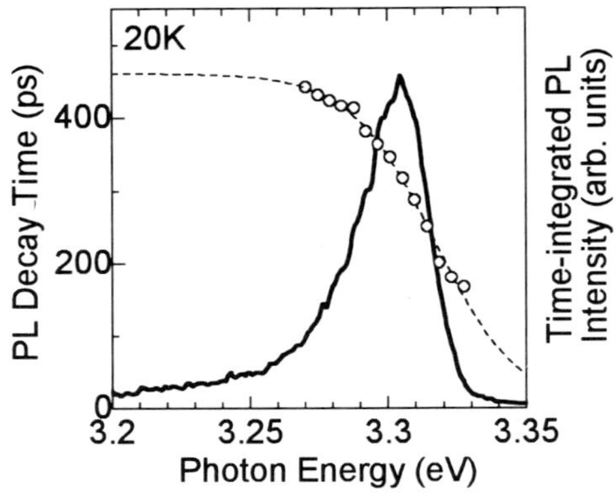

Figure 5.65 Time-integrated PL spectrum as well as PL decay times as a function of monitored emission energy at 20K taken at the near-UV LD.

of exp($-E/E_0$). The best fitting was obtained using values as τ_{rad} = 460 ps and E_0 = 15 meV as shown by the dotted curve in Figure 5.65. This E_0 value (degree of localization) fairly accounts for the result where the exciton is almost delocalized at RT due to the thermal energy.

The time-integrated PL spectrum taken at 23K in the violet LD structure (sample (b)) is shown in Figure 5.66. The PL spectrum was composed of three emission bands that are labelled E_{ex1A}, E_D and E_{lc}, respectively. The E_{lc} band consists of the main peak at 2.934eV and LO-phonon side bands. The TR, ER and PLE spectroscopy revealed that the transition at 3.216eV (E_{ex1A}) is due to excitons confined in the lowest quantized level within the well which is composed of n = 1 conduction electrons and n = 1 A (Γ_{9v}) valence holes. The PL transient monitored

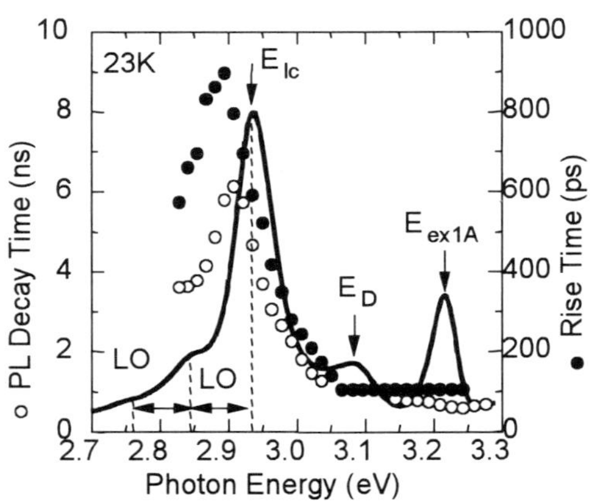

Figure 5.66 Time-integrated PL spectrum as well as decay times and rise times of PL as a function of monitored emission energy at 23K taken at the violet LD structure.

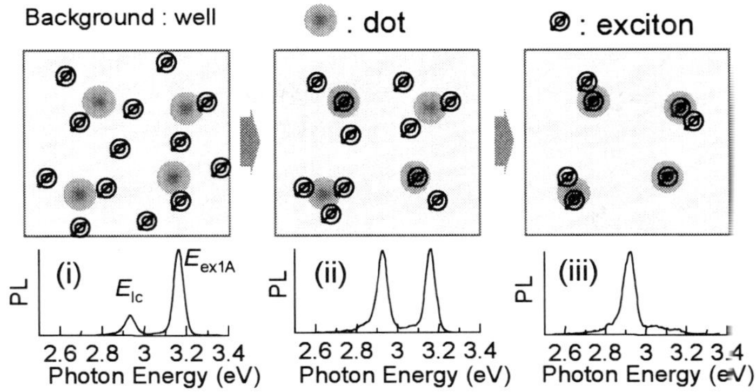

Figure 5.67 TRPL spectra of the violet LD at 23K monitored at (i) t = 0-0.6 ns, (ii) t = 0.6-2 ns and (iii) t>2 ns after pulsed photo-excitation. Schematic of recombination dynamics is also shown.

at the E_D band was well expressed by the double exponential curve whose decay times were 1 ns and 70 ns. Fast decay component is contributed from the overlapping of neighboring emission bands, while long decay time represents the intrinsic lifetime of this emission bands. Since the oscillator strength of this emission is low, this weak E_D band is probably due to the impurity-related centers. Important finding was that the E_{lc} band is attributed to the excitons localized trap center within the well. Decay times as well as the rise times of PL are also shown in Figure 5.66 as a function of monitored emission energies. The rise time is defined as the time to reach maximum PL intensity after pulsed excitation. At 23K, decay times monitored in the vicinity of the E_{ex1A} band are almost constant at 600 ps, which is much smaller than those at the E_{lc} band. The rise times of the E_{ex1A} band were constant at about 100 ps, while the rise times of E_{lc} band changed with photon energy in the range from 200 ps to 900 ps whose dependence was almost same as that of PL decay times. These features indicate that photogenerated excitons are transferred from the n = 1 quantized level (E_{ex1A}) to the localized centers. Figure 5.67 shows the TRPL spectra monitored at different time-interval after photo-excitation. As illustrated by the schematics, exciton-localization phenomena were observed from the E_{ex1A} band to the E_{lc} band.

The time-integrated PL spectra and PL decay times as a function of emission energies at 23K, 100K, 200K and 300K in the sample (b) are shown in Figure 5.68. Decay times in the main PL band (E_{lc}) increase with decreasing monitored photon energies down to about 2.90 eV at low temperature below about 150 K. In fact, TRPL spectra revealed the red-shift of the peak energies with increasing time after the pulsed photo-excitation. The red-shift observed at t = 40 ns was about 30 meV at 23K. The dependence of the emission energy on the PL decay times became less dominant with increasing temperature. Since the piezo-electric field is expected to be almost constant with temperature, the piezo-electric effect may not play an important role in this particular sample. Nevertheless, the maximum decay time observed at around the peak of the E_{lc} band almost represents the τ_{rad} value of the localized excitons. The radiative recombination lifetime (τ_{rad}) of localized excitons (E_{lc}) was estimated to be 6.3 ns at 23K, assuming the non-radiative recombination

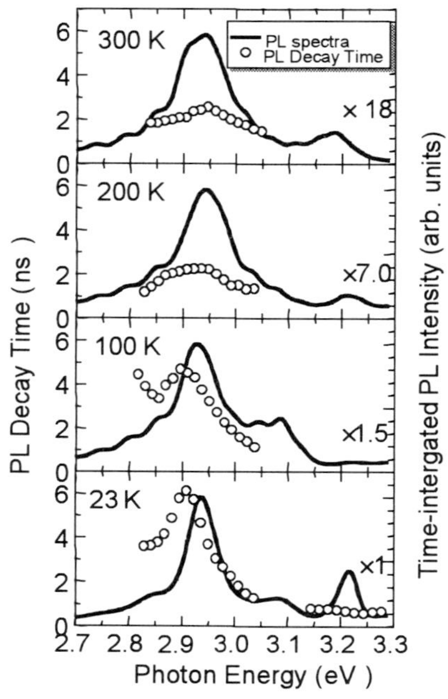

Figure 5.68 Time-integrated PL spectra and the PL decay times as a function of monitored emission energy of the violet LD structure taken at 23K, 100K, 200K and 300K, respectively. Excitation intensity was kept constant at 120 nJ/cm^2.

lifetime ($\tau_{non\text{-}rad}$) is much larger that the τ_{rad} at this temperature. This value is about one order of magnitude larger than that observed in the near-UV LED. This is probably as a result of the mesoscopic effect, where the oscillator strength of excitons is reduced with decreasing the radius as shown in Equation (5.52). Since the depth of the localization center is so large that the E_{lc} band dominates the PL spectrum even at RT.

Figure 5.69 shows temperature dependence of the integrated PL intensity (*I(T)*) of the main PL band in (i) the near UV LD and (ii) the violet LD, in which *I(T)* value at the lowest temperature is normalized to unity. The *I(T)* values in the violet LD were almost constant in the temperature region (23K-100K) and then decreased sharply with temperature above about 200K. While the *I(T)* values in the near-UV LD were constant in the lower temperature region (20K-40K) and then decreased sharply with temperature above about 120K. The activation energy (E_A) of PL quenching is estimated to be 31 meV and 75 meV for the near-UV LD and the violet LD, respectively. The difference of E_A value reflects that the temperature dependence of the capture cross-section to the non-radiative centers varies according to the degree of localization. The internal quantum efficiency at RT of the violet LD (about 8%) is slightly larger than that of the near-UV LD (about 5%).

Figures 5.70 and 5.71 show temperature dependence of τ_{PL}, τ_{rad} and $\tau_{non\text{-}rad}$ values of localized excitons in the near-UV LD and in the violet LD, respectively.

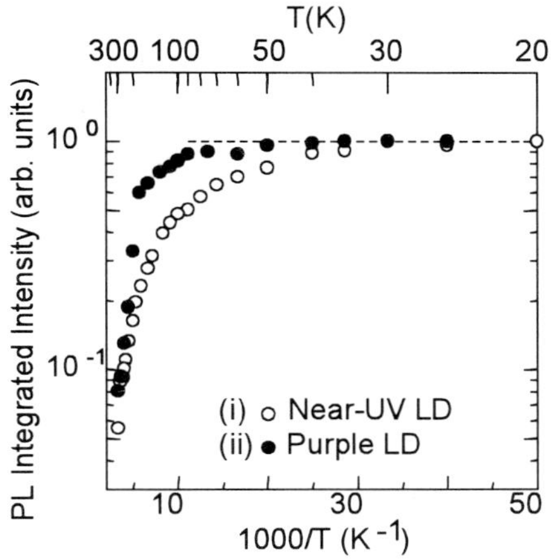

Figure 5.69 Temperature dependence of time-integrated PL intensity of (i) the main emission band in the near-UV LD, and of (ii) the E_{lc} band in the violet LD.

The τ_{rad} values in the near-UV LD were constant at 460 ps between 20K and 40K, and then increased linearly with temperature above about 100K. Such a T-linear dependence of τ_{rad} values represents the two-dimensional (2D) feature of excitons. It is considered that excitons are localized at the potential minima in the QW at low temperature, and then are delocalized by thermal energy with increasing temperature. As a result, the dimensionality of excitons in the near UV-LD exhibits nearly 2D-character at RT.

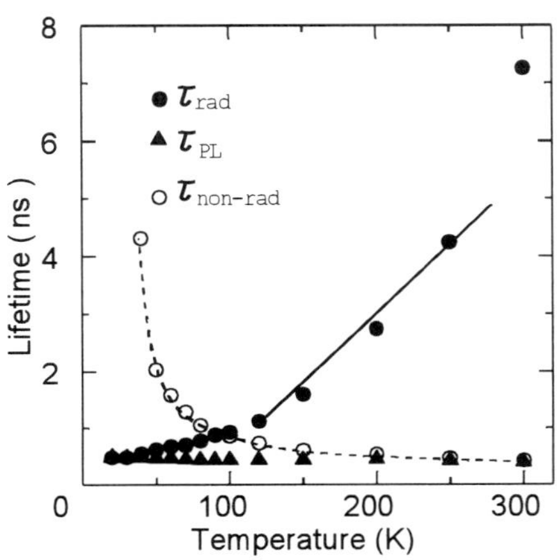

Figure 5.70 Temperature dependence of τ_{PL}, τ_{rad} and $\tau_{\mathrm{non\text{-}rad}}$ values of excitons in the near-UV LD.

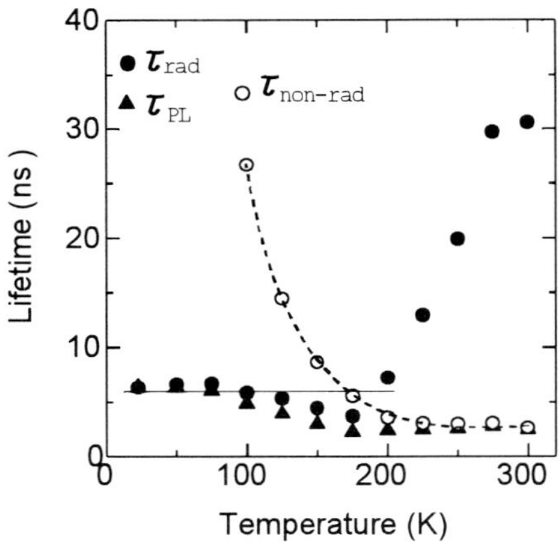

Figure 5.71 Temperature dependence of τ_{PL}, τ_{rad} and $\tau_{non\text{-}rad}$ values of localized excitons in the violet LD.

On the other hand, the τ_{rad} values in the violet LD stay almost constant around 6 ns between 23K and 200K, increased above 200K and then reached to about 30 ns at 300K. This may be because the excitons confined in the lowest quantized levels within the QD-like region are thermalized to higher sub-bands. Comparing the theoretical calculation in Figure 5.46, it is probable that the spatial extent of the localization center is about several nm within the plain of the QW.

In this section, emission mechanism was assessed for two-types of LDs. It was found that the dimensionality of excitons at RT is 2D-character for the In-poor QWs (near-UV LD), while quasi-0D-character for the In-rich QWs (violet LD).

Blue light emitting diodes

Currently, incandescent $In_xGa_{1-x}N$-based blue-LEDs are commercially available. The η_{ext} value of these LEDs are typically 10%. However, much more efficient

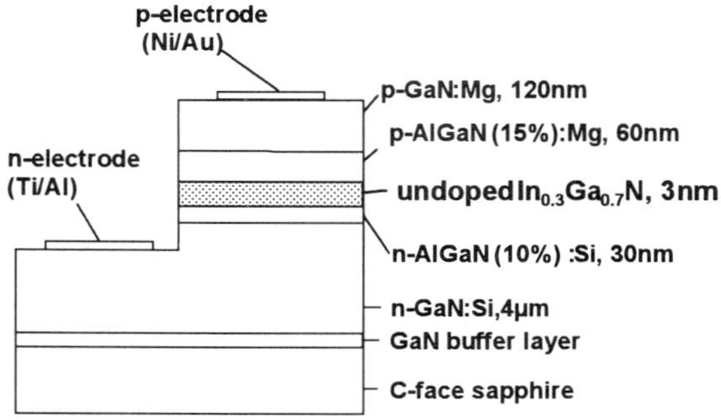

Figure 5.72 Schematic structure of UV-LED whose active layer is composed of 3nm-thick undoped $In_{0.3}Ga_{0.7}N$.

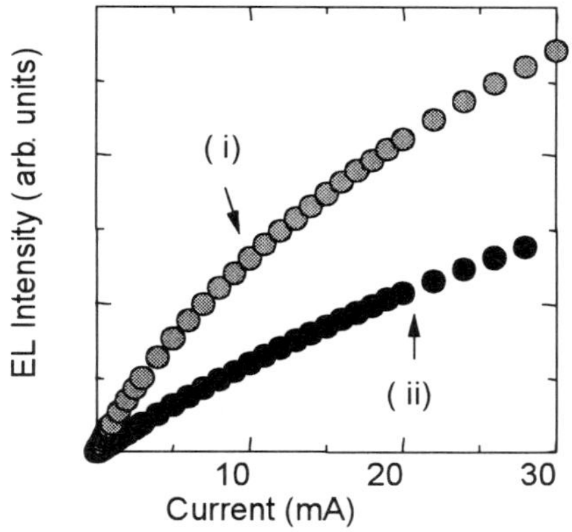

Figure 5.73 *I-L* characteristics of blue LEDs. (i) experimentally obtained NICHIA LED (SB1). (ii) commercially available NICHIA LED (B1).

blue-LEDs have recently been fabricated experimentally. In this section, the mechanism of radiative recombination is investigated in the highly efficient blue LED. The schematic of the blue LED structure is shown in Figure 5.72. The LED consists of a 30nm-thick GaN buffer layer, a 4 μm-thick layer of n-type GaN:Si, a 30nm-thick layer of n-type $Al_{0.1}Ga_{0.9}N$:Si, a 3nm-thick active layer of undoped $In_{0.3}Ga_{0.7}N$, a 60 nm-thick layer of p-type $Al_{0.15}Ga_{0.85}N$:Mg, and a 120nm-thick p-type GaN:Mg. Figure 5.73 (i) shows the EL intensity as a function of driving current of the LED (SB1). For comparison, the characteristics of commercially available blue LED (B1) are depicted in Figure 5.73 (ii). The η_{ext} value at I = 20 mA of SB1 is about two times as large as that of B1. However, EL intensity tends to saturate under high driving current. This nonlinear behavior is more pronounced in the SB1. The mechanism of this saturation is not clear at this stage. One feasible explanation may be because the filling of luminescence centers easily takes place due to the limited number of DOS.

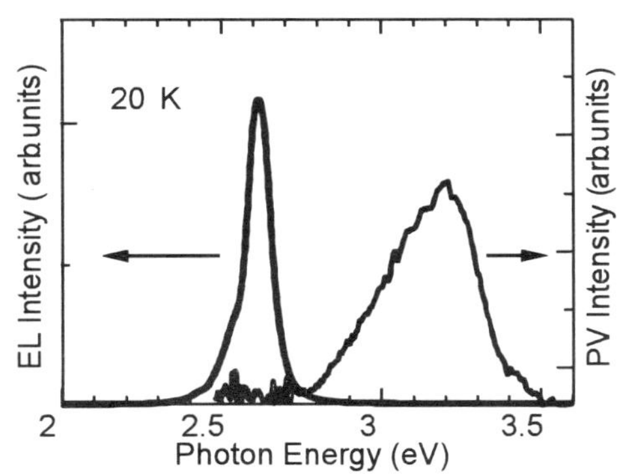

Figure 5.74 EL and PV spectra of the blue LED (SB1) at 20 K.

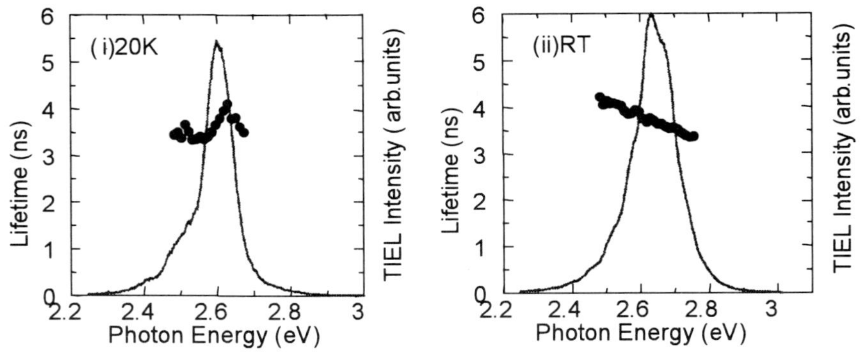

Figure 5.75 Time-integrated EL spectra and EL lifetime as a function of emission energy at (i) 20 K and at (ii) RT obtained at the blue LED (SB1).

Figure 5.74 shows the EL and PV spectra at 20K obtained at the blue LED (SB1). The EL band peaks at 2.608 eV with the linewidth of 86 meV. Two-shoulder-like structures appearing at about 2.4 eV and 2.5 eV are LO-phonon replicas of the main emission band. The PV peak is located at about 3.2 eV. The energy difference between this peak and the EL peak is about 600 meV. The PV signal in the vicinity of EL peak could not be clearly detected due to the noise level.

The EL lifetime of this blue LED (SB1) was measured using the system depicted in Figure 5.53 (Kawakami *et al.*, 1997; Narukawa *et al.*, 1998a). Obtained

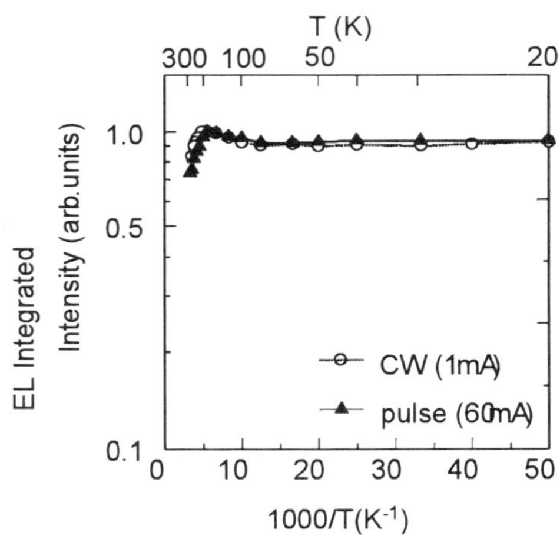

Figure 5.76 Temperature dependence of integrated EL intensity taken at the blue LED (SB1) driven under constant pulse current of 60 mA. The dependence under constant CW current of 1 mA is also plotted for comparison.

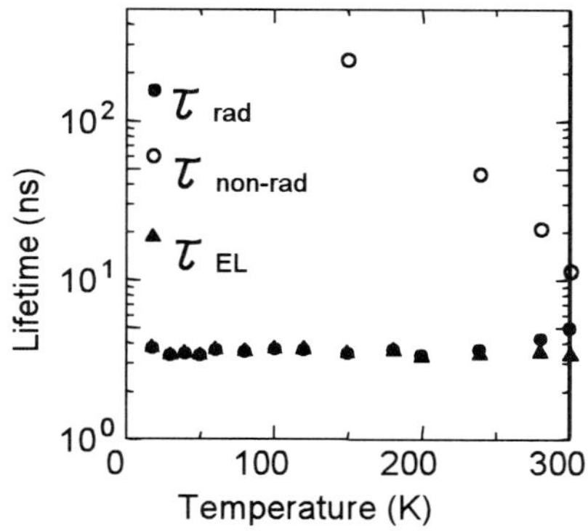

Figure 5.77 Temperature dependence of τ_{EL}, τ_{rad} and $\tau_{non\text{-}rad}$ obtained at the blue LED (SB1).

results at (i) 20K and (ii) RT are shown in Figure 5.75. At 20K, two peaks appear in the EL lifetimes within the estimated range. This is probably due to the effect of both localization and the contribution from 1-LO phonon replica. The effect of latter is namely that the low-energy-side of the main peak overlaps with high-energy-side of the 1-LO phonon replica whose temporal feature is almost same as that in the high-energy-side of the zero-phonon band. On the other hand, EL lifetimes gradually increase with decreasing energy at RT. The variation is from 3.3 ns to 4.4 ns within the estimated range.

Figure 5.76 shows the temperature dependence of integrated EL intensity obtained at the blue LED (SB1). It was found that the internal quantum efficiency (η_{int}) at RT is estimated to be as large as about 73% driven under pulsed current of 60 mA.

Temperature dependence of τ_{EL}, τ_{rad}, and $\tau_{non\text{-}rad}$ in the blue LED (SB1) is shown in Figure 5.77. The radiative lifetimes are almost constant even up to RT, indicating that the zero-dimensional feature of excitons where the spatial extent of localization centers within the QW plain is within a few nm. It is likely that the achievement of high external quantum efficiency of $In_{0.3}Ga_{0.7}$-blue LED is mostly due to the large localization of excitons because the pathways of nonradiative recombination are hindered once the excitons are captured in a small volume.

5.3.4.5 Conclusion

The dynamical behavior based on localization radiative and non-radiative recombination processes of excitons has been studied in the $In_xGa_{1-x}N$-based LEDs and the LDs, by employing the TRPL and the TREL spectroscopy. It was found that excitons were weakly localized mainly due to the small fluctuation of alloy compositions if the In alloy composition was less than about 10%. The depth of exciton localization grew with increasing with In compositions, and the self-formation of deep localization centers was observed in the sample with x>20%.

The zero-dimensional feature of excitons was observed in such a highly localized system.

If we could do the TRPL with nanoscopic spatial resolution, five dimensional (5D) data composed of space, wavelength and time would give us a clearer view on the correlation between nanoscopic structures and macroscopic optical properties. Such a subject is now under progress.

Acknowledgment
Experimental results reported in Section 5.3.4 is a collaborative work between Fujita laboratory at Kyoto University, and Dr. Shuji Nakamura's group at Nichia Chemical Industries Ltd. Thanks are due to Prof. Sg. Fujita for valuable discussion and comments. The author is grateful to Y. Narukawa, K. Sawada and S. Saijyo for their contribution to the optical characterization. A part of this measurement was done using the facility at the "Venture Business Laboratory (VBL)" in Kyoto University.

5.3.5 Optical Gain in Nitrides

5.3.5.1 Optical gain in InGaN QWs

In this subsection, after a phenomenological basic theory, on optical gain in a semiconductor is briefly reviewed according to Saleh and Teich (1991), features of the optical gain in InGaN MQWs are discussed.

Let us consider the interaction between an optical field of frequency ν and carriers of a semiconductor material of band gap Eg via band-to-band transitions for the condition of $\nu > Eg/h$, where h is the Planck's constant. In this case one has to consider two processes, i.e., the absorption and the stimulated emission processes. When the latter overcomes the former, the number of photons increases exponentially resulting in the amplification of the optical field or net optical gain. To do so, the population inversion is necessary. The population inversion in the semiconductor is a situation of the presence of a great number of electrons in the conduction band and of holes in the valence band. This is a nonequilibrium situation. Thus the population inversion is usually achieved by electric current injection in a p-n junction diode. It is of course achieved by optical pumping.

Let a monochromatic light wave traveled in the z-direction with intensity I(z) and photon-flux density, i.e., photons per unit area per unit time, $\phi(z) = I(z)/h\phi$. When the net optical gain is ensured, the number of photons traveling in this direction increases via the stimulated emission process. The emitted photons stimulate further emissions and, consequently, $\phi(z)$ will increase exponentially as increasing z. It is easy to derive an expression for the relation between the photon-flux density at z = 0 and at any value of z using the net gain coefficient $g(\nu)$ per unit length. The result is

$$\phi(z)=\phi(0)\exp[g(\nu)z]. \tag{5.66}$$

Thus, the optical intensity can be written as,

$$I(z)=I(0)\exp[g(\nu)z]. \tag{5.67}$$

It is well known in usual phenomenological semiconductor laser theory that an expression for the net gain coefficient, g(ν), of the semiconductor materials is given by,

$$g(\nu)=\frac{\lambda^2}{8\pi\tau_r}\rho(\nu)[f_c(E_2)-f_v(E_1)]. \tag{5.68}$$

where λ is the wavelength of photons, τ_r is the radiative recombination lifetime, ρ is the optical joint density-of-states. ρ is determined by the electron and hole bands or subbands structures and by the conservation of energy and momentum. The larger ρ leads to larger g(ν), consequently, the larger gain. But in this case the threshold current density becomes also large. $f_c(E_2)-f_v(E_1)$ is the Fermi inversion factor. Here $E_2-E_1 = h\nu$. It depends on both the Fermi function for the conduction band, $f_c(E) = 1/\{1+\exp(E-E_{fc})/k_BT\}$, and the Fermi function for the valence band, $f_v(E) = 1/\{1+\exp(E-E_{fv})/k_BT\}$. It is a function of the temperature and of the quasi-Fermi levels, E_{fc} and E_{fv}, which are detrmined by a pumping rate. The Fermi inversion factor represents the degree of population inversion.

From Equations (5.66)-(5.68), a semiconductor medium provides the net optical gain at the frequency ν when $f_c(E_2)>f_v(E_1)$ and, conversely, the net absorption when $f_c(E_2)<f_v(E_1)$. The former condition is equal to the requirement that the photon energy is smaller than the separation between the quasi-Fermi levels. Of course, the photon energy must be larger than E_g. Thus, the medium can acts as an amplifier for the photon energy range $E_g<h\nu<(E_{fc}-E_{fv})$. This situation is schematically illustrated in Figure 5.78 for the two dimensional case.

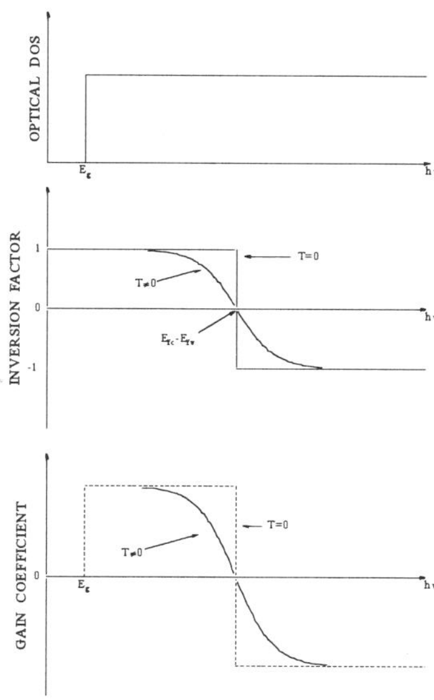

Figure 5.78 Schematic illustration of dependence of the joint optical density-of-states (DOS) in the two dimention, $\rho(\nu)$, the Fermi inversion factor, and the gain coefficient, g(ν), at T = 0 K (solid lines) and at the finite temperature (dotted lines).

As can be seen from Figure 5.78 while the medium is transparent for $h\nu<E_g$, it is an absorber for $h\nu>(E_{fc}-E_{fv})$. The amplifier bandwidth increases with $E_{fc}-E_{fv}$, i.e., with the pumping level.

The following should be pointed out. The many-body effects are important at high pumping levels and are known to pull down E_g to the lower energy region (Bohnert *et al.* 1981, Chow *et al.* 1994) and may enhance ρ (Chow *et al.* 1994). The latter effect is called Coulomb enhancement or excitonic enhancement. Furthermore, a heating effect also causes the red shift of E_g.

Experimental method to study the optical gain spectra in semiconductors bulk and their devices involve the variation of excitation-stripe length (VEL) method, (Benoit À la Guillaume *et al.* 1969, Shaklee and Leheny 1971, Shaklee *et al.* 1973) the Paoli- Hakki one,(Paoli and Hakki 1973, Hakki and Paoli 1973, 1975) and the pump-probe one (Bohnert *et al.* 1981). Among them, the most widely used method is the VEL method. Motivated from Equation (5.67), this method measures the single-pass, ℓ, amplified spontaneous emission spectra as a function of excitation-stripe length, over which the semiconductor materials are optically pumped and the net modal gain, g(ν) is determined from the following equation,

$$I(W/cm^2 steradian)=\frac{I_s\Omega}{g}(e^{g\ell}-1). \tag{5.69}$$

where I_s is the power density of the spontaneous emission (W/cm^3 steradian) and Ω is the solid angle into which the spontaneous radiation is emitted(Shaklee *et al.* 1973). Reasons why the VEL method is widely used are that the method gives not only information of the optical gain on the entire spectrum but also provides information on the spontaneous emission, loss, and saturation characteristics of the semiconductors.

In the following, experimental data obtained using the VEL method are presented on two kinds of early stage InGaN based continuous wave blue laser diodes operated at room temperature (Deguchi 1999). The difference of them arose from growth conditions of InGaN QWs. A sample called CWS1(CWS2) was considered to have larger (smaller) In compositional fluctuation. Both device structures were the same as those given in a book by Nakamura and Fasol (1997). The device life time of them was about half an hour. In both samples, no drastic blue shift of the emission peak even in increasing the excitation power was observed (Deguchi *et al.* 1998b). This is probably due to heavily doping of Si in both wells and barriers. Therefore the effect of internal field (Takeuchi *et al.* 1997b, 1998a,b, Akasaki and Amano 1997, Im *et al.* 1998) is neglected in the following discussions.

In Figure 5.79 are shown transmittance spectra and emission spectra under various excitation sources. A letter T indicates the transmittance spectra. Electroluminescence spectra are denoted by EL. The spectra denoted by A were measured under excitation by photons from Xe lamp excitation, where photons with energy larger than 3.2 eV are used with the use of a glass filter, and those denoted by B were measured under excitation by SHG photons from a Ti:sapphire laser pumped by a Ar^+ ion laser, which had a pulse width of 80 fs and a pulse repetition rate of 80 MHz. The density of photo-excited carriers is much larger in the latter case than in the former one. ASE means the amplified stimulated emission spectra measured using the VEL method. A pulsed Nd^{3+}:YAG laser was

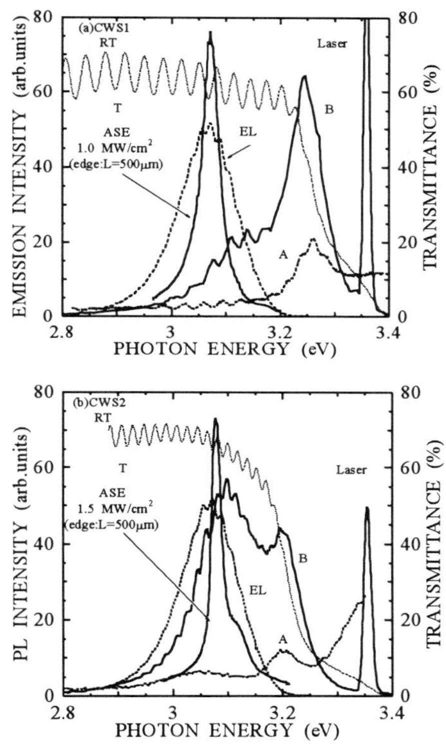

Figure 5.79: Transmission and emission spectra in samples (a) CWS1 and (b) CWS2 measured at RT. With respect to meaning of letters, see the text.

used as an excitation source, whose pulse duration and repetition rate was, respectively, 10 ns and 10 s (Deguchi *et al.* 1998a, b).

Figure 5.79 shows that the emission spectra A and B are consistent with the transmittance one. Note that no drastic change is seen in the transmittance spectra because the total thickness of the wells is thin. Among them, the spectra A appears to reflect most faithfully, at least, the optical joint density-of-sates (OJDOS) in the lower energy portions in the wells. The shape of OJDOS is different from that in typical cubic III-V compound semiconductor QWs. It seems that there exists an effective band gap inhomogeneity, probably arising from the fluctuation of In concentration from place to place in the lateral plane (Chichibu *et al.* 1997d; 1998a; 1999a; 1999b). Note that the magnitude of the lower part of the OJDOS in CWS2 is larger than in CWS1. It is interesting that EL peaks always appear in the lowest portion in the OJDOS. This appears to indicate that the resistivity of the region with the smallest band gap energy is lower than that in the other region (Deguchi *et al.* 1998b). The ASE is seen to occur apparently in the lower energy portion of the OJDOS. This is due to the fact that both the many-body effects and the heating effect dominate the band filling effect needed for achieving the population inversion.

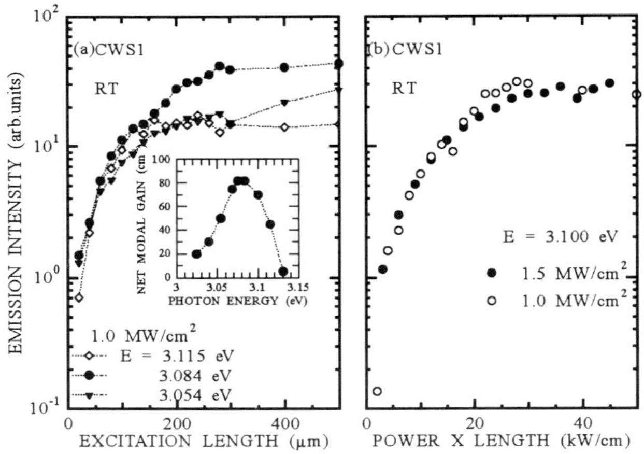

Figure 5.80 (a) Emission intensity as a function of excitation length and (b) emission intensity as a function of the product of P and corresponding to the number of excitation photons in the sample CWS1.

To examine features of the optical gain in details, the emission intensities, I, were recorded as a function of for some fixed excitation power densities, P. As an example, Figure 5.80 (a) shows the emission intensity as a function of measured in CWS1 under $P = 1.0$ MW/cm^2 for three photon energies, where dotted lines are the guide to eyes. I increases superlinearly for $\ell < 200$ μm, to indicate the onset of stimulated emission, while I saturates for $\ell > 300$ μm. Fitting Equation (5.68) to each data set, the value of $g(\nu)$ is obtained for each emitted photon energy, hν. The inset of Figure 5.80 (a) indicates the obtained gain spectrum, $g(\nu)$. $g(\nu)$ is similar to that reported for typical cubic III-V compound semiconductors bulk (Shaklee *et al.* 1973) and QWs (Saleh and Teich, 1991) and also to that reported for InGaN previously by Frankowsky *et al.* (1996) and by Kuball *et al.* (1997). Comparing $g(\nu)$ and the spectrum A in Figure 5.80 (a), it follows that $g(\nu)$ indeed reflects the OJDOS.

However, $g(\nu)$ obtained under different P no longer coincides. This is because the photo-excited carrier density is different and, consequently, observed portions of OJDOS is no longer the same for a given photon energy. Furthermore, as the difference of P leads to the different magnitude of I, the saturation properties become also different. To avoid the complexity, a good idea was proposed by Shaklee *et al.* (1973). The essence of their idea is to look again the experimental data as a function of the number of excitation photons, i.e., photo-excited carrier density. They plotted I as a function of the product of P and ℓ, and obtained a characteristic curves for a given material. Proceeding the same rearrangement of the data, a characteristic curve within experimental errors shown in Figure 5.80 (b) are obtained. This indicates that the gain and saturation mechanisms are independent of the excitation power density. The behaviour of the characteristic curve suggests that the saturation arises from the onset of self-absorption of emitted photons through the stimulated emission (Shaklee *et al.* 1973).

The same argument holds for CWS2 except for the following. That is, reflecting the shape of OJDOS in the lower energy portion, see the spectrum A and B shown in Figure 5.79 (b), g(ν) has a relatively sharp peak.

5.3.5.2 Effects of Si-doping in InGaN lasers

As written in Section 5.3.3.12, doping with Si in the barriers improves the emission properties of InGaN QWs in terms of both electronic and structural improvements. These effects may also improve the laser performance. Optical gain spectra from the InGaN QWs previously reported (Frankowsky *et al.*, 1996; Wiesmann *et al.*, 1996; Kuball *et al.*, 1997) were thought to be explained by the well-known electron-hole plasma (EHP) model with Coulomb enhancement (Haug and Koch, 1990). However, Deguchi *et al.* (1997a; 1997b) showed characteristic gain emergence in the low energy portions of the InGaN MQW LD spectra for laser material which yielded cw LDs (Nakamura and Fasol, 1997a). Therefore it is necessary to investigate mechanisms for the optical gain in more detail to understand the material physics of InGaN QWs. Here the improvement in the optical gain of InGaN/InGaN MQW LDs (Mack *et al.*, 1997; 1998) due to Si-doping in the barriers is discussed.

TE-polarized ASE spectra of LD wafers (lasing at 420nm) were measured at

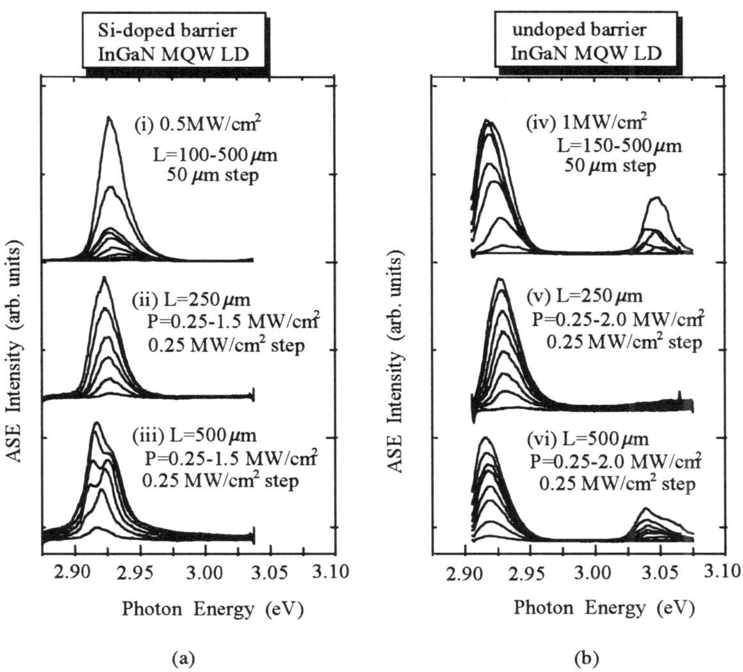

Figure 5.81 ASE spectra of InGaN /InGaN MQW LD wafers having ten QWs with (a) Si-doped and (b) undoped $In_{0.04}Ga_{0.96}N$ barriers. ASE spectra were recorded for different excitation stripe lengths (*L*) [traces (i) and (iv)] and pumping power densities (*P*) [(ii), (iii), (v), and (vi)].

RT by means of VEL method (Shaklee and Leheny, 1971) using a frequency-tripled 10 Hz Q-switched Nd^{+}:YAG laser with a pulse duration of 10 ns. The LD wafers were grown by MOCVD and had ten periods of InGaN/InGaN QWs. The LD structure is given in the literature (Mack *et al.*, 1997). The XRD analysis of the test MQW structures (Cohen *et al.*, 1998) showed that one sample contained 2.5-nm-thick $In_{0.15}Ga_{0.85}N$/7.0-nm-thick Si-doped $In_{0.05}Ga_{0.95}N$ MQW while the other had 2.5-nm-thick $In_{0.12}Ga_{0.88}N$/6.0-nm-thick undoped $In_{0.04}Ga_{0.96}N$ MQW. The Si-doping level was approximately $5\times10^{18}cm^{-3}$. To minimize light scattering, mechanically polished or reactive-ion etched facets were prepared. A 200-μm wide and 500-μm long pumping beam was directed perpendicular to the surface, but not perpendicular to the mirror facet, to obtain pure single-pass gain.

ASE spectra measured at various excitation stripe lengths (L) and power densities (P) are shown in Figure 5.81. The Si-doped barrier MQW exhibited an ASE peak at 2.937 eV, which shifted to 2.927 eV with increasing L [Figure 5.81 (a) (i)]. With increasing P an anomalous second peak (Deguchi *et al.*, 1997b) was observed for L=500 μm [Figure 5.81 (a) (iii)]. Within the EHP model, the redshift due to the increase of P is due to band gap renormalization. The apparent redshift with increasing L seems to be due combined effect of carrier depletion at the end of the pump stripe, which is due to the strong stimulated emission intensity there and the subsequent absorption of the high-energy side of the spectrum, and

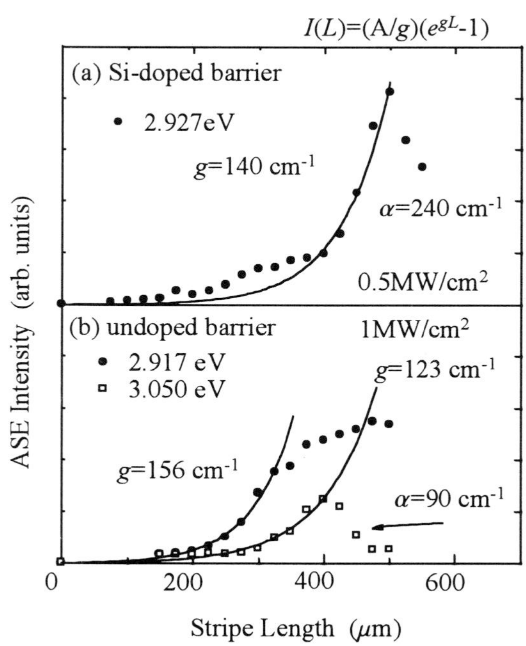

Figure 5.82 ASE intensity as a function of L for MQW LD wafers with (a) Si-doped and (b) undoped barriers measured with the excitation power densities of 0.5 MW/cm^2 and 1 MW/cm^2, respectively. Net modal gain g was obtained through fitting the data using Equation (5.73), the fitting curves being drawn by solid lines.

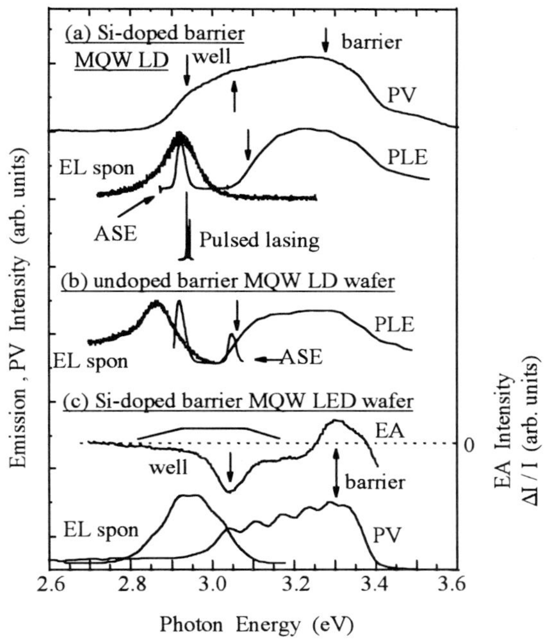

Figure 5.83 Comparison of PV, PLE, spontaneous EL, EA, ASE, and lasing EL spectra for wafers of (a) Si-doped barrier MQW LD, (b) undoped barrier MQW LD, and (c) Si-doped barrier MQW LED. Respective resonance energies in the wells and barriers are indicated by arrows or an arc.

additional absorption of the high energy side owing to the large effective bandgap inhomogeneity (Chichibu *et al.*, 1996d; 1997d; 1999a; Jeon *et al.*, 1996; Narukawa *et al.*, 1997a; 1997b).

The net modal gain g was obtained by fitting the ASE intensity $I(L)$ using the relation given by Equation (5.73). The ASE intensities of the Si-doped barrier MQW pumped at 0.5 MW/cm^2 are shown as a function of L in Figure 5.82 (a). The data were fit giving g of 140cm^{-1} at 2.927 eV. Since the pump stripe length was 500 μm, L greater than 500 μm left an unpumped region of increasing length, from which we obtained the absorption coefficient α of nearly 250cm^{-1}. These values are reasonable for InGaN MQW LDs (Im *et al.*, 1996; Frankowsky *et al.*, 1996; Wiesmann *et al.*, 1996; Kuball *et al.*, 1997; Nakamura and Fasol, 1997a; Deguchi *et al.*, 1997a; 1997b; 1998a; 1998c; Chichibu *et al.*, 1998a; 1998b; 1999a; Cohen *et al.*, 1998) according to the EHP theory. Conversely, the undoped barrier MQW structure exhibited a secondary ASE peak at 3.05 eV in addition to the primary one, as shown in Figure 5.81 (b) (iv)-(vi). Although there exists macroscopic effective bandgap inhomogeneity of the order of hundreds of micrometers within the wafer, the secondary peak was found in many portions. The secondary peak at 3.05 eV appeared for $L \geq 275$ μm at P = 1 MW/cm^2, as shown in Figure 5.81 (b) (iv). For L=250 μm, a precursor of it is noticeable for P$\geq$1.5 MW/cm^2 and only 0.5

MW/cm^2 was needed to observe it for $L = 500\ \mu m$, as shown in Figure 5.81 (b) (v) and (vi). Once the secondary peak appeared, the primary peak redshifts to 2.917 eV, which is presumably due to depletion of carriers at the light-emitting edge owing to large optical field. Typical g values are $160 cm^{-1}$ for the primary peak and $120\ cm^{-1}$ for the secondary peak, as shown in Figure 5.82 (b). The emergence of the secondary ASE peak indicates that there are at least two distinct density of states which are uncoupled with each other. The origin of this is unclear, but is most likely due to a sort of phase separation during the p-type overlayer growth, which may act as annealing.

To compare the electronic states, several optical spectra are summarized in Figure 5.83. Because the potential broadening in the wells and barriers was large, static measurements like PV or PLE could not distinguish (Jeon *et al.*, 1996) the resonance signal in the wells from that of the barriers. Therefore we measured the electroabsorption (EA) spectrum of Si-doped barrier MQW LED (Kozodoy *et al.*, 1997) having identical MQW with semi-transparent electrodes to resolve the signal from the wells. The PLE spectra of both MQW LDs exhibited tail states extending more than 100 meV to lower energy from the nominal quantized energy level, which we define the energy at which the PLE signal intensity is half the maximum. This broadening might be due to combined effects of the in-plane bandgap inhomogeneity and tailing of the energy level due to the internal electric field. Predominant resonance energies are found at 3.104 eV and 3.070 eV for the Si-doped and undoped barrier MQWs, respectively. This 34-meV blueshift could be explained by Coulomb screening of F_{PZ}, which may induce the redshift of the level due to QCSE. The EL peaks appeared at the low energy tail of absorption signals and the Stokes-like shifts (Chichibu et al., 1998a; 1998b; 1998c; 1998d; 1998e; 1999a; 1999b) are 180 meV and 210 meV for the Si-doped and undoped barrier MQW LDs, respectively. This result strongly supports the presence of effective bandgap inhomogeneity in the wells. The decrease of the Stokes-like shift in the Si-doped barrier MQW indicates screening of F_{PZ}, improved bandgap homogeneity, and filling of the localized tail states by electrons. Indeed, a Si-doped barrier MQW test structure exhibited higher order satellite peak of the (0004) XRD pattern, as compared with an undoped barrier one as seen in Figure 5.35.

It is interesting to note from Figure 5.83 that while the electronic states in the wells extended from 3.2 eV to 2.8 eV, the primary ASE peak appeared at the same energy (2.93 eV) for both MQW LDs. This energy agreed with the lowest energy structure of the PV spectra, which are sensitive to lower energy localized tail states. The appearance of the modal gain means that there exists JDOS. The energy difference between the predominant resonance and the primary ASE peak is more than 140 meV, which is larger than the expected amount of the band gap renormalization since the PLE resonance energy is already shifted to low energy by the many-body effect due to the Si modulation doping. Moreover, the Si-doped barrier MQW had a lower threshold power density ($20\ kW/cm^2$) to obtain stimulated emission than that of the undoped one ($50\ kW/cm^2$), and the net gain was very large (Cohen *et al.*, 1998). Therefore the primary ASE peak seems to be formed in the tail states (Deguchi *et al.*, 1997a; 1997b), possibly in Q-disks or segmented QWs (Chichibu *et al.*, 1997d; 1998a; 1998b; 1998c; 1998d; 1999a), like one type of the Nichia cw MQW LDs (Deguchi *et al.*, 1997b; Chichibu *et al.*,

1998a). Carriers in higher-energy states appear to relax efficiently into the lower states to provide the high gain.

Conversely, the secondary ASE peak at 3.05 eV in the undoped MQW seems to be come from higher energy states near the PLE resonance at 3.1 eV. It appears that this reservoir of carriers is isolated from the primary reservoir at 2.93 eV. The role of Si-doping in improving carrier transport or relaxation is still unclear. The improved gain might be due to the effect of modulation doping on the Fermi levels. The most probable explanation is that Si-doping modifies the growth mode to reduce the degree of potential fluctuation and to connect electronically the InGaN nanocolumns (Ponce *et al.*, 1998b), which act as the source of Q-disks or segmented QWs.

As shown, Si-doping was found to eliminate the secondary ASE peak, which is uncoupled with the primary one. The ASE threshold power density was decreased by Si-doping. In contrast to other III-V QWs, under our growth conditions, Si-doping was found to improve the structural and electronic homogeneity.

5.4 REFERENCES

Akasaki, I., Sota, S., Sakai, H., Tanaka, T., Koike, M. and Amano, H., 1996, *Electron. Lett.*, **32**, p. 1105.

Akasaki, I. and Amano, H., 1997, *Jpn. J. Appl. Phys.*, **36**, p. 5393.

Akiyama, H., Koshiba, S., Someya, T., Wada, K., Noge, H., Nakamura, Y., Inoshita, T., Shimizu, A., and Sakaki, H., 1994, *Phys. Rev. Lett.*, **72**, p. 924.

Amano, H., Hiramatsu, K., and Akasaki, I., 1988, *Jpn. J. Appl. Phys.*, **27**, p. L1384.

Amano, H., Watanabe, N., Kiode, N. and Akasaki, I., 1993, *Jpn. J. Appl. Phys.*, **32**, p. L1000.

Antonelli, A., Cen, J. and Bajaj, K.K., 1996, *Semicond. Sci. Technol.*, **11**, p. 74.

Arguello, C.A., Rousseau, D.L. and Porto, S.P.S., 1969, *Phys. Rev.*, **181**, p. 1351.

Aspnes, D.E., 1980, *Handbook on Semiconductors*, ed. Moss, T.S., North-Holland, Amsterdam, Nederland, Vol. 2, Chap. 4A, p. 109.

Aspnes, D.E., 1973, *Surf. Sci.*, **37**, p. 418.

Atanasov, R., Bassani, F., D'Andrea and Tomassini, N., 1994, *Phys. Rev. B*, **50**, p. 14381.

Azuhata, T., Sota, T., Suzuki, K. and Nakamura, S., 1995, *J. Phys. Condens. Mat.*, **7**, p. L129.

Azuhata, T., Matsunaga, T., Shimada, K., Yoshida, K., Sota, T., Suzuki, K. and Nakamura, S., 1996, *Physica B*, **219/220**, p. 493.

Azuhata, T, Sota, T., Chichibu, S., Kuramata, A., Horino, K., Yamaguchi, M., Yagi, T. and Nakamura, S., 1997, *Mater. Res. Soc. Symp. Proc.*, **468**, p. 445.

Azuhata, T., Shimada, K., Deguchi, T., Sota, T., Suzuki, K., Chichibu, S., and Nakamura, S., 1998, *Jpn. J. Appl. Phys.*, **38**, p. L151.

Bastard, G., Mendez, E.E., Chang, L.L. and Esaki, L., 1982, *Phys. Rev. B*, **26**, p. 1974.

Bebb, H.B. and Williams, E., 1972, *Semiconductors and Semimetals*, eds. Willardson, R. K. and Beer, A. C., Academic, New York, USA, Vol.8, p. 182-392.

Behr, D., Wagner, J., Schneider, J., Amano, H. and Akasaki, I., 1996, *Appl. Phys. Lett.*, **68**, p. 2404.

Benoit À la Guillaume, C., Debever, J.-M. and Salvan, F., 1969, *Phys. Rev.*, **177**, p. 567.

Bergman, J., Saksulv, N., Dalfors, J., Holtz, P., Monemar, B., Amano, H. and Akasaki, I., 1998, *Mater. Res. Soc. Symp. Proc.*, **482**, p. 631.

Bernardini, F., Fiorentini, V. and Vanderbilt, 1997, *Phys. Rev. B*, **56**, p. R10024.

Bir, G.L. and Pikus, G.E., 1974, *Symmetry and Strain-Induced Effects in Semiconductors*, John Wiley & Sons, New York, USA, p. 322.

Bohner, K., Anselment, M., Kobbe, G., Klingshirn, C., Haug, H., Koch, S.W., Schmitt-Rink, S. and Abraham, F.F., 1981, *Zeitschrift für Physik B Condens. Matt.*, **42**, p. 1, and references cited therein.

Brafman, O., Lengyel, G., Mitra, S.S., Gielisse, P.J., Plendl, J.N. and Mansur, L.C., 1968, *Solid State Commun.*, **6**, p. 523.

Bulman, G., Doverspike, K., Sheppard, S., Weeks, T., Kong, H., Dieringer, H., Edmond, J., Brown, J., Swindell, J. and Schetzina, J., 1997, *Electron. Lett.*, **33**, p. 1556.

Bulutay, C., Dagli, N. and Imamoglu, A., 1999, *IEEE J. Quantum Electron.*, QE-, to be published in April issue.

Burns, G., Dacol, F., Marinace, J.C., Scott, B.A. and Burstein, E., 1973, *Appl. Phys. Lett.*, **22**, p. 356.

Bykhovski, A., Gelmont, B., Shur, M. and Khan, M.A., 1995, *J. Appl. Phys.*, **77**, p. 1616.

Bykhovski, A., Kaminski, V.V., Shur, M.S., Chen, Q.C. and Khan, M.A., 1996, *Appl. Phys. Lett.*, **68**, p. 818.

Cardona, M., 1969, *Modulation Spectroscopy*, *Solid State Phys.* Suppl. **11**, ed. Seitz, S., Turnbull, D. and Ehrenreich, H., Academic, New York, USA.

Carlone, C., Lakin, K.M. and Shanks, H.R., 1984, *J. Appl. Phys.*, **55**, p. 4010.

Carlos, W.E. and Nakamura, S., 1997, *Appl. Phys. Lett.*, **70**, p. 2019.

Chen, G.D., Smith, M., Lin, J.Y., Jiang, H.X., Wei, S.-H., Kahn, M.A. and Sun, C.J., 1996, *Appl. Phys. Lett.*, **68**, p. 2784.

Chen, H.M., Chen, Y.F., Lee, M.C. and Feng, M.S., 1997, *Phys. Rev. B*, **56**, p. 6942.

Chichibu, S., Nakanishi, H. and Shirakata, S., 1995, *Appl. Phys. Lett.*, **66**, p. 3513.

Chichibu, S., Azuhata, T., Sota, T. and Nakamura, S., 1996a, *J. Appl. Phys.*, **79**, p. 2784.

Chichibu, S., Shikanai, A., Azuhata, T., Sota, T., Kuramata, A., Horino, K. and Nakamura, S., 1996b, *Appl. Phys. Lett.*, **68**, p. 3766.

Chichibu, S., Azuhata, T., Sota, T. and Nakamura, S., 1996c, *Proc. Int. Symp. on Blue Laser and Light Emitting Diodes*, Ohmsha, Tokyo, Japan, p. 202.

Chichibu, S., Azuhata, T., Sota, T. and Nakamura, S., 1996d, *Appl. Phys. Lett.*, **69**, p. 4188.

Chichibu, S., Okumura, H., Nakamura, S., Feuillet, G., Azuhata, T, Sota, T. and Yoshida, S., 1997a, *Jpn. J. Appl. Phys.*, **36**, p. 1976.

Chichibu, S., Azuhata, T., Sota, T., Amano, H. and Akasaki, I., 1997b, *Appl. Phys. Lett.*, **70**, p. 2085.

Chichibu, S., Azuhata, T., Sota, T. and Nakamura, S., 1997c, *Appl. Phys. Lett.*, **70**, p. 2822.

Chichibu, S., Wada, K. and Nakamura, S., 1997d, *Appl. Phys. Lett.*, **71**, p. 2346.

Chichibu, S., Sota, T., Wada, K. and Nakamura, S., 1998a, *J. Vac. Sci. Technol. B*, **16**, p. 2204.

Chichibu, S.F., Abare, A.C., Minsky, M.S., Keller, S., Fleischer, S.B., Bowers, J.E., Hu, E., Mishra, U.K., Coldren, L.A., DenBaars, S.P. and Sota, T., 1998b, *Appl. Phys. Lett.*, **73**, p. 2006.

Chichibu, S., Cohen, D., Mack, M., Abare, A., Kozodoy, P., Minsky, M., Fleischer, S., Keller, S., Bowers, J., Mishra, U., Coldren, L., Clarke, D. and DenBaars, S., 1998c, *Appl. Phys. Lett.*, **73**, p. 496.

Chichibu, S.F., Marchand, H., Keller, S., Fini, P.T., Ibbetson, J.P., Minsky, M.S., Fleischer, S.B., Speck, J.S., Bowers, J.E., Hu, E., Mishra, U.K., DenBaars, S.P., Deguchi, T., Sota, T. and Nakamura, S., 1998d, *Proc. 2nd Int. Symp. on Blue Laser and Light Emitting Diodes*, Ohmsha, Tokyo, Japan, p. 604.

Chichibu, S.F., Sota, T., Wada, K., DenBaars, S.P. and Nakamura, S., 1999a, *Mater. Res. Soc. Internet J. Nitride Semicond. Res.*, **4S1**, p. G2.7.

Chichibu, S.F., Marchand, H., Minsky, M.S., Keller, S., Fini, P.T., Ibbetson, J.P., Fleischer, S.B., Speck, J.S., Bowers, J.E., Hu, E., Mishra, U.K., DenBaars, S.P., Deguchi, T., Sota, T. and Nakamura, S., 1999b, *Appl. Phys. Lett.*, **74**, p. 1460.

Chichibu, S., Abare, A., Mack, M., Minsky, M., Deguchi, T., Cohen, D., Kozodoy, P., Fleischer, S., Keller, S., Speck, J., Bowers, J., Hu, E., Mishra, U., Coldren, L., DenBaars, S., Wada, K., Sota, T. and Nakamura, S., 1999c, *Mater. Sci. Eng. B*, **59**, p.298.

Cho, Y.-H., Song, J.J., Keller, S., Minsky, M.S., Hu, E., Mishra, U.K. and DenBaars, S.P., 1998, *Appl. Phys. Lett.*, **73**, p. 1128.

Chow, W., Koch, S. and Sargent III, M., 1994, *Semiconductor-Laser Physics*, Springer, Berlin,.

Chow, W., Wright, A. and Nelson, J., 1996, *Appl. Phys. Lett.*, **68**, p. 296.

Christensen, N.E. and Gorczyca, I., 1994, *Phys. Rev. B*, **50**, 4397.

Chtchekine, D.G., Fu, L.P., Gilliland, G.D., Chen, Y., Ralph, S.E., Bajaj, K.K., Bu, Y., Lin, M.C., Bacalzo, F.T. and Stock, S.R., 1997, *J. Appl. Phys.*, **81**, p. 2197.

Cingolani, A., Ferrara, M., Lugara, M. and Scamarcio, G., 1986, *Solid State Commun.*, **58**, p. 823.

Cohen, D., Margalith, T., Abare, A.C., Mack, M, Coldren, L.A., DenBaars, S.P. and Clarke, D., 1998, *Appl. Phys. Lett.*, **72**, p. 3267.

Davydov, S.Y. and Tikhonov, S.K., 1996, *Semiconductors*, **30**, p. 514.

Davydov, V. Yu., Averkiev, N.S., Goncharuk, I.N., Nelson, D.K.,Nikitina, I.P., Polkovnikov, A.S., Smirnov, A.N., Jacobson, M.A. and, Semchinova, O.K., 1997a, *J. Appl. Phys.*, **82**, p. 5097.

Davydov, V., Kitaev, Y., Goncharuk, I., Tsaregorotsev, A., Smirnov, M., Mirgorodsky, A. and Semchinova, O., 1997b, *Proc. 2nd Int. Conf. on Nitride Semiconductors*, October 27-31, Tokushima, Japan, p. 410.

Deger, C., Born, E., Angerer, H., Ambacher, O., Stutzmann, M., Hornstein, J., Riha, E. and Fischeruer, G., 1998, *Appl. Phys. Lett.*, **72**, p. 2400.

Deguchi, T., Azuhata, T., Sota, T., Chichibu, S. and Nakamura, S., 1997a, *Proc. 2nd Int. Conf. on Nitride Semiconductors*, Tokushima, Japan, p. 300.

Deguchi, T., Azuhata, T., Sota, T., Chichibu, S. and Nakamura, S., 1997b, *Mater. Sci. Eng. B*, **50**, p. 251.

Deguchi, T., Shikanai, A., Torii, K., Sota, T., Chichibu, S. and Nakamura, S., 1998a, *Appl. Phys. Lett.*, **72**, p. 3329.

Deguchi, T., Ichiryu, D., Sota, T., Suzuki, K., Horikoshi, Y., Azuhata, T., Yamaguchi, M., Yagi, T., Chichibu, S. and Nakamura, S., 1998b, *unpublished.*

Deguchi, T., Azuhata, T., Sota, T., Chichibu, S., Arita, M., Nakanishi, H. and Nakamura, S., 1998c, *Semicond. Sci. Technol.*, **13**, p. 97.

Deguchi, T., 1999, *Ph. D. Thesis*, Waseda University, Japan.

Demangeot, F., Frandon, J., Renucci, M.A., Briot, O., Gil, B. and Aulombard, R.L., 1996, *Solid State Commun.*, **100**, p. 207.

Demangeot, F., Frandon, J., Renucci, M.A., Meny, C., Briot, O. and Aulombard, R.L., 1997, *J. Appl. Phys.*, **82**, p. 1305.

Demangeot, F., Frandon, J., Renucci, M.A., Grandjean, N., Beaumont, B., Massies, J. and Gibart, P., 1998, *Solid State Commun.*, **106**, p. 491.

Detchprohm, T., Hiramatsu, K., Itoh, K. and Akasaki, I., 1992, *Jpn. J. Appl. Phys.*, **31**, p. L1454.

Detsch, T., 1962, *Proc. Int. Conf. on Physics of Semiconductors,* Exter, p. 505.

Deveaud, B., Clerot, F., Roy, N., Satzke, K., Sermage, B. and Katzer, D.S., 1991, *Phys. Rev. Lett.*, **67**, p. 2355.

Dingle, R., Sell, D.D., Stokowski, S.E. and Ilegems, M., 1971, *Phys. Rev. B,* **4**, p. 1211.

Domen, K., Kuramata, A. and Tanahashi, T., 1998, *Appl. Phys. Lett.*, **72**, p. 1359.

Drechsler, M., Hofmann, D.M., Meyer, B.K., Detchprohm, T., Amano, H. and Akasaki, I., 1995, *Jpn. J. Appl. Phys.*, **34**, p. L1178.

Dresselhaus, G., 1957, *Phys. Rev.*, **105**, p. 135.

Edwards, N.V., Yoo, S.D., Bremser, M.D., Weeks, Jr. T.W., Nam, O.H., Liu, H., Stall, R.A., Horton, M.N., Perkins, N.R., Kuech, T.F. and Aspens, D.E., 1997, *Appl. Phys. Lett.*, **70**, p. 2001.

Feldman, J., Peter, G., Gobel, E.O., Dawson, P., Moore, K., Foxon, C. and Elliot, R.J., 1987, *Phys. Rev. Lett.*, **59**, p. 2337.

Filippidis, L., Siegle, H., Hoffmann, A., Thomsen, C., Karch, K. and Bechstedt, F., 1996, *Physica Stat. Sol. B*, **198**, p. 621.

Flores, G.R., Contreras, H.N., Martinez, A.L., Powell, R.C. and Greene, J.E., 1994, *Phys. Rev. B*, **50**, p. 8433.

Frankowsky, G., Steuber, F., Härle, V., Scholz, F. and Hangleiter, A., 1996, *Appl. Phys. Lett.*, **68**, p. 3746.

Frenkel, J., 1931, *Phys. Rev.*, **37**, p. 17.

Giehler, M., Ramsteiner, M., Brandt, O., Yang, H. and Ploog, K.H., 1995, *Appl. Phys. Lett.*, **67**, p. 733.

Gil, B., Briot, O. and Aulombard, R.L., 1995, *Phys. Rev. B*, **52**, p. R17028.

Gil, B., Clur, S. and Briot, O., 1997a, *Solid State Commun.*, **104**, p. 267.

Gil, B. and Briot, O., 1997b, *Phys. Rev. B*, **55**, p. 2530.

Goldstein, L. Horikoshi, Y., Tarucya, S. and Okamoto, H., 1983, *Jpn. J. Appl. Phys.*, **22**, p. 1489.

Gorczyca, I., Christensen, N.E., Pelzery Blanca, E.L. and Rodriguez, C.O., 1995, *Phys. Rev. B*, **51**, p. 11936.

Gotoh, H., Ando, H., Takagahara, T., Kamada, H., Pirson, A.C. and Temmyo, J., 1997, *Jpn. J. Appl. Phys.*, **36**, p. 4204.

Gourdon, C. and Lavallard, P., 1989, *Physica Stat. Sol. B*, **153**, p. 641.

Greene, R.L., Bajaj, K.K. and Phelps, D.E., 1984, *Phys. Rev. B*, **29**, p. 1807.

Grudowski, P.A., Elting, C.J., Park, J., Shelton, B.S., Lambert, D.J.H. and Dupuis, R.D., 1997, *Appl. Phys. Lett.*, **71**, p. 1537.

Hakki, B.W. and Paoli, T.L., 1973, *J. Appl. Phys.*, **44**, p. 4113.

Hakki, B.W. and Paoli, T.L., 1975, *J. Appl. Phys.*, **46**, p.1299.

Hanamura, E., 1988, *Phys. Rev. B*, **38**, p. 1228.

Harima, H., Sakashita, H., Inoue, T. and Nakashima, S., 1997, *Proc. 2nd Int. Conf. on Nitride Semiconductors*, Tokushima, Japan, p. 86.

Harris, C.I., Monemar, B., Amano, H. and Akasaki, I., 1995, *Appl. Phys. Lett.*, **67**, p. 840.

Haug, H. (ed.), 1988, *Optical Nonlinearlities and Instabilities in Semiconductors*, Academic, San Diego, USA.

Haug, H. and Koch, S.W., 1990, *Quantum Theory of the Optical and Electronic Properties of Semiconductors*, World Scientific, Singapore.

Hayashi, K., Itoh, K., Sawaki, N. and Akasaki, I., 1991, *Solid State Commun.*, **77**, p. 115.

Haynes, J.R., 1960, *Phys. Rev. Lett.*, **4**, p. 361.

Hellman, E., 1998, *Mater. Res. Soc. Internet J. Nitride Semicond. Res.*, **3**, p. 11.

Ho, I-hsiu and Stringfellow, G.B., 1996, *Appl. Phys. Lett.*, **69**, p. 2701.

Holst, J., Eckey, L., Hoffmann, A., Broser, I., Schottker, B., As. D.J., Schikora, D. and Lischka, K., 1998, *Appl. Phys. Lett.*, **72**, p. 1439.

Hopfield, J.J., 1960, *J. Phys. Chem. Solids*, **15**, p. 97.

Hopfield, J.J., 1967, *II-VI Semiconducting Compounds*, p. 786.

Hou, H.Q., Segawa, Y., Aoyagi, Y., Namba, S. and Zhou, J.M., 1990, *Phys. Rev. B*, **42**, p. 1284.

Huang, W.H., 1973, *Phys. Status Solidi B*, **60**, p. 309.

Humphreys, T.P., Sukow, C.A., Nemanich, R.J., Posthill, J.B., Rudder, R.A., Hattangady, S.V. and Markunas, R.J., 1990, *Mater. Res. Soc. Symp. Proc.*, **162**, p. 531.

Hvam, J.M. and Ejder, E., 1976, *J. Lumin.*, **12/13**, p. 611.

Im, J., Härle, V., Scholz, F. and Hangleiter, A., 1996, *Mater. Res. Soc. Internet J. Nitride Semicond. Res.*, **1**, p. 37.

Im, J., Kollmer, H., Off, J., Sohmer, A., Scholz, F. and Hangleiter, A., 1998, *Phys. Rev. B*, **57**, p. R9435.

Iotti, R.C. and Andreani, L.C., 1995, *Semicond. Sci. Technol.*, **10**, p. 1561.

Ishibashi, T., Tarucha, S. and Okamoto, H., 1981, *Inst. Phys. Conf. Ser.*, **63**, p. 587.

Itaya, K., Onomura, M., Nishio, J., Sugiura, L., Saito, S., Suzuki, M., Rennie, J., Nunoue, S., Yamamoto, M., Fujimoto, H., Kokubun, Y., Ohba, Y., Hatakoshi, G. and Ishikawa, M., 1996, *Jpn. J. Appl. Phys.*, **35**, p. L1315.

Jeon, E., Kozlov, V., Song, Y., Vertikov, A., Kuball, M., Nurmikko, A., Liu, H., Chen, C., Kern, R., Kuo, C. and Crawford, M., 1996, *Appl. Phys. Lett.*, **69**, p. 4194.

Joyce, M.J., 1992, *Superlattices and Microstructures*, **12**, p. 293.

Kai, A., Okada, K., Yamada, Y., Taguchi, T., Sasaki, F., Kobayashi, S., Tani, T. and Taniguchi, H, 1996, *Jpn. J. Appl. Phys.*, **35**, p. 1424.

Kamiyama, S., Ohnaka, K., Suzuki, M. and Uenoyama, T., 1995, *Jpn. J. Appl. Phys.*, **34**, p. L821.

Karch, K., Wagner, J.-M. and Bechstedt, F., 1998, *Phys. Rev. B*, **57**, p. 7043.

Kawakami, Y., Peng, Z.G., Narukawa, Y., Fujita, Sz., Fujita, Sg. and Nakamura, S., 1996, *Appl. Phys. Lett.*, **69**, p. 1414.

Kawakami, Y., Narukawa, Y., Sawada, K., Saijo, S., Fujita, Sz., Fujita, Sg. and Nakamura, S., 1997, *Mater. Sci. Eng. B*, **50**, p. 256.

Keller, S., Keller, B., Minsky, M., Bowers, J.E., Mishra, U.K., DenBaars, S.P. and Seifert, W., 1998a, *J. Cryst. Growth*, **189/190**, p. 29.

Keller, S., Chichibu, S.F., Minski, M.S., Hu, E., Mishra, U.K. and DenBaars, S.P., 1998b, *J. Cryst. Growth*, **195**, p. 258.

Keyes, R.W., 1961, *Hydrogen-like impurity states in axially symmetric crystals*, *IBM J. Research and Development*, **5**, p.65-66.

Kim, K., Lambrecht, W.R.L. and Segall, B., 1996, *Phys. Rev. B*, **53**, p. 16310.

Kim, K., Lambrecht, W.R.L., Segall, B. and van Schilfgaarde, M., 1997, *Phys. Rev. B*, **56**, p. 7363.

Kimura, Y., Miyachi, M., Takahashi, H., Tanaka, T., Nishitsuka, M., Watanabe, A., Ota, H. and Chikuma, K., 1998, *Jpn. J. Appl. Phys.*, **37**, p. L1231.

Kirillov, D., Lee, H. and Harris Jr., J.S., 1996, *J. Appl. Phys.*, **80**, p. 4058.

Kisielowski, C., Kruger, J., Ruvimov, S., Suski, T. Ager III, J.W., Jones, E., Liliental-Weber, Z., Rubin, M., Weber, E.R., Bremser, M.D. and Davis, R.F., 1996, *Phys. Rev. B*, **54**, p. 17745.

Kisielowski, C., Liliental-Weber, Z. and Nakamura, S., 1997, *Jpn. J. Appl. Phys.*, **36**, p. 6932.

Kisielowski, C., 1998, *Proc. 2nd Int. Symp. on Blue Laser and Light Emitting Diodes*, Ohmsha, Tokyo, Japan, p. 321.

Kleinman, D.A., 1983, *Phys. Rev. B*, **28**, p. 871.

Knap, W., Contreras, S., Alause, H., Skierbiszewski, C., Camassel, J., Dyakonov, M., Robert, J.L., Yang, J., Chen, Q., Khan, M.A., Sadowski, M.L., Huant, S., Yang, F.H., Goiran, M., Leotin, J. and Shur, M.S., 1997, *Appl. Phys. Lett.*, **70**, p. 2123.

Kneissl, M., Bour, D.P., Johnson, N.M., Romano, L.T., Krusor, B.S., Donaldson, R., Walker, J. and Dunnrowicz, C., 1998, *Appl. Phys. Lett.*, **72**, p. 1539.

Knox, R.S., 1963, *Theory of Excitons*, Academic, New York, Suppl.5.

Kobayashi, T., Nakamura, F., Naganuma, K., Tojyo, T., Nakajima, H., Asatsuma, T., Kawai, H. and Ikeda, M., 1998, *Electron. Lett.*, **34**, p. 1494.

Koch, S.W., Haug, H., Schmieder, G., Bohnert, W. and Klingshirn, C., 1978, *Phys. Status Solidi B*, **89**, p. 431.

Korona, K.P., Wysmolek, A., Paluka, K., Stepniewski, R., Baranowski, J.M., Grzegory, I., Lucznik, B., Wroblewski, M. and Porowski, S., 1996, *Appl. Phys. Lett.*, **69**, p. 788.

Koukitsu, A., Takahashi, N., Taki, T., and Seki, H., 1996, *Jpn. J. Appl. Phys.*, **35**, p. L673.

Kozawa, T., Kachi, T., Kano, H., Taga, Y., Hashimoto, M., Koide, N. and Manabe, K., 1994, *J. Appl. Phys.*, **75**, p. 1098.

Kozawa, T., Kachi, T., Kano, H., Nagase, H., Koide, N. and Manabe, K., 1995, *J. Appl. Phys.*, **77**, p. 4389.

Kozodoy, P., Abare, A.C., Sink, R.K., Mack, M.P., Keller, S., DenBaars, S.P., Mishra, U.K. and Steigerwald, D., 1997, *Mater. Res. Soc. Symp. Proc.*, **468**, p. 481.

Kuball, M., Jeon, E., Song, Y., Nurmikko, A., Kozodoy, P., Abare, A., Keller, S., Coldren, L., Mishra, U., DenBaars, S. and Steigerwald, D., 1997, *Appl. Phys. Lett.*, **70**, p. 2580.

Kuball, M., Demangeot, F., Frandon, J., Renucci, M.A., Massies, J., Grandjean, N., Aulombard, R.L. and Briot, O., 1998, *Appl. Phy. Lett.*, **73**, p. 960.

Kubota, K., Kobayashi, Y. and Fujimoto, K., 1989, *J. Appl. Phys.*, **66**, p. 2984.

Kuo, H.-R., Feng, M.-S., Guo, J.-D. and Lee, M.-C., 1995, *Jpn. J. Appl. Phys.* **34**, p. 5628.

Kuramata, A., Domen, K., Soejima, R., Horino, K., Kubota, S. and Tanahashi, T., 1997, *Jpn. J. Appl. Phys.*, **36**, p. L1130.

Kuramata, A., Kubota, S., Soejima, R., Domen, K., Horino, K. and Tanahashi, T., 1998, *Jpn. J. Appl. Phys.*, **37**, p. L1373.

Kwon, H.-J., Lee, Y.-H., Miki, O., Yamano, H. and Yoshida, A., 1996, *Appl. Phys. Lett.*, **69**, p. 937.

Lampert, M.A., 1958, *Phys. Rev. Lett.*, **1**, p. 450.

Lee, I.-H., Choi, I.-H., Lee, C.-R., Shin, E.-J., Kim, D., Noh, S.K., Son, S.-J., Lim, K.Y. and Lee, H.J., 1998, *J. Appl. Phys.*, **83**, p. 5787.

Lemos, V., Arguello, C.A. and Leite, R.C.C., 1972, *Solid State Commun.*, **11**, p. 1351.

Leszczynski, M., Teisseyre, H., Suski, T., Grzegory, I., Bockowski, M., Jun, J., Porowski, J., Pakula, K., Baranowski, J.M., Foxon, C.T. and Cheng, T.S., 1996, *Appl. Phys. Lett.*, **69**, p. 73.

Li, X.B., Sun, D.Z., Zhang, J.P., Kong, M.Y. and Yoon, S.F., 1998, *Appl. Phys. Lett.*, **72**, p. 936.

Logothetidis, S., Petalas, J., Cardona, M. and Moustakas, T.D., 1995, *Mat. Sci. Eng.*, B **29**, p. 65.

Mack, M., Abare, A., Aizcorbe, M., Kozodoy, P., Keller, S., Mishra, U., Coldren, L. and DenBaars, S., 1997, *Mater. Res. Soc. Internet J. Nitride Semicond. Res.*, **2**, p. 41.

Mack, M., Abare, A., Aizcorbe, M., Kozodoy, P., Keller, S., Mishra, U., Coldren, L. and DenBaars, S., 1998, *J. Cryst. Growth*, **180/190**, p. 837.

Majewski, J.A., Stadele, S. and Vogl, P., 1996, *Mater. Res. Soc. Internet J. Nitride Semicond. Res.*, **1**, p. 30.

Majewski, J.A., Stadele, M. and Vogl, P., 1997, *Mater. Res. Soc. Symp. Proc.*, **449**, p. 887.

Manchon Jr., D.D., Barker Jr., A.S., Dean, P.J. and Zetterstrom, R.B., 1970, *Solid State Commun.*, **8**, p. 1227.

Marshall, R. and Mitra, S.S., 1964, *Phys. Rev.*, **134**, p. A1019.

Martin, R.W., Middleton, P.G., O'Donnell, K.P. and Van der Stricht, W., 1999, *Appl. Phys. Lett.*, **74**, p. 263.

Masselink, W.T., Pearah, P.J., Klem, J., Peng, C.K., Morkoc, H., Sanders, G.D. and Chang, Y.-C., 1985, *Phys. Rev. B*, **32**, p. 8027.

Matsuoka, T., 1997, *Appl. Phys. Lett.*, **71**, p. 105.

McCluskey, M.D., Van de Walle, C.G., Master, C.P., Romano, L.T. and Johnson, N.M., 1998, *Appl. Phys. Lett.*, **72**, p. 2725.

McNeil, L.E., Grimsditch, M. and French, R.H., 1993, *J. American Ceramics Soc.*, **76**, p. 1132.

Menniger, J., Jahn, U., Brandt, O., Yang, H. and Ploog, K., 1996a, *Phys. Rev. B*, **53**, p. 1881.

Menniger, J., Jahn, U., Brandt, O., Yang, H. and Ploog, K., 1996b, *Appl. Phys. Lett.*, **69**, p. 836.

Miller, D.A., Chelma, D.S., Damen, T.C., Gossard, A.C., Wiegmann, W., Wood, T.H. and Burrus, C.A., 1984, *Phys. Rev. Lett.*, **53**, p. 2173.

Miller, D.A., Chelma, D.S., Damen, T.C., Gossard, A.C., Wiegmann, W., Wood, T.H. and Burrus, C.A., 1985, *Phys. Rev. B*, **53**, p. 1043.

Miller, D.A., Chemla, D.S. and Schmitt-Rink, S., 1986, *Phys. Rev. B*, **33**, p. 6976.

Miller, R.C., Kleinman, D.A., Nordrand Jr., W.A. and Gossard, A.C., 1980, *Phys. Rev. B*, **22**, p. 863.

Minsky, M.S., Chichibu, S.F., Fleischer, S.B., Abare, A.C., Bowers, J.E., Hu, E.L., Keller, S., Mishra, U.K. and DenBaars, S.P., 1998, *Jpn. J. Appl. Phys.*, **37**, p. L1362.

Mirjalili, G., Parker, T.J., Shayesteh, S.F., Bulbul, M.M., Smith, S.R.P., Cheng, T.S. and Foxon, C.T., 1998, *Phys. Rev. B*, **57**, p. 4656.

Miyoshi, S., Onabe, K., Ohkouchi, N., Yaguchi, H., Ito, R., Fukatsu, S. and Shiraki, Y., 1992, *J. Cryst. Growth*, **124**, p. 439.

Miwa, K. and Fukumoto, A., 1993, *Phys. Rev. B*, **48**, p. 7897.

Mohs, G., Aoki, T., Nagai, M., Shimano, R., K-Gonokami, M. and Nakamura, S., 1997, *Solid State Commun.*, **104**, p. 643.

Monemar, B., 1974, *Phys. Rev B,* **10**, p. 676.

Monemar, B., Bergman, J.P., Amano, H., Akasaki, I., Detchprohm, T., Hiramatsu, K. and Sawaki, 1996a, N., *Proc. Int. Symp. on Blue Laser and Light Emitting Diodes*, Ohmsha, Tokyo, Japan, p. 135.

Monemar, B., Bergman, J.E., Buyanova, I.A., Li, W., Amano, H. and Akasaki, I., 1996b, *Mater. Res. Soc. Internet J. Nitride Semicond. Res.*, **1**, p. 2.

Moore, K. Duggan, G., Woodbridge, K., and Roberts, C., 1990, *Phys. Rev. B*, **41**, p. 1090.

Muensit, S. and Guy, I. L., 1998, *Appl. Phys. Lett.*, **72**, p. 1896.

Mukai, T., Morita, D. and Nakamura, S., 1998a, *J. Cryst. Growth*, **189/190**, p. 778.

Mukai, T., Narimatsu, H. and Nakamura, S., 1998b, *Jpn. J. Appl. Phys.*, **37**, p. L479.

Mukai, T., Yamada, M. and Nakamura, S., 1998c, *Jpn. J. Appl. Phys.*, **37**, p. L1358.

Mukai, T., Takekawa, K. and Nakamura, S., 1998d, *Jpn. J. Appl. Phys.*, **37**, p. L839.

Murugkar, S., Merlin, R., Botchkarev, A., Salvador, A. and Morkoc, H., 1995, *J. Appl. Phys.*, **77**, p. 6042.

Mysyrowicz, A., Grun, J.B., Levy, R., Bivas, A. and Nikitin, S., 1968, *Phys. Lett.*, **26A**, p. 615.

Nair, S. V. and Takagahara, T., 1997, *Phys. Rev. B*, **55**, p. 5153.

Nakadaira, A. and Tanaka, H., 1997, *J. Electron. Mater.*, **26**, p. 320.

Nakamura, F., Kobayashi, T., Asatsuma, T., Funato, K., Yanashima, K., Hashimoto, S., Naganuma, K., Tomioka, S., Miyajima, T., Morita, E., Kawai, H. and Ikeda, M., 1998, *J. Cryst. Growth*, **180/190**, p. 841.

Nakamura, S., 1991, *Jpn. J. Appl. Phys.*, **30**, p. L1705.

Nakamura, S., Mukai, T. and Senoh, M., 1992, *Jpn. J. Appl. Phys.*, **31**, p. 2883.

Nakamura, S., Mukai, T. and Senoh, M., 1993, *Jpn. J. Appl. Phys.*, **32**, p. L16.

Nakamura, S., Senoh, M., Iwasa, N. and Nagahama, S., 1995a, *Jpn, J. Appl. Phys.*, **34**, p. L797.

Nakamura, S., Senoh, M., Iwasa, N., Nagahama, S., Yamada T. and Mukai, T., 1995b, *Jpn, J. Appl. Phys.*, **34**, p. L1332.

Nakamura, S., Senoh, M., Nagahama, S., Iwasa, N., Yamada, T., Matsushita, T., Kiyoku, H. and Sugimoto, Y., 1996a, *Jpn. J. Appl. Phys.*, **35**, p. L74.

Nakamura, S., Senoh, M., Nagahama, S., Iwasa, N., Yamada, T., Matsushita, T., Sugimoto, Y. and Kiyoku, H., 1996b, *Appl. Phys. Lett.*, **69**, p. 4056.

Nakamura, S. and Fasol, G., 1997a, *The Blue Laser Diode*, Springer-Verlag, Heildelberg, Germany.

Nakamura, S., Senoh, M., Nagahama, S., Iwasa, N., Yamada, T., Matsushita, T., Kiyoku, H., Sugimoto, Y., Kozaki, T., Umemoto, H., Sano, M. and Chocho, K., 1997b, *Jpn. J. Appl. Phys.*, **36**, p. L1568

Nakamura, S., Senoh, M., Nagahama, S., Iwasa, N., Yamada, T., Matsushita, T., Kiyoku, H., Sugimoto, Y., Kozaki, T., Umemoto, H., Sano, M. and Chocho, K., 1998a, *Appl. Phys. Lett.*, **72**, p. 211.

Nakamura, S., Senoh, M., Nagahama, S., Iwasa, N., Yamada, T., Matsushita, T., Kiyoku, H., Sugimoto, Y., Kozaki, T., Umemoto, H., Sano, M. and Chocho, K., 1998b, *Jpn. J. Appl. Phys.*, **37**, p. L309.

Naniwae, K., Itoh, S., Amano, H., Itoh, K., Hiramatsu, K. and Akasaki, I., 1990, *J. Cryst. Growth*, **99**, p. 381.

Narukawa, Y., Kawakami, Y., Fujita, Sz., Fujita, Sg. and Nakamura, S., 1997a, *Phys. Rev. B*, **55**, p. R1938.

Narukawa, Y., Kawakami, Y., Funato, M., Fujita, Sz., Fujita, Sg. and Nakamura, S., 1997b, *Appl. Phys. Lett.*, **70**, p.981.

Narukawa, Y., Saijyo, S., Kawakami, Y., Fujita, Sz, Fujita, Sg. and Nakamura, S., 1998, *J. Cryst. Growth*, **189/190**, p. 593.

Narukawa, Y., Saijyou, S., Kawakami, Y., Fujita, Sg., Mukai, T. and Nakamura, S., 1999a, *Appl. Phy. Lett.*, **74**, p. 558.

Narukawa, Y., Kawakami, Y., Fujita, Sg. and Nakamura, S., 1999b, *Phys. Rev. B*, **59**, p. 10283.

Nido, M., 1995, *Jpn. J. Appl. Phys.*, **34**, p. L1513.

Nipko, J.C., Loong, C.-K., Balkas, C.M. and Davis, R.F., 1998, *Appl. Phys. Lett.*, **73**, p. 34.

Niwa, A., Ohtoshi, T. and Kuroda, T., 1997, *Appl. Phys. Lett.*, **70**, p. 2159.

Nozue, Y., Itoh, M. and Cho, K., 1981, *J. Phys. Soc. Jpn.*, **50**, p. 889.

O'Clock, Jr., G.D. and Duffy, M.T., 1973, *Appl. Phys. Lett.*, **23**, p. 55.

O'Donnell, K.P., Martin, R.W. and Middleton, P.G., 1999, *Phys. Rev. B*, **82**, p. 237.

Ohtoshi, T., Niwa, A. and Kuroda, T., 1996, *Jpn. J. Appl. Phys.*, **35**, p. L1566.

Okada, K., Yamada, Y., Taguchi, T., Sasaki, F., Kobayashi, S., Tani, T., Nakamura, S. and Shinomiya, G., 1996, *Jpn. J. Appl. Phys.*, **35**, p. L787.

Okumura, H., Misawa, S. and Yoshida, S., 1991, *Appl. Phys. Lett.*, **59**, p. 1058.

Okumura, H., Ohta, K., Feuillet, G., Balakrishnan, K., Chichibu, S., Hamaguchi, H., Hacke, P. and Yoshida, S., 1997, *J. Cryst. Growth*, **178**, p. 113.

Osamura, K., Naka, S. and Murakami, Y., 1975, *J. Appl. Phys.*, **46**, p. 3432.

Paisley, M.J., Sitar, X., Posthill, J.B. and Davis, R.F., 1989, *J. Vac. Sci. Technol. A*, **7**, p. 701.

Pakula, K., Wysmolek, A., Korona, K.P., Baranowski, J.M., Stepniewski, R., Grzegory, I., Bockowski, M., Jun, J., Krukowski, S., Wroblewski, M. and Porowski, S., 1996, *Solid State Commun.*, **97**, p. 919.

Pankove, J.I., Berkeyheiser, J.E., Maruska, H.P. and Wittke, J., 1970, *Solid State Commun.*, **8**, p. 1051.

Pankove, J.I., 1971, *Optical Processes in Semiconductors*, Prentice-Hall, New Jersey, USA.

Paoli, T.L. and Hakki, B.W., 1973, *J. Appl. Phys.*, **44**, p. 4108.

Perlin, P., Jauberthie-Carillon, C., Itie, J.P., Miguel, A.S., Grzegory, I. and Polian, A., 1992, *Phys. Rev. B*, **45**, p. 83.

Perlin, P., Polian, A. and Suski, T., 1993, *Phys. Rev. B*, **47**, p. 2874.

Perlin, P., Camassel, J., Knap, W., Taliercio, T., Chervin, J.C., Suski, T., Grzegory, I. and Porowski, S., 1995, *Appl. Phys. Lett.*, **67**, p. 2524.

Perlin, P., Litwin-Staszewska, E., Suchanek, B., Knap, W., Camassel, J., Suski, T., Piotrzowski, R., Grzegory, I., Porowski, S., Kaminska, E. and Chervin, J.C., 1996, *Appl. Phys. Lett.*, **68**, p. 1114.

Polian, A., Grimsditch, M. and Grzegory, J., 1996, *J. Appl. Phys.*, **79**, p. 3343.

Ponce, F.A., Steeds, J.W., Dyer, C.D. and Pitt, G.D., 1996, *Appl. Phys. Lett.*, **69**, p. 2650.

Ponce, F. and Bour, D., 1997, *Nature*, **386**, p. 351.

Ponce, F., Galloway, S., Götz, W. and Kern, R., 1998a, *Mater. Res. Soc. Symp. Proc.*, 482, p. 625.

Ponce, F., Cherns, D., Götz, W. and Kern, R., 1998b, *Mater. Res. Soc. Symp. Proc.*, 482, p. 453.

Popovici, G., Xu, G.Y., Botchkarev, A., Kim, W., Tang, H., Salvador, A., Morkoc, H., Strange, R. and White, J.O., 1997, *J. Appl. Phys.*, **82**, p. 4020.

Ramsteiner, M., Menniger, J., Brandt, O., Yang, H. and Ploog, K. H., 1996, *Appl. Phys. Lett.*, **69**, p. 1276.

Ramsteiner, M., Brandt, O. and Ploog, K.H., 1998, *Phys. Rev. B*, **58**, p. 1118.

Reithmaier, J.-P., Höger, R. and Riechert, H., *Phys. Rev. B*, **43**, p. 4933.

Reynolds, D.C. and Collins, T.C., 1981, *Excitons: Their Properties and Uses*, Academic, New York, USA.

Reynolds, D.C., Look, D.C., Talwar, D.N., Kim, W., Aktas, O., Botchkarev, A., Salvador, A. and Morkoc, H., 1996, *J. Appl. Phys.*, **80**, p. 594.

Reynolds, D.C., Look, D.C., Jogai, B. and Molnar, R.J., 1998, *Solid State Commun.*, **108**, p. 49.

Rosner, S., Carr, E., Ludwise, M., Girolami, G. and Erikson, H., 1997, *Appl. Phys. Lett.*, **70**, p. 420.

Ruvimov, S., Liliental-Weber, Z., Suski, T., Ager, J.W., Washburn, J., Krueger, J., Kisielowski, C., Weber, R., Amano, H. and Akasaki, I., 1996, *Appl. Phys. Lett.*, **69**, p. 990.

Sakai, A., Sunakawa, H. and Usui, A., 1997, *Appl. Phys. Lett.*, **71**, p. 2259.

Sakai, H., Takeuchi, T., Sota, S., Katsuragawa, M., Komori, M., Amano, H. and Akasaki, I., 1998, *J. Cryst. Growth*, **189/190**, p. 831.

Saleh, B.E.A. and Teich, M.C., 1991, *in Fundamentals of Photonics,* John Wiley & Sons, Inc. New York, p. 542.

Sanjurjo, J.A., Lopez-Cruz, E., Vogl, P. and Cardona, M., *Phys. Rev. B*, **28**, p. 4579

Sasaki, H., Morishima, S., Maruyama, T. and Akimoto, K., 1998, *Proc. 2nd Int. Symp. on Blue Laser and Light Emitting Diodes*, Ohmsha, Tokyo, Japan, p. 642.

Satake, A., Masumoto, Y., Miyajima, T., Asatsuma, T., Nakamura, F. and Ikeda, M., 1998, *Phys. Rev. B*, **57**, p. R2041.

Sato, H., Sugahara, T., Naoi, Y. and Sakai, S., 1998, *Jpn. J. Appl. Phys.*, **37**, p. 2013.

Savastenko, V.A. and Sheleg, A.V., 1978, *Phys. Status Solidi A*, **48**, p. K135.

Scholz, F., Härle,V., Steuber, F., Sohmer, A., Bolay, H., Syganow, V., Doernen, A., Im, J., Hangleiter, A., Duboz, J.Y., Galtier, P., Rosencher, E., Ambacher, O., Brunner, D. and Lakner, H., 1997, *Mater. Res. Soc. Symp. Proc.*, **449**, p. 3.

Schwarz, R.B., Khachataryan, K. and Weber, E.R., 1997, *Appl. Phys. Lett.*, **70**, p. 1122.

Shaklee, K.L. and Leheny, R.F., 1971, *Appl. Phys. Lett.*, **18**. p. 475.

Shaklee, K.C., Leheny, R.F. and Nahory R.E., 1971, *Phy. Rev. Lett.*, **26**, p. 888.

Shaklee, K.L., Nahory, R.E. and Leheny, R.F., 1973, *J. Luminescence*, **7**, p. 284 and cited therein.

Shan, W., Shmidt, T.J., Yang, X.H., Hwang, S.J., Song, J.J. and Goldenberg, B., 1995, *Appl. Phys. Lett.*, **66**, p. 985.

Shan, W., Shmidt, T.J., Yang, X.H., Song, J.J. and Goldenberg, B., 1996, *J. Appl. Phys.*, **79**, p. 3691.

Shan, W., Walukiewicz, W., Haller, E.E., Little, B.D., Song, J.J., McCluskey, M.D., Johnson, N.M., Feng, Z.C., Schuman, M. and Stall, R.A., 1998, *J. Appl. Phys.*, **84**, p. 4452.

Sheih, S.J., Tsen, K.T., Ferry, D.K., Botchkarev, A., Sverdlov, B., Salvador, A. and Morkoc, H., 1995, *Appl. Phys. Lett.*, **67**, p. 1757.

Shikanai, A., Azuhata, T., Sota, T., Chichibu, S., Kuramata, A., Horino, K. and Nakamura, S., 1997, *J. Appl. Phys.*, **81**, p. 417.

Shimada, K., Sota, T. and Suzuki, K., 1998, *J. Appl. Phys.*, **84**, p. 4951.

Shinada, M. and Sugano, S., 1966, *J. Phys. Soc. Jpn.*, **21**, p. 1936.

Shionoya, S., Saito, H., Hanamura, E. and Akimoto, O., 1973, *Solid State Commun.*, **12**, p. 223.

Shirakata, S. and Chichibu, S., 1996a, *J. Appl. Phys.*, **79**, p. 2043.

Shirakata, S., Chichibu, S. and Isomura, S., 1996b, *Cryst. Res. Technol.*, **31**, p. S717.

Shu, C.K., Ou, J., Lin, H.C., Chen, W.K. and Lee, M.C., 1998, *Appl. Phys. Lett.*, **73**, p. 641.

Siegle, H., Eckey, L., Hoffmann, A., Thomsen, C., Meyer, B.K., Schikora, D., Hankeln, M. and Lischka, K., 1995, *Solid State Commun.*, **96**, p. 943.

Siegle, H., Thurian, P., Eckey, L., Hoffman, A., Thomsen, C., Meyer, B.K., Amano, H., Akasaki, I., Detchprohm, T. and Hiramatsu, K., 1996, *Appl. Phys. Lett.*, **68**, p. 1265.

Siegle, H., Hoffmann, A. Eckey, L., Thomsen, C., Christen, J., Bertram, F., Schmidt, D, Rudloff, D. and Hiramatsu, K., 1997a, *Appl. Phys. Lett.*, **71**, p. 2490.

Siegle, H., Kaczmarczyk, G., Filippidis, L., Litvinchuk, A.P., Hoffmann, A. and Thomsen, C., 1997b, *Phys. Rev. B*, **55**, p. 7000.

Singh, R., Doppalapudi, D., Moustakas, T.D. and Romano, L., 1997, *Appl. Phys. Lett.*, **70**, p. 1089.

Skromme, B.J., Zhao, H., Wang, D., Kong, H.S., Leonard, M.T., Bulman, G.E. and Molnar, R.J., 1997, *Appl. Phys. Lett.*, **71**, p. 829.

Slak, G.A., Tanzilli, R.A., Pohl, R.O. and Vandersande, J.W., 1987, *J. Phys. Chem. Solids*, **48**, p. 641.

Sohmer, A., Off, J., Bolay, H., Härle, V., Syganow, V., Im, J.S., Wagner, V., Adler, F., Hangleiter, A., Dornen, F., Scholz, F., Brunner, D., Ambacher, O. and Lakner, H., 1997, *Mater. Res. Soc. Internet J. Nitride Semicond. Res.*, **2**, p. 14.

Song. Y.K., Kuball, M., Nurmikko, A.V., Bulman, G.E., Doverspike, K., Sheppard, S.T., Weeks, T.W., Leonard, M., Kong, H.S., Dieringer, H. and Edmond, J., 1998, *Appl. Phys. Lett.*, **72**, p. 1418.

Speck, J.S., Marchand, H., Kozodoy, P., Fini, P.T., Wu, X.H., Ibbetson, J.P., Keller, S., DenBaars, S.P., Mishra, U.K. and Rosner, S., 1998, *Proc. 2nd Int. Symp. on Blue Laser and Light Emitting Diodes*, Ohmsha, Tokyo, p. 37.

Stepniewski, K., Korona, K.P., Wysmolek, A., Baranowski, J.M., Pakula, K., Martinez, G., Gregory, I. and Porowski, S., 1997, *Phys. Rev. B*, **56**, p. 15151.

Strite, S., Ruan, J., Li, Z., Manning, N., Salvador, A., Chen, H., Smith, D.J., Choyke, W.J. and Morcoc, H., 1991, *J. Vac. Sci. Technol. B*, **9**, p. 1924.

Strite, S. and Morkoc, H., 1992, *J. Vac. Sci. Technol. B*, **10**, p. 1237.

Sugahara, T., Sato, H., Hao, M., Naoi, Y., Kurai, S., Tottori, S., Yamashita, K., Nishino, K., Romano, L.T. and Sasaki, S., 1998, *Jpn. J. Appl. Phys.*, **37**, p. L398.

Sugawara, M., 1995, *Phys. Rev. B*, **51**, p. 10743.

Suzuki, M., Uenoyama, T. and Yanase, A., 1995, *Phys. Rev. B*, **52**, p. 8132.

Suzuki, M and Uenoyama, T., 1996, *Jpn. J. Appl. Phys.*, **35**, p. 1420.

Tabata, A., Enderlein, R., Leite, J.R., da Silva, S.W., Galzerani, J.C., Schikora, D., Kloidt, M. and Lischka, K., 1996, *J. Appl. Phys.*, **79**, p. 4137.

Taguchi, T., Yamada, Y., Okada, K., Maeda, T., Sasaki, F., Kobayashi, S., Tani, T., Nakamura, S. and Shinomiya, G., 1996, *Proc. 23rd Int. Conf. on Physics of Semiconductors*, eds. Scheffler and Zimmerman, **1**, p. 541.

Tait, W.C., Cambell, D.A., Packard, J.R. and Weiher, R.L., 1967, *II-VI Semiconducting Compounds*, p. 370.

Takeuchi, T., Takeuchi, H., Sota, S., Sakai, H., Amano, H. and Akasaki, I., 1997a, *Jpn. J. Appl. Phys.*, **36**, p. L177.

Takeuchi, T., Sota, S., Katsuragawa, M., Komori, M., Takeuchi, H., Amano, H. and Akasaki, I., 1997b, *Jpn. J. Appl. Phys.*, **36**, p. L382.

Takeuchi, T. Sota, S., Sakai, H., Amano, H., Akasaki, I., Kaneko, Y, Nakagawa, S., Yamaoka, Y. and Yamada, N., 1998a, *J. Cryst. Growth*, **189/190**, p. 616.

Takeuchi, T., Wetzel, C., Yamaguchi, S., Sakai, H., Amano, H. and Akasaki, I., 1998b, *Appl. Phys. Lett.*, **73**, p. 1691.

Toyozawa, Y., 1959, *Prog. Theor. Phys. Suppl.*, **12**, p. 111.

Tsen, K.T., Joshi, R.P., Ferry, D.K., Botchkarev, A., Sverdlov, B., Salvador, A. and Morkoc, H., 1996, *Appl. Phys. Lett.*, **68**, p. 2990.

Tsen, K.T., Ferry, D.K., Botchkarev, A., Sverdlov, B., Salvador, A. and Morkoc, H., 1997, *Appl. Phys. Lett.*, **71**, p. 1852.

Tsen, K.T., Ferry, D.K., Botchkarev, A., Sverdlov, B., Salvador, A. and Morkoc, H., 1998, *Appl. Phys. Lett.*, **72**, p. 2132.

Tsubouchi, K. and Mikoshiba, N., 1985, *IEEE Trans. Sonics and Ultrasonics*, **SU-32**, p. 634.

Uenoyama, T. and Suzuki, M., 1995, *Appl. Phys. Lett.*, **67**, p. 2527.

Vacher, R. and Boyer, L., 1972, *Phys. Rev. B*, **6**, p. 639.

Van de Walle, C.G. and Neugebauer, J., 1997, *Appl. Phys. Lett.*, **70**, p. 2577.

Vertikov, A., Nurmikko, A., Doverspike, K., Bulman, G. and Edmond, J., 1998, *Appl. Phys. Lett.*, **73**, p. 493.

Volm, D., Oettinger, K., Streibl, T., Kovalev, D., Ben-Chorin, M., Diener, J., Meyer, B.K., Majewski, J., Eckey, L., Hoffman, A., Amano, H., Akasaki, I., Hiramatsu, K. and Detchprohm, T., 1996, *Phys. Rev. B*, **53**, p. 16543.

Wang, Y.J., Kaplan, R., Ng, H.K., Doverspike, K., Gaskill, D.K., Ikedo, T., Akasaki, I. and Amano, H., 1996, *J. Appl. Phys.*, **79**, p. 8007.

Wannier, G.H., 1937, *Phys. Rev.*, **52**, p. 191.

Wei, S.-H. and Zunger, A., 1996, *Appl. Phys. Lett.*, **69**, p. 2719.

Wei, G., Zi, J., Zhang, K. and Xie, X., 1997, *J. Appl. Phys.*, **82**, p. 4693.

Wetzel, C., Walukiewicz, W., Haller, E.E., Ager III, J., Grzegory, I., Porowski, S. and Suski, T., 1996a, *Phys. Rev. B*, **53**, p. 1322.

Wetzel, C., Haller, E.E., Amano, H. and Akaski, I., 1996b, *Appl. Phys. Lett.*, **68**, p. 2547.

Wetzel, C., Amano, H., Akasaki, I., Suski, T., Ager, J., Weber, E., Haller, E. and Meyer, B. K., 1998, *Mater. Res. Soc. Symp. Proc.*, **482**, p. 489.

Wieser, N., Klose, M., Dassow, R., Scholz, F. and Off, J., 1997, *Proc. 2nd Int. Conf. on Nitride Semiconductors*, Tokushima, Japan, p. 80.

Wiesmann, D., Brener, I., Pfeiffer, L., Kahn, M. and Sun, C., 1996, *Appl. Phys. Lett.*, **69**, p. 3384.

Wright, A. F., 1997, *J. Appl. Phys.*, **82**, p. 2833.

Yamada, T., Kaneko, Y., Watanabe, S., Yamaoka, Y., Hidaka, T., Nakagawa, S., Marenger, E., Takeuchi, T., Yamaguchi, S., Amano, H. and Akasaki, I., 1997, *Proc. 10th IEEE Lasers and Electro-Optics Society Annual Meeting*, San Francisco, USA, Nov. 10-13, PD1.2.

Yamada, Y., Yamamoto, T., Nakamura, S., Taguchi, T., Sasaki, F., Kobayashi, S. and Tani, T., 1996, *Appl. Phys. Lett.*, **69**, p. 88.

Yamaguchi, A.A., Mochizuki, Y., Sasaoka, C., Kimura, A., Nido, M. and Usui, A., 1997, *Appl. Phys. Lett.*, **71**, p. 374.

Yamaguchi, A.A., Mochizuki, Y., Sunakawa, H. and Usui, A., 1998a, *J. Appl. Phys.*, **83**, p. 4542.

Yamaguchi, A.A., Mochizuki, Y., Sasaoka, C., Kimura, A., Nido, M. and Usui, A., 1998b, *Mater. Res. Soc. Symp. Proc.*, **482**, p. 893.

Yamaguchi, E. and Junnarkar, M. R., 1998, *J. Cryst. Growth*, **189/190**, p. 570.

Yamaguchi, M., Yagi, T., Azuhata, T., Sota, T., Suzuki, T., Chichibu, S. and Nakamura, S., 1997, *J. Phys. Condens. Matters*, **9**, p. 241.

Yamaguchi, M., Yagi, T., Shimada, K., Deguchi, T., Sota, T., Chichibu, S. and Nakamura, S., 1998 (unpublished).

Yang, F., Wilkinsons, M., Austin, E.J. and O'Donnell, K.P., 1993, *Phys. Rev. Lett.*, **70**, p. 323.

Yeo, Y.C., Chong, T.C. and Li, M.F., 1998, *Mater. Res. Soc. Symp. Proc.*, **482**, p. 923.

Zhang, J.M., Ruf, T., Cardona, M., Ambacher, O., Stutzmann, M., Wagner, J.-M. and Bechstedt, F., 1997, *Phys. Rev. B*, **56**, p. 14399.

Zhang, R., Yang, K., Qin, L.H., Shen, B., Shi, H.T., Shi, Y., Gu, S.L., Zheng, Y.D., Huang, Z.C. and Chen, J.C., 1996, *J. Vac. Sci. Technol. A*, **14**, p. 840.

Zimmerman, R., 1990, *J. Cryst. Growth*, **101**, p. 346.

Zimmermann, R., Hofmann, M.R., Euteneuer, A., Mobius, J., Weber, D., Ruhle, W.W., Gobel, E.O., Meyer, B.K., Amano, H. and Akasaki, I., 1997, *Mater. Sci. & Eng. B*, **50**, p. 205.

Chapter 6
Life Testing and Degradation Mechanisms in InGaN LEDs

Marek Osiñski and Daniel L. Barton

6.1 INTRODUCTION

Commercial products have always been subjected to some kind of scrutiny before sale to insure that they will have a high probability of meeting or exceeding the manufacturer's stated lifetime. Without this kind of analysis, products may suffer from infant mortality failures (samples which fail earlier in life than the normal population) or from premature wearout (normal but unintended failure). Testing products helps manufacturers understand the weaknesses in their products to either adjust the operating specifications or the stated lifetime to ensure that the product would meet the stated specifications, and helps them improve the product by eliminating the causes of early failure.

Microelectronics is perhaps the ideal product area in which to study reliability, as large amounts of performance data can be collected on large populations of samples without too much difficulty. The problem with microelectronics is that each production batch of wafers is somewhat unique and may have different reliability characteristics the previous or subsequent lots. In this case, production of large numbers of lots from which reliability data are collected will help establish an average normal product lifetime and generate a complete cross section of failure types.

Digital and analog microelectronic devices are relatively easy to test in large numbers, as only electrical data need be collected. The collection and monitoring of electrical data has proven to be very easy to perform in parallel test systems, which themselves are not likely to change during the time frame of the life test. Optoelectronic devices, which are the subject of this chapter, introduce the added challenge of reliably monitoring the optical performance of many devices in a manner that will be reliable for the duration of the life test.

While the manufacturer may be interested in eliminating infant mortality issues in their products, the purpose of many independent life tests is to learn about the intrinsic reliability and degradation mechanisms of a specific technology type. In this chapter, the subject is life testing of GaN-based light emitting diodes (LEDs). Two types of tests will be discussed. In Section 6.2, life testing of blue-light emitting diodes under close-to-normal operating conditions is described. In

Section 6.3, LEDs are used to simulate laser-like operating conditions (in terms of very high current densities) in order to provide early information about possible degradation processes relevant to GaN-based diode lasers.

6.2 LIFE TESTING OF InGaN/AlGaN/GaN LEDS

6.2.1 Life Testing Primer

In its simplest form, a life test consists of a sample of product used in its intended fashion. For LEDs, the test would be to monitor the output of a number of LEDs which are being subjected to operating conditions (current, voltage, temperature, etc.) that are within the specified product limits. Given a statistically meaningful population, a failure rate, $\lambda(t)$, as shown in Figure 6.1, would be expected. The curve shown in Figure 6.1 has three distinct regions: infant mortality, constant or normal failure rate, and wearout.

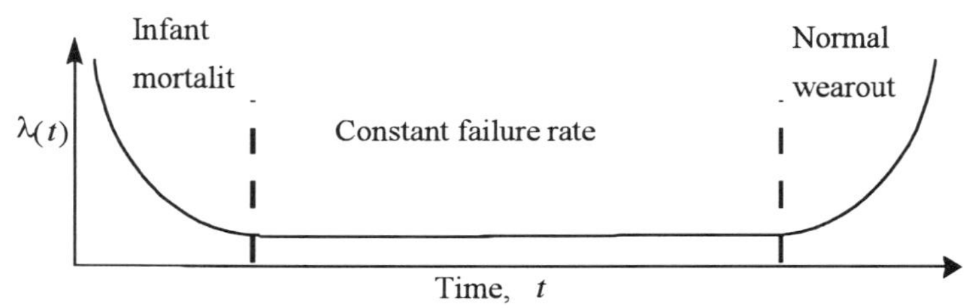

Figure 6.1 Bathtub failure rate curve showing the changes in failure rate, $\lambda(t)$, over the lifetime of a population of LEDs.

In practice, normal lifetimes of high quality devices can easily extend into the 104 to 105 hour range, making the testing under normal operating conditions rather impractical. The only way around this is to accelerate the degradation of the samples being tested. Great care must be taken when determining the conditions for the tests, as the goal is to accelerate intrinsic degradation mechanisms (those that would normally occur given a long enough amount of time) and not cause degradation by other, extrinsic mechanisms. For LEDs, the most common acceleration means are to increase the voltage applied to the LED (and hence the optical power output of the LED) and increase the ambient temperature.

Increasing the ambient temperature is perhaps the most common means to accelerate ageing in many types of microelectronic devices. Temperature accelerates many degradation mechanisms in a very predictable manner. Most thermally accelerated mechanisms can be characterised by fitting the degradation rate to an Arrhenius relation,

$$R = R_0 e^{-E_A/kT} \tag{6.1}$$

where R and R_0 represent the failure rates at elevated temperature and some nominal temperature, respectively, k is Boltzmann's constant, T is the absolute temperature, and E_A is the activation energy of the degradation mechanism. Since most degradation mechanisms in microelectronic devices follow a lognormal failure distribution, Equation (6.1) provides a simple way to predict the changes in failure rates under different temperature conditions. The activation energy, E_A, can be measured by observing the failure rates of two or more populations of LEDs being stressed at different temperatures.

Acceleration by increasing the voltage applied to LEDs serves to both increase the current density, thereby accelerating degradation by electromigration of contact metals and similar processes, and accelerate degradation due to optical processes, specifically mechanisms like recombination enhanced defect motion. These mechanisms, which are also accelerated thermally, rely on the flow of carriers and photons through the materials. Many studies have been reported which link degradation of optoelectronic devices (in material systems such as ZnSe, GaAs, InGaAsP, etc.) to the formation and motion of crystalline defects under normal and accelerated test conditions. In most cases, an Eyring equation (David *et al.*, 1992) can be used to predict the acceleration factor under increased voltage (or current) conditions. In most cases, an Erying equation (David *et al.*, 1992) can be used to predict the acceleration factor under increased voltage (or current) conditions. For the Erying equation, the constant tem, R_0, in Equation (6.1) is modified to include terms for other stress factors, such as $(V/V_0)n$ for example, for elevated voltage stress.

6.2.2 Potential Degradation Regions in LEDs

Figure 6.2 shows a generalised diagram of an LED along with the major areas responsible for device degradation. Of the four areas identified in Figure 6.2, the inner region is of the highest interest to development of GaN-based devices. The relative immaturity of manufacturing technology in this material system leaves many unanswered questions about degradation mechanisms and long-term reliability. The concern is especially serious in view of the fact that GaN-based epitaxial layers grown on sapphire or SiC are known to have very high densities of threading dislocations and other extended defects (Lester *et al.*, 1995; Liliental-Weber *et al.*, 1995; Qian *et al.*, 1995a; Qian *et al.*, 1995b; Chien *et al.*, 1996; Ning *et al.*, 1996; Pelzmann *et al.*, 1996; Wu *et al.*, 1996; Cherns *et al.*, 1997a; Cherns *et al.*, 1997b; Christiansen *et al.*, 1997; Ponce *et al.*, 1997; Vennégués *et al.*, 1997; Potin *et al.*, 1998; Wu *et al.*, 1998). On the other hand, the lack of defect motion at normal device temperatures (Sugiura, 1997a; Sugiura, 1997b) and the suggestion that dislocations do not act as recombination centres (Lester *et al.*, 1995) support the long-term reliability potential of GaN-based devices. Studies on the response of GaN-based LEDs to high current pulses with low duty cycles (to minimise device heating) have detected no measurable change in deep level trap densities even under high current densities and optical power levels (Barton *et al.*, 1995a; Zeller *et al.*, 1995, Zeller *et al.*, 1996). In light of these data, degradation mechanisms related to the inner region should not be expected during low current (smaller than the current required to reach the onset of thermal rollover), low temperature life testing. The other degradation regions shown in Figure 6.2 should thus be suspect. These, however, represent technologies that are very mature and have been studied for many years. The only exceptions are the contact metals used to provide electrical connection to the n- and p-type GaN layers. Fortunately, changes in parameters such as forward series resistance are very easy to detect and measure in LEDs.

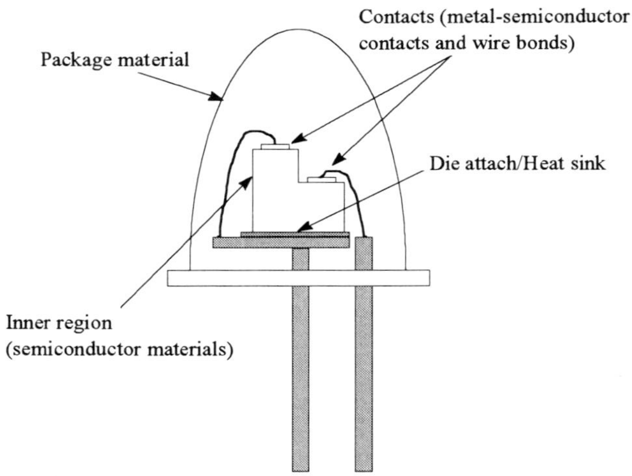

Figure 6.2 General LED structure with marked potential regions of degradation.

6.2.3 Life Test System Considerations

Because of the cost of life testing equipment and the time involved, few commercial companies are able to extract actual lifetime data from their optoelectronic products. Gathering exhaustive life testing data would not be a cost-effective or timely feedback mechanism for improving the manufacturing process of the devices. A complete, and very expensive, system for LED life testing would include the *in-situ* monitoring of output power and light spectrum for each LED, as well as full temperature control of all temperature sensitive components (such as drive circuitry) in the test.

The life test system described in this section was designed to control and monitor each LED's voltage, current, and intensity as well as the supply voltage, the time the data were recorded, and the chamber temperature (Helms *et al.*, 1996). A block diagram of the life test system is shown in Figure 6.3. All test control and measurements were fully computer automated across an IEEE-488 bus and run by a personal computer.

During the first 1,000 h of testing with the system shown in Figure 6.3, LED currents were controlled using two methods. An op-amp circuit was designed to provide constant current to one half of the LEDs, even if the resistance across an LED changed. For comparison, a second circuit based on current limiting resistors was used on the other half of the LEDs, to prevent excessive current drain by any single LED. The resistor-based circuit relied on the constant voltage from its power supply. In theory, the op-amp circuit was to provide better current regulation. After the first 1000 h of testing, though, this proved unfounded, as both designs provided adequate control of the drive current. In later segments of life testing, where higher drive currents were required, the resistor-based circuit was more easily adapted to handle the elevated currents. Hence, all current control was subsequently accomplished using the resistor-based circuit. The use of a circuit board that supplied approximately constant current to each LED allowed the use of a single power supply to power all of the 20 LEDs subjected to life testing.

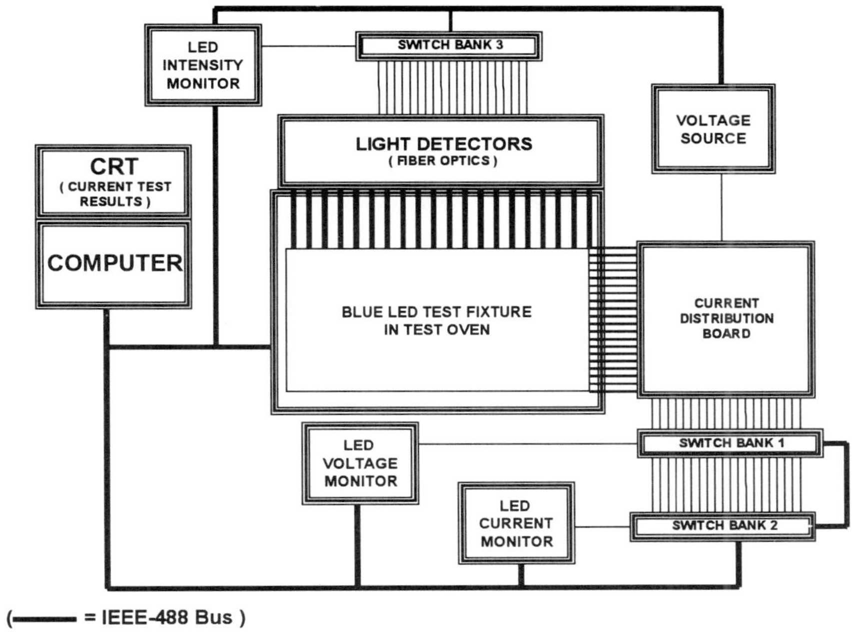

Figure 6.3 Block diagram of a life test system for optoelectronic devices.

Since temperature control is a vital parameter in life testing, all LEDs were placed inside a large temperature chamber. To monitor the light output intensity, a fiberoptic probe was secured to the output end of each LED. A metallic fixture held each fiber-LED pair in a fixed relationship to insure that variations in coupling efficiency did not contribute to measured intensity changes. The fibers traveled outside the temperature chamber, and each was coupled to a silicon photodiode using a fixture that blocked the ambient light. Three automated switch banks and one switch controller were used to obtain data from the twenty separate channels. The LED intensity was monitored by recording detector currents with a picoammeter. LED voltage and current were measured with an IEEE-controlled multimeter. LED voltage was monitored at the LED contacts, while the LED current was derived from the voltage across low value resistors (1-Ω) on the current distribution board.

Since only a portion of each LED's output would be monitored during actual testing, a well controlled pre- and post-test power measurement system was incorporated as a check on the *in-situ* intensity data. The power measurements utilized a 5-inch diameter optical integrating sphere, an optical power meter, and a detector. Forty power readings were taken during a three-minute warm-up of the LED and were computer controlled for consistent timing between LEDs. Power data were recorded as the average of 10 readings. One disadvantage of taking these power measurements was the possibility of damaging the LEDs due to the increased handling.

At one time, the measurement of both the real-time intensity and the post-test power was considered. It was found that the two sets of measurements were in quantitative disagreement, and for good reasons. Real-time intensity data showed the gradual changes in LED output over long continuous test periods and often revealed warm-up or burn-in periods of 50 to 100 h, during which the output intensity would increase. In contrast, the optical power data represented measurements on LEDs running for only a few minutes after a long rest period. Since the most meaningful measure of continuous performance would come from the periodic monitoring of LED intensities *in-situ*, it was decided to concentrate on recording the relative intensity data and not the pre- and post-test optical power. The *in-situ* monitoring of LED intensities also helped to avoid catastrophic failures that could be caused by mishandling of the devices. Intensities were monitored once per hour for the first 10 h, and then once each 12 h for the remainder of each test.

Also consider that a system for *in-situ* monitoring of optical power would require an optical integrating sphere and a dedicated, calibrated detector for each LED. The cost and volume of such a setup would prevent the testing of a large number of devices. When cost is an issue, a choice must be made between monitoring all meaningful LED performance criteria and having enough devices under test to validate the data.

6.2.4 Results of Life Tests on Nichia Blue InGaN/AlGaN/GaN Double Heterostructure LEDs

Using the life test system described in the previous section, twenty Nichia NLPB-500 double heterostructure (DH) blue LEDs (Nakamura, 1994a; Nakamura *et al.*, 1994; Nakamura, 1995) were put up for evaluation. Eighteen of the LEDs were from a newer DH production lot (April 1995), while the remaining two LEDs were from an earlier vintage production lot (March 1994). The NLPB-500 LEDs were the first nitride-based devices to appear on the market, and no data regarding their reliability were available at that time.

The characteristics of the two generations of LEDs were very similar, with the largest difference being in the output spectra. Figure 6.4 shows the output spectra for both an older generation LED and a newer generation LED. The spectra were measured at room temperature (295K) at a forward voltage of 3.5 V. The band-to-band recombination found in the older LED (manifested as the second peak at the high photon energy side of the spectrum) has been suppressed in the newer LED by an increase in the Zn doping concentration in the active InGaN layer (Nakamura, 1996).

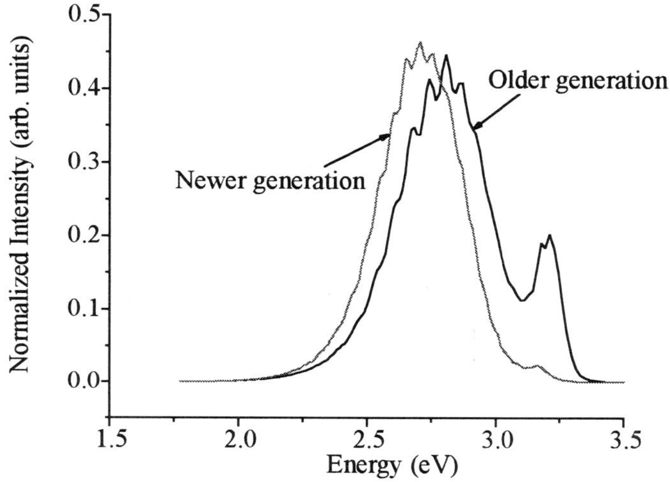

Figure 6.4 Output spectra of the old and newer generation blue DH LEDs, measured at 3.5 V.

The overall life test was broken down into segments and conducted over a period of about 7,200 hours. Figure 6.5 shows the relative luminous intensity from all 20 devices tested, normalized to their initial readings. Table 6.1 shows a summary of the various test times, shown in cumulative hours from the start of the life test, and the temperature at which each test was run. The general trend for the first 1,000 hours for the 18 newer generation LEDs was for the output intensity to increase at a faster rate within the first 50-100 h, then at a slower rate over the remainder of the test. The output intensity of the two older LEDs increased within the first 50 h, and then decreased during the remainder of the first test.

After the first 1,000 hours of testing, the drive currents of the LEDs were increased to try to accelerate the degradation process in some of the devices under test. The previously tested eighteen devices from the new batch were divided into six groups of three. Each group was driven at one of six currents: 20, 30, 40, 50, 60, or 70 mA. Of the two older devices, one (#19) was subjected to a high current of 70 mA, and the other (#20) remained driven at 20 mA. Table 6.2 shows a summary of the biasing conditions for each LED for test segments two through ten. The maximum current level of 70 mA is close to the condition producing a maximum CW output power from the LEDs. From a measurement of the output intensity as a function of drive current, the onset of thermal rollover was observed at 80 mA with slightly decreased output. A current of 70 mA was thus expected to be sufficiently high to cause measurable degradation after a few hundred hours. The changes in both current and temperature for this test were designed to provide degradation data on both of the available acceleration modes simultaneously in one extended life test run.

The relative intensity of one of the older-generation LEDs (#19) dropped to about half of its initial value (a 50% decrease in output was the initial failure criterion) after approximately 1,200 h and the device was removed from the test. In this case, the high current (70 mA) had indeed caused a rapid failure. The remaining devices driven at the same current level performed much better. After a relatively fast drop in their output (10-15% over the first 750 h), their degradation rate slowed as shown in Figure 6.5.

The remaining LEDs were returned to life testing where the temperature was subsequently increased by 5°C after each 500 hours of testing. The output from one of the newer LEDs driven at 70 mA (LED #16) degraded to 55% of its original value after 3,600 hours and a second newer LED (LED #17) degraded by a similar amount after 4,400 hours. The remaining LEDs that degraded during the life testing did not show any changes that could be attributed to degradation in the semiconductor portion of the LED.

Table 6.1. Life test summary

Test	Completion Time (cumulative hours)	Temperature [°C]
1	1004	23
2	2654	23
3	3133	30
4	3636	35
5	4138	40
6	4562	45
7	5100	55
8	5578	65
9	6137	75
10	6636	85
11	7146	95

Table 6.2 LED numbers and associated drive currents.
Note that all devices were biased at 20 mA for test 1 and LEDs 19 and 20 are "older" generation devices

Current [mA]	Device numbers
20	1,2,3,20
30	4,5,6
40	7,8,9
50	10,11,12
60	13,14,15
70	16,17,18,19

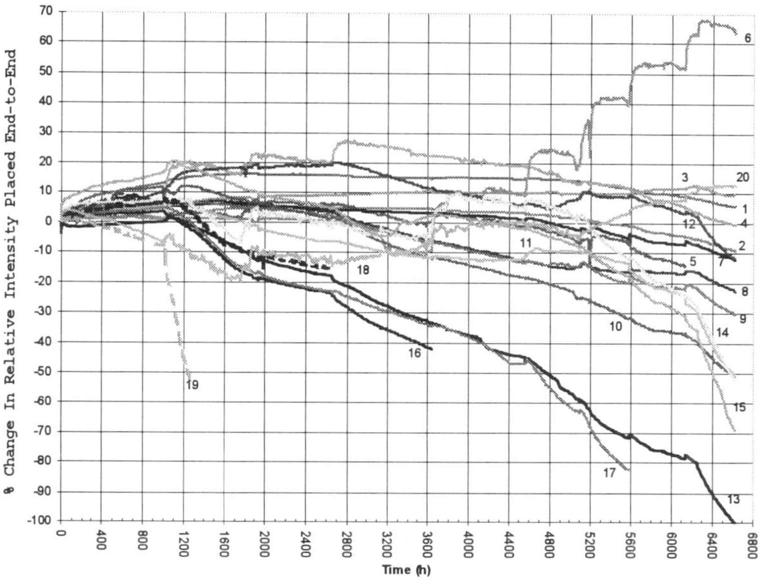

Figure 6.5 Relative change in intensity of 20 Nichia blue LEDs subjected to various currents and temperatures for 6,636 h.

The intensity values averaged over all LEDs operating under the same conditions (usually three LEDs per group) is shown in Figure 6.6 for the first four test segments. By referring to Table 6.1, the number of LEDs used to obtain these average intensities can be seen.

Because there were typically only three newer LEDs per drive-current group, taking an average did little to smooth out the effects of outlying LEDs. There were two specific examples of this. Only one LED out of three in the 30 mA group showed an increase in intensity of 10 to 15% over the first 100 to 200 h of Tests 1 through 4. Thus, the LED with the unusually high intensity dominated the average of the 30mA group. The second occurrence of an outlying LED occurred during Test 2 where one of the three LEDs in the 50 mA group did not degrade. These two instances may be the reason that the 30 mA data exceeds the 20 mA data and the 50 mA data exceeds the 40 mA data in Figure 6.6.

In Test 1, the general trend for all LEDs was for the output intensity to increase more quickly over the first 50 to 100 h. The newer LEDs then continued to slowly increase in intensity, while the older LEDs began to decrease.

In Test 2, the LEDs were separated into groups of different drive currents. The older LED at 20 mA continued its gradual decline in output intensity. The older LED at 70 mA suffered a catastrophic decline in intensity to 50% of its original value and was removed from Test 2 after about 250 h. The newer LEDs were divided into groups of three, and a group was driven at 20, 30, 40, 50, 60, or 70 mA. The intensity of the newer 20 mA LEDs continued to slowly increase then leveled out at 10% above the initial value. The intensities of the 30 mA through 70 mA groups all began to decline, with the decline being proportional to drive current. This proportionality did not apply to the 50 mA group, which contained one LED of unusually high intensity.

In Test 3 the temperature was increased to 30°C. The intensity of the older LED biased at 20 mA showed a slight increase, possibly indicating that a temperature above 23°C had reversed some of the ill effects of Test 1. The newer LEDs at 20 mA maintained their performance at 10% above their initial average intensity value. Newer LEDs driven above 20 mA continued to decline in intensity, but at a greater rate than they did during Test 1. The temperature increase to 30°C appeared to be increasing the activity of the degradation mechanisms occurring within these LEDs.

In Test 4 the temperature was increased to 35°C. The intensity of the older 20mA LED continued its slow and noisy increase. This noise may have been due to the relatively low detector currents generated by the output of this older LED. This LED may have also been most sensitive to daily changes in the temperature of the facilities, which could have had an effect on the room temperature and the equipment outside of the temperature chamber. During Test 4 the newer LEDs driven at 20 mA continued to hold an intensity 10% above their initial average value. The newer LEDs at higher currents all continued to decline in intensity, with no noticeable change in slope relative to their performances at 30°C (as in Test 3).

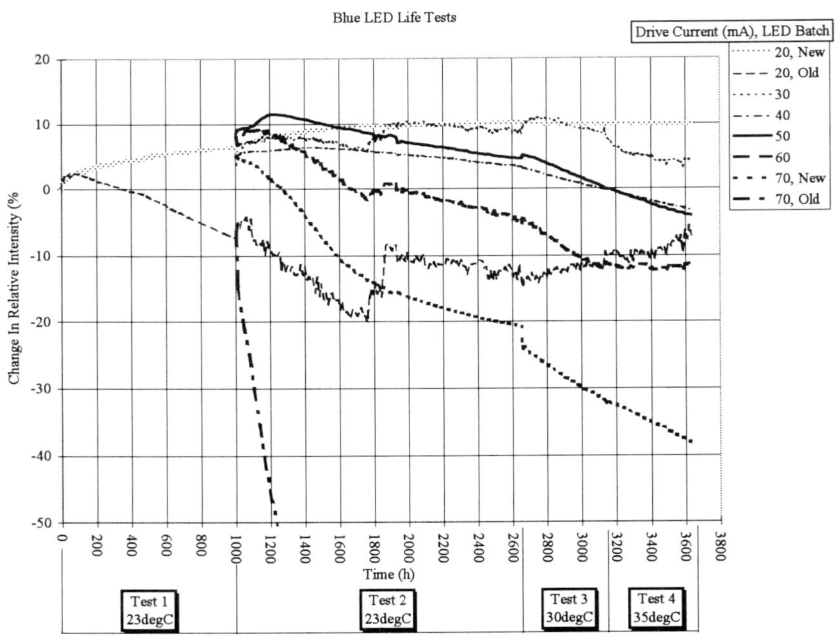

Figure 6.6 Percentage change in relative intensity for the LEDs tested, averaged according to drive current.

6.3 ANALYSIS OF EARLY TEST FAILURES

6.3.1 Analysis of LED #19

In order to determine the cause of the degradation for device #19, a series of I-V measurements were made. The measurements indicated that the junction leakage had increased during the life testing and that the light output degradation was not due to a change in contact resistance or a change in the optical transmission of the plastic encapsulation. This data is significant in that the degradation mechanism could be similar to the one identified in the high pulsed current testing done earlier (Barton *et al.*, 1995b) Figure 6.7 shows the I-V characteristics of the degraded device #19 and an unstressed LED (#52) for comparison. Figure 6.7 shows that the degraded device has an ohmic leakage path across the junction of about 600 MΩ.

To identify the cause of the degradation, device #19 was de-encapsulated and prepared for electron beam-induced voltage (EBIV) analysis. EBIV analysis differs slightly from the more common EBIC analysis in that the voltage variations created across the junction by the incident electron beam is used to create the image instead of the current collected across the junction. The EBIV analysis quickly identified that the cause of the light output degradation was a crack in the LED which isolated part of the junction area from the p-contact. With an electron beam energy of 5 keV, the beam interaction volume did not penetrate down to the n-contact. The EBIV image (Figure 6.8) clearly shows that the crack has propagated through the p-contact and the active layer thus isolating part of the LED from the electrical stimulus and reducing its light output accordingly. We therefore concluded that LED #19 did not degrade in a manner that would be considered typical for operation under normal conditions.

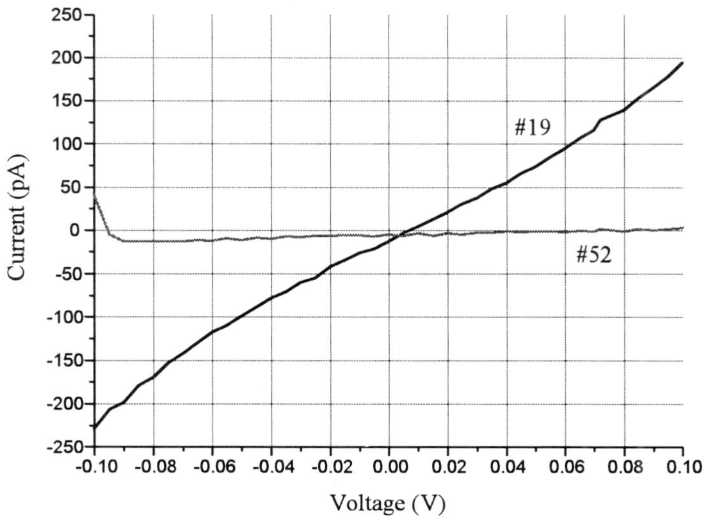

Figure 6.7 I-V characteristics of the degraded LED (#19) and an unstressed LED (#52) for comparison.

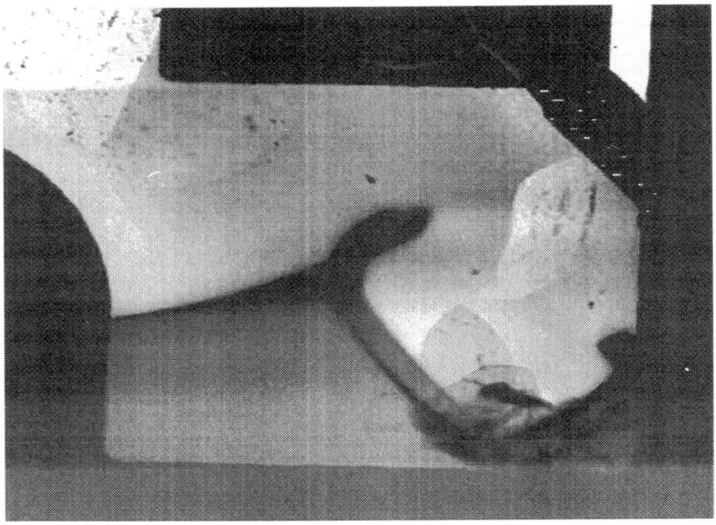

Figure 6.8 EBIV image of LED #19 at 13 keV.

6.3.2 Analysis of LEDs #16 and 17:

In order to identify the process responsible for the degradation in light output measured on LEDs #16 and 17, the first task was to carefully measure the current-voltage characteristics of each device and compare the results to a control LED of the same type. Figures 6.9–6.10 show some of the measurements made on the two degraded LEDs along with a control LED. Figures 6.9–6.10 indicate that LED #16 has not undergone a significant change in its I-V characteristics. In fact, the I-V data indicates that LED #16 has a lower forward series resistance than the unstressed control device, #120. None of the I-V data leads us to a degradation mechanism that involves the electrical performance of the LED. The only possibilities for degradation are limited to a change in the radiative versus non-radiative recombination rate or a change in the optical properties of the packaging materials. The former process would involve a 45% loss in radiation from the LED. A careful inspection of LED #16 under a low forward bias (just above the LED's turn-on voltage) revealed no significant non-radiative areas that could account for that level of loss in light output. The LED showed almost perfectly uniform illumination distribution which is unlike LED #19. The results leave the plastic encapsulation material as the most likely cause for output degradation.

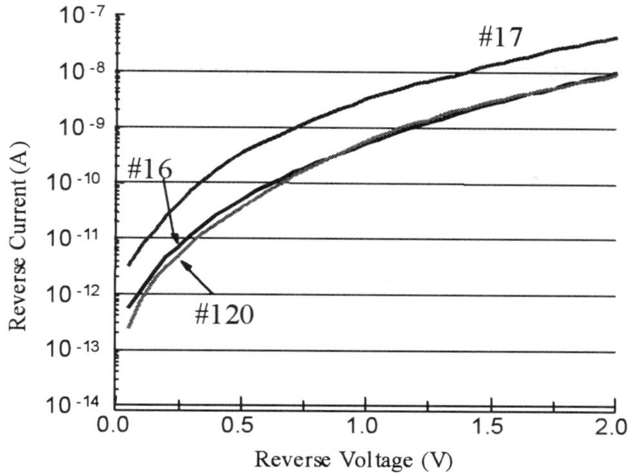

Figure 6.9 Reverse I-V characteristics.

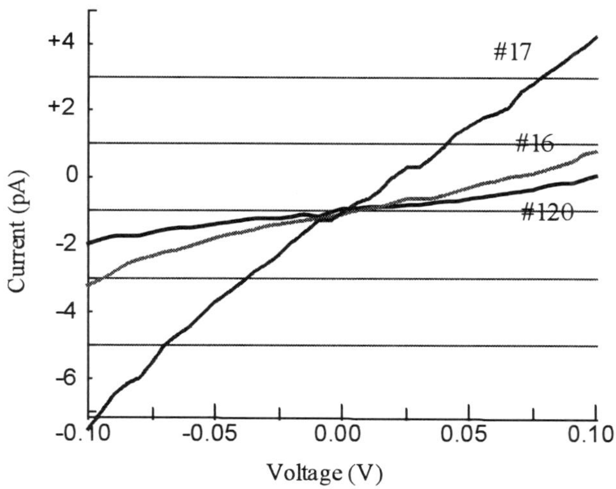

Figure 6.10 Junction leakage measurements.

The I-V data in Figures 6.9–6.10 show a different degradation mechanism for LED #17. This LED shows a significant difference in its characteristic when compared to both the unstressed and the other degraded LED. Figure 6.9 indicates that #17 has about an order of magnitude more leakage current across the junction than #120 and #16. From Figure 6.10, calculations of the ohmic leakage paths across the junctions of the three LEDs showed that LED #17 had an order of magnitude lower resistance across the junction (18 GΩ for #17 versus 113 GΩ for #120). Since the degradation of #17 was still fairly small, this LED was subsequently returned to life testing to continue the degradation process until its output reached a relative decrease in output power of 80%.

The I-V characteristics of LED #17 were measured again after the conclusion of tests 8 and 10 where it was found to have a cumulative loss in relative intensity of over 80% from its initial value. At this point, the LED was removed from the life test and subjected to failure anlaysis. The I-V characteristics of this LED were carefully measured and compared to earlier measurements. The results shown in Figure 6.11 indicate that no further degradation had occurred in the diode which would account for the additional loss in output intensity. This analysis indicates that the degradation may be again tied to the transparency of the plastic encapsulation and not to changes in the semiconductor materials themselves.

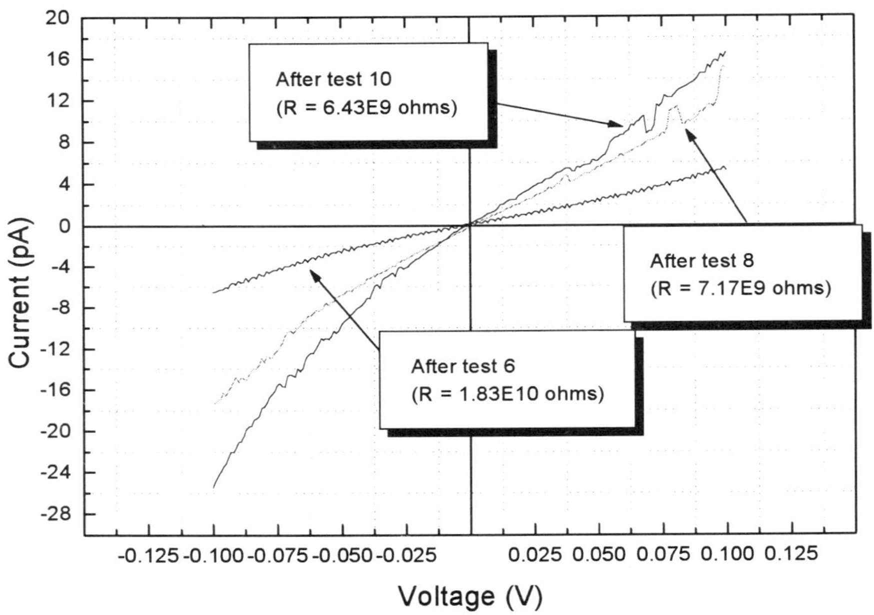

Figure 6.11 I-V measurements of LED #17 after tests 6, 8, and 10.

The data shown in Figure 6.5 was used to estimate device lifetimes and the effects of bias current on degradation acceleration. The data in Figure 6.5 was re-plotted as a function of the square root of time in to allow a linear fit to be made to estimate the time to a 50% loss in output intensity. This is a common method (Fukuda, 1991) for the prediction of LED lifetimes under accelerated aging, especially under the conditions of small sample size and limited acceleration capabilities due to device constraints. From this analysis, device lifetimes under drive currents from 30 to 70 mA were obtained. The average relative intensity changes from each of the groups of three LEDs driven at the same currents over the first 2,654 hours of testing at 23°C were used. The results are shown in Figure 6.12 where a sudden increase in acceleration factor (and a corresponding decrease in device lifetime) is observed for currents above about 50 mA. Since Figure 6.12

is based on data collected from a room temperature (23°C) life test segment, thermal degradation would not be suspect, but degradation mechanisms related to current density and optical power may be. The data in Figure 6.4 showed that the output spectrum was composed of an impurity related emission component and a band-to-band component. While the impurity related component should saturate at some point, the band-to-band component should continue to grow with increasing forward current. Degradation related to the effects of short wavelength emission from the LED could be associated with low temperature, high current life tests.

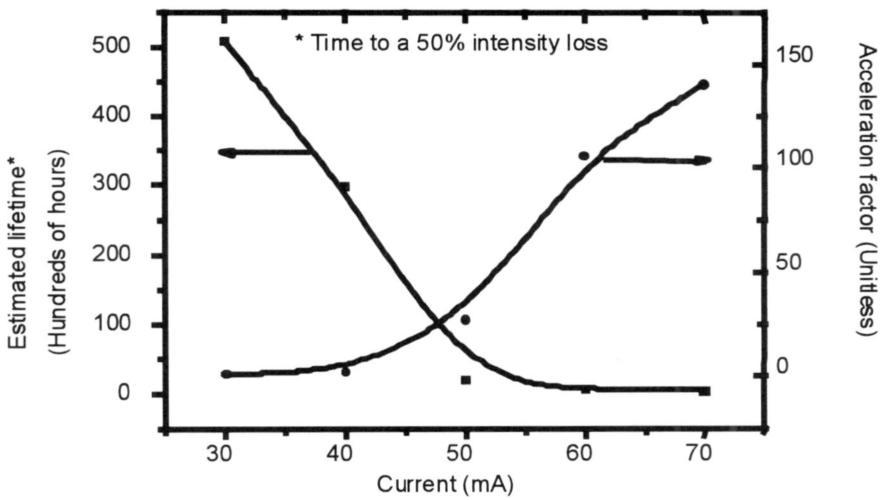

Figure 6.12 Estimated lifetime and degradation acceleration factors extracted from life test data.

In order to investigate the effect of temperature on light output of these LEDs a series of measurements were made at various currents and temperatures. Figure 6.13 shows the changes in light output intensity measured from 25 to 80°C. From Figure 6.13, it is apparent that increases in temperature will reduce the light output and thus slow any optically induced degradation mechanisms.

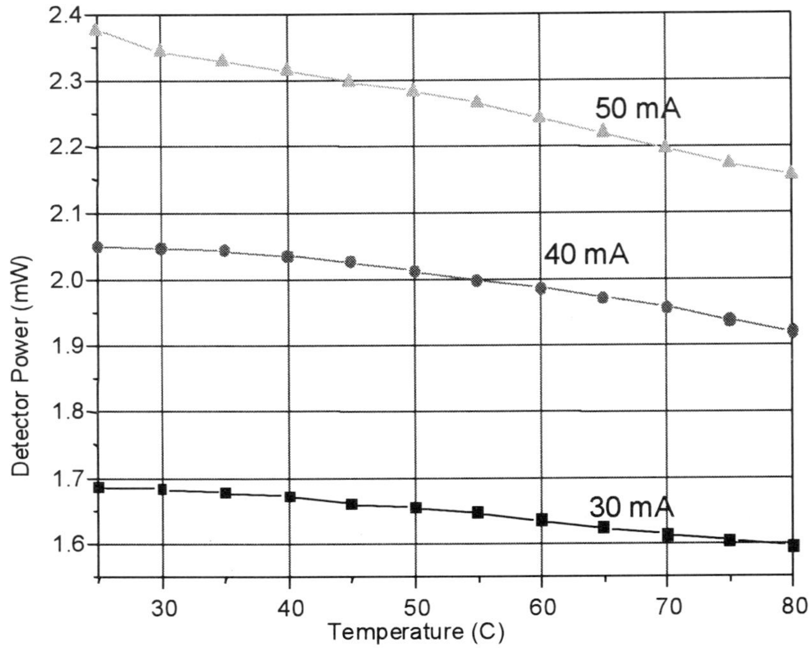

Figure 6.13 LED output power at various temperatures.

The results presented in this section clearly illustrate that there is a severe limitation imposed on accelerated life testing by the current and temperature performance of these LEDs. The current limitations are imposed by the onset of thermal rollover at approximately 80 mA limiting the acceleration factors to those shown in Figure 6.12. The decrease in output intensity with temperature indicates that room temperature (or low temperatures) will provide the fastest degradation due to optically induced mechanisms.

6.4 EFFECTS OF UV EMISSION ON PLASTIC TRANSPARENCY

It is commonly known that prolonged exposure to ultraviolet radiation can reduce the optical transparency of many types of polymers. Since band-to-band recombination in the GaN system can produce ultraviolet radiation, a degradation mechanism of this type is reasonable. Figure 6.4 shows the output spectra for both the older generation, double heterojunction Nichia LED and a newer generation Nichia LED. The spectra were measured at room temperature (295K) at a forward voltage of 3.5 V. The band-to-band recombination found in the older LED has been suppressed in the newer LED by an increase in the Zn doping concentration in the active InGaN layer (Nakamura *et al.*, 1994b). It is also well known (Barton *et al.*, 1995b) that the band-to-band recombination emission component does increase significantly with forward bias indicating that the newer generation LEDs should have a long lifetime at currents less than the 20 mA cw limit in the data sheet, but may show significant degradation by this mechanism at elevated currents.

In order to study the effects of high energy radiation on the optical transparency of the encapsulation material used on the NLPB-500 series LEDs, sections from the plastic from both degraded and un-degraded LEDs were removed. The sections were polished and were 0.05 to 0.08cm thick. Since conventional optical transmission techniques are limited to an accuracy of approximately 0.005 absorptance units, photothermal deflection spectroscopy (PDS) was used which has parts per million sensitivity (Jackson *et al.*, 1981). The PDS technique involves immersing the samples in an organic medium which has a refractive index which is sensitive to temperature. A highly stable HeNe laser beam is placed parallel to and near the surface of the sample while a chopped heating light is incident on the sample surface. With this method, any heating of the sample caused by absorption of the heating beam will bend the HeNe laser beam. Deflection of the HeNe laser beam is detected using lock-in techniques.

Samples from an unstressed LED were first exposed to light at 3.25 eV which is approximately equal to the direct bandgap of GaN for about 5 hours with no noticeable increase in absorptance. In order to speed up the test, energies of 3.9, 3.8, 3.7, and 3.6 eV were used to set a basis for extrapolation to 3.25 eV. Figure 6.14 shows typical absorbed energy data (measured by the in-phase PDS signal) for exposures at 3.9 eV for 11 hours. The upper curve was obtained after the 11 hour curve and demonstrates that even an exposure as short as a few minutes to light above the fundamental π–π* molecular transition causes a large increase in absorption at lower energies. The absorption coefficient can be obtained from the PDS data by using the relation, α(cm-1) = (1.62XPDS1/2)/Lt, assuming that the absorption is reasonably uniform as is the case for these measurements. The parameter, Lt, is the polymer thermal length and is estimated to be 50 μm for this material. The in-phase PDS signal is XPDS and is the measured parameter in this equation.

To determine the effects from exposure to light at 3.25 eV, absorption coefficients were calculated from the data in Figure 6.14 at energies from 3.6 to 3.9 eV and a linear relationship was used to calculate the change in absorption coefficient at 2.76 eV (the blue emission from these LEDs) based on the absorption coefficients measured at higher energies (which is valid based on the data collected). The proportionality constant in this case is proportional to the light intensity and, hence, the emission from the LED as a function of energy. The data suggests that the absorption coefficient at 2.76 eV will grow exponentially with time. The results are plotted in Figure 6.15 which shows the prefactors obtained from parabolic fits (a t_T dependence) to the absorption coefficients measured at different energies. The data shows that the degradation rate is exponentially dependent on exposure energy and that the degradation at 3.25 eV proceeds a decade more slowly than the lowest exposure energy used in this study.

This data can be used to predict the LED lifetime based on a 50% loss in blue light output failure criterion. By calculating the light output and the plastic area exposed, the lifetime can be bounded by using either a linear or parabolic dependence of α(t). Figure 6.16 shows the final result from these calculations using the spectral information for the older generation LED in Figure 6.4. A comparison of the life test data collected and the calculated plastic degradation for these LEDs yields a measured lifetime that is around the mean of these two bounds.

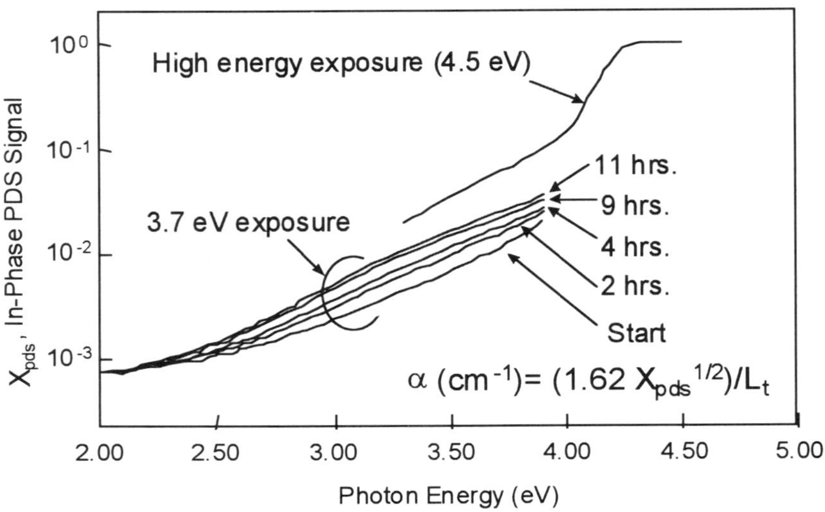

Figure 6.14 In-phase PDS signal from exposure to light at 3.9 eV up to 11 hours.

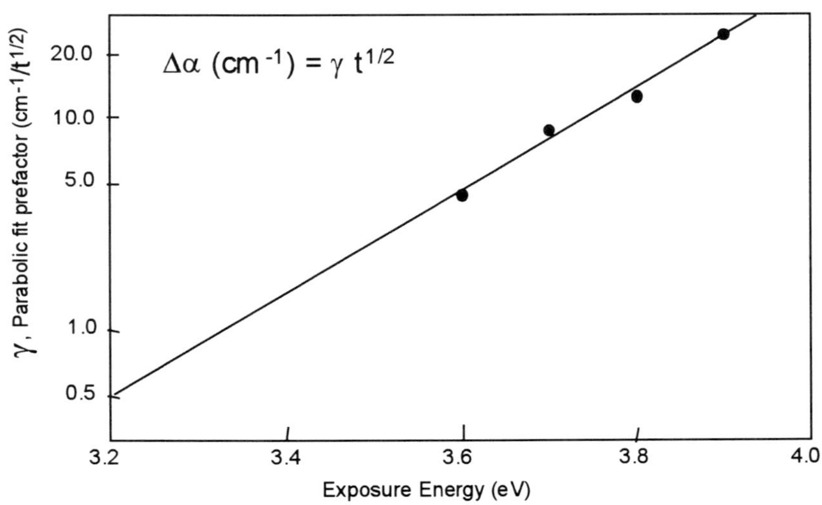

Figure 6.15 Prefactors from parabolic fits to data in Figure 6.14.

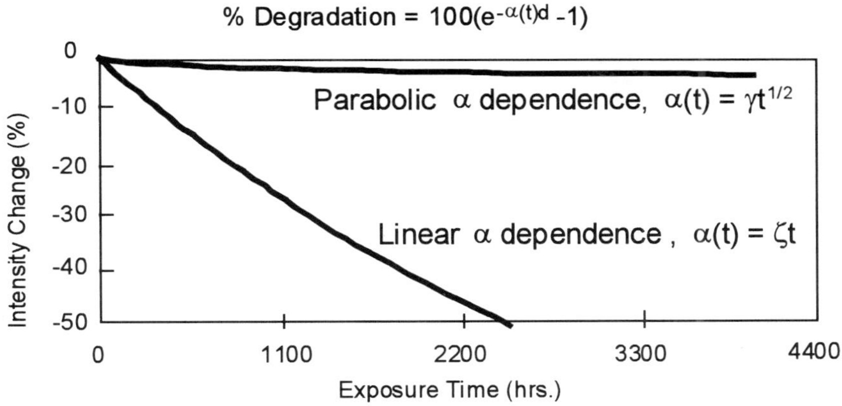

Figure 6.16 LED lifetime based on PDS measurements.

6.5 THERMAL DEGRADATION OF PLASTIC PACKAGE TRANSPARENCY

A comparison between the life test data shown in Figure 6.5 for the early stage of the life test (at low ambient temperatures) and the data shown in Figure 6.17 for test 11 (at 95°C) shows a distinct difference in the degradation rates for the same LEDs. The obvious explanation is that the increase in the degradation rate in the higher temperature test is an acceleration of the same degradation mechanism because of the increase in temperature. The degradation at lower ambient temperatures has been shown to be related to the changes in the plastic packaging material with exposure to the short wavelength emission from the LED. The problem with this degradation mechanism being responsible for the degradation at high ambient temperatures was identified in Figure 6.13. Figure 6.13 shows how the light output is reduced at higher temperatures which would cause degradation mechanisms related to light output to be reduced as the temperature is raised. Thus, a different mechanism must be responsible.

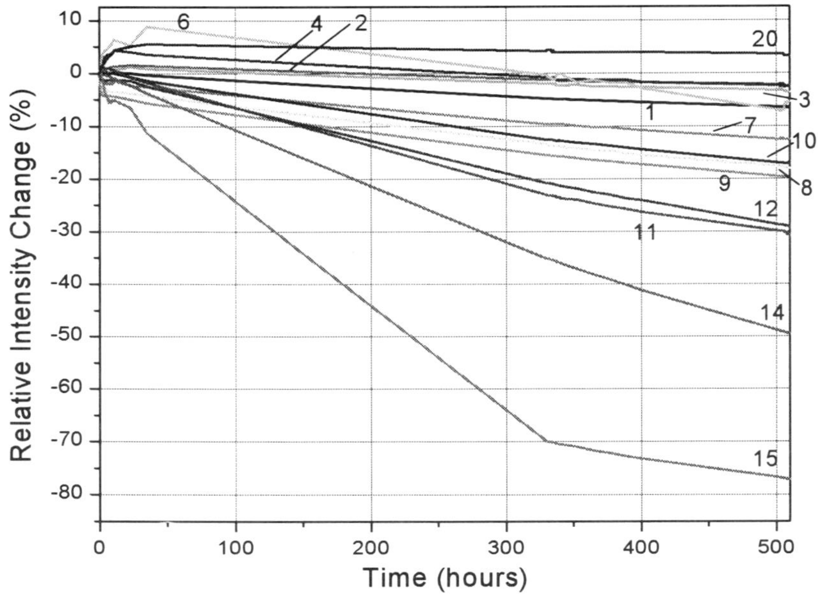

Figure 6.17 Relative change in intensity of remaining Nichia blue LEDs at 95°C for the final 510 hours.

In order to identify the high temperature mechanism, the degradation rates or the slopes of the intensity versus time curves shown in Figure 6.17 were extracted from the life test data from test segment 11, the final, 95°C test. The data, as shown in Figure 6.18, indicates that there is a dramatic increase in the degradation rate for currents greater than or equal to 40 mA at an ambient temperature of 95°C.

In order to understand the cause of this change, the effects of forward current on the temperature of the LED were investigated. The first test was to measure the change in the LED temperature with forward current with the LED in a room temperature environment. For this analysis, a thermocouple was attached to the heat sink of the LED. The results are shown in Figure 6.19 where the LED temperature and the applied power are plotted with forward current. Figure 6.19 shows that the temperature of the LED (as measured external to the package) can increase by over 40°C at the current levels used in the life test. The use of an external point on the package to measure these temperatures means that the temperature of the diode could be significantly higher than what is measured.

Since the data suggests that the degradation may be related to a combination of ambient temperature and LED self-heating, the changes in the plastic with temperature were investigated. Figure 6.20 shows three sections of plastic removed from an Nichia LED control sample. The piece on the left has not been exposed to more than normal room temperature and has a clear appearance. The sample in the middle was placed in a burn-in oven at 150°C in air for a 133 hours. This sample has turned to a deep brown color. The sample in the right was placed in the same oven at 200°C for 130 minutes and shows a similar change to the sample exposed to 150°C only to a lesser degree. This test indicated that at temperatures of around 150 °C (or slightly lower) the plastic can change by a purely thermal effect in such a manner that the light output of the LED could be attenuated. While the LEDs used in the life test did not show plastic that had browned that was visible on the outside of the package, the plastic in contact with the LED may have. This evidence showed that a more accurate junction temperature measurement was needed.

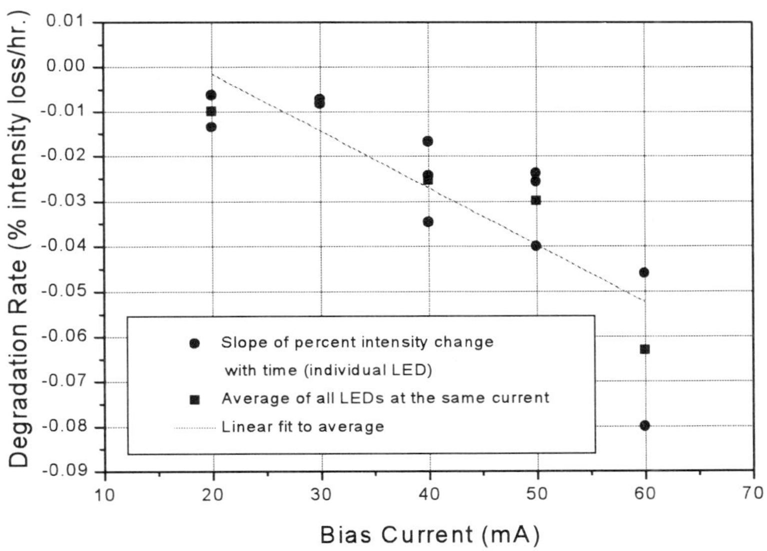

Figure 6.18 Slopes extracted from test 11 data showing a sharp change in degradation rate at 95°C for currents around 40 mA.

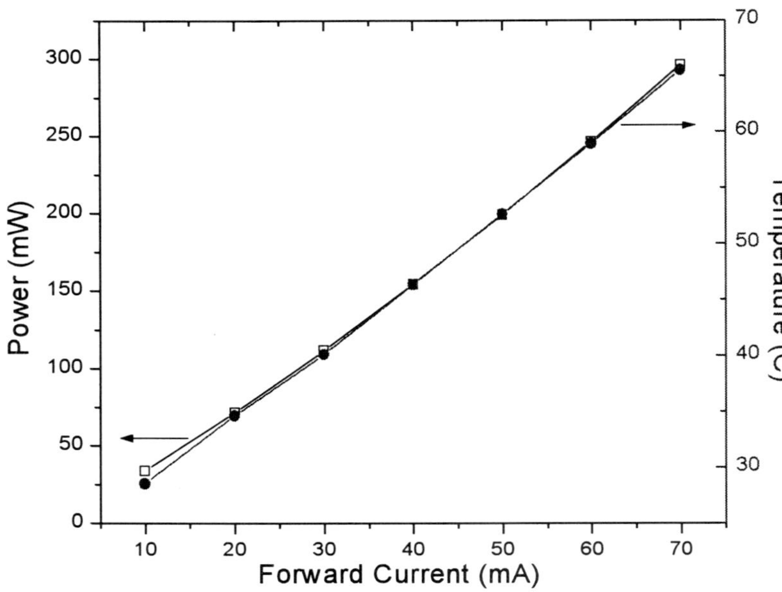

Figure 6.19 LED Power dissipation and measured temperature rise at various currents at room temperature.

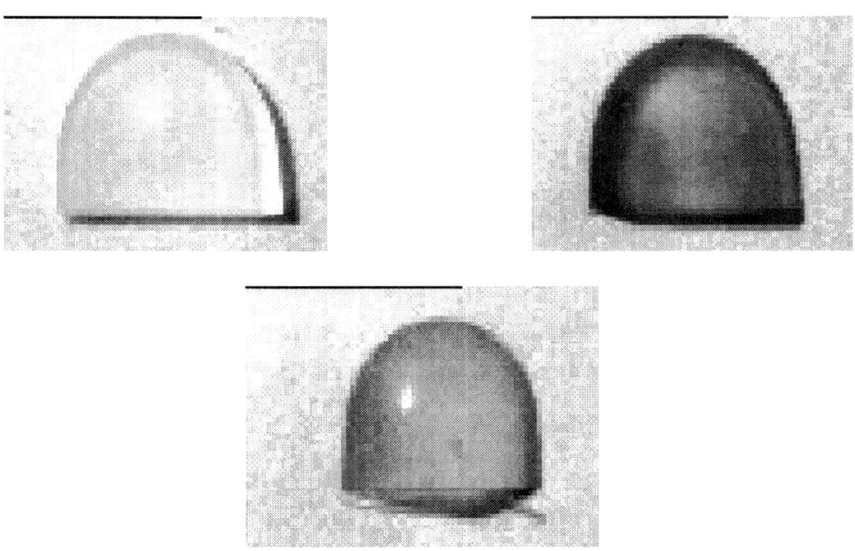

Figure 6.20 Sections of Nichia LED package material: (left) unstressed, (middle) 133 hours at 150°C, (right) 130 minutes at 200°C.

The only way to accurately measure the temperature at the diode is by measuring the junction voltage. To make this measurement, a control LED was placed in a burn-in oven and allowed to come to thermal equilibrium with the oven temperature. The junction voltage was measured when the LED was first turned on to minimize changes in the voltage across the junction due to self-heating. A semiconductor parameter analyzer was used for the measurement allowing the voltage to be accurately measured within 10 ms after the current was applied. This process was repeated at currents from 10 mA to 70 mA (coinciding with the range of currents used in the life tests) and at temperatures from 30°C to 150°C. The resulting data is plotted in Figure 6.21. The data in Figure 6.21 shows that there is a linear relationship between the voltage measured across the junction and the temperature of the junction. From this data, a better measure of the temperature that the plastic in contact with the diode can be made.

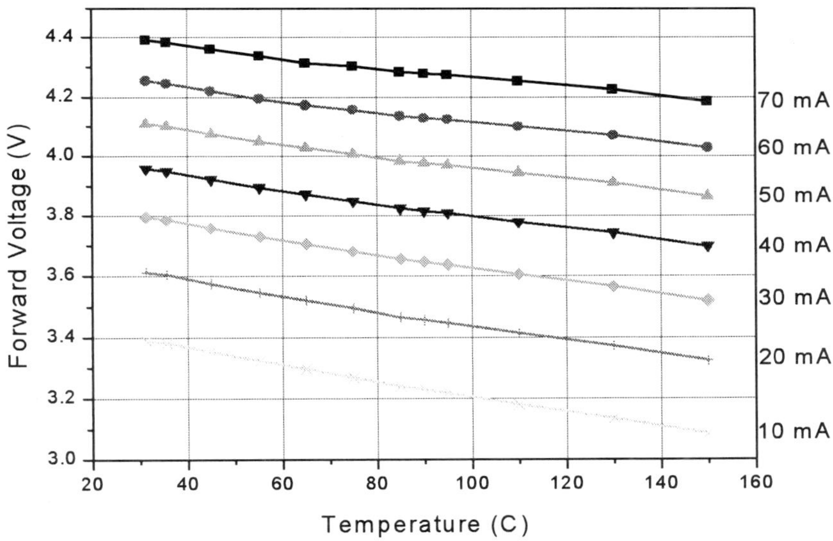

Figure 6.21 Forward junction voltage measured at turn-on for various currents and temperatures.

To calculate the diode temperatures experienced by the LEDs in test 11 (95°C), a control LED from the same production lot was placed in an oven at 95°C and allowed to come to thermal equilibrium. The same forward currents that were used in the life tests were applied and allowed to stabilize meaning that the final temperature resulting from the ambient temperature and self-heating had been reached. This point was determined by observing the junction voltage and waiting for it to stop decreasing. The results along with the temperature measured on the external thermocouple are shown in Table 6.3.

Table 6.3 LED forward voltage at various currents in a 95°C ambient.

Forward Current (mA)	Voltage (V)	Thermocouple Temperature (°C)
10	3.1689	96.5
20	3.3669	97.5
30	3.538	98.4
40	3.7019	99.6
50	3.8679	100.9
60	4.0378	102.2

Using the linear fit parameters from the data in Figure 6.21 along with the data in Table 6.3, the junction temperatures were calculated for each forward current. The results are shown in Figure 6.22. The calculations showed that for forward currents of 40 mA and higher, the junction temperature was in excess of 145°C which is very close to the 150°C temperature where the plastic has been shown to discolor in Figure 6.20. For currents smaller than 40 mA, the temperature is less than 135°C where, according to the data in Figure 6.18, the degradation rate is significantly lower. While the exact temperature at which the plastic begins to discolor has not been determined, the temperature where the degradation begins to affect this material in a time frame important to LED lifetimes has been shown to be between 135 and 145°C.

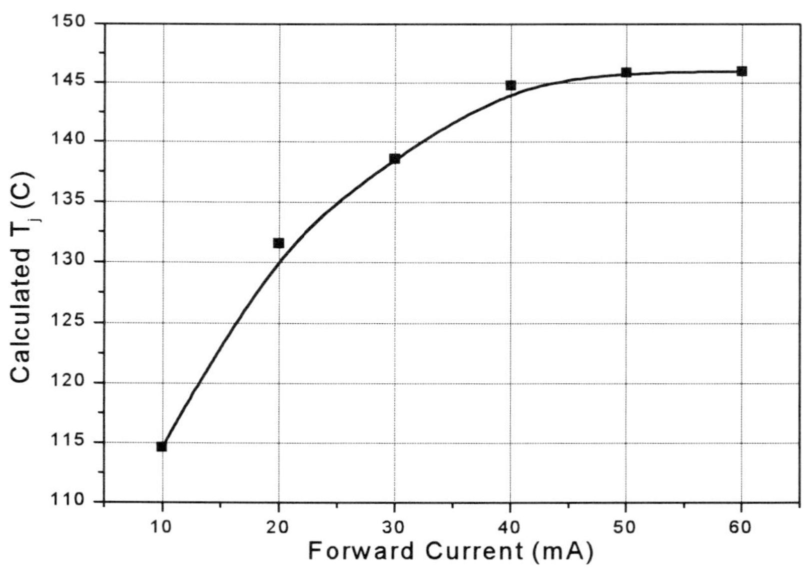

Figure 6.22 Calculated junction temperature (Tj) at various forward currents (ambient temperature 95°C).

6.6 DEGRADATION OF GAN-BASED LEDS UNDER HIGH CURRENT STRESS

The testing results detailed in this section focus on the performance of double heterostructure (DH) and quantum well GaN LEDs under high electrical stress conditions. The earlier, DH, LEDs were subjected to pulsed currents with amplitudes up to 1,540 mA. Analysis of stressed samples showed no significant changes in DLTS and admittance spectroscopy results indicating that trap levels or densities had not been significantly changed by the current pulse stress. LEDs stressed to failure showed a tendency for contact metals to migrate along defect tubes eventually producing shorts across the junction. Similar tests performed to the newer quantum well LEDs showed significant improvement in their resistance to the failure modes observed in the DH devices. These LEDs were subjected to current pulse amplitudes from 1 to 7 A with a very short duty cycle to minimize heating effects. In this case, the reduction in the defect density in the quantum well LEDs showed an increased tolerance to the current stress with the observed failures attributed to weaknesses in the plastic package material and not the diode itself.

6.7 DOUBLE HETEROSTRUCTURE DEVICE TESTING

The impetus for this section was failures that occurred while output spectrum measurements were being made on LEDs under high current conditions. The absolute maximum rating (as stated in the data sheet) for forward current are 30 mA for continuous wave (cw) and 100 mA for pulsed conditions. Under standard operating conditions, the diodes emit in a broad (~70nm) impurity-related band, centered at 450-460nm. The output power reaches a maximum at ~75 to 80 mA and then gradually decreases to about one-third of its peak value at 140 mA (the maximum forward current we have used for cw measurements). The impurity-related contribution peaks at about 70 mA and then decreases due to joule heating. In contrast, the band-to-band component continues to increase beyond 70 mA. This suggests that the current density, rather than temperature, is the main factor determining the efficiency of the band-to-band transitions, while elevated temperature suppresses the impurity-related transitions. This behavior is fully recoverable and no deterioration of device characteristics was observed during the measurements.

Similar changes in output spectra were observed under high-current pulsed conditions. Figure 6.23 shows the output spectra under pulsed conditions (100 ns rectangular pulses at 1 kHz repetition rate) with amplitudes ranging from 40 mA to 1,540 mA. The band-edge emission around 380nm increases at a faster rate than the impurity-related transitions around 460nm, and becomes dominant at very large current densities, while the impurity-related emission appeared to saturate.

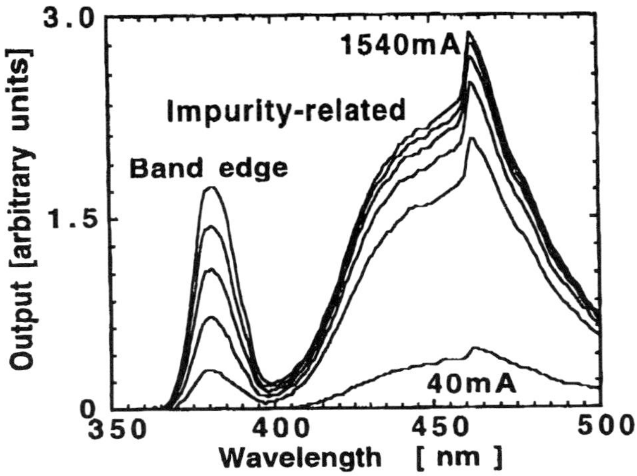

Figure 6.23 Output spectra for various pulsed currents

During the high pulsed current measurements, several of the LEDs showed a degradation in I-V characteristics; some showed a low resistance ohmic short (40 to 800 Ω). The shorts were interesting in that they could be removed by applying a modest (approximately 5 V) reverse bias and re-created by applying a pulsed forward current. This process could be repeated so long as the LED remained encapsulated, i.e. the heat sink for the die was intact.

While the LED was in the shorted state, light output under cw conditions was not possible. The LED would still emit light under pulsed conditions. After a reverse bias was applied to remove the short, cw operation was once again possible. The effects of the stress were not completely removable, i.e. the LED could be placed in a shorted state or in a degraded state, not back into its original, unstressed state. The degraded state, not the shorted state was used for subsequent deep level transient spectroscopy (DLTS) and admittance spectroscopy measurements.

Figures 24 and 25 show I-V characteristics of both unstressed and stressed devices. The forward characteristics shown in Figure 6.24 show both the I-V data for forward bias as well as the resistance (forward voltage divided by forward current). The I-V curves show a small difference, mostly a stress induced increase in conductivity at voltages below the threshold voltage (~2.8 V). The resistance curves support this observation by showing a fairly constant resistance of about 10 kΩ for the degraded device versus several orders of magnitude greater resistance for the unstressed device. As expected, the difference between the two resistance curves is reduced after turn-on as the resistance across the junction will be very small compared to any other parallel resistance path. The reverse bias curves in Figure 6.25 show a more dramatic difference. In this case, the normally small leakages found across reverse biased junctions are easily distinguishable from the ohmic path found across the stressed LED.

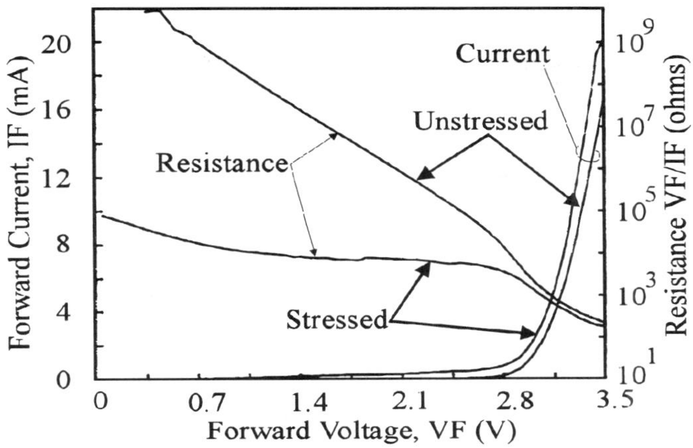

Figure 6.24 Forward I-V characteristics and resistances of a stressed and an unstressed LED.

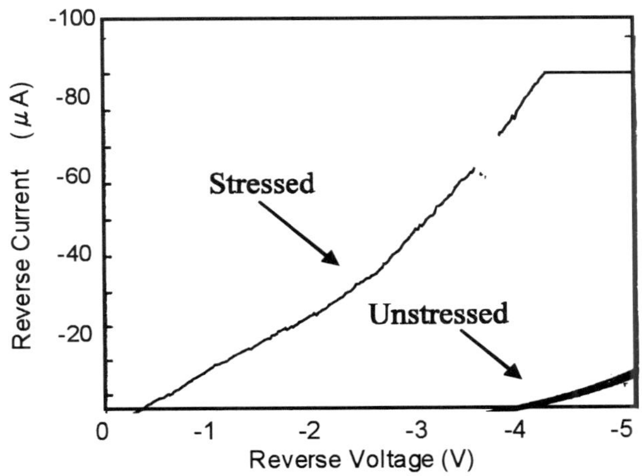

Figure 6.25 Reverse I-V characteristics of one stressed and three unstressed LEDs.

In order to understand the physical phenomenon responsible for the changes in the characteristics of LEDs that have been subjected to high current stresses, several different tests were done.

DLTS was used to compare stressed samples with unstressed, control samples to see if the degradation was due to vacancies, interstitials, impurity atoms, etc. Figure 6.26 compares the DLTS spectrum from a stressed sample to an unstressed one. The DLTS measurements were made on a spectrometer using a 50 ms rate window. The system uses a cooling stage capable of cooling down to 80K. Figure 6.26 shows that for the scanned temperature range, (T > 80K), there were no distinguishable peaks for either sample. The data suggests that there is a peak for temperatures less than 80K. The data does show that the low temperature peak would be indicative of the presence of shallow impurities and not from deep levels. The similarity between the stressed and unstressed samples suggests that the stress did not cause a significant change in any impurity level, especially deep level impurities.

This data contradicts reports on deep-level defects observed in n-type GaN (Hacke *et al.*, 1994 and Gotz *et al.*, 1994). The DLTS results on the LEDs tested in this study show only an unresolved low temperature peak associated with minority carrier traps.

Admittance spectroscopy was done to determine if there was a detectable concentration of deep levels and, if so, what the activation levels of those levels

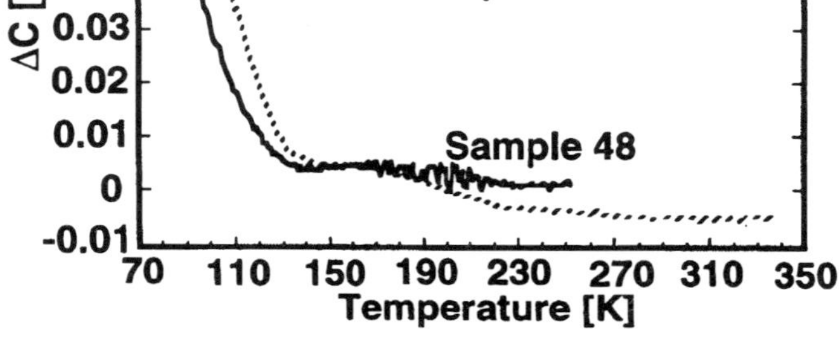

Figure 6.26 DLTS spectra for stressed and unstressed LEDs.

were. The measurements were made at several different frequencies and temperatures using the same temperature control equipment as was used for the DLTS measurements. The measurements were made over a temperature range of 80K to 340K and a frequency range from 2 MHz to 11 MHz.

Figure 6.27 shows a conductance versus temperature plot for an unstressed LED at several different frequencies. Figure 6.28 shows the same plot for a stressed LED. Both figures show that, in accordance with admittance spectroscopy theory, a peak could be found in the conductance and that the peak shifted to higher temperatures at higher frequencies. From these data, an Arrhenious plot can be made whose slope gives the activation energy ΔE = ET-EV, where ET is the energy level of the trap and EV is the valence band energy. Figure 6.29 shows the Arrhenious plots for both the stressed and unstressed LEDs. From this figure, activation energies of 91 meV for the stressed LED and 77 meV for the unstressed LED are obtained. These results also indicate that no deep levels were detected, only shallow ones. The difference in activation energies is insignificant as the difference is within the standard deviation of the measurement on this system. These activation energies most likely correspond to the low temperature DLTS peak from Figure 6.26 and are virtually unaffected by electrical stress.

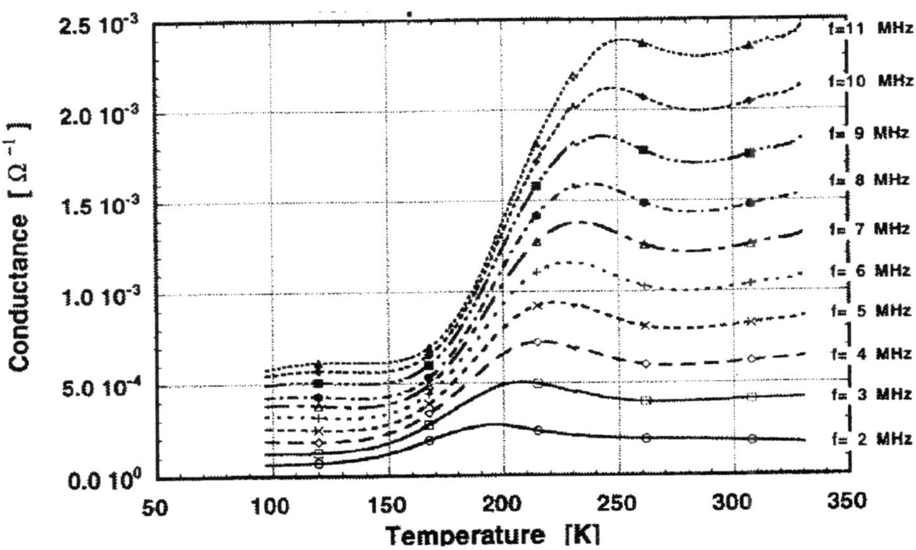

Figure 6.27 Conductance versus temperature plot for an unstressed LED at several different frequencies.

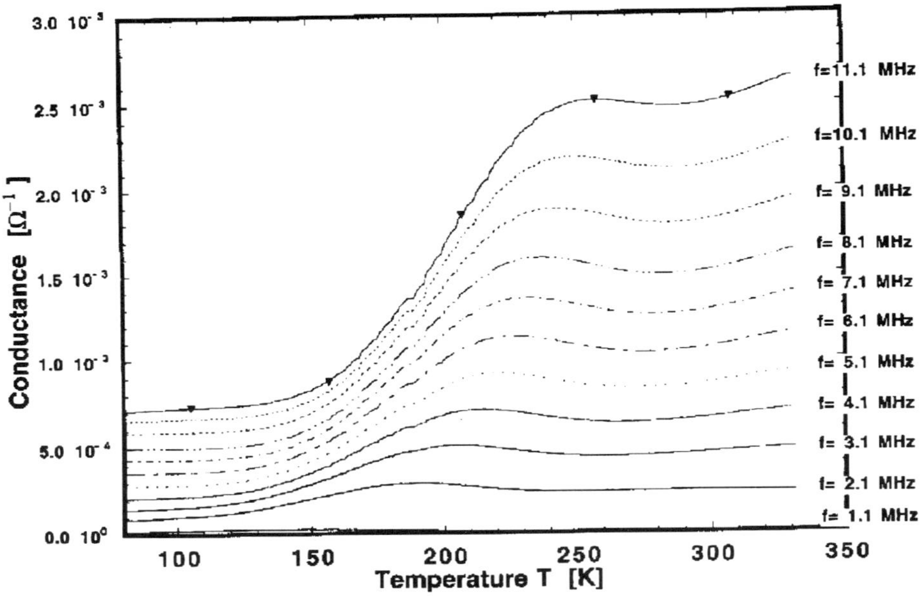

Figure 6.28 Conductance versus temperature plot for a stressed LED at several different frequencies.

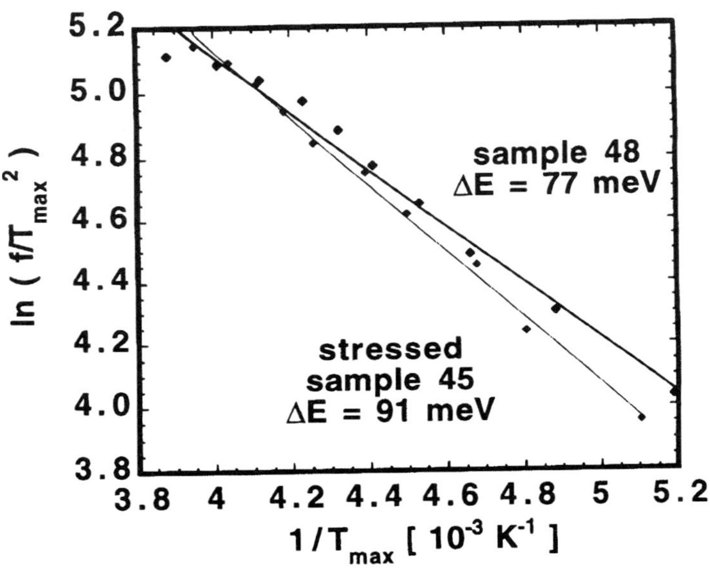

Figure 6.29 Arrhenious plots from admittance spectroscopy data for both stressed and unstressed LEDs.

6.8 EBIC ANALYSIS

Electron beam induced current (EBIC) imaging was used to identify the location of a low resistance path across an LED that had been stressed to failure. Figure 6.30 is an EBIC image taken with a 7 keV primary electron beam and a 109 current gain. Figure 6.31 is a secondary electron image of the area imaged in Figure 6.10. The EBIC image shows two distinct areas of increased current collection. The first is the edge of the p-contact area. Since this area allows the beam to come into contact with the junction, its presence in the EBIC images is expected. The only other signal is from the short across the junction. Figures 6.32 and 6.33 show higher magnification EBIC and secondary images, respectively, of the same defect region.

Figure 6.30 EBIC image of stressed LED (54X, 109 gain) at 7 keV.

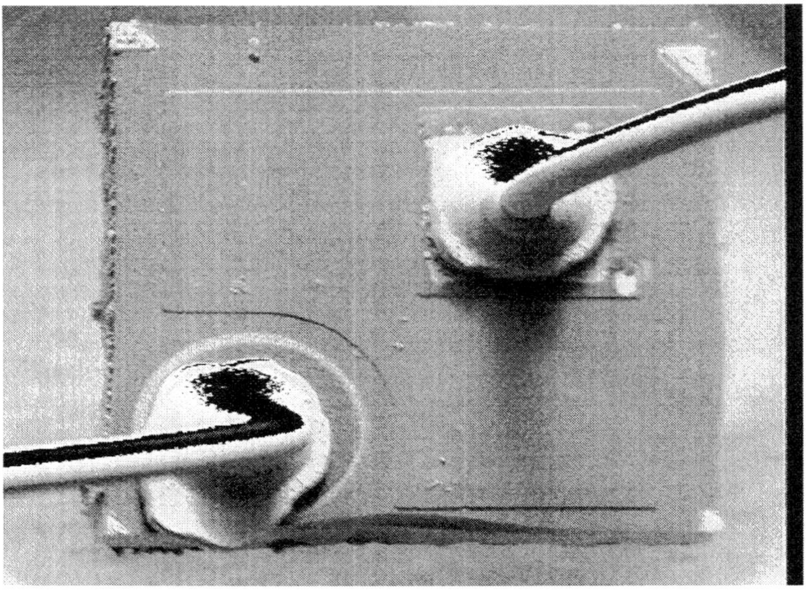

Figure 6.31 Secondary electron image of area in Figure 6.30. (54X)

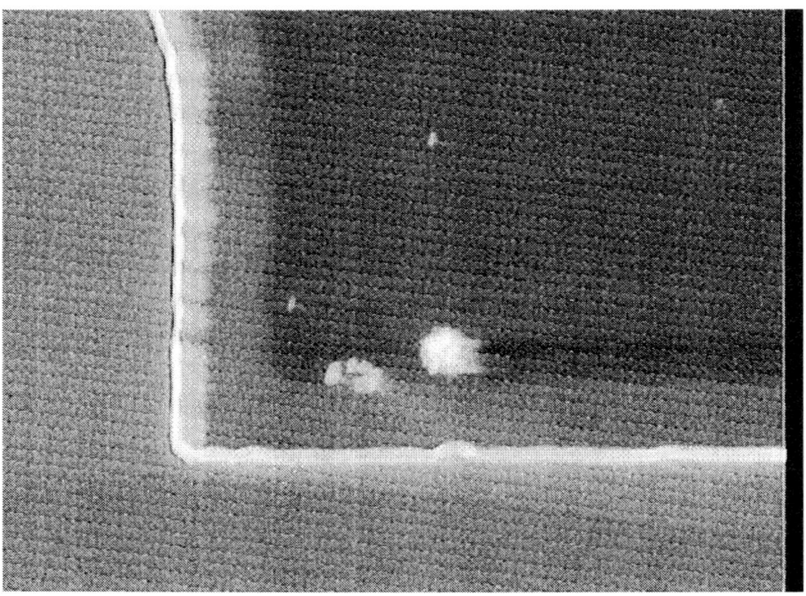

Figure 6.32 Higher magnification EBIC image of stressed LED (220X, 109 gain) at 7 keV.

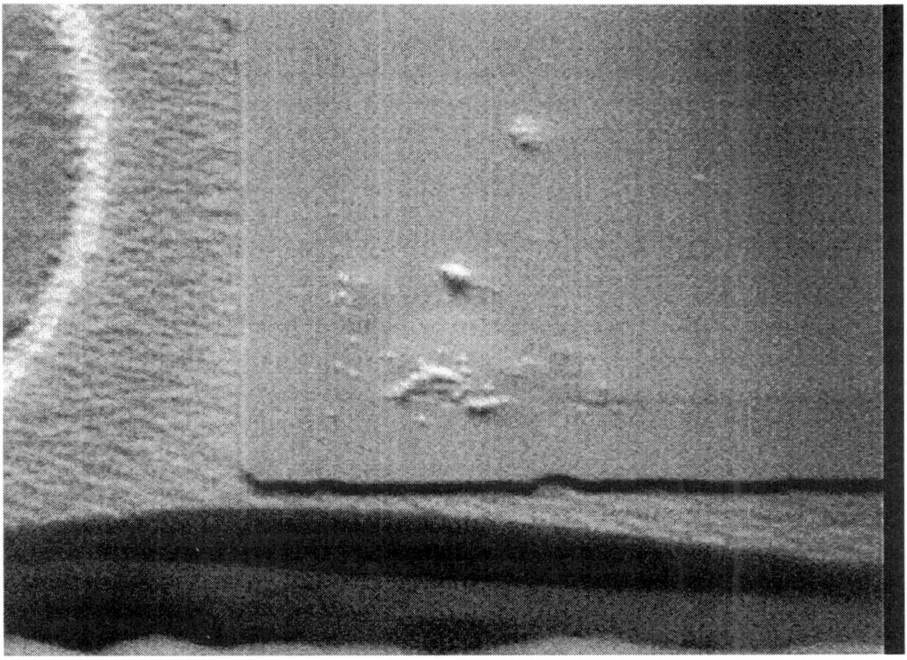

Figure 6.33 Secondary electron image (220X) of area in Figure 6.32.

Figure 6.34 shows a detail of the signal collected from the shorted region of the LED. This image shows that the high forward currents applied to this devices have caused metal from the p-contact to migrate across the junction. This figure also shows that the short consists of multiple metal fingers rather than a single thread. This observation supports the electrical data in that when the junction is reverse biased, the current would be carried almost entirely by the shorted area allowing these fingers to be fused open at relatively low voltages. Conversely, in forward bias, the junction would carry most of the current load.

Figure 6.34 EBIC image (905X) of stressed LED detailing shorted region.

Figure 6.35 is an EBIC image from the same LED using a 10 keV beam. This image shows several dark regions caused by debris from the decapsulation process attenuating the electron beam as well as a strong signal from the entire junction area. At this energy, the electron beam penetrates deep enough to allow the entire p-n junction area to collect the generated electron-hole pairs. This energy does not, however, allow the location of the short to be imaged because of the strong signal from the p-n junction. At 7 keV, as shown in Figures 6.30, 32, and 34, the beam only penetrates deep enough to allow current collection in the vicinity of the short.

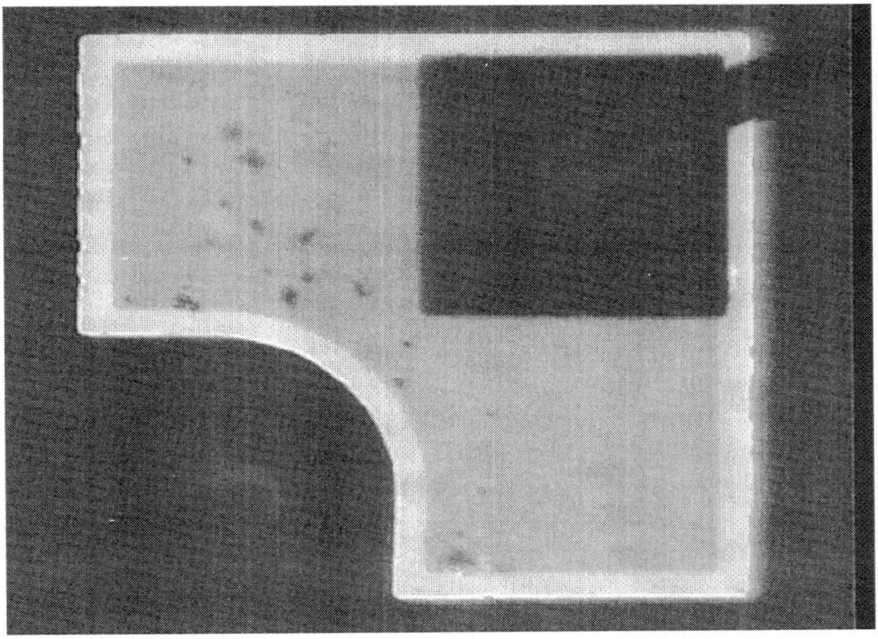

Figure 6.35 EBIC image of stressed LED (430X, 109 gain) at 10 keV.

The LED shown in Figures 6.30-6.35 was deprocessed to analyze the physical phenomenon causing the EBIC signal. Because of its small size and transparency, the LED had to be handled with care during the deprocessing operations. The gold p-contact was removed using aqua regia. Since the p-contact metallization is Au/Ni, the aqua regia successfully removed the gold and left a thin layer of nickel behind. Figure 6.36 shows two secondary images of the area identified using EBIC. The images clearly show that metal has migrated into the GaN under electrical stimulation.

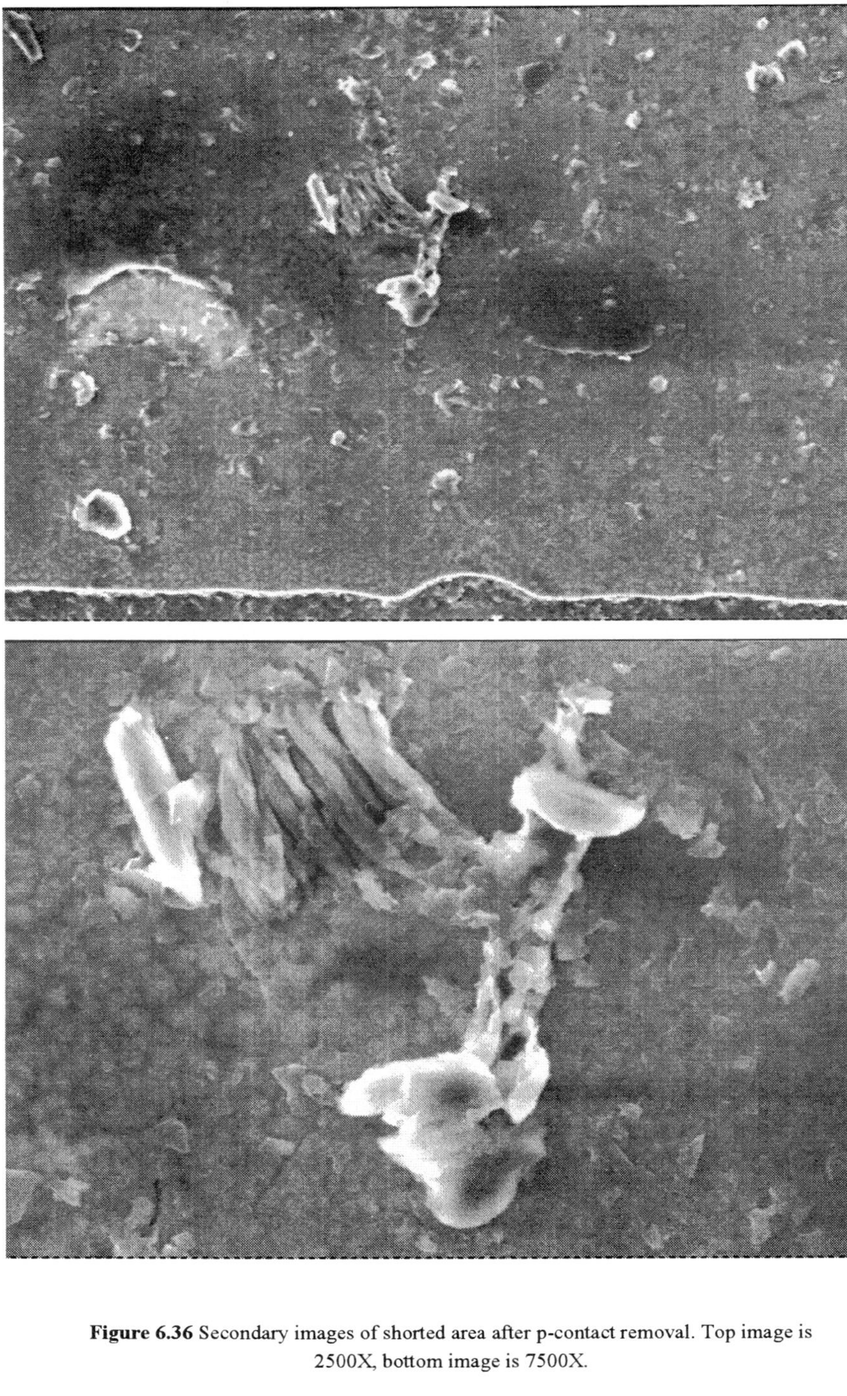

Figure 6.36 Secondary images of shorted area after p-contact removal. Top image is 2500X, bottom image is 7500X.

6.9 PULSED CURRENT STRESS EXPERIMENTS AND RESULTS ON QUANTUM WELL LEDS

The diodes under study were stressed using a pulse generator with a 100 ns pulse width at a repetition rate of 1 kHz in order to minimize joule heating. The light output was monitored by a power meter with a silicon diode detector. The I-V characteristics were measured with a pico-amp meter/voltage source.

Early in this study it was observed that the LEDs would degrade immediately if they were subjected to current pulses with amplitudes above 6 A. The degradation took about 1 s to occur and was accompanied by a visible electrical discharge between the electrodes. At these currents, the voltage drop across the diode is above 70 V. In order to investigate the effects of longer term exposure to reduced currents, it was found that using pulsed current amplitudes less than 6 A would cause a very slow degradation allowing the diode to continue working for between 24 and 90 hours. Several LEDs were stressed at a forward current of 5 A with a duty cycle of 10-4 per cent.

In Figure 6.37, the intensity of light emission from a diode subjected to high current pulses at 4.5 A and 5.8 A is shown as a function of time. Two different regions can be distinguished in Figure 6.37. First, in the first approximately 5 hours, a fast degradation is observed with the slope –0.17%/h. This is followed by a second region for times longer than 5 hours where the slope is –0.03%/h. By using a simple extrapolation from this data, the lifetime of the diode should be as long as 300 hours, but in reality, the diodes degraded rapidly after 24 to 90 hours. Analysis of the damage on these LEDs revealed the same kind of damage that was found on LEDs which had been subjected to currents above 6 A. This implies that the same degradation mechanism is responsible for the failure but the time to failure has been significantly increased by stressing the LEDs at lower currents.

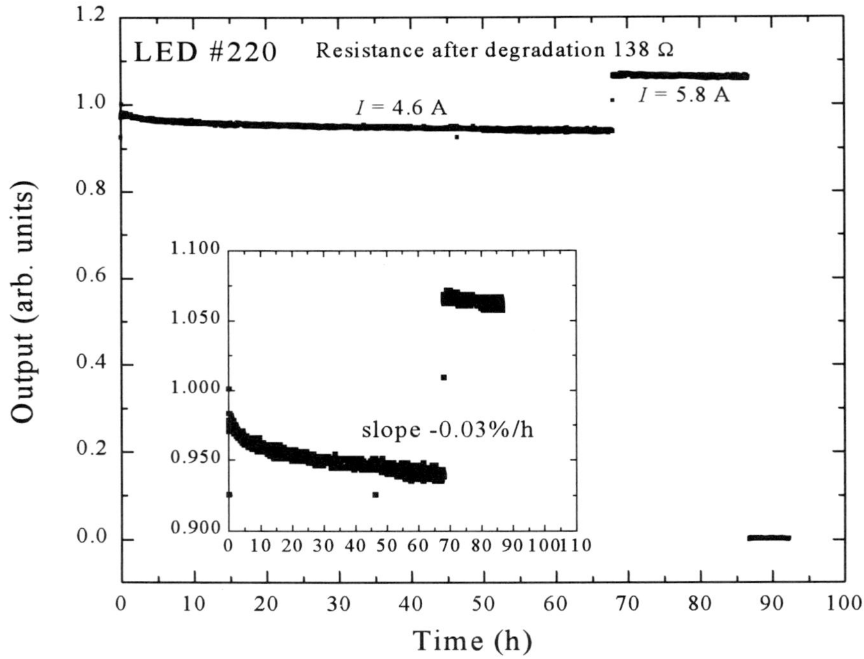

Figure 6.37 Light output of a stressed LED (#220) as a function of time.

In Figure 6.38, the I-V characteristics of an LED (#219) both before and after being subjected to high current stress are shown. It is characteristic of these LEDs that the reverse current leakage increases quickly when first subjected to stress but slows considerably as the stress is continued. This behavior is similar to that that observed in the optical power (faster degradation at the beginning and slower degradation after). Interestingly, the I-V curve shown in Figure 6.38 at 22 hours was measured just before the diode failed demonstrating that neither the electrical nor the optical measurements could predict that catastrophic degradation was about to occur. This observation was consistent in all of the LEDs subjected to pulsed currents less than 6 A and indicates that the device parameters which describe the GaN-based portion of these LEDs are not affected by the pulsed current stress.

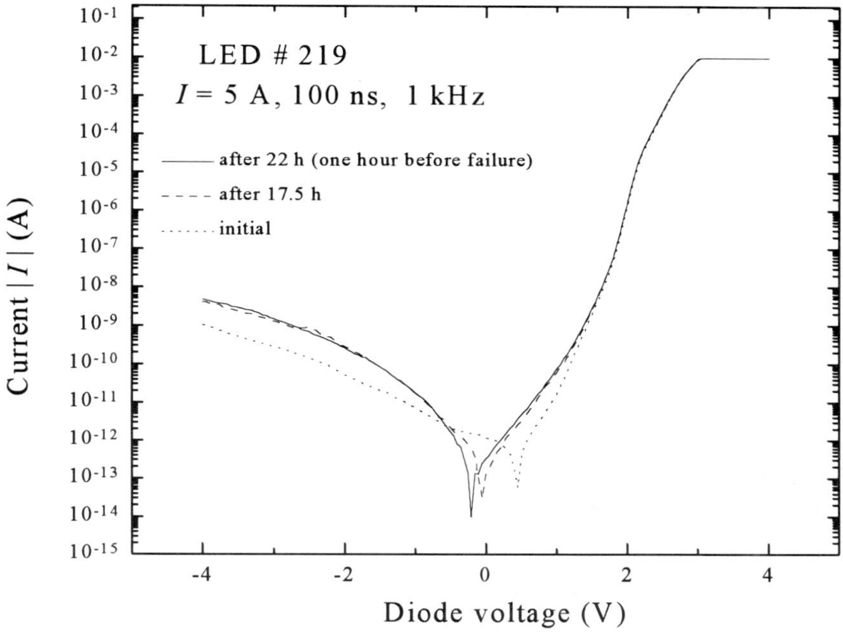

Figure 6.38 I-V characteristic of an LED (#219) before and after degradation.

6.10 FAILURE ANALYSIS OF DEGRADED QUANTUM WELL LEDS

Three of the devices (numbers 202, 219, and 220) which were stressed to the point of failure (identified by a sudden and complete loss in light output) were analyzed to determine the cause of the degradation. These LEDs lasted several hundred hours under stress before failure. The I-V characteristics of the three devices were all linear with slopes indicating that the LEDs had resistive shorts in the 18 Ω to 140 Ω range. A fourth device, #221, was subjected to the same stress, but it was removed from the test prior to failure. The I-V characteristics of #221 showed no significant changes but the LED was decapsulated anyway for comparison to the other three devices. Figure 6.39 shows an optical micrograph of LED #202 after decapsulation. It is apparent from Figure 6.39 that the LED has been severely damaged by the pulsed current stress. Further inspection of the device indicated that the observed damage was mostly the plastic encapsulation material whose material properties had been altered by the testing to the point where it was not removed during the decapsulation process. Since hot sulfuric acid was used to remove the encapsulant, this indicated that the material close to the LED had changed. This type of insolubility is frequently observed in failure analysis on devices that have been exposed to high temperatures. In this degradation process, the packaging material burns leaving a conductive, carbon film against the die.

Thus, in this case, most of the damage may be contained in the plastic and not in the LED. The observations were consistent with those made on the other two samples that were stressed to failure, LEDs #219 and 220. A fluorescent microthermal image (FMI) (Barton, 1994) of LED #202 shows a hot spot that indicates the location of the short is shown in Figure 6.40.

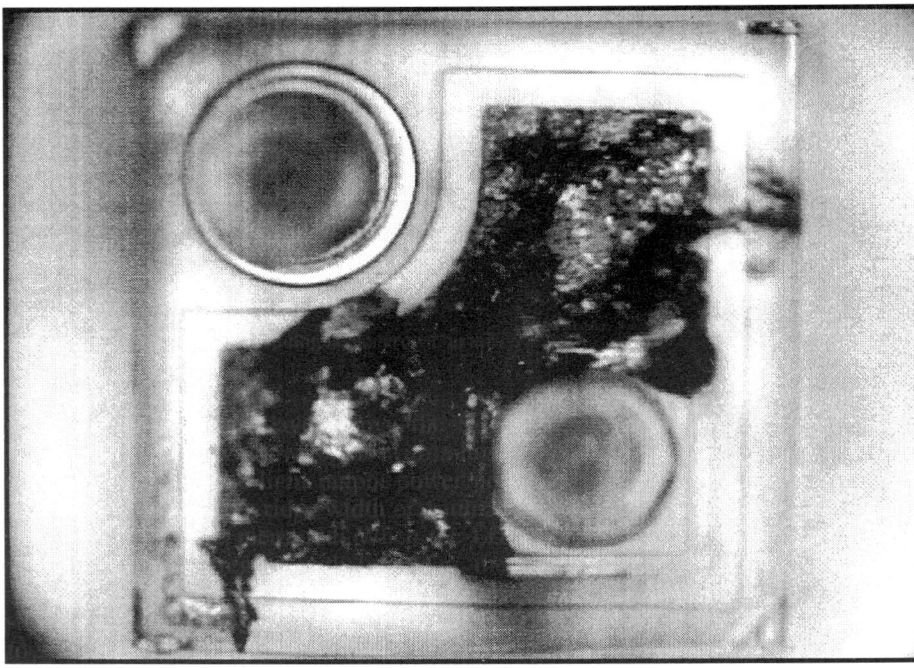

Figure 6.39 Optical micrograph of LED #202 after plastic removal (the darkened areas over the p-contact area are damaged plastic which could not be removed).

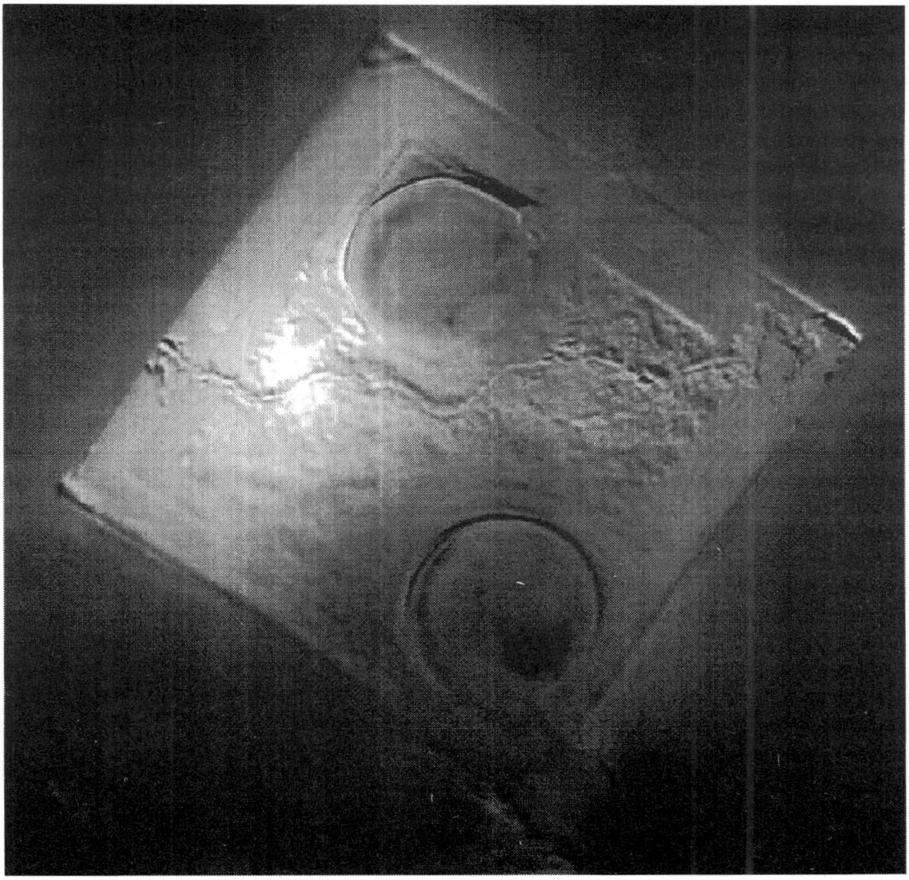

Figure 6.40 FMI thermal image localizing the short in LED #202.

Figure 6.41 shows an optical micrograph of LED #221 that was not stressed to failure. This image shows several very important details. First, as in the LEDs stressed to failure, some of the plastic packaging material was left behind after decapsulation. This means that even though there were no significant changes in light output or in I-V characteristics, there was a change in the plastic composition. Second, there are several black spots in the image, mostly at the edges of the semi-transparent metallization on the p-contact layer. These spots are damaged (burnt) plastic areas created when the junction below went into a non-permanent breakdown under the applied electrical stress. Under continued stress, enough of the damaged plastic sites will form to create a conductive layer that forms a short circuit across the LED. Further application of current to the LED causes catastrophic device failure where the contact metal is extruded across the junction like that located in Figure 6.40 with FMI.

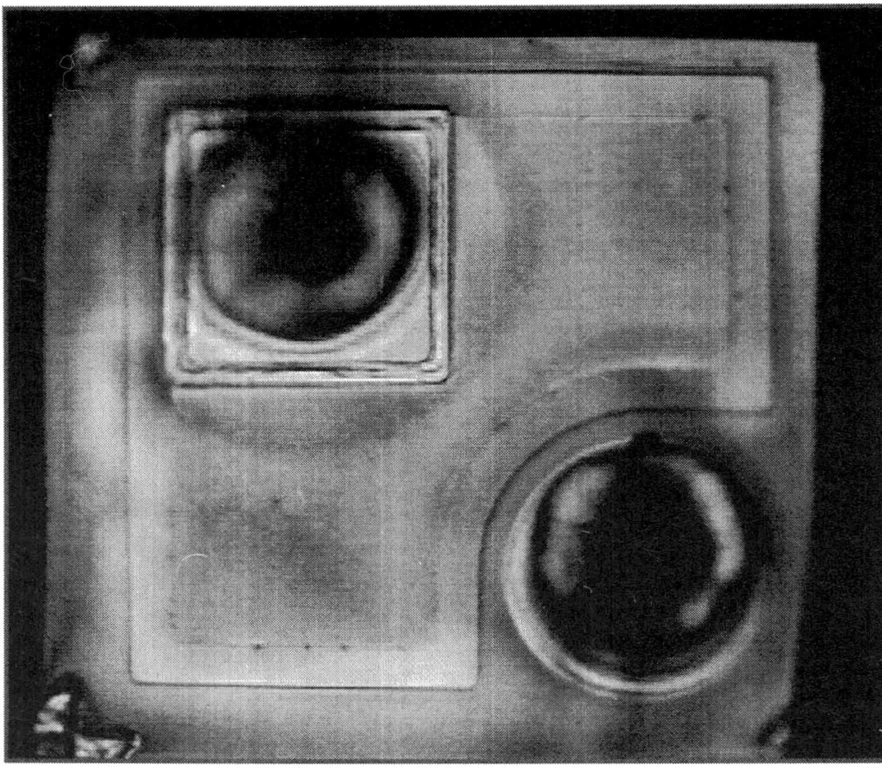

Figure 6.41 Optical micrograph of damaged plastic areas on LED #221.

6.11 DISCUSSION

Under normal operating conditions, Nichia diodes do not show any signs of rapid degradation. Lifetimes in excess of 10,000 hours are projected by the manufacturer. In the DH LED tests, the electromigration process is enhanced by the high voltage (tens of volts) required to drive high-current pulses through the device. Reversibility of the degradation of diode characteristics supports this interpretation. DLTS and admittance spcectroscopy data suggest that the applied voltage stress does not change the density or population of deep levels. Sugiura's work (Sugiura 1997a and 1997b) agrees with this conclusion by showing that unless the junction temperature exceeds about 400°C, defect motion in GaN-based systems will be inconsequential compared to that observed in other optial materials. Degradation in GaAs and InGaAsP systems can be attributed to defect motion even under normal operating conditions at relatively low temperatures. Egawa, *et.al.* (1997) has observed a defect motion-based degradation mechanism in GaN LEDs on sapphire substrates. Egawa's data is difficult to compare to the data in this review because the data in (Egawa *et al.*, 1997) was not based on Nichia grown LEDs. Also, the tests described in (Egawa *et al.*, 1997) were carried out under very high currents (up to 250 mA) and under dc conditions where very high junction temperatures can be reached (in spite of the copper headers that were used) where the defect mobilities can be significant and in accordance with Sugiura's data. The tests summarized in this review were carefully designed to minimize the effects of joule heating by using only short duty cycle pulsed current stress. The experiments on quantum well LEDs have supported this by demonstrating that the packaging material used on the commercial LEDs was the weak link (the LEDs tested in (Egawa *et al.*, 1997) were not packaged) and was responsible for the observed failures. Furthermore, the junction temperature measurements showed that heat generated by the LED could degrade the transparency of the package material well before any significant defect motion was achieved.

Diode lasers will require very high current densities which may, in turn, cause metal migration or other degradation related to heating effects at the contacts. In order to prevent this from occurring, it is essential to design the lasers to operate at low voltages and to minimize all contact resistances.

6.12 SUMMARY

In this chapter, the topic of life testing and pulsed stress testing of GaN-based DH LEDs have been reviewed. The results from over 7,000 hours of life testing of commercially available Nichia LEDs has confirmed the long lifetime potential of devices made in this material system. The results have identified thermal degradation of the package material to be a concern for high current, cw operation at high ambient temperatures where the combination of ambient temperature and applied power raise the junction temperature to around 140°C. At low ambient temperatures and forward currents above about 50 mA, the UV output around 3.25 eV from the LED can adversely affect the transparency of the packaging material at the energies of desired emission, around 2.75 eV. The reported life test did successfully identify a single infant mortality failure during the second segment of the life test. This LED had latent mechanical damage which caused a sudden loss of light output under a 70 mA current stress.

The data collected on the earlier production, double heterojunction LEDs indicated a possible connection between the large number of crystalline defects and a tendency for metal to migrate from the p-contact across the junction and short out the device. The newer generation, quantum well LEDs showed a significant improvement in resistance to this type of stress and revealed that the limitation may be in the plastic packaging material and not in the diode itself. This work has demonstrated that the slow degradation in light output intensity observed on the quantum well LEDs is likely not due to the creation of internal defects but rather to changes in the transparency of the plastic package material caused during high current pulses which are not destructive to the LED. The data collected from earlier, constant current life tests has shown that the plastic can change in transparency due to a combination of joule heating and elevated ambient temperatures. The conditions in the final stage of the CW life test showed that the package would be adversely affected by prolonged exposure to temperatures around 140°C. The analysis of LEDs subjected to pulsed current stress showed that regions on the LED may have gone above 140°C during a single pulse thereby damaging the plastic in that region. The observed rapid degradation occurs when the plastic has degraded to the point where its conductivity increases and facilitates burn-out of the LED. At this point low resistance, ohmic shorts are observed across the LED.

6.13 REFERENCES

Barton, D.L., 1994, *Proc. 20th ISTFA*, Los Angeles, CA, p. 87.

Barton, D.L., Zeller, J., Phillips, B.S., Chiu, P.-C., Askar, S., Lee, D.-S., Osiñski, M. and Malloy, K.J., 1995a, *Proc. 33rd Annual IEEE Int. Reliability Phys. Symp.*, Las Vegas, Nevada, 4-6 April 1995, Paper 3B.3, p. 191.

Barton D. L., *et.al.*, 1995b, *IRPS.*

Cherns, D., Young, W.T., Steeds, J.W., Ponce, F.A. and Nakamura, S., 1997a, *J. Cryst. Growth*, **178**, p. 201.

Cherns, D., Young, W.T. and Ponce, F.A., 1997b, *Mater. Sci. Eng. B, Solid State Mater. Adv. Technol.*, **50**, p. 76.

Chien, F.R., Ning, X.J., Stemmer, S., Pirouz, P., Bremser, M.D. and Davis, R.F., 1996, *Appl. Phys. Lett.*, **68**, p. 2678.

Christiansen, S., Albrecht, M., Dorsch, W., Strunk, H.P., Pelzmann, A., Mayer, M., Kamp, M., Ebeling, K.J., Zanotti-Fregonara, C. and Salviati, G., 1997, *Mater. Sci*, B - *Solid State Mater. Adv. Technol.*, **43,** p. 296.

David, P.K. and Montarni, G.C., 1992, *European Transactions on Electrical Power Engineering*, **2**, p. 187.

Egawa, T., Jimbo, T. and Umeno, M., 1997, *J. Appl. Phys.*, **82**, No. 11, p. 5816.

Fukuda, M., 1991, *Reliability and Degradation of Semiconductor Lasers and LEDs*, Artech House, Boston , Ch. 3.

Gotz, W., Johnson, N., Amano, H. and Akasaki, I., 1994, *Appl. Phys. Lett.*, **65**, No. 4, p. 463.

Hacke, P., Detchprohm, T., Hiramatsu, K., Sawaki, N., Tadatomo, K. and Miyake, K., 1994, *J. Appl. Phys.*, **76**, No. 1, p. 304.

Helms, C.J., Berg, N.H., Barton, D.L. and Osiñski, M., 1996, *Testing, and Reliability of Semiconductor Lasers*, San Jose, California, 31 Jan. – 1 Feb. 1996, eds. Fallahi, M., and Wang, S. C., *SPIE Proc.*, **2683**, p. 74.

Jackson, W.B., Amer, N.M., Boccara, A.C. and Fourner, D., 1981, *Appl. Optics*, Vol. 20, No. 8, p. 1333.

Lester, S.D., Ponce, F.A., Craford, M.G. and Steigerwald, D.A., 1995, *Appl. Phys. Lett.*, **66**, p. 1249.

Liliental-Weber, Z., Sohn, H., Newman, N. and Washburn, J., 1995, *J. Vac. Sci. Technol. B: Microelectronics and Nanometer Structures*, **13**, p. 1578.

Nakamura, S., 1994a, *J. Cryst. Growth*, **145**, p. 911.

Nakamura, S., 1994b, *2nd Workshop on Wide Bandgap Nitrides*, St. Louis, MO., October 17-18.

Nakamura, S., Mukai, T., and Senoh, M., 1994, *Appl. Phys. Lett.,* **64**, p. 1687.

Nakamura, S., 1995, *J. Vac. Sci. Technol. A - Vacuum, Surfaces, and Films*, **13**, p. 705.

Nakamura, S., 1996, *Third St. Louis Workshop on Wide Bandgap Nitrides*, St. Louis, Missouri, 13-15 March 1996.

Ning, X.J., Chien, F.R., Pirouz, P., Yang, J.W. and Khan, M.A., 1996, *J. Mater. Res.*, **11**, p. 580.

Pelzmann, A., Mayer, M., Kirchner, C., Sowada, D., Rotter, T., Kamp, M., Ebeling, K.J., Christiansen, S., Albrecht, W., Strunk, Holländer, B. and Mantl, S., 1996, *Mater. Res. Soc. Internet J. Nitride Semicond. Res.*, **1**, Art. 40.

Ponce, F.A., Cherns, D., Young, W.T., Steeds, J.W. and Nakamura, S., 1997, *III-V Nitrides, Boston, Massachusetts*, 2-6 Dec., 1996, eds. Ponce, F. A., Moustakas, T. D., Akasaki, I. and Monemar, B. A., *Mater. Res. Soc. Symp. Proc.*, **449**, p. 1173.

Potin, V., Vermaut, P., Ruterana, P. and Nouet, G., 1998, *J. Electron. Mater.*, **27**, p. 266.

Qian, W., Skowronski, M., Doverspike, K., Rowland, L.B. and Gaskill, D.K., 1995a, *J. Cryst. Growth*, **151**, p. 396.

Qian, G.S. Rohrer, W., Skowronski, M., Doverspike, K., Rowland, L.B. and Gaskill, D.K., 1995b, W. Qian, M. Skowronski, K. Doverspike, L.B. Rowland, and D.K. Gaskill, *Appl. Phys. Lett.*, **67**, p. 2284.

Sugiura, L., 1997a, *J. Appl. Phys.*, **81**, p. 1633.

Sugiura, L., 1997b, *Appl. Phys. Lett.*, **70**, p. 1317.

Vennéguès, P., Beaumont, B., Vaille, M. and Gibart, P., 1997, *Appl. Phys. Lett.*, **70**, p. 2434.

Wu X.H., Brown L.M., Kapolnek D., Keller S., Keller B., DenBaars S.P., and Speck J. S., 1996, *J. Appl. Phys.*, **80**, p. 3228.

Wu, X.H., Fini, P., Tarsa, E.J., Heying, B., Keller, S., Mishra, U.K., DenBaars, S. P. and Speck, J.S., 1998, *J. Cryst. Growth* **189/190**, p. 231.

Zeller, J., Barton, D.L., Askar, S., Lee, D.-S., Chu, A.-S., Osiñski, M. and Malloy, K.J., 1995, *CLEO '95 Conference on Lasers and Electro-Optics*, Baltimore, Maryland, 22-26 May, 1995, Paper CMH6, p. 35.

Zeller, J., Eliseev, P.G., Sartori, P., Perlin, P. and Osiñski, M, 1996, *Gallium Nitride and Related Materials*, Boston, Massachusetts, 27 Nov. - 1 Dec., 1995, eds. Ponce, F.A., Dupuis, R.D., Nakamura, S. and Edmond, J.A., *Mater. Res. Soc. Symp. Proc.*, **395** p. 937.

Chapter 7
Development and Future Prospects of InGaN-based LEDs and LDs

Shuji Nakamura

7.1 PROPERTIES OF InGaN-BASED LEDs

7.1.1 Introduction

There has been much interest in light-emitting diodes (LEDs), which emit from ultraviolet (UV) to red wavelengths, fabricated from III-V nitride compound semiconductors. Major developments in wide-bandgap III-V nitride semiconductors have recently led to the commercial production of high-efficiency UV/blue/green/amber LEDs (Nakamura *et al.*, 1995a; 1995b; Mukai *et al.*, 1998a; 1998b). All of these light-emitting devices use an InGaN active layer instead of a GaN active layer because of the difficulty in fabricating highly efficient light-emitting devices using a GaN active layer (Morkoç *et al.*, 1994; Ponce and Bour, 1997; Mukai *et al.*, 1998b) Several groups recently reported that the dislocations are nonradiative recombination centers in GaN (Rosner *et al.*, 1997; Sugahara *et al.*, 1998). In InGaN, however, the role of dislocations has not yet been clarified. Highly efficient blue/green InGaN single-quantum-well (SQW) or multi-quantum-well (MQW) structure LEDs and LDs have been fabricated directly on a sapphire substrate in spite of a high dislocation density of 1-10 x $10^{10}cm^2$ originating from a large lattice mismatch between GaN and the sapphire substrate (Nakamura *et al.*, 1995b; Nakamura and Fasol, 1997; Mukai *et al.*, 1998a). Due to the high efficiency of the LEDs grown on the sapphire substrates, the dislocations in InGaN do not appear to work as a nonradiative recombination center (Lester *et al.*, 1995). Here, the present performance of UV/blue/green/amber LEDs and the role of the dislocations in the InGaN are described.

7.1.2 Amber LEDs

The longest-wavelength InGaN-based LEDs were amber InGaN single-quantum-well (SQW) structure LEDs with an emission wavelength of 590 nm and an output power of 0.5 mW, reported previously by the present author's group (Nakamura *et al.*, 1995a). However, this output power was lower than that of commercially available amber AlInGaP LEDs (0.7 mW) (Kish and Fletcher, 1997). Amber AlInGaP LEDs are used for many applications such as roadway traffic light signals, automobile signal lighting, optical fiber communications and various displays (Hodapp, 1997). Using current AlInGaP LEDs, however, the output power is reduced dramatically due to a poor band offset between the active layer and cladding layers when the ambient temperature is increased (Kish and Fletcher, 1997). Here, amber InGaN LEDs with a high luminous efficiency and an almost temperature-independent output power are described (Mukai *et al.*, 1998a). Here, we call InGaN SQW structure LEDs InGaN LEDs.

III-V nitride films were grown by the two-flow metalorganic chemical vapor deposition (MOCVD) method (Nakamura, 1991). The growth was conducted at atmospheric pressure. Sapphire with (0001) orientation (C face) and two-inch diameter was used as a substrate. The amber InGaN LED device structures consist of a 300-Å-thick GaN buffer layer grown at a low temperature (550°C), a 0.7-μm-thick layer of undoped GaN, 3.3-μm-thick layer of n-type GaN:Si, a 400-Å-thick layer of undoped GaN (current spreading layer), a 25-Å-thick active layer of undoped InGaN, a 300-Å-thick layer of p-type $Al_{0.2}Ga_{0.8}N$:Mg and a 0.2-μm-thick layer of p-type GaN:Mg. Figure 7.1 shows the structure of the amber LED.

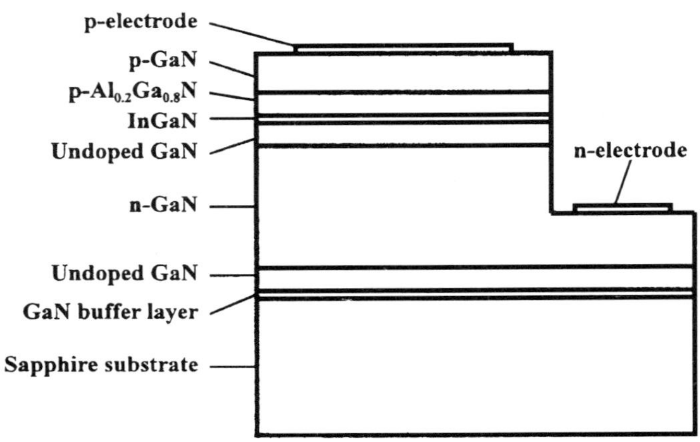

Figure 7.1 The structure of amber InGaN LEDs.

It was difficult to determine the exact In composition of the InGaN active layer due to the weak signal intensity in X-ray diffraction and photoluminescence measurements. Unlike the previous InGaN-based LEDs (Nakamura *et al.*, 1995a; 1995b; Nakamura and Fasol, 1997), the n-type GaN:Si layer was replaced by undoped GaN and n-GaN:Si. A low resistivity Si-doped GaN layer with a resistivity of 6×10^{-3} Ω·cm and carrier concentration of 4×10^{18}cm^{-3} was used as a contact layer for the n-type electrode. In order to spread the current uniformly throughout the active layer, a high-resistivity undoped GaN layer was inserted between the active layer and the contact layer. Using this structure, the operating voltage was reduced resulting from the lower resistivity and the higher carrier concentration of the contact layer in comparison with the previous structure. Fabrication of LED chips was accomplished as previously reported (Nakamura *et al.*, 1995a; 1995b; Nakamura and Fasol, 1997). The characteristics of the LEDs were measured under direct-current (DC)-biased conditions at room temperature, except for the measurement of the temperature dependence of the output power.

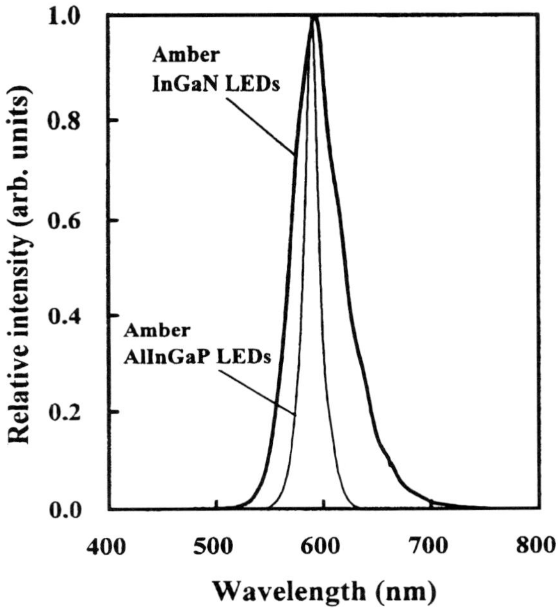

Figure 7.2 The emission spectra of amber InGaN and AlInGaP LEDs at a forward current of 20 mA.

Figure 7.2 shows the emission spectra of amber InGaN and AlInGaP LEDs at a forward current of 20 mA at RT (Mukai *et al.*, 1998a). For comparison, the spectrum of commercially available amber AlInGaP LED fabricated on an absorbing substrate of GaAs (Type: HLMP-DL32, Hewlett Packard) is also shown (Kish and Fletcher, 1997). The peak wavelength and the full-width at half-maximum (FWHM) of the emission spectra of the amber InGaN LEDs were 594 nm and 50 nm, respectively, and those of amber AlInGaP LEDs were 591 nm and 17 nm, respectively. The spectrum width of the InGaN LEDs is relatively broad mainly due to a phase separation of InGaN during growth, which was reported recently (Chichibu *et al.*, 1996; 1997a; Narukawa *et al.*, 1997a; 1997b). In view of the spectrum width, the amber AlInGaP LEDs seem to have superior color purity. With regard to the application of LEDs, however, in the color range from red to yellow, it is hard for the human eye to detect the difference in color purity caused by a spectrum width difference of 16 nm to 50 nm, as shown in Figure 7.2.

Figure 7.3 shows the output power of amber InGaN and AlInGaP LEDs as a function of the ambient temperature from –30°C to +80°C. The output powers of amber InGaN and AlInGaP LEDs at 25°C were 1.4 mW and 0.66 mW, respectively. The output power of InGaN LEDs was about twice as high as that of AlInGaP LEDs. The output power of each LED was normalized to 1.0 at 25°C. When the ambient temperature was increased from RT to 80°C, the output power of amber AlInGaP LEDs decreased dramatically to half that at RT due to a carrier overflow caused by a small band offset between the active layer and cladding layers (Kish and Fletcher, 1997). In the AlInGaP system, the band offset is small under a lattice-matched condition between the AlInGaP epilayer and GaAs substrate. On the other hand, the temperature dependence of amber InGaN LEDs is relatively weak. When the ambient temperature is increased from RT to 80°C, the output power of amber InGaN LEDs only decreases to 90% of that at RT, probably due to a small carrier overflow caused by a large band offset between the active layer and cladding layers. In terms of the temperature dependence of the LEDs, InGaN LEDs are superior to the AlInGaP LEDs.

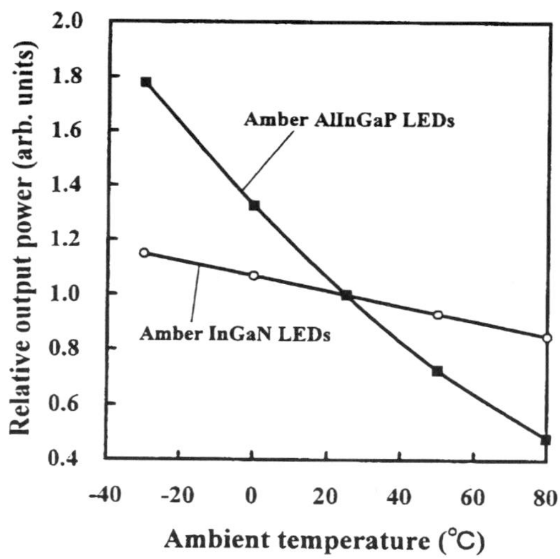

Figure 7.3 The output power of amber InGaN and AlInGaP LEDs as a function of the ambient temperature from – 30°C to + 80°C. The output power of each LED was normalized to 1.0 at 25°C.

The lifetime test of the amber InGaN LEDs was performed under a forward current of 20 mA at RT. After 1,000 hours of operation, no degradation of the LEDs was observed. The typical forward voltage was 3.3 V at a forward current of 20 mA. Luminous efficiency is the luminous flux emitted by the device per input power (lm/W). This parameter is an important measure of performance of visible LEDs and other light sources. The luminous efficiency of amber InGaN LEDs (10 lm/W) was almost the same as that of amber AlInGaP LEDs (8 lm/W). Optical response time was also measured. The rise time and fall time of the amber InGaN LEDs were 20 ns and 30 ns, respectively, almost the same as those of amber

AlInGaP LEDs (27 ns and 24 ns). In summary, amber InGaN LEDs with an emission wavelength of 594 nm and luminous efficiency of 10 lm/W were developed. These values were almost the same as those for commercially available amber AlInGaP LEDs. However, there was a large difference in the temperature dependence of the output power between InGaN and AlInGaP LEDs. When the ambient temperature was increased from RT to 80°C, the output power of AlInGaP LEDs decreased dramatically. On the other hand, the output power of InGaN LEDs remained almost constant. In addition, the output power of InGaN LEDs was about twice as high as that of AlInGaP LEDs at RT. For InGaN-based emitting devices, a sapphire substrate is used, which is a safe, nontoxic material. For AlInGaP-based emitting devices, however, the GaAs substrate, which is a toxic material, is used. Taking the environmental issues into consideration, only InGaN-based LEDs, which are safe, would be used for any kind of applications in the future. In addition, the technology of InGaN-based emitting devices is still at an early stage. Thus, InGaN-based LEDs with improved characteristics will be obtained very soon.

7.1.3 UV/Blue/Green LEDs

Because of low symmetry, the wurtzite system such as GaN-based materials, displays pyroelectric and piezoelectric behavior (Chichibu *et al.*, 1997b; Takeuchi *et al.*, 1997; Nardelli *et al.*, 1997; Im *et al.*, 1998). The macroscopic polarization in the material comprising the active region of the SQW or multi-quantum-well (MQW) structure gives rise to a net electric field perpendicular to the plane of the well. This field, if strong enough, will induce a spatial separation of the electron and hole wave functions in the well. As the charge density becomes concentrated near the walls of the well, the wave function overlap decreases and the interband recombination rate decrease (the quantum-confined Stark effect (QCSE)). The blueshift of the electroluminecence (EL) of the InGaN blue/green SQW-structure LEDs with increasing operating current may be explained by the QCSE resulting from piezoelectric fields induced by the lattice mismatch (Im *et al.*, 1998). However, a higher efficiency of the LEDs with increasing strain in the SQW upon increasing the In content in the InGaN well layers, was observed (Nakamura and Fasol, 1997; Mukai *et al.*, 1998b). These phenomena cannot be explained by only the QCSE. It may be that the localization effects induced by composition fluctuations (Chichibu *et al.*, 1996; 1997a; Narukawa et al., 1997a; 1997b) must overcome these intrinsic limitations due to the piezoelectric field. The localization induced by In composition fluctuation seems to be a key factor in the high efficiency of the InGaN-based LEDs. However, the emission mechanism of these InGaN-based LEDs is still not clear at present. Here, current and temperature dependencies of EL of InGaN UV/blue/green SQW-structure LEDs were measured to study these emission mechanisms. We call InGaN UV/blue/green SQW structure LEDs as UV/blue/green LEDs for a simplicity.

The UV/blue/green LED device structures were almost identical to that of amber LED in Figure 7.1 except for the InGaN well layer. The In composition of the InGaN well layer was nearly zero for UV LEDs (Mukai *et al.*, 1998b), 0.2 for blue LEDs, 0.45 for green LEDs (Nakamura *et al.*, 1995a; 1995b). Here, the In composition was determined assuming that the bowing parameter of InGaN was 1 eV (Osamura *et al.*, 1975). However, recent studies revealed that the bowing parameter of InGaN is much higher than 1 eV (Amano *et al.*, 1997; McCluskey *et al.*, 1998). In that case, the In content in the InGaN layer should be much smaller than the above values. For UV LEDs, the thickness of the undoped InGaN well layer was changed to 55 Å and $Al_{0.2}Ga_{0.8}N$ barrier layer was used for both the n- and p-type barrier layers. The characteristics of UV/blue/green LEDs were measured under pulsed current-biased conditions with a frequency of 200 Hz and a duty ratio of 1% to reduce heat generation.

Figure 7.4 shows the electroluminescence (EL) of green LEDs. Figure 7.4 (a) shows the operating current dependence of the EL. The blueshift is observed with increasing current due to a band-filling effect of the localized energy states caused by In composition fluctuation in the InGaN well layer. Figure 7.4 (b) shows the ambient temperature dependence of the EL. There is no change of the EL on changing the ambient temperature. This is a peculiar result because the band-gap energy of semiconductor materials should become smaller with increasing temperature. Then, the emission peak wavelength should becomes longer with increasing temperature. The blue LEDs with an emission peak wavelength of 475 nm showed the same change of the EL as did green LEDs. A blue shift of the emission peak wavelength with increasing current and no change of the EL with increasing temperature were observed.

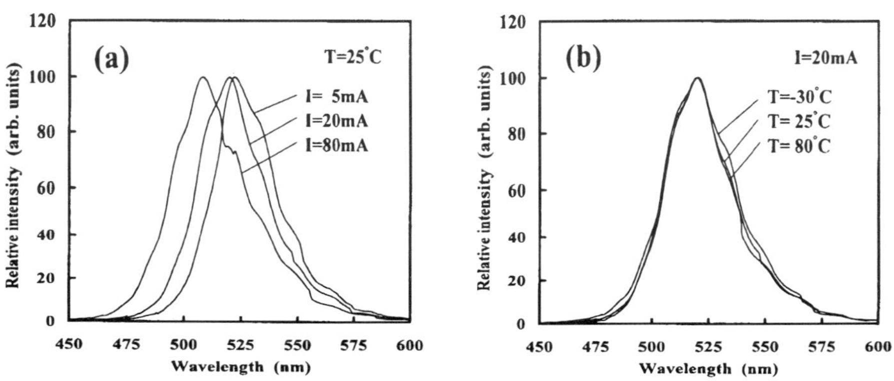

Figure 7.4 Operating current (a) and ambient temperature (b) dependence of the EL of InGaN green SQW-structure LEDs.

Figure 7.5 shows results for the conventional AlInGaP red MQW LEDs (Toshiba TLRH 157P). With increasing temperature, a redshift of the emission peak wavelength is observed due to the temperature dependence of the band-gap energy, as shown in Figure 7.5 (b). On the other hand, there is no change of the emission peak wavelength with increasing operating current due to the lack of localized energy states, as shown in Figure 7.5 (a). Based on these results, the band-gap energy of InGaN seems to be independent of temperature. In order to investigate these results further, we studied UV LEDs.

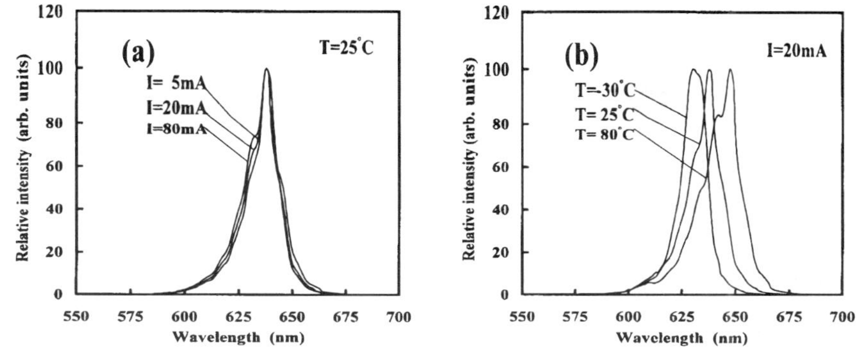

Figure 7.5 Operating current (a) and ambient temperature (b) dependence of the EL of conventional AlInGaP red MQW LEDs (Toshiba TLRH 157P).

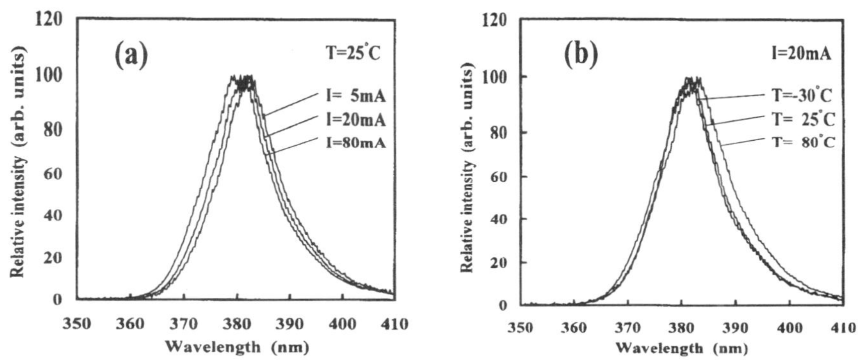

Figure 7.6 Operating current (a) and ambient temperature (b) dependence of the EL of InGaN UV SQW-structure LEDs with an emission peak wavelength of 380 nm.

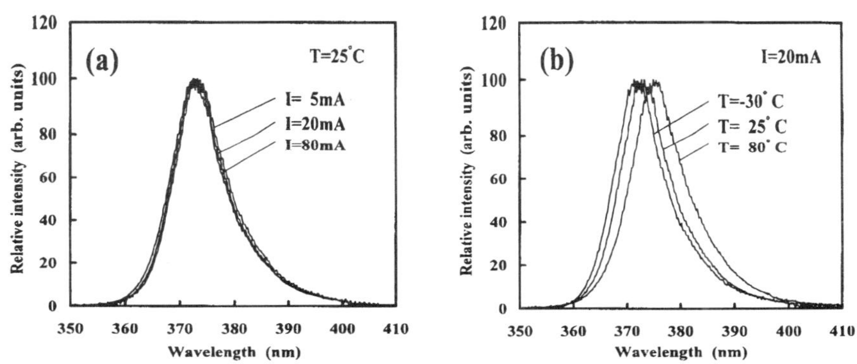

Figure 7.7 Operating current (a) and ambient temperature (b) dependence of the EL of InGaN UV SQW-structure LEDs with an emission peak wavelength of 375 nm.

Figure 7.6 shows the EL of UV LEDs with an emission peak wavelength of 380 nm at a current of 20 mA. A small blueshift of the emission peak wavelength with increasing current is observed due to the small localization energy of the carriers caused by a small In composition fluctuation in the InGaN well layer, as shown in Figure 7.6 (a). In Figure 7.6 (b), almost no change of the emission peak wavelength is observed, the same as those of green and blue LEDs.

Figure 7.7 shows the EL of UV LEDs with an emission peak wavelength of 375 nm. At this emission peak wavelength, the EL showed the drastic change. The emission peak wavelength did not show any change with increasing current, as shown in Figure 7.7 (a). On the other hand, a redshift of the emission peak wavelength was observed with increasing temperature. These emission peak wavelength changes are almost the same as those of conventional AlInGaP LEDs, as shown in Figure 7.5. When we measured the EL of UV LEDs with an emission peak wavelength of 370 nm and a small amount of In (nearly zero), the change of the EL on current and temperature was almost the same as that in Figure 7.7. Thus, there is an abrupt change of EL at the emission peak wavelength of 375 nm. In other words, there is a change of the emission mechanism at the emission peak wavelength of 375 nm. The UV LEDs with an emission peak wavelength of 375 nm which have small localized energy states resulting from small fluctuation in In composition. This means that emission mechanism is dominated by a conventional band-to-band emission as in AlInGaP LEDs. However, the results in Figures 7.4 and 7.7 of green/blue/UV LEDs cannot be explained by the simple band-to-band emission mechanism. The large localized energy states due to the large In composition fluctuation seem to be a key factor in the blueshift with increasing forward current. However, the constant emission peak wavelength with increasing temperature cannot be explained by only the localized energy states. An additional mechanism is required to explain this constant emission peak wavelength with increasing temperature. When the temperature is increased, the energy distribution of carriers at the localized energy states moves to higher energy levels due to a small density of the states. These high-energy carriers may compensate the redshift of the emission peak wavelength due to the bandgap narrowing caused by increasing temperature.

Figure 7.8 shows the relative output power as a function of the emission peak wavelength of UV LEDs. The output power of the UV LEDs containing a small amount of In in the active layer, with the emission wavelength of 380 nm was about 20 times higher than that containing no In with the emission wavelength of 368 nm. With decreasing In composition in the active layer, which means that the emission peak wavelength become shorter, the output power decreases dramatically. Thus, high-power UV LEDs can be obtained only when using the InGaN active layer with relatively high In composition. This is related to the large localized energy states caused by In composition fluctuation in the InGaN well layer (Chichibu *et al.*, 1996; 1997a; Narukawa *et al.*, 1997a; 1997b). When electrons and holes are injected into the InGaN active layer of the LEDs, these carriers are captured by the localized energy states before they are captured by the nonradiative recombination centers caused by the large number of threading dislocations (TDs). At these large localized energy states, localized excitons with a relatively high binding energy due to a quantum well structure are formed to recombine radiatively. Without In in the active layer, there are no In composition fluctuations that form the large localized energy state in the InGaN active layer. Thus, the QCSE resulting from the piezoelectric field due to the strain becomes dominant (Takeuchi *et al.*, 1997; Nardelli *et al.*, 1997; Im *et al.*, 1998). This field, if strong enough, will induce a spatial separation of the electron and hole wave functions in the well. Then, the wave function overlap decreases and the interband recombination rate is reduced. Also, there is no In composition fluctuation where carriers are captured by the localized energy states before they are captured by the nonradiative recombination centers. In addition, the lifetime of carriers becomes much longer due to the QCSE, which helps carriers recombine easily at the non-radiative recombination centers originating from a large number of TDs. Thus, the efficiency of the UV LEDs becomes extremely low when the active is GaN or InGaN with a small amount of In, as shown in Figure 7.8. When the emission peak wavelength becomes shorter than 371 nm, the output power decreases dramatically probably due to an additional reason of a self absorption of the light by n- and p-type GaN contact layers.

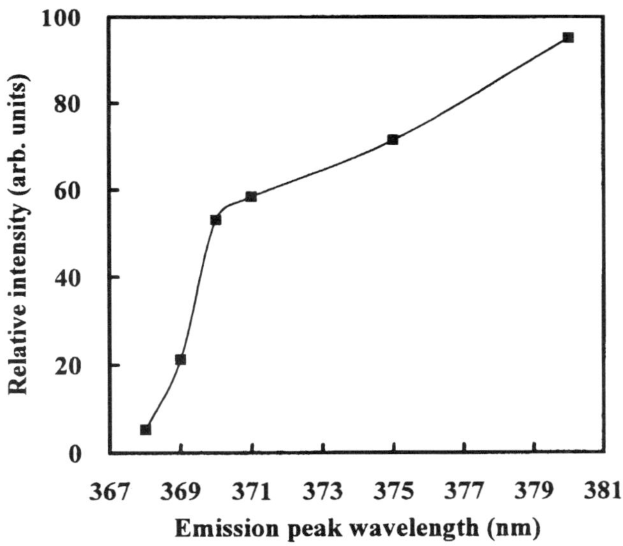

Figure 7.8 Relative output power of UV LEDs as a function of emission peak wavelength with different In mole fractions in the active layer.

Figure 7.9 shows the relative output power of UV LEDs as a function of ambient temperature. When the temperature is decreased from the 50°C to –30°C, the output power of UV LEDs decreases dramatically. The output power of each UV LED was normalized to be one at a temperature of 25°C. The change of the output power with the temperature is totally different from those of blue, green and amber LEDs as shown in Figure 7.3 (Mukai *et al.*, 1998a). The output power of blue, green and amber LEDs increases gradually when the temperature is decreased. This difference in the UV LEDs is probably due to the QCSE caused by the piezoelectric field resulting from the strain. In Figure 7.8, the output power of the UV LEDs is relatively low due to the QCSE or poor carrier localization. When the temperature is decreased, the strain between InGaN and AlGaN layers become large due to the difference in thermal expansion coefficients. Then the QCSE becomes highly dominant with decreasing temperature. As a results, the radiative recombination rate drops sharply, as discussed in Figure 7.8. However, there is a problem with this assumption because no blueshift of the emission peak wavelength with increasing current due to the QCSE is observed in Figure 7.7 (a). If the QCSE is dominant in the UV LED, a blueshift with increasing current should be observed due to carrier screening of the piezoelectric field. This may be due to the small blueshift of the QCSE resulting from a thin InGaN well layer of 55 Å.

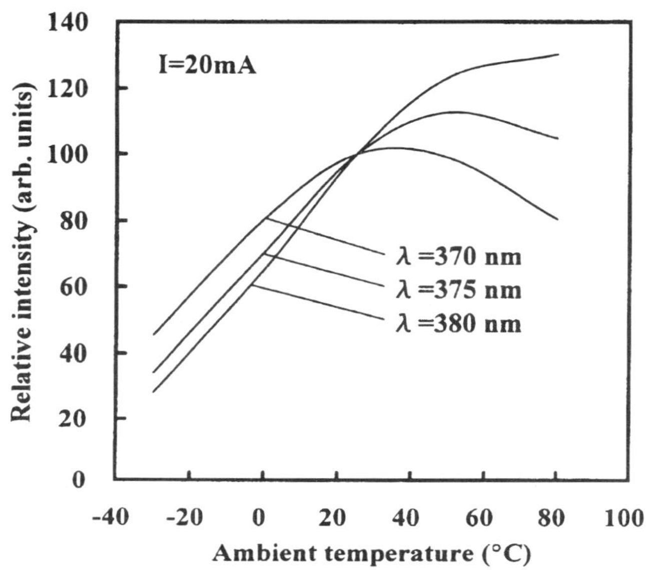

Figure 7.9 Relative output power of UV LEDs with different emission peak wavelengths as a function of ambient temperature. The output powers of the UV LEDs were normalized to be one at an ambient temperature of 25°C.

Using the QCSE, the blueshift with increasing current in Figures 7.4 (a) and 7.6 (a) may be explained by carrier screening of the piezoelectric field. Also, the constant emission peak wavelength with increasing temperature in Figures 7.4 (b) and 7.6 (b) may be explained by the compensation of both the blueshift due to the decreased QCSE resulting from the reduced strain and the redshift due to band-gap narrowing. Using the QCSE, however, it is impossible to explain why the blueshift with increasing current disappears suddenly at an emission peak wavelength of 375 nm in Figure 7.7 in spite of the relatively large strain, and why the output power of UV LEDs decreases dramatically with decreasing emission peak wavelength, as shown in Figure 7.8.

Figure 7.10 shows the output power and external quantum efficiency of a blue-green InGaN MQW LED as a function of the forward current. The emission peak wavelength of the LED is 507 nm at a current of 20 mA. The external quantum efficiency has the highest value of 19% at a current of 0.1 mA. When the forward current is lower or higher than 0.1 mA, the external quantum efficiency decreases gradually. If the QCSE would be a dominant effect on the emission mechanism of the LED, the external quantum efficiency has to increase with increasing the current due to the carrier screening effects of the piezoelectric field. However, only the blueshift of the emission peak wavelength with increasing current was observed as same as blue/green InGaN SQW LEDs, as shown in Figure 7.4 (a).

Thus, the emission mechanism of InGaN-based LEDs with an emission peak wavelength longer than 375 nm is dominated by carrier recombination at the localized energy states caused by In composition fluctuation in the InGaN well layer. The QCSE due to the large strain between the InGaN and AlGaN layer is negligible at these emission peak wavelengths due to the much greater effects of the large localized energy states. When the emission peak wavelength becomes shorter than 375 nm, the conventional band-to-band emission mechanism becomes dominant due to poor carrier localization resulting from the small In composition fluctuation. In addition, the QCSE also becomes dominant, which causes the low output power of the UV LEDs.

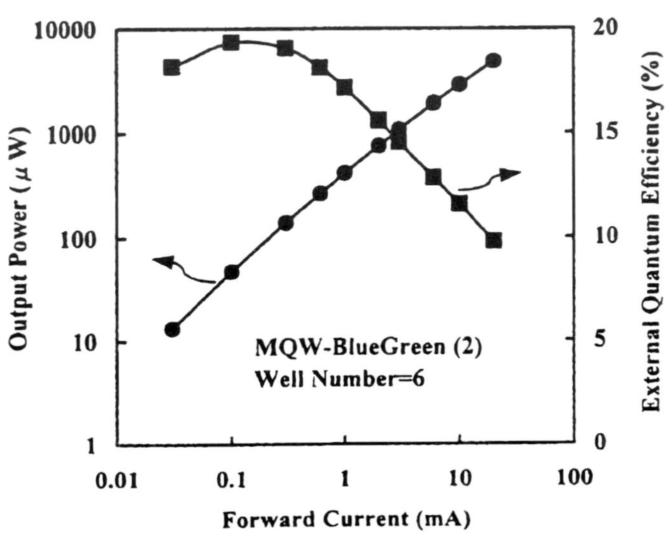

Figure 7.10 Output power and external quantum efficiency of InGaN MQW LEDs as a function of the forward current.

7.1.4 Roles of Dislocations in InGaN-Based LEDs

All of III-V nitride-based LEDs and LDs use an InGaN active layer instead of a GaN active layer because of the difficulty in fabricating highly efficient light-emitting devices using a GaN active layer (Mukai *et al.*, 1998b). Highly efficient UV/blue/green/amber InGaN quantum-well structure LEDs have been fabricated directly on a sapphire substrate in spite of a high dislocation density of 1-$10x10^{10}cm^{-2}$ originating from a large lattice mismatch between GaN and the sapphire substrate (Nakamura and Fasol, 1997). Due to the high efficiency of the LEDs grown on the sapphire substrates, the dislocations in InGaN do not appear to work as a nonradiative recombination center (Lester *at al.*, 1995). Epitaxially laterally overgrown GaN (ELOG) on sapphire was developed recently to reduce the number of threading dislocations in the GaN epitaxial layers (Nam *et al.*, 1997; Sakai et al., 1997; Usui *et al.*, 1997; Zheleva *et al.*, 1997; Marchand *et al.*, 1998). Using the ELOG, the number of threading dislocations was reduced significantly to almost zero in the GaN grown on the SiO_2 stripe mask. Thus, there is a great interest in fabricating LEDs using the ELOG substrate which has a small number of dislocations in order to study the role of these dislocations in the InGaN-based LEDs. For the study, blue InGaN SQW structure LEDs were fabricated on the ELOG and sapphire substrates (Mukai *et al.*, 1998c).

First, selective growth of GaN was performed on a 2-μm-thick GaN layer grown on a (0001) C-face sapphire substrate. The 0.1-μm-thick silicon dioxide (SiO_2) mask was patterned to form 3-μm-wide stripe windows with a periodicity of 9 μm in the GaN <1-100> direction. Following the 7-μm-thick GaN growth on the SiO_2 mask pattern, the coalescence of the selectively grown GaN enabled the achievement of a flat GaN surface over the entire substrate. This coalesced GaN was designated the ELOG. We examined the defect density by plan-view transmission electron microscopy (TEM) observations of the ELOG substrates.

Figure 7.11 Plan-view TEM image of the surface of the ELOG substrate with a thickness of 7 μm.

Figure 7.11 shows plan-view TEM image of the surface of the ELOG substrate with a thickness of 7 μm. The number of TDs on the SiO_2 mask area was lower than $1x10^{6}cm^{-2}$, and that on the window area was approximately $3x10^{7}cm^{-2}$. The TD density of conventional GaN which was grown directly on the sapphire substrate without the ELOG was of the order of $1x10^{10}cm^{-2}$ (Lester *et al.*, 1995), the number of the TDs was reduced considerably when the ELOG substrate was used. After obtaining a 15-μm-thick ELOG substrate, an InGaN SQW blue LED structure was grown on the substrate in order to study a role of the dislocations. The LED structure was already described in Figure 7.1 and Section 7.1.3. The characteristics of the LEDs were measured under a direct current (DC) at room temperature. For a comparison, blue LED with the same structure was also fabricated directly on the sapphire substrate without the ELOG.

Figure 7.12 show the photograph of real LED on ELOG under a forward current of 3 mA. This photograph was taken using a microscope. The size of each LED chip is as large as 350 μm x 350 μm. Each LED chip includes many window and SiO_2 stripe regions periodically. No inhomgeneties of blue emission intensity over the entire surface of the p-type transparent Ohmic contact layer was observed. There was no emission intensity fluctuation on the surface of the LED depending on the TD density fluctuation periodically across the window and SiO_2 stripe regions. Thus, it is concluded that a large number of TDs of $3x10^7 cm^{-2}$ on the window region do not serve to decrease the efficiency of the LEDs. The ELOG and GaN on sapphire had average TD densities of $1x10^7 cm^{-2}$ and $1x10^{10} cm^{-2}$, respectively. Here, average TD density of the ELOG on sapphire was obtained by dividing the TD density of $3x10^7 cm^{-2}$ on the window region by the ratio of (stripe periodicity of 9 μm)/ (window width of 3 μm) because the TD density on the SiO_2 stripe region was lower than $1x10^6 cm^{-2}$, as shown in Figure 7.11.

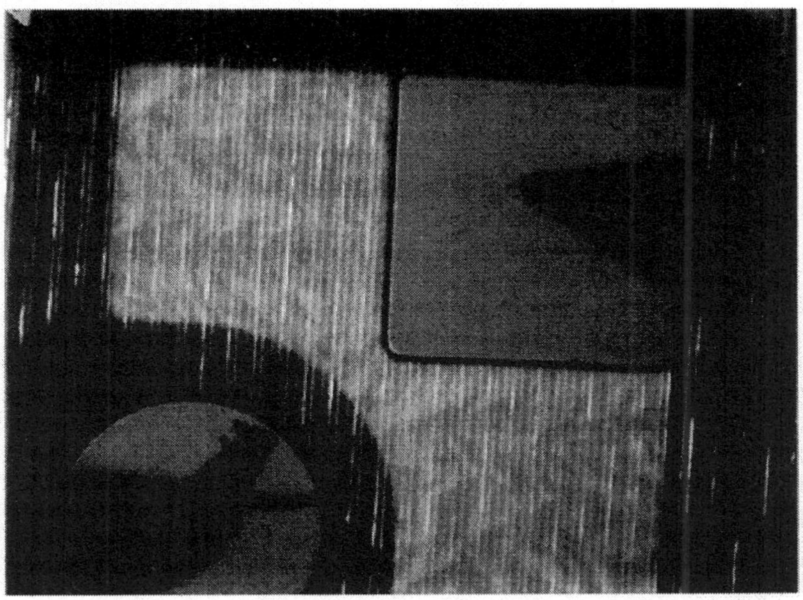

Figure 7.12 Photograph of blue InGaN SQW LED on ELOG under a forward current of 3 mA.

Figure 7.13 shows the relative output power of both LEDs as a function of the forward current. The output power of both LEDs at a current of 20 mA was 6 mW, which is equal to the external quantum efficiency of 11%. In spite of a large number of dislocations, the LED on sapphire had the same output power as that on ELOG. Thus, the dislocation seems not to work as nonradiative recombination centers in the InGaN active layer. If the TDs work as nonradiative recombination centers, the output power of the LED on ELOG have to be much higher than that on sapphire due to a small number of average TD density.

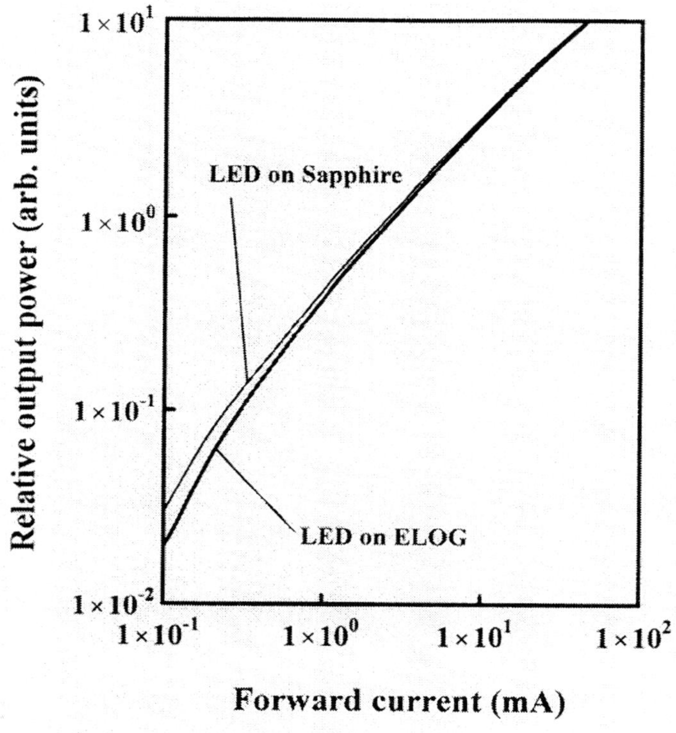

Figure 7.13 Relative output power of blue InGaN SQW LEDs grown on ELOG and on sapphire substrates as a function of the forward current.

Figure 7.14 shows the emission spectra of the blue InGaN SQW LEDs on ELOG at forward currents from 0.1 mA to 20 mA. At a current of 20 mA, the peak wavelength and the full-width at half maximum (FWHM) of the emission spectra of the LED on ELOG were 470 nm and 27 nm, respectively, and those on sapphire were 464 nm and 26 nm, respectively. The peak wavelength of the LED on ELOG is somewhat longer than that on sapphire probably due to a growth fluctuation. There is no significant difference between in the spectra of both LEDs. The broad spectral width is mainly due to an In composition fluctuation in the InGaN well layer (Chichibu *et al.*, 1996; 1997a; Narukawa *et al.*, 1997a; 1997b). This means that the size of the fluctuation is the same in both LEDs in spite of a large difference in the dislocation density. Thus, the In composition fluctuations are not related to the dislocations.

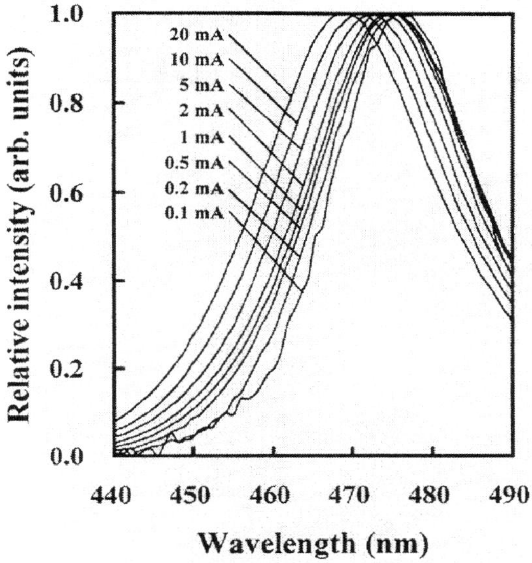

Figure 7.14 Emission spectra of blue InGaN SQW LED grown on ELOG substrate at various forward currents.

In the spectra, the blueshift is easily observed due to a band-filling effect of the deep localized energy states. The blue LED on sapphire also showed the same blueshift with increasing the forward current. We measured the absorption spectra and EL of blue/green InGaN SQW LEDs grown on sapphire substrate in previous studies (Chichibu *et al.*, 1996; 1997a; Narukawa *et al.*, 1997a; 1997b). The entire EL appeared at the lower energy tail of the absorption spectra. The EL of blue and green LEDs originated from the carrier recombination at the deep localized energy states with a localization energy of 290 meV and 570 meV, respectively. The blueshift of the EL of the green SQW LEDs with increasing forward current was due to a band-filling effect of the localized energy states. Both LEDs showed the same peak wavelength dependence on the current, which implies the same blueshifts. Thus, the TDs are not directly related to the formation of the localized energy states. These localized states are probably formed by the In composition fluctuation in the InGaN well layer due to a natural phase separation of InGaN during growth. Therefore, assuming that the localized states are formed by the In composition fluctuation in the InGaN well layer, the TD related composition fluctuation is eliminated because the TD density of the epilayer on the ELOG is relatively small.

Thus, the localization induced by the In composition fluctuations seem to be a key role of the high efficiency of the InGaN-based blue LEDs. When the electrons and holes are injected into the InGaN active layer of the LEDs, these carriers are captured by the localized energy states before they are captured by the nonradiative recombination centers caused by the large number of TDs (Nakamura, 1998). It was reported that the TDs served as a nonradiative recombination center in GaN and InGaN (Rosner *et al.*, 1997; Sugahara *et al.*, 1998). These localized energy states can be formed only in InGaN films during the growth due to a phase separation of the InGaN. Assuming that the lateral spacing of the effective bandgap (potential) minimum due to the In composition determines the carrier diffusion length in InGaN, the diffusion length was estimated to be less than 60 nm from the spatially resolved CL spectrum mapping measurement (Chichibu *et al.*, 1997a). It was also concluded that the efficiency of light emission was high as long as the minority carrier diffusion length was shorter than the dislocation spacing (Sugahara *et al.*, 1998). Considering about these previous results, the carrier diffusion length determined by the potential fluctuation due to InGaN phase separation must be less than the TD spacing in the InGaN layer in order to obtain high-efficient InGaN-based LEDs.

Figure 7.15 shows the reverse biased I-V characteristics of both LEDs. At a reverse bias voltage of –20 V, the reverse current of the LED on the ELOG was 0.009 μA, and that on sapphire was 1 μA. Under the reverse biased condition, the LED on sapphire had a considerable amount of leakage current. This leakage current is probably caused by the TDs because the LEDs on sapphire have a considerable amount of leakage current and many TDs (Sasaoka *et al.*, 1998).

For summary, InGaN SQW blue LEDs were grown on ELOG and sapphire substrates. The emission spectra showed the same blueshift with increasing forward currents between both LEDs. The output power of both LEDs was as high as 6 mW at a current of 20 mA. The LED on sapphire had a considerable amount of leakage current in comparison with that on ELOG. These results indicate that In composition fluctuation is not caused by TDs, free carriers are captured by radiative recombination centers before they are captured by nonradiative recombination centers in InGaN, and that the dislocations form the leakage current pathway in InGaN.

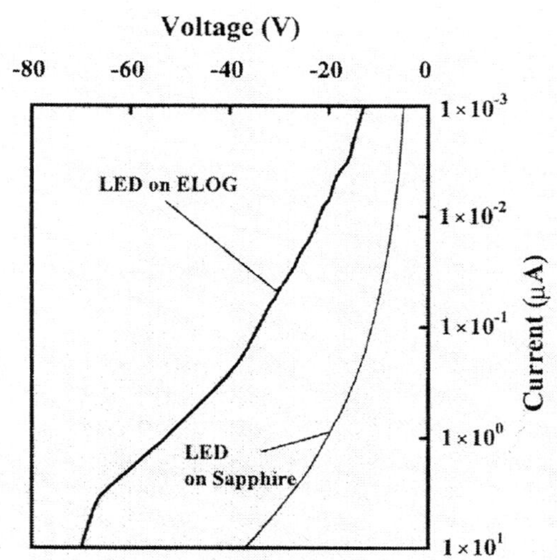

Figure 7.15 I-V characteristics of blue InGaN SQW LEDs grown on ELOG and on sapphire substrates under reverse biased conditions.

7.2 LDs GROWN ON SAPPHIRE SUBSTRATE

7.2.1 Introduction

First demonstration of RT violet laser light emission in InGaN/GaN/AlGaN-based heterostructures under pulsed operations were achieved in 1995 (Nakamura *et al.*, 1996). Since this report of pulsed operation, many groups have reported pulsed operation of the laser diodes (LDs) using the same structure (Itaya *et al.*, 1996; Bulman *et al.*, 1997; Mack *et al.*, 1997; Kuramata *et al.*, 1997; Kneissl *et al.*, 1998; Nakamura *et al.*, 1998a; Katoh *et al.*, 1998). In 1996, first continuous-wave (CW) operation of the LDs with a lifetime of 27-300 hours were reported (Nakamura *et al.*, 1997a; 1997b). Since the success in the CW operation of the LDs, various characteristics of the LDs were studied under CW operation, including emission mechanism, gain, far-field patterns (FFPs) and others. Here, the present performance of InGaN/GaN/AlGaN separate confinement heterostraucture (SCH) LDs grown on sapphire substrate with a lifetime of more than 300 hours are described (Nakamura *et al.*, 1997b).

7.2.2 LDs Grown on Sapphire Substrates

First, the laser structure was grown directly on sapphire substrate. Therefore, the TD density in the InGaN active layer is as high as $1x10^8$ to $1x10^{12}cm^{-2}$, as shown in Figure 7.16. The area of the ridge-geometry LD was 4 μm x 450 μm. High-reflection facet coatings for both dry-etched facets were used to reduce the threshold current density (Nakamura and Fasol, 1997).

Figure 7.17 shows the light output power-current (L-I) per coated facet of the LD and the voltage-current (V-I) characteristics under RT-CW and pulsed (pulse period of 1 ms and pulse width of 0.5 μs) operations (Nakamura *et al.*, 1997b). No stimulated emission was observed up to a threshold current of 75 mA, which corresponded to a threshold current density of 4.2 kA/cm^2, as shown in Figure 7.17(a). The operating voltage at the threshold current was 4.3 V, as shown in Figure 7.17(b). The differential quantum efficiency of 13% per facet and output power of 50 mW per facet were obtained at a current of 220 mA under RT-CW operation. The threshold current is 75 mA for both CW and pulsed operations. However, the threshold voltage is 5 V under pulsed operation, which is higher than that under CW operation (4.3 V), probably due to a small amount of heat generated under pulsed operation. The activation ratio of holes from Mg acceptors of p-type GaN become smaller when the temperature of the LD becomes lower due to the deep acceptor levels. Thus the resistivity of the p-type GaN layer and the voltage increase when the temperature of the LD decrease.

The temperature dependence of the L-I curves of the LDs was measured under CW operation, as shown in Figure 7.18(a). Lasing was achieved between 20°C and 100°C. The threshold current increases gradually with increasing temperature. The characteristic temperature T_0, which was used to express the temperature dependence of the threshold current in the form $I_{th}(T) = I_0exp(T/T_0)$,

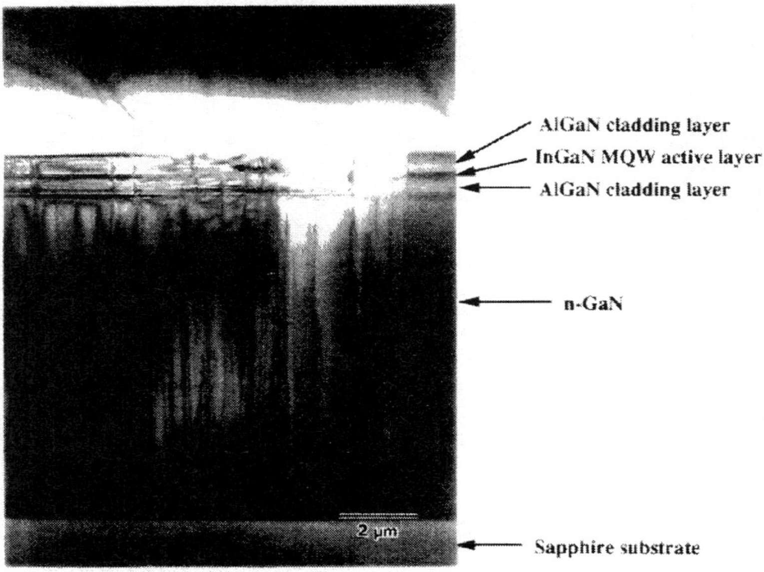

Figure 7.16 Cross-sectional transmission electron micrographs (TEM) of the InGaN-MQW/GaN/AlGaN SCH LD grown directly on the sapphire substrate. The LD structure consisted of a 300-Å-thick GaN buffer layer grown at a low temperature of 550°C, a 3-μm-thick layer of n-type GaN:Si, a 0.1-μm-thick layer of n-type $In_{0.05}Ga_{0.95}N$:Si, a 0.6-μm-thick layer of n-type $Al_{0.08}Ga_{0.92}N$:Si, a 0.1-μm-thick layer of n-type GaN:Si, an $In_{0.15}Ga_{0.85}N/In_{0.02}Ga_{0.98}N$ MQW structure consisting of three 35-Å-thick Si-doped $In_{0.15}Ga_{0.85}N$ well layers forming a gain medium separated by 70-Å-thick Si-doped $In_{0.02}Ga_{0.98}N$ barrier layers, a 200-Å-thick layer of p-type $Al_{0.2}Ga_{0.8}N$:Mg, a 0.1-μm-thick layer of p-type GaN:Mg, a 0.6-μm-thick layer of p-type $Al_{0.08}Ga_{0.92}N$:Mg, and a 0.3-μm-thick layer of p-type GaN:Mg. The 0.1-μm-thick n-type and p-type GaN layers were light-guiding layers. The 0.5-μm-thick n-type and p-type $Al_{0.08}Ga_{0.92}N$ layers acted as cladding layers for confinement of the carriers, and the light was emitted from the active region of the InGaN MQW structure.

was estimated to be 170K for this LD. Here, I_0 is a constant, T is the absolute temperature and $I_{th}(T)$ is the threshold current. Figure 7.18 (b) shows the temperature dependence of the V-I curve. With increasing temperature, the operating voltage decreases due to the increase of the hole carrier concentration resulting from the increased activation ratio of holes from Mg acceptors.

Figure 7.19 shows the results of a lifetime test on CW-operated LDs carried out at RT, in which the operating current is shown as a function of time under a constant output power of 2 mW per facet controlled using an autopower controller (APC). The operating current increases gradually due to the increase in the threshold current from the initial stage and increases sharply after 300 hours. Breakdown of the LDs occurred after a period of more than 300 hours due to the formation of a short circuit in the LDs. It is clarified later that the short lifetime of the LD is due to a large number of TDs.

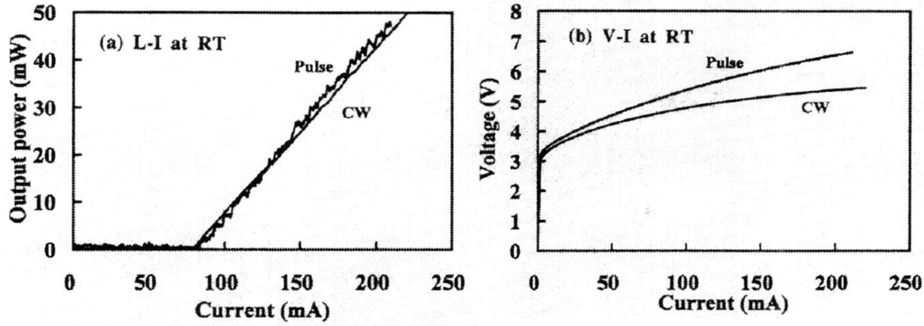

Figure 7.17 (a) L-I and (b) V-I characteristics of InGaN MQW LDs measured under CW and pulsed operations at RT. The pulse period and pulse width were 1 ms and 0.5 μs, respectively.

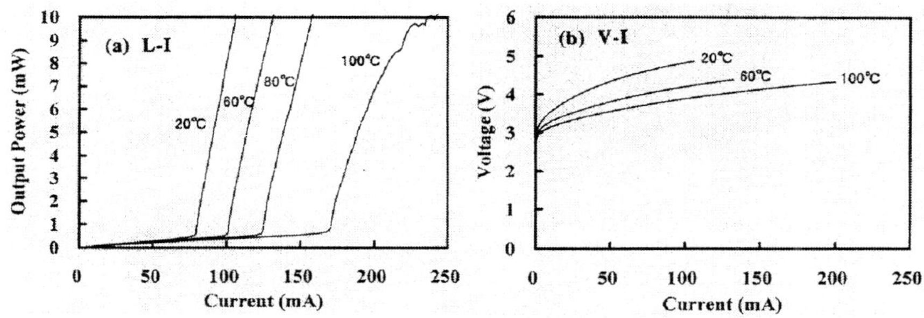

Figure 7.18 Temperature dependence of (a) L-I and (b) V-I characteristics of InGaN MQW LDs measured under CW operation.

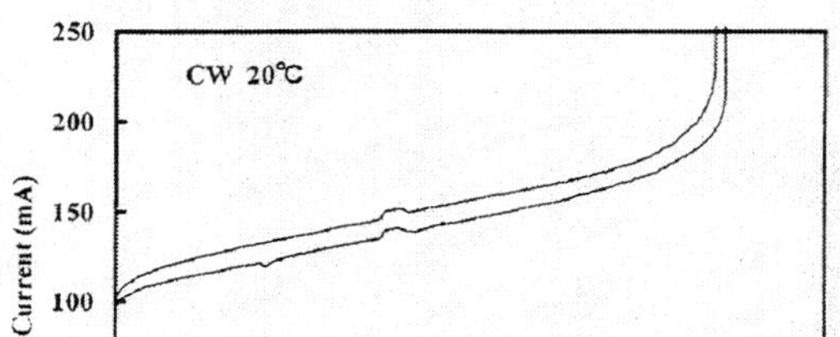

Figure 7.19 Operating current as a function of time under a constant output power of 2 mW per facet controlled using an autopower controller. The LD was operated under DC at RT.

The emission spectra measured under RT-CW operation are shown in Figure 7.20. At the forward current of 20 mA, the spontaneous emission with an emission peak wavelength of 417 nm and a spectrum width of 20 nm was observed. Above the threshold current, the single mode emission was observed. The peak wavelength of the stimulated emission was 416 nm at the forward current of 85 mA.

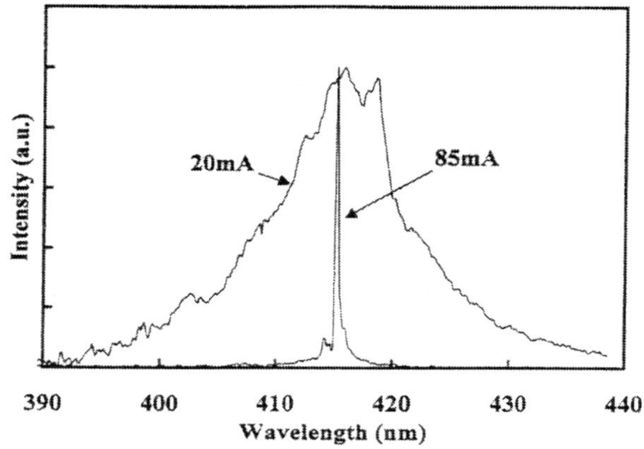

Figure 7.20 Emission spectra measured under RT-CW operation with currents of 80 mA and 86 mA.

7.2.3 ELOG Substrate

In order to lengthen the lifetime of the LD to more than 300 hours, the number of TD density had to be decreased. For that purpose, the ELOG substrate which was described in Section 7.1.4 was used. Figure 7.21 (a) shows the etched surface morphology of the ELOG substrate, and Figure 7.21 (b) shows that of the 12-μm-thick GaN film grown under the same conditions as the ELOG substrate without the SiO_2 mask patterns for comparison (Nakamura *et al.*, 1997c; 1998b; 1998c; 1998d). In this case, The 0.1-μm-thick silicon dioxide (SiO_2) mask was patterned to form 4-μm-wide stripe windows with a periodicity of 12 μm in the GaN <1-100> direction. The etching of the GaN was performed by reactive ion etching with Cl_2 plasma (Nakamura and Fasol, 1997). The etching depth was as great as 2 μm in order to reveal the etch pit clearly. Many hexagonal etch pits with the number of $1x10^8 cm^{-2}$ were observed over the entire region of the GaN film grown directly on the sapphire substrate, as shown in Figure 7.21 (b). On the ELOG substrate, a small number of etch pits were observed on the 4-μm-wide stripe window, as shown in Figure 7.21 (a). The etch pit density was about $2x10^7 cm^{-2}$ in the region of the 4-μm-wide stripe window. However, the etch pit density was almost zero in the region of the 8-μm-wide SiO_2 stripe.

Figure 7.22 shows cross-sectional TEM image of the ELOG substrate. TDs, originating from the GaN/sapphire interface, propagate to the regrown GaN layer within the window regions of the mask. It is important to note that the TDs extend only to just above the window areas. In contrast, there were no observable TDs in the overgrown layer. However, a few short edge-on dislocation segments parallel to the interface plane were observed in the GaN layer on the SiO_2 mask area. These dislocations were parallel to the (0001) plane via the extension of the vertical threading dislocations after a 90° bend in the regrown region. These dislocations did not subsequently propagate to the surface of the overgrown GaN layers. We examined the defect density by plan-view TEM observation of the surface of the ELOG substrates, as shown in Figure 7.11. The number of dislocations on the SiO_2 mask area was lower than $1x10^6cm^{-2}$, and that on the window area was approximately $1x10^7cm^{-2}$. As the dislocation density of conventional GaN was of the order of $1x10^{10}cm^{-2}$, the number of the TDs was reduced considerably when the ELOG substrate was used. These results are almost identical to those obtained through the measurements of etch pit density in Figure 7.21 (a).

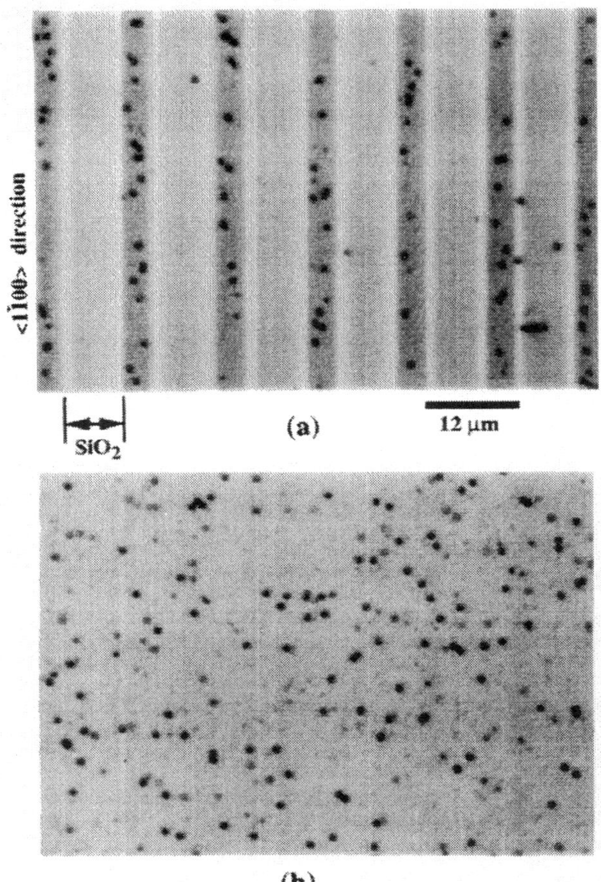

Figure 7.21 Etched surface morphologies of (a) the ELOG substrate and (b) the 12-μm-thick GaN film grown under the same conditions as the ELOG substrate without the SiO_2 mask patterns. The SiO_2 stripe width and window width of the ELOG substrate were 8 μm and 4 μm, respectively. The etch pit density was approximately $2\ x10^7cm^2$ in the region of the 4-μm-wide stripe window.

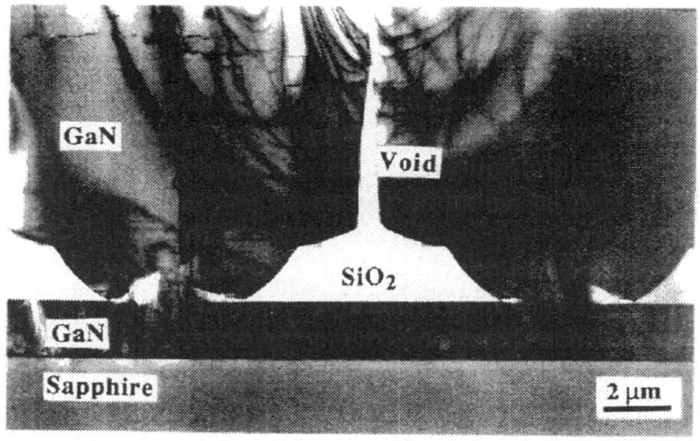

Figure 7.22 Cross-sectional TEM micrograph of the laterally overgrown GaN layer on a SiO_2 mask and window area.

7.2.4 InGaN-Based LDs Grown on ELOG Substrates

The LDs with an InGaN MQW/GaN/AlGaN SCH was grown on above-mentioned ELOG substrate. To grow the thick AlGaN cladding layer with a high Al content required for optical confinement without any cracks, GaN/AlGaN modulation doped strained-layer superlattices (MD-SLSs) within the range of critical thickness instead of thick AlGaN layers were used as cladding layers (Nakamura *et al.*, 1997c; 1998b; 1998c; 1998d). The LD structure was grown on the ELOG substrate with a thickness of 15 μm. The InGaN MQW LD device consisted of a 3-μm-thick layer of n-type GaN:Si, a 0.1-μm-thick layer of n-type $In_{0.1}Ga_{0.9}N$:Si, a $Al_{0.14}Ga_{0.86}N$/GaN MD-SLS cladding layer consisting of 120 25-Å-thick undoped GaN separated by 25-Å-thick Si-doped $Al_{0.14}Ga_{0.86}N$ layers, a 0.1-μm-thick layer of Si-doped GaN, an $In_{0.15}Ga_{0.85}N/In_{0.02}Ga_{0.98}N$ MQW structure consisting of four 35-Å-thick Si-doped $In_{0.15}Ga_{0.85}N$ well layers forming a gain medium separated by 105-Å-thick Si-doped $In_{0.02}Ga_{0.98}N$ barrier layers, a 200-Å-thick layer of p-type $Al_{0.2}Ga_{0.8}N$:Mg, a 0.1-μm-thick layer of Mg-doped GaN, a $Al_{0.14}Ga_{0.86}N$/GaN MD-SLS cladding layer consisting of 240 25-Å-thick undoped GaN separated by 25-Å-thick Mg-doped $Al_{0.14}Ga_{0.86}N$ layers, and a 0.05-μm-thick layer of p-type GaN:Mg. The 0.1-μm-thick n-type and p-type GaN layers were light-guiding layers. The n-type and p-type $Al_{0.14}Ga_{0.86}N$/GaN MD-SLS layers acted as cladding layers for confinement of the carriers and the light emitted from the active region of the InGaN MQW structure. The structure of the ridge-geometry InGaN MQW LD was almost the same as that described previously (Nakamura and Fasol, 1997). First, the surface of the p-type GaN layer was partially etched until the n-type GaN layer and the p-type $Al_{0.14}Ga_{0.86}N$/GaN MD-SLS cladding layer were exposed, in order to form a ridge-geometry LD. The laser cavity was formed parallel to the direction of the SiO_2 stripe. A mirror facet was also formed by dry etching, as reported previously (Nakamura and Fasol, 1997). High-reflection facet coatings (90%) consisting of 2 pairs of quarter-wave TiO_2/SiO_2 dielectric multilayers were used to reduce the threshold current. A Ni/Au contact was evaporated onto the p-type GaN layer, and a Ti/Al contact was evaporated onto the n-type GaN layer. The electrical characteristics of the LDs fabricated in this way were measured under a direct current (DC). The structure of the LD is shown in Figure 7.23.

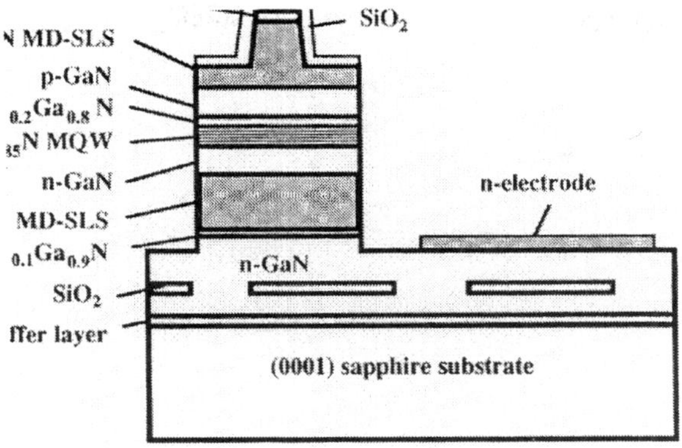

Figure 7.23 Structure of the InGaN MQW-structure LDs with MD-SLS cladding layers grown on the ELOG substrate.

The ridge-geometry LDs were formed on the GaN layer above the SiO_2 region with low TDs of lower than $1x10^6$ cm^{-2} and the window region with a high TD density of $3x10^7 cm^{-2}$. The V-I characteristics and the light output power per coated facet of the LD with a cavity length of 450 µm and a ridge width of 4 µm as a function of the forward DC current (L-I) at RT were measured. When the LD was formed on the GaN layer above the SiO_2 mask region with low TDs, the threshold current was 53 mA, which corresponded to a threshold current density of 3 $kAcm^{-2}$. When the LD formed on the window region with the high TD density had a threshold current density of 6 to 9 $kAcm^{-2}$, which was much higher than that of the LD formed on the SiO_2 mask. The higher threshold current density is probably caused by the large number of TD density of $3x10^7$ cm^{-2} at the window region. It was confirmed that the dislocation served as a leakage current pathway in InGaN as shown in Figure 7.15. Thus, there is a possibility that a leakage current due to a large number of TDs caused the high threshold current density on the window region. Further studies are required to determine the excat reasons of the high threshold current density caused by TDs.

Figure 7.24 shows the results of a lifetime test for CW-operated LDs formed on the GaN layer above the SiO_2 mask region carried out at 20°C, in which the operating current is shown as a function of time under a constant output power of 2 mW per facet controlled using an autopower controller (APC). The LDs is survived 8,000 hours of operation. The LDs formed on the window region showed the lifetimes of 1,000–3,000 hours due to the high threshold current density of 6 to 9 $kAcm^{-2}$.

The emission spectra of the LDs were measured under RT CW operation at currents of 50 and 60 mA, as shown in Figure 7.25. At a current of 50 mA, longitudinal modes with a mode separation of 0.04 nm due to the cavity were observed. At a current of 60 mA, a single-mode emission was observed at an emission wavelength of 396.6 nm.

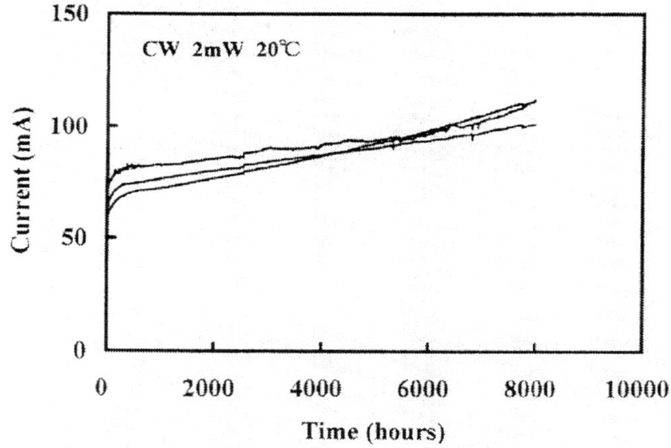

Figure 7.24 Operating current as a function of time under a constant output power of 2 mW per facet controlled using an autopower controller. The InGaN MQW LDs with MD-SLS cladding layers grown on the ELOG substrate were operated under DC at 20°C.

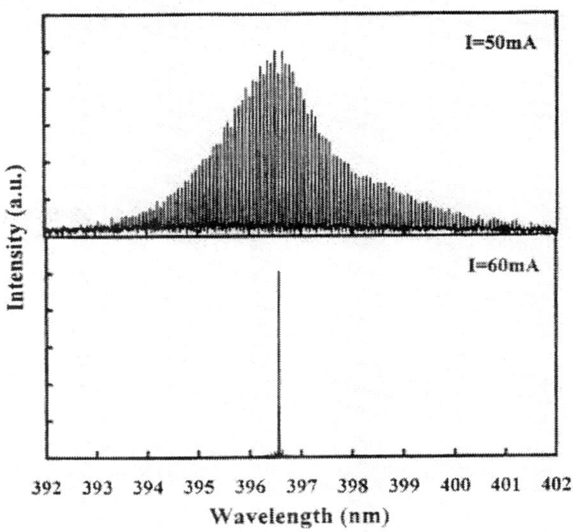

Figure 7.25 Laser emission spectra measured under RT CW operation with currents of 50 mA and 60 mA.

Next, high-power LDs were also fabricated on the ELOG substrate. After obtaining a 20-μm-thick ELOG substrate, the InGaN MQW LD structure was grown on the surface. The InGaN MQW LD structure was the same as described in Figure 7.23 except for the active layer. The InGaN MQW structure was changed to an $In_{0.15}Ga_{0.85}N/In_{0.02}Ga_{0.98}N$ MQW structure consisting of two 40-Å-thick undoped

$In_{0.15}Ga_{0.85}N$ well layers forming a gain medium separated by 100-Å-thick Si-doped $In_{0.02}Ga_{0.98}N$ barrier layers. The area of the ridge-geometry LD was 3 μm x 450 μm. A laser cavity was formed by cleaving the facets along {1-100} of the LD grown on the ELOG substrate after polishing the sapphire substrate to 70 μm. A facet coating (90%) consisting of 2 pairs of quarter-wave TiO_2/SiO_2 dielectric multilayers was formed on one side of the facet. The output power of the LD was measured from the other side of uncoated facet.

Figure 7.26 shows the V-I characteristics and the light output power per uncoated cleaved facet of the LD grown on the ELOG substrate with a number of well of 2 as a function of the forward DC current (L-I) at RT. No stimulated emission was observed up to a threshold current of 40 mA, which corresponds to a threshold current density of 3 kA/cm^2. The slope efficiency was as high as 1.2 W/A per facet because the output power was measured from the uncoated cleaved mirror facet having low reflectivity. The differential quantum efficiency per facet was 39%. The output power of the LDs was as high as 420 mW at an operating current of 490 mA. This output power is the highest ever reported for III-V nitride-based LDs under RT-CW operation. At an output power of 100 mW, a kink was observed in the L-I curve because the transverse mode change occurred at an output power of 100 mW. The emission spectra of the LDs were measured under RT-CW operation at output powers of 10 mW and 30 mW. At output powers of 10 mW and 30 mW, stimulated emission was observed at wavelengths of around 408.2 and 408.8 nm, respectively.

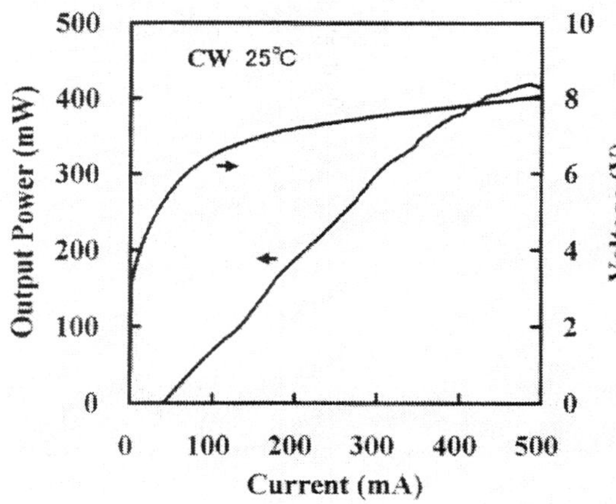

Figure 7.26 Typical L-I and V-I characteristics of InGaN MQW LDs measured under CW operation at RT.

Figure 7.27 shows the temperature dependence of the peak emission wavelength of the LDs under CW operation at an output power of 30 mW. During this measurement, the LDs were placed on a Peltier-type cooler to maintain the temperatures of the LDs between 25°C and 50°C. The average wavelength drift caused by the temperature change was estimated to be 0.06 nm/K from this figure. Assuming that the wavelength drift of the longitudinal mode is determined by the temperature dependence of band gap energy of the active layer, the wavelength drift is determined as

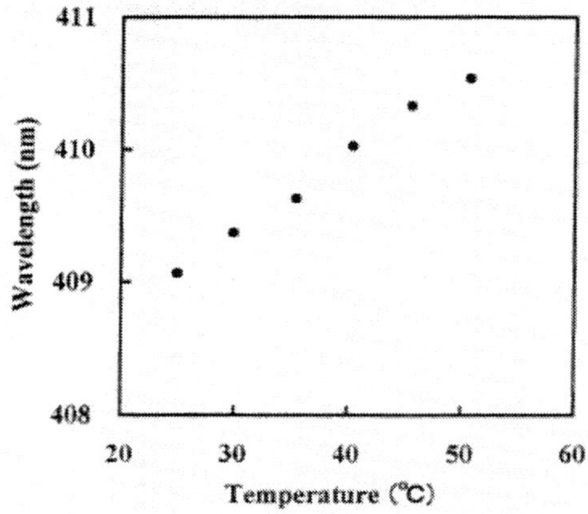

Figure 7.27 Temperature dependence of the peak emission wavelength of the InGaN MQW LDs under CW operation with an output power of 30 mW.

$$d\lambda_p/dT=(d\lambda_p/dE_g)(dE_g/dT) \tag{7.1}$$

$$\lambda_p=1240/E_g \tag{7.2}$$

Taking the derivative of Equation (7.2), we obtain

$$d\lambda_p/dE_g=-1240/E_g^2 \tag{7.3}$$

Therefore, Equation (7.1) becomes

$$d\lambda_p/dT=-(1240/E_g^2)(dE_g/dT) \tag{7.4}$$

where λ_p (nm) is the emission wavelength of the LD, E_g (eV) is the band gap energy of the active layer and T (K) is absolute temperature. Other groups have already estimated the temperature coefficient $dE_g/dT=-6.0\times10^{-4}$ eV/K around RT from the temperature dependence of the principle emission peak of GaN (Strite and Morkoç, 1992; Morkoç *et al.*, 1994). Using this value for GaN and Equation (7.4), we were able to estimate a value of 0.06 nm/K for the wavelength drift. This value agrees well with the experimentally obtained one of 0.06 nm/K, even though the actual active layer is an InGaN MQW layer whose temperature dependence of the band gap energy has not been obtained. From Equation (7.4), the wavelength drift becomes much larger when the band gap energy becomes smaller because $d\lambda_p/dT$ is proportional to $1/E_g^2$ assuming that dE_g/dT is almost constant between different materials, such as InGaN, and AlInGaP. For this purpose, the wavelength drift of commercially available strained AlInGaP MQW red LDs (SANYO: DL-3038-023) with an emission wavelength of 640 nm was measured. The wavelength drift was as large as 0.3 nm/K due to the small band gap energy. Thus, based on the wavelength drift, wide band gap III-V nitride-based LDs are suitable for various applications due to a large band gap energy.

Figure 7.28 shows the results of a lifetime test of CW-operated LDs carried out at an ambient temperature of 50°C, in which the operating current is shown as a function of time under a constant output power of 30 mW controlled using an autopower controller. The initial operating current was as high as 100 mA due to the high-temperature operation. A small increase in the operating current is observed with increasing operation time. After 250 hours of operation, the operating current increases dramatically. Thus, the lifetime of the LD was 250 hours.

The InGaN MQW LDs grown on the ELOG substrate showed an output power as high as 420 mW under RT-CW operation. The wavelength drift caused by the temperature change was estimated to be 0.06 nm/K. The lifetime of the LDs at a constant output power of 30 mW was 250 hours under CW operation at an ambient temperature of 50 °C. In order to improve the lifetime under such high power and high temperature conditions, the threshold current density has to be decreased further by optimizing the structure of the LDs. The longest lifetime of 8,000 hours at a constant output power of 2 mW under RT-CW operation was achieved.

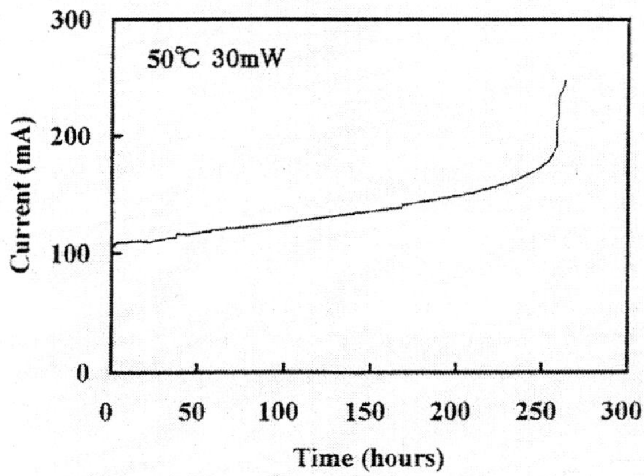

Figure 7.28 Operating current of InGaN MQW LDs as a function of time under a constant output power of 30 mW at an ambient temperature of 50 °C controlled using an autopower controller.

7.3 LDs GROWN ON GAN SUBSTRATE

Ultimate substrate for GaN growth is GaN. Here, LDs grown on GaN substrate are described (Nakamura *et al.*, 1998e; 1998f). It means a homoepitaxy of GaN. In the previous sections, we described InGaN MQW-structure LDs grown on sapphire or ELOG substrates. Using both the MD-SLSs and the ELOG substrates, LDs with an estimated lifetime of more than 10,000 hours were developed. In the structures, sapphire substrates were used. However, when using a sapphire substrate, it is difficult to obtain cleaved mirror facets which were used for the cavities of conventional LDs. In addition, the thermal conductivity of sapphire (0.5 W/cm·K) is not as high as that of GaN (1.3 W/cm·K) for the heat dissipation generated by the LDs (Strite and Morkoç, 1992; Morkoç *et al.*, 1994). Here, the LDs grown on pure GaN substrates which are easily cleaved and have a high thermal conductivity, are described.

7.3.1 Free-Standing GaN Substrates

After obtaining a 20-μm-thick ELOG substrate, the wafer was removed from the MOCVD reactor. Next, GaN growth was carried out using a conventional hydride vapor phase epitaxy (HVPE) system using a horizontal quartz reactor because the growth rate of GaN is much higher when using the HVPE method. In the HVPE system, the Ga source was GaCl formed by the reaction between Ga metal and HCl gas. The Ga metal was positioned in the upstream region of the reactor. The temperature of the Ga metal was maintained at 900°C. The substrate holder was set in the downstream region of the reactor, where GaCl and NH_3 were mixed. NH_3 was supplied, as the N source, at a flow rate of 3 l/min. The total flow rate of the H_2 carrier gas was 3 l/min. The flow rate of the HCl gas was 50 cc/min. The substrate temperature was kept constant at 1000°C during the growth. The growth rate of GaN was approximately 50 μm/hour under these conditions. Undoped GaN growth was continued up to 200 μm. Then, the sapphire substrate was removed by polishing, in order to obtain a pure GaN substrate with a thickness of approximately 150 μm.

Then, the InGaN MQW LD structure was grown on the surface of the 150-μm-thick GaN substrate by MOCVD. The InGaN MQW LD structure was the same as described in the section 7.2.4 except for the active layer. The InGaN MQW structure was changed to an $In_{0.15}Ga_{0.85}N/In_{0.02}Ga_{0.98}N$ MQW structure consisting of two 40-Å-thick undoped $In_{0.15}Ga_{0.85}N$ well layers forming a gain medium separated by 100-Å-thick Si-doped $In_{0.02}Ga_{0.98}N$ barrier layers. The area of the ridge-geometry LD was 3 μm x 400 μm. A laser cavity was formed by cleaving the facets along {1-100} of the LD grown on the GaN substrate. A facet coating (90%) consisting of 2 pairs of quarter-wave TiO_2/SiO_2 dielectric multilayers was formed on one facet. The output power of the LD was measured from another uncoated facet.

7.3.2 Characteristics of LDs

Figure 7.29 shows the V-I characteristics and the light output power per uncoated cleaved facet of the LD grown on the GaN substrate by the HVPE as a function of the forward DC current (L-I) at RT. No stimulated emission was observed up to a threshold current of 80 mA, which corresponds to a threshold current density of 7 kA/cm^2. The slope efficiency was as high as 0.8 W/A per facet because the output power was measured from the uncoated cleaved mirror facet having low reflectivity. The differential quantum efficiency per facet was 26%. The output power of the LDs was as high as 160 mW at an operating current of 300 mA. At an output power of 80 mW, a kink was observed in the L-I curve because the transverse mode change occurred at an output power of 80 mW.

Figure 7.30 shows near-field patterns (NFP) of the LDs taken by a charge-coupled device (CCD) camera through the optical microscope at output powers of

70 mW and 100 mW. The fundamental transverse mode was observed below the output power of 80 mW, as shown in Figure 7.30 (a). With increasing output power above 80 mW, the first-order transverse mode appeared, as shown in Figure 7.30 (b). This transverse mode change at an output power of 80 mW caused the kink in the L-I curve. In order to suppress the change, we must control the transverse mode to be a fundamental mode, by further narrowing the ridge width.

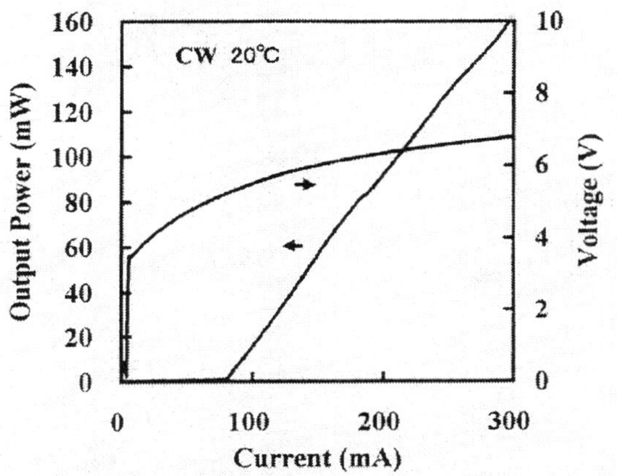

Figure 7.29 Typical L-I and V-I characteristics of InGaN MQW LDs grown on GaN substrate measured under CW operation at RT.

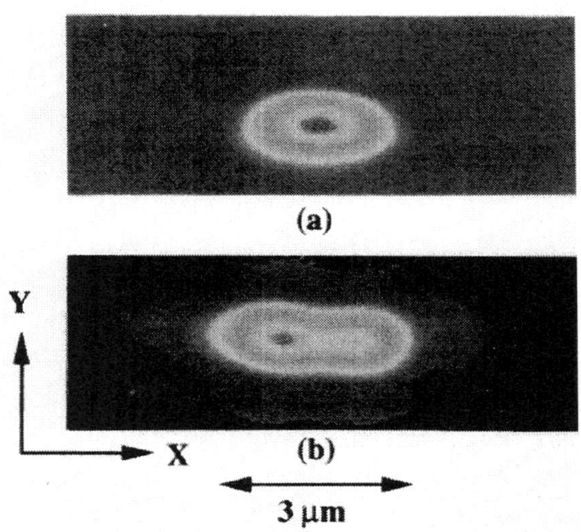

Figure 7.30 Near-field pattern of InGaN MQW LDs at output powers of (a) 70 mW and (b) 100 mW under RT-CW operation. X and Y directions are parallel and perpendicular to the junction of the LD, respectively

The far-field pattern (FFP) of the LDs with the cleaved mirror facet was measured in the planes parallel and perpendicular to the junction, as shown in Figure 7.31. The FFP revealed single-mode emission at the output power up to 80 mW. Above 80 mW, multimode emission occurred due to the change of the transverse mode from fundamental to first order. The beam full-width at half-power levels for the parallel and perpendicular FFP were 8° and 30°, respectively.

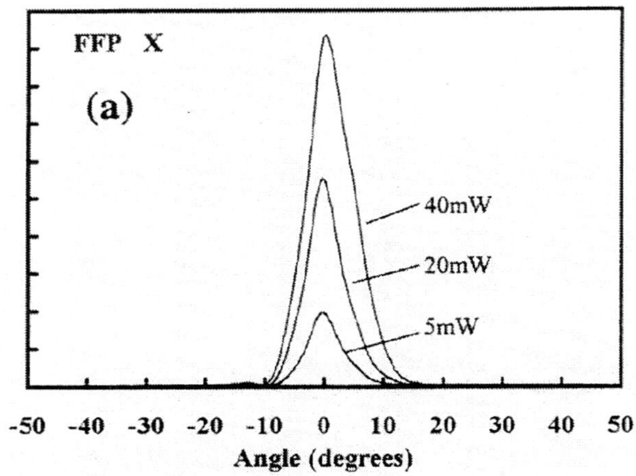

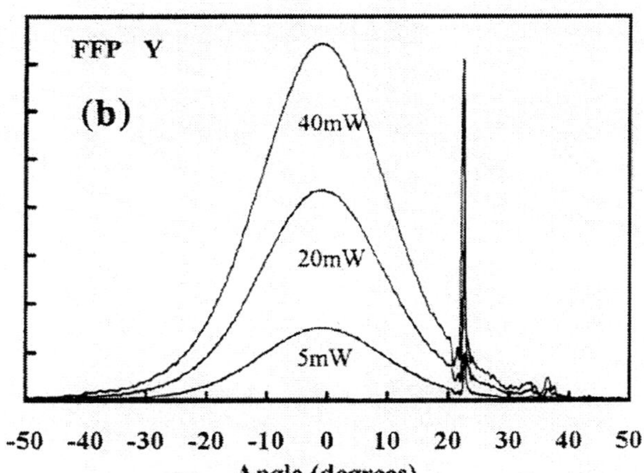

Figure 7.31 Far-field pattern of InGaN MQW LDs in the plane (a) parallel and (b) perpendicular to the junction at output powers of 5 mW, 20 mW and 40 mW.

In the FFP perpendicular to the junction, strong emission was observed at an angle of 23 degrees. This laser emission originates from the GaN substrate because a small portion of the laser beams penetrated from the waveguide region to the GaN substrate region due to poor optical confinement of the cladding layer

(Hofstetter *et al.*, 1997). We could suppress this emission by increasing the thickness or Al content of the cladding layers.

The emission spectra of the LDs were measured under RT-CW operation at output powers of 3 mW, 10 mW and 40 mW. At an output power of 3 mW, longitudinal modes with many sharp peaks due to a cavity with a length of 400 μm, were observed. At output powers of 10 mW and 40 mW, multimode emission was observed at wavelengths of around 410.5 and 411.3 nm, respectively.

Figure 7.32 shows the results of a lifetime test of CW-operated LDs carried out at an ambient temperature of 50°C, in which the operating current is shown as a function of time under a constant output power of 5 mW controlled using an autopower controller. The initial operating current was as high as 160 mA due to the high-temperature operation. A small increase in the operating current is observed with increasing operation time. After 180 hours of operation, the operating current increased dramatically. The lifetime was about 180 hours due to a high threshold current density of 14 kA/cm^2.

InGaN MQW LDs were fabricated on a GaN substrate grown by MOCVD followed by HVPE. The LDs with cleaved mirror facets showed an output power as high as 160 mW under RT-CW operation. The fundamental transverse mode was observed up to an output power of 80 mW. The emission spectra indicated multimode emission with an emission wavelength of around 410 nm. The lifetime of the LDs at a constant output power of 5 mW was about 180 hours under CW operation at an ambient temperature of 50°C, due to a high threshold current density. Thus, GaN substrates grown by the HVPE method can be used for the GaN growth. If the threshold current density of the LDs grown on the GaN substrate could be reduced to approximately 1-4 kAcm^{-2}, the lifetime of the LDs would be much longer.

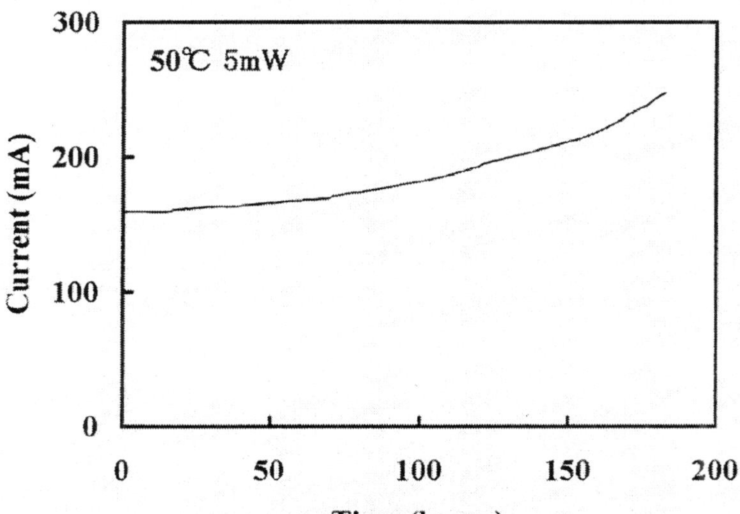

Figure 7.32 Operating current of InGaN MQW LDs as a function of time under a constant output power of 5 mW at an ambient temperature of 50 °C controlled using an autopower controller.

7.4 FUTURE PROSPECTS OF InGaN-BASED EMITTING DEVICES

RT CW operation of bluish-purple InGaN MQW LDs was demonstrated with a lifetime of 8,000 hours in 1998. The laser emission wavelength was 390-440 nm which was the shortest ever reported for semiconductor laser diodes. Using these LDs with an emission wavelength of 400 nm, there would be many applications that would cause a big impact on the optoelectronic industry, in the way that blue and green LEDs did. The shorter wavelength of 400 nm means that the light can be focused more sharply, which would increase the storage capacity of magnetic and optical disks.

Here, using III-V nitride compound semiconductors, short wavelength emitters, such as UV/blue/green/amber LEDs and LDs have been described as one example. However, these III-V nitride materials can also be used for other applications of UV detectors, solar battery and electronic devices such as high-power and high-frequency FETs the same as other semiconductors (such as Si, Ge, GaAs, InP and GaAlAs). At present, our challenge is that all light sources would be changed to III-V nitride based LEDs or LDs from conventional bulb and fluorescent lamps in the future considered from environmental issues (especially from saving energy and natural resources). When nitride based LED are used for traffic lights, energy consumption is reduced to 1/10 in comparison with conventional bulb lamp traffic lights. Also, the lifetime of LED traffic light is more than ten years (almost forever). It means that maintenance of the LED traffic light is not required. Also, we can save a lot of the natural resources due to a long lifetime. The lifetime of the conventional bulb lamp traffic light is only one year. They need a lot of maintenance work for replacement of the burned-out traffic light lamps. It is expensive. If the LEDs were used for other applications instead of conventional bulb lamps, the same effects would happen, and we could reduce a lot of energy consumption and save a lot of natural resources. As a result, we could reduce number of nuclear power plants. These are very important issues for human beings considered from an environmental issue. In addition, the technology of III-V nitride-based devices is still at an early stage. For example, very recently, GaN substrates have become available. Using these substrates, III-V nitride-based devices with improved characteristics could be performed soon.

7.5 REFERENCES

Amano, H., Takeuchi, T., Sota, S., Sakai, H. and Akasaki, I., 1997, *Mater. Res. Soc. Symp. Proc.*, **449**, p. 1143.

Bulman, G.E., Doverspike, K., Sheppard, S.T., Weeks, T.W., Kong, H.S., Dieringer, H.M., Edmond, J.A., Brown, J.D., Swindell, J.T. and Schetzina, J.F., 1997, *Electronics Letters*, **33**, p. 1556.

Chichibu, S., Azuhata, T., Sota, T. and Nakamura, S., 1996, *Applied Physics Letters*, **69**, p. 4188.

Chichibu, S., Wada, K. and Nakamura, S., 1997a, *Appl. Phys. Lett.*, **71**, p. 2346.

Chichibu, S., Azuhata, T., Sota, T. and Nakamura, S., 1997b, *Appl. Phys. Lett.*, **70**, p. 2822.

Hodapp, M.E., 1997, *In Semiconductors and Semimetals*, Vol. **48**, eds. Stringfellow, G.B. and Craford, M.G. San Diego: Academic Press p. 227.

Hofstetter, D., Bour, D.P., Thornton, R.L. and Johnson, N.M., 1997, *Appl. Phys. Lett.*, **70**, p. 1650.

Im, J.S., Kollmer, H., Off, J., Sohmer, A., Scholz, F. and Hangleiter, A., 1998, *Mater. Res. Soc. Symp. Proc.*, **482**, p. 513.

Itaya, K., Onomura, M., Nishino, J., Sugiura, L., Saito, S., Suzuki, M., Rennie, J., Nunoue, S., Yamamoto, M., Fujimoto, H., Kokubun, Y., Ohba, Y., Hatakoshi, G. and Ishikawa, M., 1996, *Jpn. J. Appl. Phys.*, **35**, p. L1315.

Katoh, H., Takeuchi T., Anbe, C., Mizutmoto, R., Yamaguchi, S., Wetzel, C., Amano, H., Akasaki, I. and Yamada, N., 1998, *Jpn. J. Appl. Phys.*, **37**, p. L444.

Kish, F. A. and Fletcher, R. M., 1997, *In Semiconductors and Semimetals*, Vol. **48**, eds. Stringfellow, G. B. and Craford, M. G. (San Diego: Academic Press) p. 149.

Kneissl, M., Bour, D.P., Johnson, N.M., Romano, L.T., Krusor, B.S., Donaldson, R., Walker, J. and Dunnrowicz, C., 1998, *Appl. Phys. Lett.*, **72**, p. 1539.

Kuramata, A., Domen, K., Soejima, R., Horino, K., Kubota, S. and Tanahashi, T., 1997, *Jpn. J. Appl. Phys.*, **36**, p. L1130.

Lester, S.D., Ponce, F.A., Craford, M.G. and Steigerwald, D.A., 1995, *Appl. Phys. Lett.*, **66**, p. 1249.

Mack, M.P., Abare, A., Aizcorbe, M., Kozodoy, P., Keller, S., Mishra, U.K., Coldren, L. and DenBaars, S., 1997, *Mater. Res. Soc. Internet J. Nitride Semicond. Res.*, **2**, No. 41 (Available from http://nsr.mij.mrs.org/2/41/)

Marchand, H., Wu, X.H., Ibbeston, J.P., Fini, P.T., Kozodoy, P., Keller, S., Speck, J.S., DenBaars, S.P. and Mishra, U.K., 1998, *Appl. Phys. Lett.*, **73**, p. 747.

McCluskey, M.D., Van de Walle, C.G., Master, C.P., Romano, L.T. and Johnson, N.M., 1998, *Appl. Phys. Lett.*, **72**, p. 2725.

Morkoç, H., Strite, S., Gao, G.B., Lin, M.E., Sverdlov, B. and Burns, M., 1994, *J. Appl. Phys.*, **76**, p. 1363.

Mukai, T., Narimatsu, H. and Nakamura S., 1998a, *Jpn. J. Appl. Phys.*, **37**, p. L479.

Mukai, T., Morita, D. and Nakamura, S., 1998b, *J. Cryst. Growth*, **189/190**, p. 778.

Mukai, T., Takekawa, K. and Nakamura, S., 1998c, *Jpn. J. Appl. Phys.*, **37**, p. L839.

Nakamura, S., 1991, *Jpn. J. Appl. Phys.*, **30**, p. L1705.

Nakamura, S., 1998, *Science*, **281**, p. 956-961.

Nakamura, S. and Fasol, G., 1997, *The Blue Laser Diode*, Heidelberg: Springer-Verlag.

Nakamura, S., Senoh, M., Iwasa, N. and Nagahama, S., 1995a, *Jpn. J. Appl. Phys.*, **34**, p. L797.

Nakamura, S., Senoh, M., Iwasa, N., Nagahama, S., Yamada, T. and Mukai, T., 1995b, *Jpn. J. Appl. Phys.*, **34**, p. L1332.

Nakamura, S., Senoh, M., Nagahama, S., Iwasa, N., Yamada, T., Matsushita, T., Kiyoku, H. and Sugimoto, Y., 1996, *Jpn. J. Appl. Phys.*, **35**, p. L74.

Nakamura, S., Senoh, M., Nagahama, S., Iwasa, N., Yamada, T., Matsushita, T., Sugimoto, Y. and Kiyoku, H., 1997a, *Appl. Phys. Lett.*, **70**, p. 1417.

Nakamura, S., Senoh, M., Nagahama, S., Iwasa, N., Yamada, T., Matsushita, T., Sugimoto, Y. and Kiyoku, H., 1997b, *Jpn. J. Appl. Phys.*, **36**, p. L1059.

Nakamura, S., Senoh, M., Nagahama, S., Iwasa, N., Yamada, T., Matsushita, T., Kiyoku, H., Sugimoto, Y., Kozaki, T., Umemoto, H., Sano, M. and Chocho, K., 1997c, *Jpn. J. Appl. Phys.*, **36**, p. L1568.

Nakamura, F., Kobayashi, T., Tojo, T., Asatsuma, T., Naganuma, K., Kawai, H. and Ikeda, M., 1998a, *Electronics Letters*, **34**, p. 1105.

Nakamura, S., Senoh, M., Nagahama, S., Iwasa, N., Yamada, T., Matsushita, T., Kiyoku, H., Sugimoto, Y., Kozaki, T., Umemoto, H., Sano, M. and Chocho, K., 1998b, *J. Cryst. Growth*, **189/190**, p. 820.

Nakamura, S., Senoh, M., Nagahama, S., Iwasa, N., Yamada, T., Matsushita, T., Kiyoku, H., Sugimoto, Y., Kozaki, T., Umemoto, H., Sano, M. and Chocho, K., 1998c, *Appl. Phys. Lett.*, **72**, p. 211.

Nakamura, S., Senoh, M., Nagahama, S., Iwasa, N., Yamada, T., Matsushita, T., Kiyoku, H., Sugimoto, Y., Kozaki, T., Umemoto, H., Sano, M. and Chocho, K., 1998d, *Jpn. J. Appl. Phys.*, **37**, p. L627.

Nakamura, S., Senoh, M., Nagahama, S., Iwasa, N., Yamada, T., Matsushita, T., Kiyoku, H., Sugimoto, Y., Kozaki, T., Umemoto, H., Sano, M. and Chocho, K., 1998e, *Jpn. J. Appl. Phys.*, **37**, p. L309.

Nakamura, S., Senoh, M., Nagahama, S., Iwasa, N., Yamada, T., Matsushita, T., Kiyoku, H., Sugimoto, Y., Kozaki, T., Umemoto, H., Sano, M. and Chocho, K., 1998f, *Appl. Phys. Lett.*, **72**, p. 2014.

Nam, O.H., Bremser, M.D., Zheleva, T. and Davis, R.F., 1997, *Appl. Phys. Lett.*, **71**, p. 2638.

Narukawa, Y., Kawakami, Y., Fujita, Sz., Fujita, Sg. and Nakamura, S., 1997a, *Physical Review*, **B55**, p. 1938R.

Narukawa, Y., Kawakami, Y., Funato, M., Fujita, Sz., Fujita, Sg. and Nakamura, S., 1997b, *Appl. Phys. Lett.*, **70**, p. 981.

Nardelli, M.B., Rapcewicz, K. and Bernholc, J., 1997, *Appl. Phys. Lett.*, **71**, p. 3135.

Osamura, K., Naka, S. and Murakami, Y., 1975, *J. Appl. Phys.*, **46**, p. 3432.

Ponce, F.A. and Bour, D.P., 1997, *Nature*, **386**, p. 351.

Rosner, S.J., Carr, E.C., Ludowise M.J., Girolami, G. and Erikson, H.I., 1997, *Appl. Phys. Lett.*, **70**, p. 420.

Sakai, A., Sunakawa, H. and Usui, A., 1997, *Appl. Phys. Lett.*, **71**, p. 2259.

Sasaoka, C., Sunakawa, H., Kimura, A., Nido, M., Usui, A. and Sakai, A., 1998, *J. Cryst. Growth*, **189/190**, p. 61.

Strite, S. and Morkoç, H., 1992, *J. Vac. Sci. Technol.*, **B10**, p. 1237.

Sugahara, T., Sato, H., Hao, M., Naoi, Y., Kurai, S., Tottori, S., Yamashita, K., Nishino, K., Romano, L.T. and Sakai, S., 1998, *Jpn. J. Appl. Phys.*, **37**, L398.

Takeuchi, T., Sota, S., Katsuragawa, M., Komori, M., Takeuchi, H., Amano, H. and Akasaki, I., 1997, *Jpn. J. Appl. Phys.*, **36**, p. L382.

Usui, A., Sunakawa, H., Sakai, A. and Yamaguchi, A., 1997, *Jpn. J. Appl. Phys.*, **36**, p. L899.

Zheleva, T.S., Nam, O.H., Bremser, M.D. and Davis, R.F., 1997, *Appl. Phys. Lett.*, **71**, p. 2472.

Appendix A
Parameters Tables

A.1 GaN

A.1.1 Electronic Properties

Energy Gap : Eg (in eV)

WZ			
	2K		Dingle *et al.* 1971
3.50	1.6K $Eg=[3.503+(58.08\times10^{-4}T^2)/(T-996)]$		Monemar 1974
3.44	295K	($T<295K$)	Monemar 1974
3.504	10K		Chichibu *et al.* 1997
ZB			
3.302	10K		Flores *et al.* 1994
3.3			Okumura *et al.* 1994

Free Exciton Energy : Ex (in eV)

WZ					
Ex(A)	*Ex(B)*	*Ex(C)*	*Ex* $(A_{n=2})$		
3.474	3.480	3.501		2K	Dingle *et al.* 1971
3.4751	3.4815	3.493		1.6K	Monemar 1974
3.4776	3.4827	3.5015		4.2K	Korona *et al.* 1996
3.478	3.485	3.506	3.498	10K	Chichibu *et al.* 1997
3.421	3.429			RT	Chichibu *et al.* 1997
ZB					
Ex					
3.272				5K	Menniger *et al.* 1996
3.27				4.2K	Okumura *et al.* 1997
3.201				RT	Nakadaira *et al.* 1997

A-Exciton Binding Energy E_b *(in meV)*

28	1.6K	Monemar 1974
20	40K	Chen *et al.* 1996
25	10K	Chichibu *et al.* 1997

Temperature and Pressure Dependence of Energy Gap : dEg/dT, dEg/dp

dEg/dT	dEg/dp	
0.67 meV/K	42 meV/GPa	Camphausen *et al.* 1971
	39.0 meV/GPa	Shan *et al.* 1995
0.370 meV/K		Chichibu *et al.* 1996
0.11 meV/K	42 meV/GPa	Korona *et al.* 1996

Splitting Energies of Valence Band Maximum : Δ_{cr}, Δ_{so} (in meV)

WZ

Δ_{cr}	Δ_{so}	
22.0	11.0	Dingle *et al.* 1971
72.9	15.6	Suzuki *et al.* 1996a
22	15	Chichibu *et al.* 1997
10	16.5	Yamaguchi *et al.* 1998

ZB

Δ_{so}	
17	Flores *et al.* 1994
17.0	Fan *et al.* 1996
15	Wei *et al.* 1996

Effective Masses (in m_0)

WZ electron

m_e	$m_{e\perp}$	$m_{e//}$	
0.20	0.20	0.20	Barker *et al.* 1973
0.22			Pankove *et al.* 1975
0.236			Meyer *et al.* 1995
	0.19	0.17	Chen *et al.* 1996
0.18	0.18	0.20	Uenoyama Chapt.2

WZ hole

m_h	$m_{hh\perp}$	$m_{lh\perp}$	$m_{ch\perp}$	$m_{hh//}$	$m_{lh//}$	$m_{ch//}$	
0.8							Pankove *et al.* 1975
0.54							Merz *et al.* 1996
	2.09	0.74	0.18	0.37	0.39	0.94	Majewski *et al.* 1997
	2.00	1.17	0.15	0.33	0.35	1.53	Kim *et al.* 1997
	1.96	1.96	0.14	1.87	0.14	1.96	Yeo *et al.* 1998
	1.76	0.76	0.16	1.61	0.14	1.04	Uenoyama Chapt.2

ZB electron

m_e	
0.15	Fan *et al.* 1996
0.17	Suzuki *et al.* 1996a

ZB hole

	m_{hh}	m_{lh}	m_{sh}	
[*k*00](Δ)	0.86	0.86	0.17	Uenoyama Chapt.2
[*kk*0](Γ)	4.89	0.84	0.15	
[*kkk*](Λ)	1.74	1.74	0.15	

Luttinger-Kohn Parameters : A

WZ

A_1	A_2	A_3	A_4	A_5	A_6	A_7	
-6.4	-0.5	5.9	-2.55	-2.56	-3.06	0.0015	Kim *et al.* 1997
-6.4	-0.8	5.93	-1.96	-2.32	-3.02	0.0026	Majewski *et al.* 1997
-7.24	-0.51	6.73	-3.36	-3.35	-4.72		Yeo *et al.* 1998
-6.56	-0.91	5.65	-2.83	-3.13	-4.85	0.00	Uenoyama Chapt. 2

ZB

γ_1	γ_2	γ_3	
2.46	0.65	0.98	Kim *et al.* 1997
2.94	0.89	1.25	Wang *et al.* 1998
2.70	0.76	1.07	Uenoyama Chapt. 2

A.1.2 Lattice Properties

Lattice Constant : a,c (in Å)

WZ

a	*c*	*c/a*	*u*	
3.1879	5.1856	1.6267		Leszczynski *et al.* 1995
3.182	5.189	1.631		Takeuchi *et al.* 1997
3.17	5.13	1.62	0.379	Kim *et al.* 1996
3.190	5.189	1.627	0.377	Schulz *et al.* 1977

ZB

a	
4.50	Lei *et al.* 1992
4.51	Leszczynski *et al.* 1996
4.460	Wright 1997
4.537	Shimada *et al.* 1998

Linear Thermal Expansion Coefficient : α (in $10^{-6}K^{-1}$)

α_a	α_c			
5.59	3.17	300-700K	on Sapp.	Maruska *et al.* 1969
5.59	7.75	700-900K		
3.1	2.8	300-350K	bulk	Leszczynski *et al.* 1995
6.2	6.1	700-750K		
3.8	2.0	300-350K	on Sapp.	Leszczynski *et al.* 1995
7.9	5.8	700-750K		

Elastic Constant : C

WZ							
C_{11}	C_{12}	C_{13}	C_{33}	C_{44}	C_{66}	B	
396	144	100	392	91	126	207	Kim *et al.* 1996
367	135	103	405	95	-	202	Wright 1997
365	135	114	381	109	115	-	Ymaguchi *et al.* 1997
377	160	114	209	81.4	109	-	Schwarz *et al.* 1997
350	140	104	376	101	115	197	Shimada *et al.* 1998

ZB						
C_{11}	C_{12}	C_{44}	$C_{44,0}$	ζ	B	
293	159	155	200	0.61	203	Wright 1997
296	154	206	225	0.5	201	Kim *et al.* 1996
285	161	149	202	0.67	202	Shimada *et al.* 1998

Phonon Wave Numbers : $\bar{\nu}$ *(in* cm^{-1}*)*

WZ								
A1		*E1*		*E2*		*B1*		
LO	*TO*	*LO*	*TO*	*high*	*low*	*high*	*low*	
735	533	743	561	569	144			Azuhata *et al.* 1995
710	533	741	559	568	143			Cingolani *et al.* 1986
734	546	739	554	565	153	728	319	Davydov *et al.* 1998
748	541	757	568	579	143	720	337	Karch *et al.* 1998

ZB		
LO	*TO*	
740	555	Siegle *et al.* 1995
739	552	Giehler *et al.* 1995
750	560	Karch *et al.* 1997a

Effective Charge : Z^* *(in e)*

WZ			
$Z^*_{\perp}$	$Z^*_{//}$	Z^*	
2.65	2.82		Barker *et al.* 1973
2.60	2.74	2.64	Karch *et al.* 1998
	2.74		Shimada *et al.* 1998

ZB	
Z^*	
2.65	Karch *et al.* 1997a
2.67	Shimada *et al.* 1998

A.1.3 Optical Properties

Index of Refraction : n

WZ

2.33	@1.0eV	Ejder 1971
2.22	@3.4eV	Torvik *et al.* 1998
2.65	@3.38eV	
$n(E)=[1+\frac{B}{E^2}\{2E_0^{\frac{1}{2}}-(E_0+E)^{\frac{1}{2}}-(E_0-E)^{\frac{1}{2}}\}+\frac{G}{E_p^2-E^2}]^{\frac{1}{2}}$	E_0=3.4eV, E_p=7.8eV, B=41.55, G=158.8	Shokhovets *et al.* 1998

ZB

2.25	@0.8eV	$n(E)=[1+148/(38.3-E^2)]^{1/2}$	Vidal *et al.* 1996
2.50	@3.1eV	(0.7-3.3eV)	

Dielectric Constants : ε

WZ

$\varepsilon(0)$	$\varepsilon(0)^{\perp}$	$\varepsilon(0)^{//}$	$\varepsilon(\infty)$	$\varepsilon(\infty)^{\perp}$	$\varepsilon(\infty)^{//}$	
	12.2				5.8	Manchon *et al.* 1970
9.8	9.5	10.4		5.35		Barker *et al.* 1973
	9.28	10.1		5.29	5.29	Azuhata *et al.* 1995

ZB

$\varepsilon(0)$	$\varepsilon(\infty)$	
9.7	5.41	Karch *et al.* 1997a
8.9	5.0	McGill *et al.* 1998

A.1.4 Impurities and Defects

Binding Energy of Accepter : ΔE (in meV)

WZ

$Impurity_{site}$	$\Delta E^{\dagger}$	$\Delta E^{\dagger\dagger}$	ΔE	
Be_{Ga}	193	204		Kim *et al.* 1997
	233	241		Suzuki *et al.* 1995
	195	208		Majewski *et al.* 1997
	185	193		Yeo *et al.* 1998
			250	Salvador *et al.* 1996
			700	Strite *et al.* 1992
Mg_{Ga}	204	215		Kim *et al.* 1997
	245	253		Suzuki *et al.* 1995
	208	221		Majewski *et al.* 1997
	197	204		Yeo *et al.* 1998
			250	Strite *et al.* 1992

Binding Energy of Accepter (continued)

$Impurity_{site}$	$\Delta E^{\dagger}$	$\Delta E^{\dagger\dagger}$	ΔE	
Zn_{Ga}	321	331		Kim *et al.* 1997
	411	419		Suzuki *et al.* 1995
	394	406		Majewski *et al.* 1997
	352	360		Yeo *et al.* 1998
			340	Strite *et al.* 1992
Ca_{Ga}	248	259		Kim *et al.* 1997
	297	305		Suzuki *et al.* 1995
	264	276		Majewski *et al.* 1997
	247	255		Yeo *et al.* 1998
Li_{Ga}			750	Strite *et al.* 1992
Cd_{Ga}			550	Strite *et al.* 1992
Hg_{Ga}			410	Strite *et al.* 1992
C_N	220	230		Kim *et al.* 1997
	264	272		Suzuki *et al.* 1995
	228	240		Majewski *et al.* 1997
	214	223		Yeo *et al.* 1998
			860	Strite *et al.* 1992
			230	Fischer *et al.* 1995
Si_N	192	203		Kim *et al.* 1997
	231	239		Suzuki *et al.* 1995
	193	205		Majewski *et al.* 1997
	183	191		Yeo *et al.* 1998
			225	Strite *et al.* 1992
Zn_N			480	Monemar *et al.* 1980

ZB

Impurity	$\Delta E^{\dagger}$	$\Delta E^{\dagger\dagger}$	
Be	117	125	Suzuki *et al.* 1995
	124	133	Kim *et al.* 1997
	126	133	Wang *et al.* 1998
Mg	123	130	Suzuki *et al.* 1995
	130	139	Kim *et al.* 1997
	133	140	Wang *et al.* 1998
Zn	155	162	Suzuki *et al.* 1995
	170	178	Kim *et al.* 1997
	177	184	Wang *et al.* 1998
Ca	143	151	Suzuki *et al.* 1995
	153	162	Kim *et al.* 1997
	157	164	Wang *et al.* 1998
C	130	138	Suzuki *et al.* 1995
	138	147	Kim *et al.* 1997
	141	148	Wang *et al.* 1998
Si	117	125	Suzuki *et al.* 1995
	123	132	Kim *et al.* 1997
	125	132	Wang *et al.* 1998

Note : $\Delta E^{\dagger}$ and $\Delta E^{\dagger\dagger}$ denote the estimated binding energies with and without the polaron correction, respectively. $\Delta E^{\dagger}$ and $\Delta E^{\dagger\dagger}$ are calculated by Mireles and Ulloa (Mireles *et al.* 1998). The binding energies are obtained with band parameters from several references arranged in descending order for each impurity.

A.1.5 Further Properties

Deformation Potential Constants : D (in eV)

D_1	D_2	D_3	D_4	D_5	
-13.9	-13.7	2.99	-1.63	-2.04	Suzuki *et al.* 1996b
		8.82	-4.41		Shikanai *et al.* 1997
		5.7	-2.85		Kim *et al.* 1997
-18.0	-15.1	2.82	-1.61	-2.04	Shimada *et al.* 1998

Melting Point : T_m (in K)

2791	Vechten 1973

Thermal Conductivity : (in $Wcm^{-1}K^{-1}$)

1.3	Sichel *et al.* 1977

A.2 AlN

A.2.1 Electronic Properties

Energy Gap : Eg (in eV)

WZ			
6.2	300K		Yim *et al.* 1973
6.28	5K		Perry *et al.* 1978
ZB			
5.11	300K	indirect	Lambrecht *et al.* 1991

Temperature Dependence of Energy Gap : dEg/dT

dEg/dT	
0.51meV/K	Tang *et al.* 1998

Splitting Energy of Valence Band Maximum : Δ_{cr} , Δ_{so} (in meV)

WZ

Δ_{cr}	Δ_{so}	
-58.5	6.8	Suzuki *et al.* 1995
-217	19	Wei *et al.* 1996

ZB

Δ_{so}	
17	Suzuki *et al.* 1995
19	Wei *et al.* 1996

Effective Masses : m (in m_0)

WZ electron

m_e	$m_{e\perp}$	$m_{e//}$	
	0.18	0.25	Suzuki *et al.* 1995
	0.35	0.35	Kim *et al.* 1997
	0.33	0.32	Majewski *et al.* 1997
0.27	0.25	0.33	Uenoyama Chapt.2

WZ hole

m_h	$m_{hh\perp}$	$m_{lh\perp}$	$m_{ch\perp}$	$m_{hh//}$	$m_{lh//}$	$m_{ch//}$	
	11.14	0.26	4.05	3.53	3.53	0.26	Kim *et al.* 1997
	20.00	0.37	4.54	3.57	3.57	0.26	Majewski *et al.* 1997
	10.42	0.24	3.81	3.53	3.53	0.25	Uenoyama Chapt.2

ZB electron

m_e	
0.20	Fan *et al.* 1996

ZB hole

	m_{hh}	m_{lh}	m_{sh}	
[*k*00](Δ)	1.40	1.40	0.32	Uenoyama Chapt.2
[*kk*0](Γ)	>10	1.41	0.27	
[*kkk*](Λ)	3.72	3.72	0.25	

Luttinger-Kohn Parameters : A

WZ

A_1	A_2	A_3	A_4	A_5	A_6	A_7	
-4.06	-0.26	3.78	-1.86	-2.02	-3.04	-	Suzuki *et al.* 1995
-3.86	-0.25	3.58	-1.32	-1.47	-1.64	-	Kim *et al.* 1997
-3.82	-0.22	3.54	-1.16	-1.33	-1.25	0	Majewski *et al.* 1997
-3.95	-0.27	3.68	-1.84	-1.95	-2.92	0.00	Uenoyama Chapt.2

ZB

γ_1	γ_2	γ_3	
1.40	0.35	0.59	Kim *et al.* 1997
1.50	0.39	0.62	Uenoyama Chapt.2

A.2.2 Lattice Properties

Lattice Constant : a, c (in Å)

WZ

a	*c*	*c/a*	*u*	
3.112	4.982			Yim *et al.* 1973
3.110	4.980	1.601	0.3821	Schulz *et al.* 1977
3.112	4.995	1.605	0.3811	Shimada *et al.* 1998

ZB

4.38	Petrov *et al.* 1992
4.376	Shimada *et al.* 1998

Linear Thermal Expansion Coefficient : α *(in* $10^{-6}K^{-1}$*)*

α_a	α_c	
4.2	5.3	Yim *et al.* 1974

Elastic Constant : C

WZ

C_{11}	C_{12}	C_{13}	C_{33}	C_{44}	C_{66}	*B*	
345	125	120	395	118	110		Tsubouchi *et al.* 1985
398	140	127	382	96	129	218	Kim *et al.* 1996
396	137	108	373	116		207	Wright 1997
398	142	112	383	127	128	212	Shimada *et al.* 1998

ZB

C_{11}	C_{12}	C_{44}	$C_{44,0}$	ζ	*B*	
304	152	199	230	0.6	203	Kim *et al.* 1996
304	160	193	237	0.55	208	Wright 1997
294	160	189	233	0.57	214	Karch *et al.* 1997b
313	168	192	236	0.56	216	Shimada *et al.* 1998

Phonon Wave Numbers : $\bar{\nu}$ *(in* cm^{-1}*)*

WZ

A1		*E1*		*E2*		*B1*		
LO	*TO*	*LO*	*TO*	*high*	*low*	*high*	*low*	
	601		350	638	228	703	534	Miwa et al. 1993
893	614	916	673	660	252			McNeil *et al.* 1993
	668		734	704	301	772	553	Ruiz *et al.* 1994
893	619	919	677	667	237	765	580	Karch *et al.* 1997b
	612		679	672	247	645	636	Shimada *et al.* 1998
881	615	922	673	667	252	778	571	Davydov *et al.* 1998

ZB

LO	*TO*	
907	662	Karch *et al.* 1997b
902	655	Harima *et al.* 1999

Effective Charge : Z^* (in e)

WZ

$Z^*_\perp$	$Z^*_{//}$	Z^*	
		2.57	Sanjurjo *et al.* 1983
2.53	2.69	2.58	Karch *et al.* 1997b
	2.67		Shimada *et al.* 1998

ZB

Z^*	
2.56	Karch *et al.* 1997b
2.55	Shimada *et al.* 1998

A.2.3 Optical Properties

Index of Refraction : n

2.15	Pastrnak et al. 1966

Dielectric Constants : ε

WZ

$\varepsilon(0)$	$\varepsilon(0)^\perp$	$\varepsilon(0)^{//}$	$\varepsilon(\infty)$	$\varepsilon(\infty)^\perp$	$\varepsilon(\infty)^{//}$	
8.5			4.68			Akasaki *et al.* 1967
8.5			4.77			Chin *et al.* 1994

A.2.4 Impurities and Defects

Binding Energy of Accepter : ΔE (in meV)

WZ

$\text{Impurity}_{\text{site}}$	$\Delta E^\dagger$	$\Delta E^{\dagger\dagger}$	
Be_{Al}	446	472	Suzuki *et al.* 1995
	223	262	Kim *et al.* 1997
	283	253	Majewski *et al.* 1997
Mg_{Al}	758	795	Suzuki *et al.* 1995
	465	514	Kim *et al.* 1997
	721	789	Majewski *et al.* 1997
Zn_{Al}	438	464	Suzuki *et al.* 1995
	219	255	Kim *et al.* 1997
	273	343	Majewski *et al.* 1997
Ca_{Al}	376	402	Suzuki *et al.* 1995

	204	240	Kim *et al.* 1997
	203	273	Majewski *et al.* 1997
Si_N	415	441	Suzuki *et al.* 1995
	214	250	Kim *et al.* 1997
	245	315	Majewski *et al.* 1997

ZB

Impurity	$\Delta E^{\dagger}$	$\Delta E^{\dagger\dagger}$	
Be	248	273	Suzuki *et al.* 1995
	265	292	Kim *et al.* 1997
Mg	305	330	Suzuki *et al.* 1995
	333	360	Kim *et al.* 1997
Zn	245	269	Suzuki *et al.* 1995
	261	288	Kim *et al.* 1997
Ca	227	252	Suzuki *et al.* 1995
	242	268	Kim *et al.* 1997
C	320	345	Suzuki *et al.* 1995
	353	380	Kim *et al.* 1997
Si	239	264	Suzuki *et al.* 1995
	255	281	Kim *et al.* 1997

Note : $\Delta E^{\dagger}$ and $\Delta E^{\dagger\dagger}$ denote the estimated binding energies with and without the polaron correction, respectively. $\Delta E^{\dagger}$ and $\Delta E^{\dagger\dagger}$ are calculated by Mireles and Ulloa (Mireles *et al.* 1998). The binding energies are obtained with band parameters from several references arranged in descending order for each impurity.

A.2.5 Further Properties

Melting Point : T_m (in K)

3487	Vechten 1973

Thermal Conductivity : (in $Wcm^{-1}K^{-1}$)

2.8	Slack *et al.* 1987

A.3 InN

A.3.1 Electronic Properties

Energy Gap : Eg (in eV)

WZ

1.89	RT	Zetterstrom 1970
2.11	78K	Osamura *et al.* 1975
2.05	RT	Tyagai *et al.* 1977

ZB

2.2	RT	Jenkins *et al.* 1987

Temperature Dependence of Energy Gap : dEg/dT

0.18meV/K	Slack *et al.* 1976

Splitting Energy of Valence Band Maximum : Δ_{cf}, Δ_{so} (in meV)

WZ

Δ_{cf}	Δ_{so}	
41	1	Wei *et al.* 1996
17	-	Yeo *et al.* 1998

ZB

Δ_{so}	
6	Wei *et al.* 1996

Effective Masses : m (in m_0)

WZ electron

m_e	$m_{e\perp}$	$m_{e//}$	
0.11			Zetterstrom 1970

WZ hole

m_h	$m_{hh\perp}$	$m_{lh\perp}$	$m_{ch\perp}$	$m_{hh//}$	$m_{lh//}$	$m_{ch//}$	
	1.61	0.11	1.67	1.67	1.67	0.10	Yeo *et al.* 1998

Luttinger-Kohn Parameters : A

WZ

A_1	A_2	A_3	A_4	A_5	A_6	A_7	
-9.28	-0.60	8.68	-4.34	-4.32	-6.08		Yeo *et al.* 1998

A.3.2 Lattice Properties

Lattice Constant : a, c (in Å)

WZ

a	*c*	*c/a*	*u*	
3.540	5.705			Martin *et al.* 1996
3.548	5.760			Zetterstrom 1970
3.53	5.72		0.378	Vogel *et al.* 1997

ZB

4.98	Strite *et al.* 1993
4.92	Kim *et al.* 1996
4.97	Tabata *et al.* 1999

Linear Thermal Expansion Coefficient : α (in $10^{-6}K^{-1}$)

α_a	α_c		
2.9	3.8	300K	Sheleg *et al.* 1976
3.5	5.2	500K	

Elastic Constant : C

WZ

C_{11}	C_{12}	C_{13}	C_{33}	C_{44}	C_{66}	B	
190	104	121	182	10		139	Sheleg *et al.* 1979
271	124	94	200	46		147	Kim *et al.* 1996
243	72	53	263	72			Azuhata *et al.* 1996
223	115	92	224	48		141	Wright 1997

ZB

C_{11}	C_{12}	C_{44}	$C_{44,0}$	ζ	
184	116	177	209	0.7	Kim *et al.* 1996
187	125	86	138	0.80	Wright 1997

Phonon Wave Numbers : $\overline{\nu}$ (in cm^{-1})

WZ

A1		*E1*		*E2*		*B1*		
LO	*TO*	*LO*	*TO*	*high*	*low*	*high*	*low*	
596					495			Kwon *et al.* 1996
590	491				491			Lee *et al.* 1998

ZB

LO	*TO*	
540		Kim *et al.* 1996
457	588	Tabata *et al.* 1999

A.3.3 Optical Properties

Index of Refraction : n

3.05	Zetterstrom 1970
2.9	Hovel *et al.* 1972
2.85	Natarajan *et al.* 1980

Dielectric Constants : ε

WZ

$\varepsilon(0)$	$\varepsilon(0)^{\perp}$	$\varepsilon(0)^{//}$	$\varepsilon(\infty)$	$\varepsilon(\infty)^{\perp}$	$\varepsilon(\infty)^{//}$	
15.4			8.4			Chin *et al.* 1994

A.3.4 Further Properties

Melting Point : T_m (in K)

2146	Vechten 1973

Thermal Conductivity : (in $Wcm^{-1}K^{-1}$)

0.8	Slack *et al.* 1987

A.4 REFERENCES

Akasaki, I. and Hashimoto, H., 1967, *Solid State Commun.*, **5**, p. 851.
Azuhata, T., Sota, T., Suzuki, K. and Nakamura, S., 1995, *J. Phys. Condens. Matter.*, **7**, p. L129.
Azuhata, T., Sota, T. and Suzuki, K., 1996, *J. Phys. Condens. Matter.*, **8**, p. 3111.
Barker, A.S. and Ilegems, M., 1973, *Phys. Rev.*, B, **7**, p. 743.
Camphausen, D.L. and Connell, G.A.N., 1971, *J. Appl. Phys.*, **42**, p. 4438.
Chen, G.D., Smith, M., Lin, J.Y., Jiang, H.X., Wei, S.H., Khan, A.A. and Sun, C.J., 1996, *Appl. Phys. Lett.*, **68**, p. 2784.
Chichibu, S., Okumura, H., Nakamura, S., Feuillet, G., Azuhata, T., Sota, T. and Yoshida, S., 1977, *Jpn. J. Appl. Phys.*, **36**, p. 1976.
Chichibu, S., Okumura, H., Azuhata, T., Sota, T. and Nakamura, S., 1996, *J. Appl. Phys.*, **79**, p. 2784.
Chin, V.W.L., Tansley, T.L. and Osotchan, T., 1994, *J. Appl. Phys.*, **75**, p. 7365.
Cingolani, A., Ferrara, M., Lugara, M. and Scamarcio, G., 1986, *Solid State Commun.*, **8**, p. 1227.
Davydov, V.Y., Kitaev, Y.E., Goncharuk, I.N., Smirnov, A.N.. Graul, J., Smirnov, O., Mirgorodsky, A.P. and Evarestov, R.A., 1998, *Phys. Rev.*, B, **58**, p. 12899.
Dingle, R., Sell, D.D., Stokowski, S.E. and Ilegems, M., 1971, *Phys. Rev.*, B, **4**, p.1211.
Ejder, E., 1971, *Phys. Status Solidi*, A, **6**, p. 445.
Fan, W.J., Li, M.F., Chong, T.C. and Xia, J.B., 1996, *J. Appl. Phys.*, **79** p. 188.
Fischer, S., Wetzel, C., Haller, E.E. and Meyer, B.K., 1995, *Appl. Phys. Lett.*, **67**, p. 1298.
Flores, G.R., Contreras, H.N., Martinez, A.L., Powell, R.C. and Greene, J.E., 1994, *Phys. Rev.*, B, **50**, p. 8433.
Giehler, M., Ramsteiner, M., Brandt, O., Yang, H. and Ploog, K.H., 1995, *Appl. Phys. Lett.*, **67**, p. 733.
Harima, H., Inoue, T., Nakashima, S., Okumura, H., Ishida, Y., Yoshida, S., Koizumi, T., Grille, H. and Bechstedt, F., 1999, *Appl. Phys. Lett.*, **74**, p. 191.
Hovel, H.J., Ishii, T. and Cuomo, J.J., 1972, *Appl. Phys. Lett.*, **20**, p. 71.
Jenkins, D.W., Hong, R.D. and Dow, J.D., 1987, *Superlatt. Microstruc.*, **3**, p. 365.
Karch, K., Wagner, J.M. and Bechstedt, F., 1998, *Phys. Rev.*, B, **57**, p. 7043.

Karch, K., Bechstedt, F. and Pletl, T., 1997a, *Phys. Rev.*, B, **56**, p. 3560.
Karch, K. and Bechstedt, F., 1997b, *Phys. Rev.*, B, **56**, p. 7404.
Kim K., Lambrecht, W.R.L., Segall, B. and Schilfgaarde, M.V., 1997, *Phys. Rev.*, B, **56**, p. 7363.
Kim, K., Lambrecht, W.R.L. and Segall, B., 1996, *Phys. Rev.*, B, **53**, p. 16310.
Korona, K.P., Wysmolek, A., Pakula, K., Stepniewski, R., Baranowski, M., Grzegory, I., Lucznik, B., Wroblewski, M. and Porowski, S., 1996, *Appl. Phys. Lett.*, **69**, p. 788.
Kwon, H.J., Lee, Y.H., Miki, O., Yamano, H. and Yoshida, A., 1996, *Appl. Phys. Lett.*, **69**, p. 937.
Lambrecht, W.R.L. and Segall, B., 1991, *Phys. Rev.*, B, **43**, p. 7070.
Lee, M.C., Lin, H.C., Pan, Y.C., Shu, C.K., Ou, J., Chen, W.H. and Chen, W.K., 1998, *Appl. Phys. Lett.*, **73**, p. 2606.
Lei, T., Moustakas, T.D., Graham, R.J., He, Y. and Berkowitz, S.J., 1992, *J. Appl. Phys.*, **71**, p. 4933.
Leszczynski, M., Suski, T., Perlin, P., Teisseyre, H., Grzegory, I., Bockowski, M., Jun, J., Porowski, S. and Major, J., 1995, *J. Phys. D: Appl. Phys.*, **28** p. A149.
Leszczynski, M., Teisseyre, H., Suski, T., Grzegory, I., Bockowski, M., Jun, J., Porowski, S., Pakula, K., Baranowski, J.M., Foxon, C.T. and Cheng, T.S., 1996, *Appl. Phys. Lett.*, **69**, p. 73.
Majewski, J.A., Stadele, M. and Vogl, P., 1997, *MRS Sym. Proc.*, **449**, p. 887.
Manchon, D.D., Barker, A.S., Dran, J.P. and Zetterstrom, R.B., 1970, *Solid State Commun.*, **8**, p. 1227.
Martin, G.T., Botchkarev, A., Rockett, A. and Morkoc, H., 1996, *Appl. Phys Lett.*, **68**, p. 2541.
Maruska, H.P. and Tiietjen, J.J., 1969, *Appl. Phys. Lett.*, **15**, p. 327.
McNeil, L.E., Grimsditch, M. and French, R.H., 1993, *J. Am. Ceram. Soc.*, **76**, p. 1132.
McGill, S.A., Fowler, K.C.B. and DeLeo, G.G., 1998, *Phys. Rev.*, B, **57**, p. 8951.
Menniger, J., Jahn, U., Brandt, O., Yang, H. and Ploog, K., 1996, *Phys. Rev.* B, **53**, p. 1881.
Merz, C., Kunzer, M. and Kaufmann, U., 1996, *Semicond. Sci. Technol.*, **11**, p. 712.
Meyer, B.K., Volm, D., Graber, A., Alt, H.C., Detchprohm, T., Amano, H. and Akasaki, I., 1995, *Solid State Commun.*, **95**, p. 597 .
Mireles, F. and Ulloa, S.E., 1998, *Phys. Rev.*, B, **58**. p. 3879.
Miwa, K. and Fukumoto, A., 1993, *Phys. Rev.*, B, **48**, p. 7897.
Monemar, B., 1974, *Phys. Rev.*, B, **10**, p. 676.
Monemar, B., Lagerstedt, O. and Gislason, H.P., 1980, *J. Appl. Phys.*, **51**, p. 625.
Nakadaira, A. and Tanaka, H., 1997, *J. Electron. Mater.*, **26**, p. 320.
Natarajan, B.R., Eltoukhy, A.H., Greene, J.E. and Barr, R.J., 1980, *Thin Solid Films*, **69**, p. 201.
Okumura, H., Yoshida, S. and Okahisa, T., 1994, *Appl. Phys. Lett.*, **64**, p. 2997.
Okumura, H., Ohta, K., Ando, K., Ruhle, W.W., Nagamoto, T. and Yoshida, S., 1997, *Solid State Electron.*, **41**, p. 201.
Osamura, K., Naka, S. and Murakami, Y., 1975, *J. Appl. Phys.*, **46**, p. 3432.
Pankove, J.I., Bloom, S. and Harbeke, G., 1975, *RCA Rev.*, **36**, p. 163.
Pastrnak, J. and Roskovcova, L., 1966, *Phys. States Solodi*, **14**, p. K5.

Perry, P.B. and Rutz, R.F., 1978, *Appl Phys. Lett.*, **33**, p. 319.

Petrov, I., Mojab, E., Powell, R.C., Greene, J.E., Hultman, L. and Sundgren, J.E., 1992, *Appl. Phys. Lett.*, **60**, p. 2491.

Ruiz, E., Alvarez, S. and Alemany, P., 1994, *Phys. Rev.*, B, **48**. p. 7115.

Salvador, A., Kim, W., Aktas, O., Botchkarev, A., Fan, Z. and Morkoc, H., 1996, *Appl. Phys. Lett.,* **69**, p. 2692.

Sanjurjo, J.A., Cruz, E.L., Vogl, P. and Cardona, M., 1983, *Phys. Rev.*, B, **28**, p. 4579.

Schwarz, R.B., Khachaturyan, K. and Weber, E.R., 1997, *Appl. Phys. Lett.*, **70**, p. 1122.

Schulz, H. and Thiemann, K.H., 1977, *Solid State Commun.*, **23**, p. 815.

Shan, W., Schmidt, T.J., Hauenstein, R.J. and Song, J.J., 1995, *Appl. Phys. Lett.*, **66**, p. 3492.

Sheleg, A.U. and Savastenko, V.A., 1979, *Izv. Akad. Nauk SSSR, Neorg. Mat.*, **15**, p. 1598.

Sheleg, A.U. and Savastenko, V.A., 1976, *Vestsi Akad. Nauk BSSR, Ser. Fiz. Mat. Nauk*, **3**, p. 126.

Shikanai, A., Azuhata, T., Sota, T., Chichibu, S.,Kuramata, A., Horino, K. and Nakamura, S., 1997, *J.Appl. Phys.*, **81**, p. 417.

Shimada, K., Sota, T, and Suzuki, K., 1998, *J. Appl. Phys.*, **84**, p. 4958.

Shokhovets, S., Goldhahn, R., Cimalla, V., Cheng, T.S. and Foxon, C.T., 1998, *J. Appl. Phys.*, **84**, p. 1561.

Sichel, E. and Pankove, J., 1977, *J. Phys. Chem. Solids*, **38**, p. 330.

Siegle, H., Eckey, L., Hoffmann, A., Thomsen, C., Meyer, B.K., Schikora, D., Hankeln, M. and Lischka, K., 1995, *Solid State Commun.*, **96**, p. 943.

Slack, G., Tanzilli, R., Pohl, R. and Vandersande, J., 1987, *J. Phys. Chem. Solids*, **48**, p. 641.

Slack, G. and McNelly, T., 1976, *J. Cryst. Growth*, **34**, p. 263.

Strite, S. and Morkoc, H., 1992, *J. Vac. Sci. Technol.*, B, **10**, p. 1237.

Strite, S., Chandrasekhar, D., Smith, D.J., Sariel, J., Chen, H., Teraguchi, N. and Morkoc, H., 1993, *J. Cryst. Growth*, **127**, p. 204.

Suzuki, M., Uenoyama, T. and Yanase, A., 1995, *Phys. Rev.*, B, **52**, p. 8132.

Suzuki, M. and Uenoyama, T., 1996a, *Jpn. J. Appl. Phys.*, **35**, p. 1420.

Suzuki, M. and Uenoyama, T., 1996b, *J. Appl. Phys.*, **80**, p. 6868.

Tabata, A., Lima, A.P., Teles, L.K., Scolfaro, L.M.R., Leite, J.R., Lemos, V., Schottker, B., Frey, T., Schikora, D. and Lischka, K., 1999, *Appl. Phys. Lett.*, **74**, p. 362.

Takeuchi, T., Takeuchi, H., Sota, S., Sakai, H., Amano, H. and Akasaki, I., 1997, *Jpn. J. Appl. Phys.*, **36**, p. L177.

Tang, X., Hossain, F., Wongchotigul, K. and Spencer, G., 1998, *Appl. Phys. Lett.*, **72**, p. 1501.

Torvik, J.T., Pankove, J.I., Iliopoulos, E., Ng, H.M. and Moustakas, D., 1998, *Appl. Phys. Lett.*, **72**, p. 244.

Tsubouchi, K. and Mikoshiba, N., 1985, *IEEE Trans. Sonic Ultrason.*, **SU-32**, p. 634.

Tyagai, V.A., Evstigneev, A.M., Krasiko, A.N., Adreeva, A.F. and Malakhov, V.Y., 1977, *Sov. Phys. Semicond.*, **11**, p. 1257.

Vechten, J.V., 1973, *Phys. Rev.*, B, **7**, p. 1479.

Vidal, M.A., Ramirez-Flores, G., Navarro-Contreras, H., Powell, R.C. and Greene, J.E., 1996, *Appl. Phys. Lett.*, **68**, p. 441.
Vogel, D., Kruger, P. and Pollmann, J., 1997, *Phys. Rev.*, B, **55**, p. 12836.
Wang, R., Ruden, P.P., Kolnik, J., Oguzman, I. and Brennan, K.F., 1998, *MRS Sym. Proc.*, **449**, p. 935.
Wei, S.H. and Zunger, A., 1996, *Appl. Phys. Lett.*, **69**, p. 2719.
Wright, A.F., 1997, *J. Appl. Phys.*, **82**, p. 2833.
Yamaguchi, M., Yagi, T., Azuhata, T., Sota, T., Suzuki, K., Chichibu, S. and Nakamura, S., 1997, *J. Phys.: Condens. Matter.*, **9**, p. 241.
Yamaguchi, A., Mochizuki, Y., Sunakawa, H. and Usui, A., 1998 *J. Appl. Phys.*, **83** p. 4542.
Yeo, Y.C., Chong, T.C. and Li, M.F., 1998a, *J. Appl. Phys.*, **83**, p. 1429.
Yeo, Y.C., Chong, T.C. and Li, M.F., 1998b, *MRS Sym. Proc.*, **482**, p. 923.
Yim, W.M., Stofko, E.J., Zanzucchi, P.J., Pankove, J.I., Ettenberg, M. and Gilbert, S.L., 1973, *J. Appl. Phys.*, **44**, p. 292.
Yim, W.M. and Paff, R.J., 1974, *J. Appl. Phys.*, **45**, p. 1456.
Zetterstrom, R.B., 1970, *J. Mater. Sci.*, **5**, p. 1102.

Subject Index

Physical Constants

velocity of light in vacuum	c	2.9979250×10^{8}	m/s
electron charge	e	1.6021892×10^{-19}	C
mass of free electron	m_0	9.109558×10^{-31}	kg
Planck's constant	h	6.626196×10^{-34}	J·s
		4.135708×10^{-15}	eV·s
	$\hbar$	1.05459×10^{-34}	J·s
Boltzmann's constant	k	1.380622×10^{-23}	J/K
		8.617097×10^{-5}	eV/K
Bohr radius	a_B	5.291772×10^{-9}	cm
Avogadro's number	N_A	6.022169×10^{23}	mol^{-1}
dielectric permittivity in vacuum	ϵ_0	8.854188×10^{-12}	F/m
magnetic permittivity in vacuum	μ_0	1.256637×10^{-6}	H/m